Information
Development

Information Development
Managing Your
Documentation Projects,
Portfolio, and People

JoAnn T. Hackos, PhD

Wiley Publishing, Inc.

Information Development: Managing Your Documentation Projects, Portfolio, and People

Published by
Wiley Publishing, Inc.
10475 Crosspoint Boulevard
Indianapolis, IN 46256
www.wiley.com

Copyright © 2007 by JoAnn T. Hackos

Published by Wiley Publishing, Inc., Indianapolis, Indiana
Published simultaneously in Canada

ISBN-13: 978-0-471-77711-3
ISBN-10: 0-471-77711-0

Manufactured in the United States of America

10 9 8 7 6 5 4 3 2 1

1MA/SY/RR/QW/IN

For general information on our other products and services or to obtain technical support, please contact our Customer Care Department within the U.S. at (800) 762-2974, outside the U.S. at (317) 572-3993 or fax (317) 572-4002.

Library of Congress Cataloging-in-Publication Data
Hackos, JoAnn T.
  Information Development : Managing Your Documentation Projects, Portfolio, and People / JoAnn T. Hackos.
     p. cm.
  Includes bibliographical references and index.
  ISBN-13: 978-0-471-77711-3 (paper/website)
  ISBN-10: 0-471-77711-0 (paper/website)
  1. Information resources management. 2. Information technology—Management. I. Title.
  T58.64.H32 2006
  658.4'038—dc22
                            2006030049

Wiley also publishes its books in a variety of electronic formats. Some content that appears in print may not be available in electronic books.

About the Author

Dr. JoAnn Hackos is President of Comtech Services, a content-management and information-design firm based in Denver, which she founded in 1978. She directs the Center for Information-Development Management (CIDM), a membership organization focused on content-management and information-development best practices. Dr. Hackos is called upon by corporate executives worldwide to consult on strategies for content management, information design and development, organizational management, customer studies, information architecture, and tools and technology selection.

For more than 25 years, Dr. Hackos has addressed audiences internationally on subjects ranging from content management, project management, structured writing and minimal information products, usability studies, and online and Web-based information to managing the information design and development process. Her seminars are dedicated to enhancing the practices and products that will best promote customer satisfaction and increase productivity.

She has authored *Content Management for Dynamic Web Delivery* (Wiley 2002), *Managing Your Documentation Projects* (Wiley 1994), co-authored with Dawn Stevens *Standards for Online Communication* (Wiley 1997), and co-authored with Ginny Redish *User and Task Analysis for Interface Design* (Wiley 1998). JoAnn is a Fellow and Past President of the International Society for Technical Communication (STC). She is a founder with IBM of the OASIS Technical Committee for the DITA standard (Darwin Information Typing Architecture). Her latest book, *Information Development: Managing Your Documentation Projects, Portfolio, and People,* is expected to be available late in 2006.

Recent clients include The International Monetary Fund, The Board of Governors of the Federal Reserve, Siemens Medical, Hewlett-Packard, The American Red Cross, Network Appliance, Varian Oncology Systems, Kone Elevators and Escalators, Dell Computer, Cadence Design Systems, SAP, Avaya, Lucent Technologies, Nokia, Motorola, Nortel, Federal Express, Compaq Computer, and more.

Credits

EXECUTIVE EDITOR
Robert Elliott

DEVELOPMENT EDITOR
Ami Frank Sullivan

PRODUCTION EDITORS
Eric Charbonneau
Pamela Hanley

COPY EDITORS
Foxxe Editorial
Kristi Bullard

EDITORIAL MANAGER
Mary Beth Wakefield

PRODUCTION MANAGER
Tim Tate

VICE PRESIDENT AND EXECUTIVE GROUP PUBLISHER
Richard Swadley

VICE PRESIDENT AND EXECUTIVE PUBLISHER
Joseph B. Wikert

PROJECT COORDINATORS
Kristie Rees
Ryan Steffen

GRAPHICS AND PRODUCTION SPECIALISTS
Joni Burns
Carrie A. Foster
Brooke Graczyk
Barbara Moore
Alicia B. South

QUALITY CONTROL TECHNICIANS
John Greenough
Brian H. Walls

PROOFREADING
Kristi Bullard
Techbooks

INDEXING
Techbooks

Contents at a Glance

Contents

Preface

In 1994, with the release of *Managing Your Documentation Projects*,[1] I put together many of the concepts and processes about managing technical documentation development that had been percolating in the field. The book has been well received, indicating that people involved in developing documentation needed a codified approach to the process. Many people tell me that *Managing Your Documentation Projects* continues to be their industry bible, providing them with a step-by-step process from planning and design through development and production. Most of the processes described in that book have changed little because they represent the basics of sound project management techniques. Except for some of the information associated with print product, little about the basics of documentation plans, project estimates and schedules, project tracking, and project completion has changed.

Nonetheless, much has changed for information development. As information-development managers, you are under considerable pressure to reduce costs and project time, to do the same or more work with fewer resources, send more projects to lower cost economies, and, in general, to increase the value of the information you deliver. I have designed this book to help you do so, in part by aiding you to make strategic decisions about information development, moving yourself squarely into the ranks of a professional mid-management leader. I have directed the discussion of project management toward smarter decision making there as well.

I hope that you find that by pursuing innovation in the design of projects, information, people, and organizations that your work is recognized as valuable to your organization as a whole.

Innovation in project management

This book brings the documentation project management ideas up to date. Although planning, estimating, tracking, and managing projects remains fundamentally the same, the new information on project management in Part III of this book looks beyond the structured project of the 1980s and 1990s to the rapidly changing projects of the 2000s. Managers and information developers find themselves challenged by shorted schedules and the adoption of agile product development techniques that rapidly iterate design ideas until the customer identifies what is needed. Consequently, this book introduces agile information development to the mix without forsaking the central focus of planning information design and development around the needs of information users.

---

[1] JoAnn T. Hackos, *Managing Your Documentation Projects,* Hoboken, NJ: John Wiley & Sons, 1994. *Managing Your Documentation Projects* remains in print and available. This new books expands upon the originally ideas presented there but does not supersede them.

The project management best practices in Part III include new attention to topic-based design as a significant new design principle, replacing the development of monolithic documents that owed more to the conventions of printing than to an understanding of user needs. Topic-based design assumes that users are looking for standalone, brief, and specific information to help them complete tasks and use products and systems quickly and efficiently.

Managing topic-based development introduces complexities in estimating, scheduling, and tracking that were not part of book-oriented development. Topics must be carefully planned, estimated in terms of scope and complexity, assigned to information developers with subject-matter expertise, and tracked carefully through myriad changes through the life of the project. The business advantage they provide far outweighs any complications of project management. Topics give you specific, standalone solutions to deliver to customers, allow you to reconfigure content to suit the demands of customers and product configuration, help you increase your ability to update as soon as needed, and assist you in decreasing the cost of producing and maintaining content in multiple languages.

Thus the best practices for project management in Part III have been rewritten to foster a topic-based approach and promote efficiency in content management and delivering content in multiple deliverables through single sourcing.

Innovation in information development

The innovations in project management are, however, only a small part of the changes that 21st century information-development managers face. Since the 1994 publication, documentation management has been transformed into information management. The term "documentation" has within it an underlying assumption that has had a negative connotation in the industry. Documentation refers to information that describes product or process and how it was developed. Product requirements and specifications, engineering drawings, manufacturing instructions, and others all explain the intricacies of a product's genesis and construction. In the same way, documentation is used to explain complete processes internal to organizations, including contractual agreements and statements of policy. Many times, such documentation includes procedures that codify the policies.

Documentation is by its connotations inward looking—tasked with explaining what is. It serves the needs of those that originated the policy, the process, and the procedures, including those processes defined during the development of hardware and software products.

Unfortunately, process and product documentation is not defined as meeting the needs of people who must use the processes or products to perform functions. People who need information to learn and be productive at work and at home are not well served by content that is focused on how a product was developed or how it is intended to work. Nor are they served by formal legal agreements or statements of contract or policy in learning to perform a procedure efficiently and effectively.

People need, of course, information that is developed with their learning and performance as a central goal, not an accident. And, more often than ever, people need information that is packaged and delivered in media that is most easily accessible. Before the 1990s, few options existed to deliver information in anything but print. Most technical and procedural information was packaged as books. Now, multiple media delivery of information, including websites, embedded and online help systems, knowledge bases, CDs, and others, is the norm.

As a result, many organizations dedicated to supporting people who need to learn and perform tasks with products or without have redefined their work. What was once documentation is now regarded as any type of information that guides users. What was documentation writing is now referred to as information development. Many technical writers today are referred to as information developers.

In the nearly 20 years since I wrote *Managing Your Documentation Projects*, information development has sought to focus on developing effective information for users rather than documenting how products were designed and developed. Although this transformation is by no means complete, information developers and managers are increasingly aware that describing product features and functions or writing legally correct policies and procedures does not promote good performance. If they want to ensure that customers and employees are productive, they must directly address their information needs and develop solutions that are more innovative and effective than shipping out an 800-page binder of incomprehensible detail.

Innovation in technology

Innovations in how to design information are influenced by better understanding of how information is used by its consumers. Innovations in how to manage projects are influenced by those design changes. Not only do the innovations increase customer satisfaction, but they encourage managers and information architects to invest in new technologies. At present, those new technologies include moving to topic-based authoring supported by XML tools and content management systems. The new technologies allow information developers to increase quality while decreasing the cost of development. Technologies that reduce time spent on formatting text encourage information developers to spend more time on planning, design, and development of sound content. Technologies that reduce production time for multiple media (print, PDF, HTML, help, and so on) increase the time devoted to ensuring that information is accurate and complete.

Technology innovations extend the information development life cycle into localization and translation. Content management systems allow you to deliver topics to translation as soon as they are ready, rather than waiting until entire books are complete. Translation memory tools preserve the asset of previous translations, and machine translations allow critical content to be delivered in a timelier manner.

Technology helps managers, and staff, reduce the number of resources required to produce a unit of content. Information planning and design encourages you to reduce the content to only what is needed by the user. Technology innovations further allow you to update content and respond to changing user needs more quickly.

Innovation in staffing

Information-development managers are generally enthusiastic about applying new technologies to information development. They are increasingly supportive of design innovations that reduce the volume of unnecessary content that must be managed. Both minimalism and user-centered design encourage new approaches to managing content rather than simply documenting the product.

However, innovations in design and technologies are still not sufficient to decrease resource requirements to the levels demanded by senior management. Consequently, information-development managers seek additional ways to reduce costs without decreasing quality.

One solution is to move a percentage of information development to lower cost countries. When you can hire five information developers for the cost of one in the US or Western Europe, you can maintain staff size while reducing development costs. Even if the cost of offshore development is not as low as you may be encouraged to believe, the overall effect on total cost can be significant, as long as the lower cost staff remain inexpensive and the cost of training and managing them does not exceed their employment costs.

However, offshore development does nothing to encourage innovation. In fact, it allows you to continue to be inefficient and to produce content that no one needs.

Innovation in portfolio management

If your responsibility is to increase productivity, decrease development costs, and maintain value for the customer in the information you deliver, innovation in managing your portfolio of projects and responsibilities becomes essential. Many times, information-development managers see themselves in one of two ways: they are either people managers, keeping everyone motivated and skilled, or they are super project managers, either managing all the projects themselves or overseeing the project managers. Certainly, people and project management is an important part of the information-development manager's job. However, both are tactical responsibilities and can easily result in spending considerable time and effort going in the wrong direction and doing the wrong thing, albeit doing it well.

Such a manager quickly becomes an order-taker from others, including product and development managers or business-line managers. You are told, "Here is your set of projects for the next quarter or next year. Figure out what resources you need to meet the deadlines. And, by the way, do the work with half the resources you calculate."

Of course, you can employ technology to make your people resources more productive or find less expensive people and let them continue developing in the same old way. Or, you can choose a strategic direction for your organization, deciding which projects are most important for the organization and applying your resources there.

By aligning your strategy with overall corporate objectives, you can apply your best resources to the most critical projects, provide average support for less important projects, and relegate the end-of-life or the going-nowhere projects to maintenance or less. Actively managing your project portfolio is never easy. You will no doubt experience a great deal of opposition from product and process managers who believe each of their projects is most important. But, by reducing resources, senior management is conveying the message that you must keep spending under control while supporting the corporate strategy. Like every other line manager responsible for manufacturing a product, in this case an information product, you must make difficult choices about what gets full attention and what is relegated to the back.

Part II of this book helps you understand the tradeoffs required for innovative management of your project portfolio, including an examination of technologies and staff growth and development. You will find the chapters of Part II organized to correspond to the four quadrants of the Balanced Scorecard, a management measurement scheme described in Chapter 3.

Reading this book

Part I of this book introduces the concepts I describe in this Preface. However, I begin in Chapter 2 with an update of the original 1994 Information Process Maturity Model (IPMM), an innovation that has become an industry standard. This 2006 IPMM gives you a method for comparing the state of your organization to others, from immature organizations indulging ad-hoc behaviors to well-organized departments led by innovative and professional managers. Use the IPMM descriptions of the five levels of process maturity and the eight existing and two new key characteristics to evaluate your present state. Consider what is needed to move to the next level.

Most of what you need to enhance the maturity of your organization is covered here. In Part II on portfolio management, I present many of the ideas I have been developing and sharing in the past 10 years on making strategic decisions about the direction of information development. In Part III on project management, I expand traditional project management to include techniques of agile project development coupled with innovations in information design.

I hope you enjoy the ride.

Acknowledgments

Many people have contributed over the past 10 years to the development of the concepts in this book. Many of them are publications managers with whom I have worked on Information Process Maturity assessments, benchmark studies, customer studies, and content management projects. The members of The Center for Information-Development Management (CIDM) have been closely involved in reviewing the content and adding to my understanding of the challenges they face in managing enterprises that are increasingly global.

In particular, I want to thank those who have worked with me by reviewing chapters of the book as I struggled through them, helping me to clarify my thinking and adding examples from their own experience. Those contributors who read chapters and added their insights include Julie Bradbury, retired as director of information development at Cadence Design Systems; Diane Davis, senior director of information development at Synopsys; Sue King, information management consultant; Vesa Purho of Nokia Networks; Susan Harkus, information architect; Amy Witherow of Cadence Design Systems; Waldemar Frank of LUZ, Inc.; and Ben Jackson of Microsoft Corporation. I want to thank them most for their continuing encouragement of my ideas.

Beth Thomerson of BMC Software; Ann Teasley of CheckFree; Monica Lake, formerly of Dell Corporation; John Russell of Oracle; and Charlotte Robidoux and Patrick Waychoff of Hewlett-Packard were kind enough to share examples of their work that have broadened considerably the examples of creative and effective management best practices throughout the book.

The most significant contributor to this book is Bill Hackos, my husband and business partner. He and I have developed the concepts and techniques together. Bill is chiefly responsible for the project estimating and tracking processes and the metrics analysis. He brings his 30 years' experience managing projects and teaching our project management workshops. We talked over every chapter as I was writing, usually providing new insights that have strengthened the discussions. He has also been patient through yet another year of writing.

Part 1

The Framework

"I thought management was going to be easy."

Chapter

An Introduction to Information-Development Management

> "Management" means, in the last analysis, the substitution of thought for brawn and muscle, of knowledge for folklore and superstition, and of cooperation for force. . . .
>
> —Peter Drucker, *People and Performance*[1]

Managing information development has never been simple. Information-development organizations are frequently orphans looking for a permanent home. In many high-tech companies, information developers work within the product development structure, reporting either to product teams or other business units or to a central development organization. In other companies, information developers report to more senior managers in marketing, marketing communications, operations, or customer support and service. In service-oriented companies and nonprofit organizations, information developers report into diverse management structures, often associated with human-resources management or operations. In many of these structures, senior managers have little knowledge, and sometimes little interest, in what information developers contribute to the organization or what they might contribute, given an effective managerial direction.

---

[1] Peter F. Drucker. *People and Performance.* Burlington, MA: Butterworth-Heinemann, New edition, 1995.

Despite the difficulties they face in winning respect and appreciation, information-development managers are generally quite adept at focusing on the details of developing publications and meeting deadlines. I often hear managers say that they never miss a deadline. They believe in the necessity of getting the information products out the door, usually on the same schedule that products are released. Meeting development schedules requires a devotion to project management, especially the task of estimating and staffing each project so that it is done on time and with the level of quality demanded by the customers and the organization.

At the same time, I believe that information-development managers must be equally adept at strategic management. The portfolio of projects is often more than can be done with quality by the existing staff. That means making hard decisions about the priorities of projects and how many resources should be devoted to each project. It also means being constantly alert to opportunities to pursue a minimalist agenda, providing only the content that users need to achieve their goals. It means pursuing content management and reusing content among related deliverables. It means striving toward a higher level of process maturity and ensuring that staff are well educated and directed toward both efficient performance and the development of effective, customer-oriented information.

If you don't already have enough to do in running an efficient and effective organization, you are responsible for reporting to your senior management and educating them about the value provided by your organization. You must develop strong professional relationships with peer managers in your organization, including those in engineering and software development, education and training, service, marketing, sales, and any others who affect customers and might benefit from your support and collaboration. You must develop your own staff, focusing on building skills and knowledge, as well as investing in the activities that bring the highest value. You need to be alert to changing strategies, especially as you move into a collaborative work model that includes global teams and outsourcing. You must be skilled at bringing together team members who are geographically distributed or who come from companies acquired through acquisition and merger. You may yourself lead a team that becomes part of another organization and be required to adjust your business methods to accommodate changing expectations.

If you are a new manager, you have much to learn about managing information development and supporting your organization's recognition as a key contributor in a larger organization. I highly recommend learning from more experienced managers by joining organizations that specialize in information-development management or provide access to a community of managers. The Society for Technical Communication (STC), the Professional Communications Society of the IEEE (IEEE-PCS), and The Center for Information-Development Management (CIDM) all provide opportunities for new managers to learn their art.

If you are an experienced manager, you have the opportunity to share your expertise with newcomers and to become part of the community of information-development managers that is growing globally. By taking part in conferences, workshops, and electronic communities, you can not only provide information yourself but also learn from the experience of others.

I hope that you view your development as a more mature and secure manager with enthusiasm. Although you may experience pitfalls along the way, the journey is rewarding.

When your organization is recognized for its contributions and you are viewed as an effective leader, you have succeeded not only in advancing your own career but also in building the profession as a whole.

Best Practices in Information-Development Management

In each chapter of this book, I have included a set of best practices for managing information development in your organization. Some of the best practices are focused on how you manage your organization as a whole, beginning with an examination of your organization's process maturity. Some are focused on strategic planning, allowing you to manage your portfolio of projects effectively. Some of the best practices are focused on the management of projects, ensuring that you develop and deliver the information that your customers most need and that satisfies the requirements of your particular business environment.

In this chapter, you learn about the importance of your own management role and those who assist you in that role, whether inside or outside your own department. The best practices in this chapter help you form an understanding of overall strategy with regards to information development. They help you better serve the needs of internal and external customers and employ tactics to ensure that you are delivering information in an efficient and cost-effective manner.

The four best practices in Chapter 1 introduce you to the four themes you will find throughout this book:

✔ Understanding your many roles as an information-development manager

✔ Recognizing the need to build a mature organization

✔ Developing an information-management strategy

✔ Ensuring that your projects are managed efficiently and effectively

The best practices in this chapter provide you with an introduction and overview of the issues I discuss in more depth in the subsequent chapters.

Best Practice—Understanding your many roles as an information-development manager

As an information-development manager, you have many responsibilities to your organization, your profession, and yourself. Each of these responsibilities represents a unique and challenging role that you assume when you join an organization. As the organization itself changes, your roles change with it.

In 1994, in *Managing Your Documentation Projects (Wiley 1994)*, I divided the roles into four critical areas. They are as relevant today as they were at that time with a bit of modification.

Your four key roles, illustrated in Figure 1-1, are

- ✔ communicator
- ✔ resource manager
- ✔ leader
- ✔ visionary

Figure 1-1: The many roles of the information-development manager

As a communicator, you are responsible for keeping the lines open to your senior management, to peer managers throughout the organization, and to your team members. A central goal of your communication activities should be to develop an understanding of and support for the information-development process and the information products you produce to meet customer needs.

As a resource manager, you are responsible for ensuring that your staff is able to meet the demands of your projects and engage in activities that advance the maturity of your organization and introduce new, innovative practices. You are responsible for prioritizing the portfolio of projects that your team manages and ensuring that you have the resources to meet the requirements of deadlines and quality.

As a leader, you need to be thoroughly engaged with your team members and understand their activities. You cannot stand on the sidelines as an administrator but must be involved in designing and implementing effective processes, information architectures, and tools. You must know how your customers think and learn so that you can guide your team to meet their needs with new ideas and best practices.

As a visionary, you need a clear picture of what you want your organization to become, one that is carefully aligned with the business objectives of your larger organization. You need to understand and appreciate business goals and objectives so that the work of your organization is never viewed as merely clerical. You need to communicate your vision of what your team can provide effectively to the decision makers.

Develop as a middle manager

When you were hired as a departmental manager or you moved into a management position from another position in the same organization, you took on the responsibilities of middle management. You report to a more senior manager in marketing, product development, operations, support, or some other part of the larger organization to which you belong. In that capacity, you are responsible for understanding the strategic objectives of your manager and the corporate management and translating those objectives to the day-to-day activities of your team and to the direction you set for your own organization. You are responsible for communicating corporate strategy and direction to your team members, even if you don't always agree with the strategy.

As a department manager, you also have a relationship to other managers in departments with which your team interacts. Those generally include managers responsible for marketing and selling products, directing operations in various parts of the organization, developing and testing products, providing service and training to customers after sales have been completed, and others appropriate to the role your organization plays.

You may also have relationships with other peer managers in parts of the organization that have different business directions. If the corporation has grown through mergers and acquisitions, you may build relationships to other technical publication managers or others responsible for writing operational or technical information elsewhere in the larger entities of the corporation. You may also be asked to establish relationships with managers and staff in partner organizations, including those reselling, servicing, or distributing your products or those who maintain technical information for products your organization uses, sells, services, or distributes.

Finally, you have a significant role to play with your own staff members. They may be located in the same facility that you work in, or they may be located anywhere in the world. As a manager, you are responsible for ensuring their success and engaging them in the active development of your organization's products and services.

Operate as a professional

Outside of your immediate organization, you may have other obligations to the profession of which you are a part. As a professional communicator, you may be a member of a trade organization that promotes the field. As a professional manager, you may be part of groups that facilitate communication among peer managers. You are responsible for knowing the state of the art and the best practices in your industry so that you can bring them into play in your own organization. You are responsible for subscribing to industry standards and deciding if they apply to your enterprise. You are also responsible for offering your own expertise and experience back to your professional colleagues, in the form of publications and presentations on a local, regional, national, or international scale.

You also have responsibilities to yourself for professional growth. If you have come up through the ranks of technical communicator to a management role, you have grown and changed from being an individual contributor to someone who takes responsibility for the contributions of others. You have gone from being a colleague to being the boss, which is often a dramatic change in direction.

You may be engaged in promoting education in the field, either through teaching opportunities informally in professional organizations or formally through programs at local colleges and universities. You may have chosen to lead a professional group in your community to increase your exposure to ideas and develop your management skills. You may yourself attend educational activities or pursue an advanced degree in your profession or in a related management area.

In all of these circumstances, you have a wide range of influence and responsibilities. In fact, you may feel that you are being pulled in too many different directions, each of them demanding a degree of commitment and loyalty that may be in direct contradiction to other demands. You may need to balance professional demands on your time and attention with responsibilities to family and other parts of your community. That balancing act is never easy and seems to become more complicated every day.

Handle the balancing act

One of the most difficult aspects for managers in this complex act of balanced loyalties is how to represent your senior management's goals and objectives to your staff members. This part of the balancing act is made more difficult if you have moved from individual contributor to manager in the same organization. Former colleagues are now your staff members. They expect you to maintain your loyalty to them and support their needs.

At the same time, you have taken on a role in the larger organization that brings a new set of expectations. Your management expects you to represent the larger organization to your staff even when you may disagree with the actions of that organization. You may know about plans that you cannot reveal to staff members, even though they will be adversely affected. You may have to refuse requests for funding and support that you find legitimate because you have other priorities that must be addressed first. You may have to take actions you find extremely unpleasant and face criticism from your staff for doing so (see Figure 1-2).

The best practice to consider in the face of a balancing act is open and honest communication. Your staff needs to know that they can count on you to tell them what their roles should be with respect to the larger organization. For example, you have been asked to reduce the amount of time and money spent on end-user information development in the form of help systems. Your staff has spent a great deal of time developing a help system and takes great pride in the help design and content they have created. They have even won an award for the help system in an international competition. At the same time, on-site studies reveal that the help system is not being used by the customers for whom it was intended. Their roles in their work environment, the training they receive, the low turnover, and the standard nature of the tasks may make the help system irrelevant, no matter how well crafted it may be. Your management has asked that the help be discontinued and effort put into other information needs.

Figure 1-2: The management balancing act

You know that your staff will be disappointed in the plans and will try to convince you to push back on management. How do you proceed?

The best practice is open and honest communication. You tell your staff about the outcome of the studies and ask them how they might react. You explain that you understand their disappointment but ask them to see the change as a challenge for doing more valued work. You explain that management doesn't want them to spend valuable time and resources on a help product that isn't meeting customer needs. You ask for ideas for new initiatives that are better aligned with what you have learned about the customers. With a combination of understanding and honesty, you communicate the message from senior management and help your team move to a new level and respond to the challenge effectively.

You face the balancing act in the other direction when your management or your peer managers ask you and your team to do work that is not appropriate. For example, consider the product developer or product manager who wants information included in the documentation that, in the best judgment of you and your staff, is not appropriate for the customers. The information may be more detailed than customers are prepared to understand or need to know to be successful. The information may be written inappropriately for the audience, with too much industry jargon or a poor writing style. The information may be irrelevant for the customer. A developer may be more interested in sounding impressive than in communicating with those who need unbiased information written in language they can understand.

Most information-development managers face this conflict continuously in their relationships with other managers and their staff members. A best practice is to clearly state your assumptions about responsibilities toward the customers. As the information developers, you and your staff are responsible for ensuring that customers are successful and interpreting their needs for information. Your organization is, in effect, the owner of the information and best situated to make decisions about content, format, and style. Although you remain open to suggestions about the information, in the end, you make the decision about what should be included, what should not, and how the information should be best presented.

Unfortunately, you may face situations in which you have no political power to enforce your position. The CEO demands that you remove all instances of contractions in a document intended for naïve consumers who will succeed better if information is not intimidating. Despite all arguments, the CEO is adamant. In such instances, you are likely to comply with the demands although you may find it safe to register your dismay and reiterate your position as the keeper of the information.

Develop a Balanced Scorecard

You will learn more about developing a Balanced Scorecard for your organization in Chapter 3: Introduction to Portfolio Management. However, understanding how to balance the demands made on you includes knowing how to focus on a larger view of your role. The Balanced Scorecard reminds managers that every part of the larger organization is responsible for the four parts of the scorecard: financial success and profitability, customer satisfaction, effective operations, and efficient and knowledgeable employees. Best practices in each of the four areas help you to ensure that you concentrate on a strategy that will produce success.

As part of your balancing act, you need to

✔ understand what your organization values most so that you can ensure that the focus of your department contributes to the organizational goals. If your organization is devoted to winning market share and developing satisfied customers, your mission will be different than if your organization is concentrating on reducing costs. How you contribute to the profitability of the organization as a whole may be difficult to measure. But learning everything you can about how financial success is defined will increase your effectiveness as a manager.

✔ understand who the real customers are, the ones who write the checks that pay your salary and allow the corporation to meet its financial goals. You need to clearly differentiate between internal and external customers, reminding the internal customers that their needs are second to those of the people who pay the bills. You need to ensure that your team members have opportunities to know customers directly, especially with reference to their information needs. When you advocate for customers, your advocacy must be based on real information, not opinions. Your information plans must ensure that information helps make your customers more successful and helps reduce the cost of ownership of your company's products and services.

✔ understand that you are involved in managing both an operations and a product-development function. It's often difficult to remember that both objectives need to be fulfilled equally. You need to run an efficient organization, one that meets its deadlines and gets the information products into the hands of customers. But you must also run an effective organization, developing information products that are genuinely useful and usable. It doesn't matter much if you meet deadlines and keep costs under control, if your customers are ready to complain that they have no tools to perform successfully with your company's products.

✔ understand that you have an obligation to your team members that goes beyond creating a pleasant working environment. You must ensure that they grow and continue to learn and innovate. Without growing, they are likely to stagnate, doing the same thing today that they did 10 or 20 years ago. Without a focus on continued growth, your staff will descend into a clerical function that is little valued and ripe for outsourcing to a lower-cost resource. Your communication to your team members must make the priorities clear. It won't be sufficient to continue producing the same old information products. The products must change to meet changing demands in the customers' workplace and must change to compete with others who produce better information and happier customers.

Learn more about the Balanced Scorecard as you progress and develop a scorecard that you can use directly to measure your progress toward increasing your organizational maturity.

Best Practice—Recognizing the need to build a mature organization

Consider what it means to have a more mature organization. The details of the Information Process Maturity Model (IPMM) are presented in Chapter 2: The Information Process Maturity Model. At this point, you should find it important to recognize that you need a mature organization if you are to meet business and professional goals and maintain an effective and efficient department. I heard recently from a colleague that her manager discouraged her from pursuing a customer contact because they were not mature enough as an organization (as measured by the IPMM) to consider customers. The manager appeared satisfied to run an immature organization, quite possibly viewing that immaturity as inevitable.

Nothing could be further from the truth. I believe that every information-development manager should strive for a higher level of process maturity because staying as a Level 1: Ad hoc or a Level 2: Rudimentary organization invites devaluing and outsourcing. Certainly, more mature organizations at Levels 3, 4, or even 5 may be wrecked by an ignorant or malevolent senior manager, but you will find it much more likely that an organization that performs primarily at a basic operational level is at risk.

What exactly is the difference between a mature and an immature organization, and why are immature organizations at risk for dissolution or outsourcing? Overall operational quality and sustained innovation is a product of a mature organization. Although it may be possible for the individual contributors who dominate Level 1 to produce exciting new ideas and efficient methods, they do so in isolation. Level 1 describes an organizational pattern that is decidedly isolated. Individuals work independently, often prized for their ability to be unmanaged or unmanageable. They thrive on reaching personal goals with little interest in collaboration or even cooperation. Some such individual contributors may indeed be very talented and produce superb information products. Others may lack motivation, skill, and professionalism, producing lackluster results.

If you look at an immature organization from a 30,000-foot perspective, you find wide differences in quality and initiative. Everyone works for him- or herself, pursuing personal agendas and without regard for corporate objectives. In an immature organization,

I see writers continue to produce the same manuals year after year, making them longer and increasingly unwieldy. I see individuals unwilling to devote any business or personal time to learning and professional growth. I see people for whom information development is an 8 to 5 job and who basically engage in a clerical function. Of course, in the next cubicle is someone who does care about the quality of the work and continually searches for better methods to decrease costs and improve quality.

As a manager, which team member do you want working for you? If you are satisfied in managing an immature organization, you are likely not interested in organizational growth. I believe, however, that such managers are not the majority. Most of the information-development managers with whom I have been engaged are devoted to building more successful and recognized organizations. They want staff who are motivated to learn and grow the organization as a team. They prefer people who are innovative and devote time to pursuing new ideas and better practices. They build into the organization time for innovations, especially those focused on knowing the customer better.

In a mature organization, sound processes are in place, projects are well planned and managed, schedules and budgets are maintained, changes are made rationally and deliberately, and everyone knows what is expected of them. The information products developed in such organizations are designed to meet both the quality expectations of customers and the business objectives of the larger organization.

Determine your current process-maturity level

Look around your organization. Evaluate the level of process maturity that your team has achieved. Then, decide what you need to do to progress to the level you want to be. In Chapter 3: Introduction to Portfolio Management, you will find suggestions for increasing your maturity level. Look carefully at the characteristics of each process-maturity level and think about your organization. Do you have standard processes in place? Do you measure those processes to judge their effectiveness? If not, you are most likely at a Level 1 or 2.

The Software Engineering Institute (SEI), the developer of the Integrated Capabilities Maturity Model (CMMI), has demonstrated that higher levels of maturity result in significant increases in efficiency of an organization. Increases in efficiency typically result in productivity gains and reduced costs of operations. Mature organizations spend less time on unproductive activities and those that add little value to the larger organization. They spend more time optimizing those activities that meet the strategic goals of the larger organization and help to increase customer satisfaction.

If you identify your current maturity level and find it too low, you need to begin a project to improve. Improvement usually means looking closely at your processes and deciding which add clear value to customers and the business and which add little or nothing. Those that produce little value need to be eliminated or drastically minimized. Those that add clear value need emphasis and close attention.

For example, you may discover that your team members spend a high percentage of their time formatting engineering specifications and labeling them user manuals. At the same time, the team spends little or no time learning about what customers need to know. The overall value of their activities is low and easily outsourced. By shifting the team's work to higher-value activities, you not only improve performance but you also gain knowledge about customers and information design that is not easily replicated by outsiders.

Process maturity is a tool for withstanding the competition. High levels of process maturity center on

- ✔ measuring performance
- ✔ learning from your mistakes and shortcomings
- ✔ preventing the same problems from occurring again
- ✔ optimizing your performance in the future

With these characteristics in place, you can improve and gain approval for your efforts.

Best Practice—Developing an information-management strategy

Peter Drucker, noted management guru, captures the importance of an information-management strategy when he writes, "There is nothing so useless as doing efficiently that which should not be done at all." It's all together too easy for us to focus on the deadlines and the press of everyday work rather than to set priorities and understand what should be done. Most people, as managers and contributors, find comfort in getting something out the door. But if the document you "get out the door" is useless to the customer, you have only succeeded in increasing the customer's level of frustration and decreasing your value to the larger enterprise.

The more information you publish that does not meet customer needs, the more it continues to cost more than it's worth. It also allows you to be overtaken by competitors who get it right by providing the best information just in time and suited to the customer's goals.

Having mature processes in place will increase your organization's competitive edge, but without a strategy for information development, you won't get the support you need from senior management. You need to develop an information-management strategy for your organization and ensure that your strategy is aligned effectively with the goals of your larger organization. The details of strategy development, from understanding budgets and customers to improving processes and encouraging employee growth and development are discussed in Part 1, "Portfolio Management."

The core of portfolio management for information development is strategic planning. In your portfolio of work, you have a variety of projects to conduct that will produce information for product releases, update information that has changed, provide operating procedures for internal departments, and create a myriad of information types for your organization. Your portfolio probably contains more than product- or service-related projects. You may be engaged in conducting user studies, designing delivery methods such as information websites, managing localization and translation, designing instructional materials, and delivering training. Depending on the nature of your organization and your responsibilities, your portfolio may include highly diverse activities or it may be focused on producing a well-defined set of documents for internal and external customers.

In addition to the types of projects that engage your team's best efforts, you may be involved with new activities, including developing requirements for new tools and technologies, designing new information solutions, soliciting customer feedback, collaborating with other parts of the organization, and engaging in external professional development activities. All of these activities need to be managed effectively. But first you must decide what resources you can afford to devote to each of them. Those decisions should be governed by strategy objectives, and those objectives must be carefully aligned with corporate objectives.

Align with corporate objectives

Consider the possible objectives that may impact your strategic planning for information development. In this first case study, the emphasis is on reducing the cost of operations and the cost of goods.

ABC Corporation: A Case Study

ABC Corporation's senior management is most concerned with reducing the cost of goods for its hardware products. The manufacturing costs include the costs of technical manuals, particularly the cost of translation. Many of ABC's product lines have long histories, with 40-year-old products still being sold to additional global markets, requiring new translations. The existing product line is updated occasionally with new parts that require small changes to the technical manuals.

As publications manager, you recognize the need to reduce operational costs and to make translations less expensive. Your strategic planning centers on efficiency and cost reductions. You decide to pursue a content management solution with a seamless integration into a translation management system maintained by your localization service provider. You also decide to implement a controlled language system based on a terminology database in English so that you get more language consistency in your source documents. This strategy requires some rewriting to standardize language and a minimalist agenda to eliminate unnecessary content.

A strategic direction like this one is common in industries in which cost reductions for traditional products is paramount. In the next case study, the corporation is attempting to move into new markets and is pursuing a "best in class" corporate strategy.

TAX Corporation: A Case Study

TAX Corporation develops hardware and software in a highly competitive telecommunications market. They recognize the need to develop superior products that meet customer requirements at a lower cost than their competitors. With manufacturing locations in low-cost economies, they are able to deliver the best prices in the industry, but customers reject their offers because their technical information is undecipherable. Their strategy of having the information authored by the engineers in a local language and then translated into English by translators who do not know the technology has backfired, even if it is the lowest cost.

As the new US publications manager, you recognize that your primary task is to improve information quality for the customers while maintaining cost efficiencies. You have engineering developers whose first language is not English, making it difficult to use American writers to learn about the new products. You decide on a global strategy of developing a first-class information development team in the new country and supporting them with your writers and editors in the US.

Because you find it difficult to hire people with any technical writing experience, you focus on strong English skills and some technical training in engineering or programming in your hiring strategy. A manager on your staff who is bilingual expresses a strong interest in leading the new team in her home country. You create a mentoring relationship between your US writers and editors and the new employees, and you provide for technical training, as well as training in information development. You know you have to make a significant investment if this enterprise will succeed, but your alignment with corporate objectives ensures that you get the funding required.

Another set of objectives requires that the information-development manager set difficult priorities to identify which projects can be supported and which cannot. He is faced with severely restrictive budgets and headcounts, which means that he does not have enough people to do all the projects that the product managers would like him to support.

BIG TIME Corporation: A Case Study

BIG TIME Corporation has a wide variety of products in its portfolio. Some have been in place for years with few sales to new customers. Others are in mid-life, with updates occurring regularly. A few are the emerging stars, solutions-based products that the company hopes will change its fortunes in the foreseeable future. The product managers for every one of these products wants a large set of technical manuals that are regularly updated. You have a continually decreasing staff in the US and an inexperienced small group recently hired in a low-cost economy. Together, your team still cannot handle the workload.

As the director of information development, you develop a bold strategy in which you prioritize the products in terms of their promise for the company and in alignment with what you have learned from senior management. The top new products don't get standard documentation sets. Their users are innovators and early adopters, closely supported during installation and implementation by engineering and programming experts. These customers, you decide, need technical white papers that explain how the products work and how they can be best combined to produce innovative solutions.

The legacy products, you recognize, have very few updates and are primarily used by long-time customers with few staff changes. They are more familiar with the products than your engineering team. You decide that these products should be supported by basic files that describe the updates and present changes to procedures, but you will not update the legacy manuals at all. You assign this task to the least experienced members of your team, including the offshore writers.

continued

BIG TIME Corporation: A Case Study *continued*

The mid-range products require more attention. However, the documentation set is clearly overblown, with unnecessary marketing fluff and technical details added over 10 or more years. You create a group that is devoted to a minimalist agenda. Their first task is to reduce the volume of documentation by at least 50% if not more. Then, they will be charged with updating the newly redesigned information.

Your strategy won't make you popular with the product managers who aren't getting their usual full set of content, but without the resources to make that possible, you believe you have a viable solution. In fact, you're convinced that the minimalist documentation will be better used by the customers than the old massive tomes.

As you can see from these examples, creating a strategy plan and setting priorities does not always make you popular among all your internal constituencies. You will find yourself defending your strategy with product and marketing managers who have agendas of their own and want your attention to their needs. You should have strong senior management support for your plan to help you withstand the blast of disapproval you may experience. Nonetheless, that disapproval, if it occurs, is easier to handle than overworking everyone on your team to the point that they quit or producing work that satisfies no one, including the customers.

I recognize that many companies do not establish a strategic direction for making decisions even at the highest level. Projects are funded because their advocates are strong politically and make the most noise, even when the projects seem unlikely to produce positive or profitable results. Projects are canceled because their advocates move on, leaving them with insufficient political support. If you find yourself in such a situation, in which you are constantly fighting to respond to rapidly changing priorities with too few resources and too many demands, you will find that setting your own priorities is even more critical. If you were in a well-managed company, the priorities would be clear from the first. In a poorly managed company, you need to set priorities on your own to enable your team members to be successful.

At some point, you might consider the "nut grass" story. We were living in West Texas and battling an influx of nut grass in our backyard. Nut grass has dagger-like spines that penetrate children's sneakers easily. Nothing we did seemed to help. Finally, we had an opportunity to have dinner with old friends who had been in West Texas for many years. Much to our surprise, Al remarked after hearing our plight that he knew how to get rid of nut grass. "How?" we implored. Al answered deliberately—"Move."

Strategic planning enables you to manage your portfolio to best effect. Projects that require information to be successful must come first in your planning. You may discover, for example, that a maturing product needs different information as it is adopted by more conservative customers than it did when it remained in the hands of early adopters. Early adopters might have succeeded with support from engineering or product developers and a few key white paper discussions of system architecture. Conservative customers whose purchases often move a product into a profitable position often require information resources that allow them to use the product successfully with fewer talented specialists. For insight into the roles that various customers play in adopting technology, see Chapter 5: Understanding the Technology Adoption Life Cycle.

A product information set that was suitable for advanced users, focusing on concepts rather than tasks, will be entirely unsuitable for beginning users who simply want to be told what to do. Conservative buyers expect technology products to improve productivity and reduce the worker cost, not increase it. Instructional information and quick reference devices assist mainstream users to come up to speed quickly and find the information they need to solve problems.

A strategic information plan that allows you to manage your portfolio creatively takes into account key customer differences, as well as indicators that emerge from identifying where a particular product is in the technology adoption life cycle. You will find that some information products need to be supported by high-quality processes that ensure usability, others need merely to be maintained at minimal levels, and still others need minimal information to support early adopters who are themselves defining how the product functionality will be used.

One solution does not fit all projects

An information-development manager will be most successful by recognizing that one product does not fit all. One information-development solution does not fit all environments. Multiple strategies are required to manage a diverse portfolio, especially when resources are scarce. Scarce resources must not be squandered on unproductive information products or unproductive work for any information product.

A strategy must not only define how to evaluate projects in the portfolio and decide upon a course of action but also define the information-development life cycle in terms of value to the customer. Your strategic planning should include not only the value associated with each project you support but also the value associated with every action to which your team members devote time and effort.

My process maturity assessments with individual companies often reveal that a significant percentage of the time spent by information developers has low value to the customers. Most dramatic is the time spent on formatting text for final deliverables, using desktop publishing systems. We know organizations in which a third or even a half of information-development time is spent on formatting. Although documents delivered to customers should certainly be readable, the customer rarely or ever benefits from the degree of control that fanatical page designers want to exercise on the output. The goal of a strategic plan should be to reduce the time spent on low-value activities and to increase the time spent on activities that provide substantial customer value. If you could trade the time spent formatting for time spent in customer studies, you would increase productivity among your team and add a base of customer knowledge that cannot be reproduced by an outsource organization. If you move your authors to tools like XML authoring, which move formatting to an automated process at the end of the information-development life cycle, you gain time for them to understand customers better and minimalize the content you deliver. If you develop a single-sourcing strategy in which you establish a database of content that is common to all or common to many, you gain time to work closely with your training or support organizations and gain customer understanding from their experiences.

Strategic planning is a critical pursuit for a successful information-development management. It leads to more effective management of your portfolio of work and the resources you need to support the portfolio. It is the first step toward recognition for the value you and your team provide to the customers and to your organization as a whole.

Best Practice—Ensuring that your projects are managed efficiently and effectively

With a strategic plan in place and the sound decisions that follow on the goals and objectives of individual projects, you are ready to approach project management as both an art and a science. The art of project management requires that you continually explore and introduce design innovations that better meet customer needs. The science of project management requires that you control project costs and deliver effective products at the time they are required.

In Part 2, "Project Management," you learn the details of effective project management, beginning with project initiation and planning, through estimating and tracking, and finally into project delivery and evaluation. As an information-development manager, you are unlikely to be managing each project yourself unless you have a very small organization. However, even if you do not manage projects directly, you are responsible for ensuring that the project management process is effective and followed successfully by all your project managers. You must also monitor the progress of projects so that you are able to assist in making trade-offs of resources and priorities.

In too many organizations, project management is assumed not to be an information-development responsibility at all. Projects are managed by other people—engineering project managers, subject-matter experts, other department heads, or anyone with whom writers are assigned to work. Unfortunately, when outsiders attempt to manage information-development projects, they underestimate the work required to produce effective information products. Engineering project managers routinely underfund information development or forget to include information developers early enough in a project to understand the functionality being developed. As a result, information developers often begin a project too late to be successful, especially when they might suggest improvements in product usability that could reduce the need for information. Subject-matter experts often believe that they should write policies and procedures themselves, often without understanding the needs of the information users. Information developers are brought in at the end of the project, with the assumption that their role is to correct the grammar and spelling and add the proper formatting.

If others in the larger organization believe they should manage the information-development project, they also assume that they should dictate the content to be produced. In most cases, the content to be produced is exactly the same content that has always been produced. Little, if any, innovation is pursued, because the project managers are not interested in innovations or are unaware of new developments in the field. It is far easier, if not necessarily less expensive, to maintain the status quo.

In addition, outside project management results in peculiar decisions about content, often to the detriment of the users. Developers and subject-matter experts use product documentation as a repository for all product information, including information that should have been part of the product requirements or specifications. That information inflates the content delivered to customers, increasing the volume and reducing accessibility. Much of the information that we find in technical manuals are artifacts of the product-development life cycle, providing no value to most customers.

Take ownership of your projects

The first responsibility of a new information-development manager is to reestablish ownership of project management. Even if an engineering or subject-matter project manager exists, you should always manage the information-development activities yourself or through lead writers or designed project managers in your organization.

However, if others are accustomed to dictating to information developers and getting what they want, you may face an uphill battle to establish ownership of your projects. If at all possible, you need the support of a senior management champion who agrees that time and money will be saved and better quality delivered less expensively if experts manage this part of the larger projects.

To gain management support, you must develop a business case for ownership. The best way to establish the business case is through customer information. If you can demonstrate for even one project that the information being delivered increases costs and reduces customer satisfaction, you have the foundation for your case.

Take for example the case for minimalism. In most organizations, you can easily establish that the customers are receiving information that is irrelevant to their goals. A simple review of one documentation set reveals content that has little or no customer value. Discussions with customers generally pinpoint the superfluous content and indicate what is necessary to their successful performance.

Pointing out the superfluous content or even the overly technical content is only the starting point. Add to your business case the cost of publishing superfluous content that no one needs or uses. The publishing costs might include the cost of localization and translation, printing costs, the costs of editing and formatting, and the costs of maintaining useless content in inventory. All of these costs can easily add up to significant expenditures.

If the information is also producing inferior performance, you have an even stronger business case to establish. You can establish a direct link between performance and information quality by surveying customers, tracking calls to customer service, and opening discussions with classroom trainers, installers, application engineers, or anyone else that has regular, direct contact with customers. Staff in these functions are often aware of customer productivity issues in learning and using your products effectively in the field.

CIDM members have often found that customer service and support personnel are often aware of areas where documentation is weak because it leads to frequent and lengthy customer calls. Comtech's consultants learned during one customer study that the instructions for installing the operating system on the company's computers was fraught with useless content, leading to an enormous number of service calls during installation. By working closely with the service organization and understanding the nature of the problems that customers were experiencing, the Comtech team was able to restructure and rewrite the installation procedures and virtually eliminate the service calls in that area. The cost savings were spectacular.

After collecting the information you need, use it to build your business case for making your organization responsible for managing information development and making the best decisions about the content that customers really need. Present your business case to your senior management to win approval. If you can show that cost savings will occur and customers are likely to be happier, you usually will succeed in making your case.

Once you have gotten senior management's support, you will need to persuade the project teams that your staff works with on the larger projects. You might begin by meeting individually with each project manager for the larger teams, explaining your business case and how the new process will make his or her job easier. Stress the fact that your staff is taking on responsibility for its own work.

One presentation is not likely to be sufficient. Some project managers will be easy to convince. They are usually the ones who have already turned responsibility for publications over to your lead writers. Others will forget what they agreed to and continue to try to dictate how the information-development projects are run and what is produced. Still others will fight the process, wanting to maintain control of the content. Succeeding among the forgetful and the recalcitrant will require constant vigilance. Even after you have established your ownership of information-development projects, new product developers, marketing managers, and subject-matter experts will appear on the scene, thinking that they own the information. You will find yourself continually making the case for your ownership clear.

Perhaps the most difficult group to convince about project ownership will be your own staff, especially those who have grown accustomed to following the dictates of engineering or other project managers. They are content to take orders because it requires little thinking on their part. They also have an exaggerated view of the importance of the technical professionals and consider them more expert about everything related to the product than they are themselves. This lack of confidence is often the most insidious form of resistance you will meet among your staff.

Some of your staff will be eager to take responsibility, having wanted that responsibility from the start. They also will be brimming with ideas for improving the information and the process used to develop the information. Your role will be to ensure that their ideas are in keeping with your strategic direction and that of the larger organization.

Other staff members will have difficulty learning to push back on the stronger, more forceful developers or subject-matter experts. They have learned to do what they are told and have little confidence in their own knowledge and judgment. Perhaps the best way to support them is to provide education to bolster their confidence. Training in information design, user studies, task analysis, and minimalism can help them feel more like experts in their field and less like scribes doing what they're told.

You will have to assume responsibility for intervening in projects that are going in the wrong direction. At the beginning of the ownership transition, it would be wise to keep close watch for information developers who are too willing to capitulate. You will need to demonstrate how to push back, helping them to build confidence. You will need to provide them with the data they need to support the position that you want them to take in designing the information deliverables. You may be aided by changing tools away from desktop publishing to XML, tools that are more difficult for others to access. You may use customer studies as evidence for better decisions about content. You may focus on cost reductions. All of these tactics will help reluctant developers and writers find their footing.

In the worst cases, you may have to escalate the problem to your champion in senior management. Discussions among the more senior managers may filter down to the project or product manager level, suggesting that they make better use of their time than in dictating technical content.

Put strong project management practices in place

Strong project management practices are essential to ensuring that your team members are adequately planning and controlling their projects. By following the recommendations in Part 2 of this book, you will have an excellent, tested process for managing your information-development projects. You need to ensure that everyone involved in projects understands the process, even if they are not the project managers themselves. Educating team members may begin by involving them in deciding how they will follow best practices in the profession and adapt them to their particular project circumstances.

I recommend using a process-development methodology that I have used successfully with many organizations. First, outline the large process functions by identifying the primary phases of your information-development process. Figure 1-3 shows the five phases of a project identified by one organization.

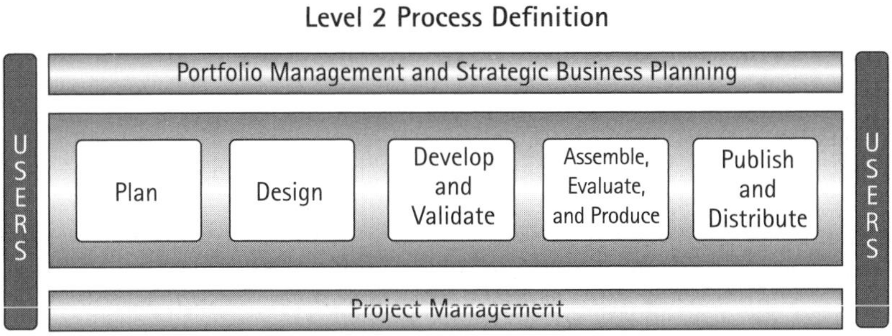

Level 2 Process Definition

Figure 1-3: The five phases of information-development projects

For each of the five or so phases you have identified, define the primary goals and activities that take place. For each of the primary phases, add more detail. Typically, a phase will have four or five basic functions that must be performed. In Figure 1-4 you see the breakdown of the Planning Phase of a project.

Finally, develop a swimlane diagram for each phase that shows who performs each activity and what relationships are required from one action to another. A swimlane diagram outlines the required actions of the information-development process in a step-by-step fashion. In developing your diagrams, you will often find the basic handoffs are not well defined and actions are duplicated that should be done once. Figure 1-5 shows the swimlane diagram related to the Project Planning Phase of the information-development process.

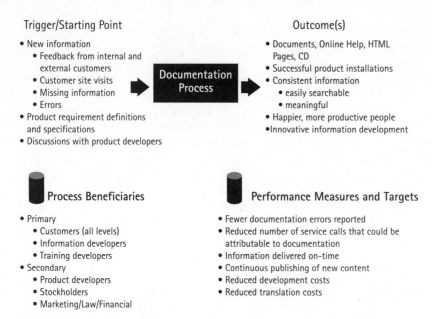

Figure 1-4: **Figure 1-4:** Project Planning Phase activities

Once your team members have agreed upon the process they hope to follow, ask them to test it in a pilot project. Undoubtedly, they will find parts of the process that don't work exactly as required and will suggest changes. However, be aware that many teams try to force the new process into their old ways of working. Be careful about changing all the activities to support an environment in which there is little collaboration and most writers work independently. If indeed, you want to move your organization to a more collaborative, global structure, you will need new processes.

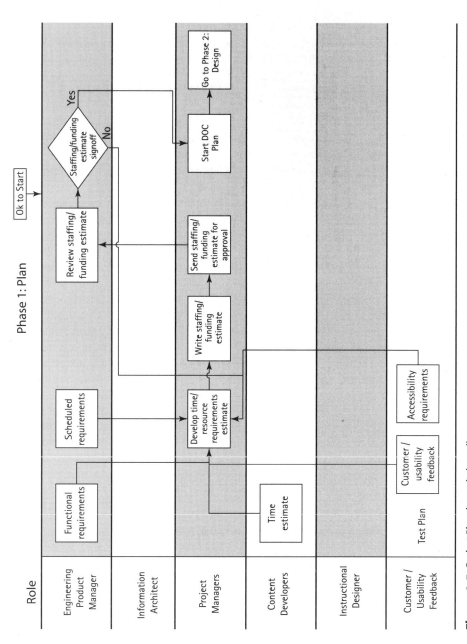

Figure 1-5: Project Planning swimlane diagram

Prioritize levels of management by project type

The swimlane diagrams usually describe the complete process that is followed for the most complex projects you conduct. They describe the "ultimate" set of activities that should occur if the projects are to be handled effectively. Some projects, however, require an abbreviated set of procedures, either because they are simpler than average or because they are emergencies.

Simple projects may include those that are quick maintenance fixes for information that is already completed and approved. They may include projects that are nearing end-of-life or have very stable information sets and an experienced user community. Such projects require fewer process steps to complete, often requiring only minimal planning and few if any design activities.

You may find it helpful to define swimlane diagrams for simple, quick-turnaround projects. Include only those steps in the process that are necessary to ensure accuracy and usability.

Emergency projects or those with very short deadlines may require yet another approach to project activities and project management. It is easy to be tempted to eliminate steps in emergency or very short projects because there is so little time. However, in my experience, emergency or short projects require more careful planning than other projects because mistakes are so damaging. Comtech has done projects that had to be turned around in less than one week, but we have included planning, design, development, and usability testing to ensure that we were headed in the right direction and able to meet customer requirements. In such a project, several steps may be done simultaneously but are done nevertheless. In a project in which we had one day to test the procedures for usability, we discovered that the procedures could not be performed at all under real-life conditions. The engineers had assumed a customer environment that simply did not exist. When we found the problem, we re-invented the procedures, called the engineer on Sunday afternoon to verify our ideas, and had the draft ready by Monday morning for review.

We often find that emergency projects are best handled with great care, but that we can often save time by reducing the number of reviewers and approvers or using a standard design pattern for the content.

Short Projects Require Advanced Planning

Comtech had a project that required a four-day turnaround for entirely new content. The project consisted of a set of instructions for parents of preschoolers setting up educational software. We had between 30 and 40 separate educational programs to document in a few days.

Planning and design were critical phases and could not be left to individuals. We met with the team working on the emergency project to define the customers' needs and consider our approach to the content design. Then, we assigned the software to team members and asked them to try it out and come up with a design for the instructions. A few hours later, the team members met again to present their design concepts and reach an agreement about a standard design pattern. With the standard design pattern in hand, the writers were able to complete the parent instructions ahead of the deadline.

All the phases of a project are important to the project's success. Be careful about agreeing that planning or design isn't required on a project because "there is not enough time." Planning makes a project shorter and more successful rather than longer.

Consider, as you plan your variances for simple, emergency, and very short projects, what is really needed to ensure success. If you don't have time for a customer study, define the "typical" customer so that all the writers understand the requirements for level of detail and writing style. If you don't have time for testing instructions with customers, consider having writers test each other's instructions. Overlap processes rather than eliminating them. In short, find the critical path through your procedures, the path that ensures accuracy and quality in the deliverable. The steps in this path are those that are required and can never be omitted. Other "nice to have" activities that you would prefer to do if you have the time are those that are easiest to eliminate to make a deadline. You may have to accept a few less than optimal page breaks rather than sacrifice accuracy and usability.

Make time for innovation

Information developers, in my experience, can easily get into a rut of following the same processes for years and updating documents without considering their usefulness. Yet, your user communities change, the nature of the information requirements change, and you need to find ways to change, too.

Making time for innovation is vitally important to maintaining a competitive edge. That competitive edge may represent competition for resources within your own company. An information-development organization that never changes how they design and deliver information may be easily outsourced or moved offshore. An information-development organization that simply regurgitates information from engineering specifications may be easily eliminated and formatting work done automatically or given to lower-cost clerical workers.

Without taking the time to innovate, you may find your staff at risk. Innovation, however, requires decisive action. It doesn't happen by keeping heads down in the cubicles. In the IPMM, innovation is one of the hallmarks of a Level 4 or 5 organization. It requires that you plan to allow innovation to happen, and it cannot occur in a vacuum.

Some managers actively pursue funding to send team members to outside activities. Attendance at conferences and workshops can stimulate new ideas through exposure to experts and colleagues. Participation in industry competitions can provide feedback from judging panels. Hiring new staff from outstanding professional education programs can introduce new ideas to a staid environment.

Note that a Balanced Scorecard, described in Chapter 3: Introduction to Portfolio Management, has as its foundation Employee Growth and Development. Too often, however, funds for professional education and training are eliminated in budget cuts or never granted. If your department is not viewed as a professional contributor to product success and is viewed as a clerical function, you are unlikely to obtain funds for education and training. If you are viewed as an important contributor, you can argue that innovations come from knowledge of the profession, the community, and the customer.

Some managers find ways close to home to pursue innovation and bring new ideas to their team members. Palmer Pearson, senior manager at Cadence Design Systems, instituted a program within his team to promote innovation. He also sponsored an Innovation Council that brought together managers from local area companies and academic institutions to share ideas and explore industry innovations. As a result, his team members began to influence innovative thinking in their larger organization, increasing the respect that product managers and software developers have for technical communication.

Monitor project management successes and failures

You can learn a great deal from monitoring the projects in which your team is engaged. Even simple accounts of the time required to complete projects can alert you to differences that may reflect management problems. You may discover, for example, that some projects consistently take more time than others, as a senior manager in the telecommunications industry discovered when she reviewed the relative time of projects. It became clear that projects done for one engineering team were always well over budget and considerably more expensive than all the other projects. As she continued to monitor these projects, she learned that the engineering project manager was weak in planning his projects and keeping them on track, negatively affecting the information-development part of the project. When senior management learned of the cost differentials, which were supported by high engineering costs as well, they quickly replaced the project manager.

As you prepare your project evaluations at the end of projects, always ask for the difference between estimates and actuals. If certain projects always take more time to complete than estimated, you may have to revise your estimating algorithms. If some projects are always under budget, you may also have inaccurate estimates. You may find that some team members take significantly more time to complete projects than others. Perhaps their projects are more difficult technically or fraught with schedule and product changes. Or, they may be adding complexities to the projects that are not justified.

Project failures often come when team members fail to respect each other's contributions. They often are poorly planned at the highest levels, with constantly changing requirements. They are often led by people who seem unresponsive to concerns about cost overruns or unnecessary changes that are products of personal opinion rather than good project management.

Project successes come from sound planning, conscientious staff, and cooperation among all the various members of the larger development team. They come when people work together effectively and respect each other's responsibilities. Successes frequently are the product of good management. These projects are well planned, estimated, and tracked. When changes inevitably happen, the lead writers or your project managers assess the affect on the team's work and suggest ways to keep the project moving toward a sound conclusion.

Knowing the difference between failures and successes and managing for success is part of your responsibility as information-development manager. Once again, your active involvement in and awareness of the projects is critical.

Summary

No one ever told us that management is easy. As you move into information-development management and gain experience in your role, you find that you have much to learn—about managing people, projects, and the business of your organization. You play a balancing act in allocating resources, helping employees grow, introducing sound management practices, inviting innovation in your design work, and managing projects that require different approaches to be successful and to keep costs under control.

In the two parts of this book, you learn about

- ✔ managing the portfolio of projects for which you are responsible

- ✔ guiding successful project management

- ✔ ensuring that you have the tools and insights to run a professional organization that aligns with larger corporate goals

The best practices in each chapter lead you through the processes and decisions you need to make to become an effective information-development manager. Each chapter ends with a summary of the key concepts discussed in the chapter. I hope you profit from the information and use the ideas to your advantage. Best wishes.

"I think we are at level 2.
How do we get to the next level?"

Chapter 2

The Information Process Maturity Model

> You can often tell an organization is in crisis by the attitudes expressed toward its customers and other outsiders. An organization in crisis is so entangled in its internal problems that it forgets its fundamental reason for existence.
>
> —Gerald Weinberg, *Quality Software Management*[1]

"How do we compare with other information-development organizations?"

"Are we following best practices?"

These questions are typical of the most Frequently Asked Questions (FAQs) that information-development managers focus on. The FAQs continue:

- ✔ How do our processes match with those of best-in-class companies?

- ✔ Are we doing as well as or better than our competitors?

- ✔ Is our information as good as everyone else's?

- ✔ What might we do to improve?

---

[1] Gerald M. Weinberg, *Quality Software Management: Systems Thinking,* New York: Dorset House Publishing, 1991.

31

Given the pressure on every manager to improve the productivity of staff members, reduce the costs of information development, and, at the same time, deliver a quality product to customers, the ability to compare a department with leaders in the field is essential. The Information Process Maturity Model (IPMM) was designed to provide standard methodology for making such comparisons. Managers worldwide have used the IPMM to demonstrate to their larger organizations that their processes follow industry standards, that they exceed industry standards in some areas, and that they need support to improve processes in other areas.

During a recent IPMM assessment, the senior manager to whom the publications manager reported stated that he expected the manager and the department to move from their current Level 2 of process maturity to a Level 4 within two years. At Level 4, the vice president felt that the department would be operating at an optimal level and be capable of meeting the changing needs of the customers.

Because many engineering managers are already familiar with process maturity assessments in their own fields of software or hardware development, they understand the IPMM's role in providing a baseline of comparison with mature and immature organizations. They recognize that activities to improve process maturity will also help to reduce costs and improve the quality of information products because they have experienced the same gains in the groups they manage today.

Today, the IPMM is fully supported by The Center for Information-Development Management (CIDM), which sponsors the IPMM assessments that I conduct.

The Information Process Maturity Model

The Information Process Maturity Model (IPMM) has grown as an increasingly reliable and informative response to the needs of information-development managers. Managers are asked to demonstrate that they are at least as efficient and cost-effective as others in the field. They are asked by senior management if their work measures up to others' work in the industry. They may even be asked if other departments have found ways to produce information that meets customer needs at lower costs by learning from the experience of similar successful organizations.

What is the IPMM?

The IPMM describes the practices that make an information-development organization successful. Since I first defined the IPMM in 1992, I have regularly updated the model based on a continuing analysis of organizations that exhibit best practices in the information-development industry. I seek out organizations and managers that have a strategic vision aligned with that of their larger organizations, are focused on supporting the profitability of their companies, know how to develop information customers truly need, and are skilled at running effective and efficient business operations.

Although I continue to update the model on which the IPMM is based, I have found that remarkably little has changed in our assessment of the characteristics of an effective information-development organization. However, I have seen changes in the opportunities for education and training of staff members, especially inhouse training. I have seen changes in organizational structure that require new management skills, including the management of remote writing groups, more often today located in countries with few information-development traditions or educational opportunities. I have also seen a considerable increase in the amount of outsourcing in certain industries, including telecommunications and computer hardware. Outsourcing has been extended to include groups in countries that offer low-cost labor.

Capable managers leading best-in-class organizations have had to learn to cope with the challenges by embracing technical and design innovations at the same time that they have had to pursue more effective ways to reduce costs. I believe that the IPMM continues to be an extremely valuable tool for managers seeking to better understand their own organizations and their relationship with others inside and outside their companies.

How did it get started?

I initially defined the IPMM in *Managing Your Documentation Projects.* However, work on the IPMM had begun some years before. Throughout the 1980s, I was engaged in many consulting projects with a wide variety of organizations, all engaged in producing information for customers and employees. I worked with software development companies that needed technical documentation and training materials for their products; with mining, energy, and manufacturing companies that were developing new business proposals and reporting on field research projects to clients; and with numerous government departments and research associations looking for ways to communicate their work to the public. In each of these encounters, I found both immature organizations unable to organize simple process flows or control costs and highly successful organizations that seemed to do everything well. And, of course, I found most organizations somewhere in between, doing some things very well and others barely adequately. As a result, I began to form a picture of the range of characteristic behaviors that seemed to make a difference in the organization's success.

At the same time, I became interested in the work of Gerald Weinberg in systems engineering. A physicist by education, Weinberg studied the success of organizations engaged in software engineering. Weinberg postulated five levels of organizational maturity in the software industry, from Pattern 0: Oblivious to Pattern 5: Congruent.

Also, during that time, the Software Engineering Institute (SEI) was organized by the Department of Defense through Carnegie Mellon University. I followed their development of a five-level model of maturity in software development with interest, although I found Weinberg's work to be much less government-focused. The problem I saw with the original SEI model was its disregard for information development. Technical information or customer documentation, it seems, was viewed as a byproduct of the software-development life cycle, without a set of mature processes of its own. The underlying assumption seemed to be that a software-development organization was successful if it was able to meet deadlines, stay within budget, and deliver product that met the specifications.

Usability, a user-centered focus, or a recognition that people needed effective training and information to perform successfully using the new products was not addressed.

The emerging problem, as I saw it, was that software developers, many of whom employed growing legions of technical writers in the early 1990s, were not going to learn about responsible information development or usability from the SEI model. In fact, a few information-development managers had called to report how concerned they were when their companies brought in assessors for what became known as the Capabilities Maturity Model (CMM). These assessors either were uninterested in looking at information development or usability or were oblivious to the special requirements of information development in effectively supporting customer performance and success.

Quite obviously, there was a need for a model to define process requirements for information development. Hence the development of the IPMM. Despite the impetus provided by the unfortunate assumptions in the CMM, the IPMM owes more to Weinberg than to the SEI. Weinberg provided a thoughtful, customer-oriented structure that I could emulate.

An IPMM assessment

The CIDM regularly conducts IPMM assessments of information-development organizations. During an assessment, the information development staff and managers complete a detailed questionnaire on the eight characteristics that form the core of the IPMM. The questionnaire results are tabulated and used as a basis for the on-site interview with the staff and managers. The interviews help to understand in depth how the organization actually functions. In addition, interviews are held with staff members, managers, and stakeholders in peer organizations and with senior management to gather their perspectives about the processes and qualities of the information-development organization. The assessment gathers data on current practices in the eight key areas described in Table 2-1. The assessors also review sample documents produced during the information-development life cycle, such as information plans, project plans, project estimates, actual project data, and so on. The assessors also review examples of information products produced by the organization.

In the comprehensive assessment report, the assessors provide detailed findings describing the current level of process maturity and recommendations for moving to the next highest level. They also provide benchmark data that compares the organization with similar organizations, based either on product area, size, or organizational direction. The assessment report recommends actions that conform to the Six Sigma continuous process-improvement model.

What are the five levels?

In parallel to the CMM and Weinberg's model, the IPMM defines five levels of process maturity. The five levels are described in Table 2-1 with basic recommendations for the transition from one level to the next.

The five levels of the IPMM provide you with a model both to assess your current organization and to set your sights on a move to the next level. The IPMM gives you a blueprint for change by capturing the characteristics of successful organizations that routinely meet or exceed customers' expectations.

Table 2-1: The Five Levels of Process Maturity

IPMM Level	Description	Transition to the Next Level
Level 1: Ad hoc	Ad-hoc organizations are characterized chiefly by a lack of structure and uniform practices. Information developers generally work alone, most often hired and managed by someone from another field, such as engineering or software development. As a result of working alone, each individual follows a unique process and applies standards independently. The quality of the final product is highly dependent on the professionalism and expertise of the individual. No quality assurance activities take place except for reviews for technical accuracy. There is little opportunity to understand the needs of the customer.	To move to Level 2, the organization usually needs to build cooperation among individual communicators. In most cases, a management position is created and a department organized. The information developers report to the publications manager and work together in a department. The manager and department members understand the need for common processes and design standards for the publications or other information products of the department.
Level 2: Rudimentary	Rudimentary organizations are in the process of putting their structures and standards in place. Initially, the group of information developers collaborates to establish style standards and institute uniform practices. At a management level, a new manager and a new department bring together formerly isolated information developers. The new manager must work to create a unified organization in the face of opposition from staff who were once autonomous. The manager and staff begin to institute quality assurance practices, including copyediting, developmental editing, and peer reviews. Despite good intentions, the rudimentary new practices are often abandoned under pressure of deadlines and constantly changing requirements, as well as lack of commitment among the staff to changing individual practices. Level 2 can be a difficult and awkward transition period.	To move to Level 3, the organization and its leadership must make a firm commitment to following the processes and standards put into place. They need a standard set of templates, a style guide, a project workflow, and sound processes in place to plan, estimate, and track projects.

Continued

Table 2-1: The Five Levels of Process Maturity (Continued)

IPMM Level	Description	Transition to the Next Level
Level 3: Organized and Repeatable	Organized and repeatable organizations have come of age after passing through he fire of Level 2. Now the majority of the staff support and are committed to following uniform processes, templates, and standards. They are convinced that the best practices they have put in place constitute the right way to run an information-development department. Their move has been supported by a strong leader who has a vision for the organization and its future and is helping the staff realize that vision. The leader and staff recognize the importance of sound planning and quality assurance activities, and they are incorporated into every project. Attention is given to hiring qualified individuals and providing them with opportunities for continuing education. Because processes work so well, staff begin to find opportunities for improvement, including redesign of legacy information, customer studies, and benchmarking with other organizations.	To move to Level 4 requires a firm commitment to follow high-quality practices, not only within the organization but also in relationship to peer organizations. Everyone needs to commit to project planning, estimating, and scheduling, and editing and reviews, even when it's difficult and they are pressed for time. If not already begun, customer studies need to be vigorously pursued.
Level 4: Managed and Sustainable	A managed and sustainable organization has made a strong and consistent commitment to the mature practices of a Level 3 organization. In fact, the leadership may change without a loss of commitment to planning, quality assurance, hiring and training, and budgetary controls. Level 4 organizations become increasingly sophisticated in handling customer studies, assessing and meeting customer needs (including regular usability analysis), and managing return on investment. Level 4 organizations are often recognized as effective by the larger organization and often play a leadership role in the development of new teams within the organizations or in the assimilation of teams acquired through mergers and acquisitions. In many cases, staff members participate in a matrixed structure in which they represent the interests and goals of information development regarding product design, support, training, human factors, and other parts of the organization. Managers are often directors or vice presidents and are recognized for their business acumen. Frequently, they serve on business leadership teams.	To move to Level 5, the leadership needs to increase their business understanding. They need to strengthen their commitment to increasing productivity, controlling and reducing costs, focusing on customer satisfaction, and aligning strategically with overall business goals and objectives.

IPMM Level	Description	Transition to the Next Level
Level 5: Optimizing	Level 5 organizations are actively involved in promoting information development throughout the organization. From an outside viewpoint, staff members may seem similar to those at Level 1 because they are completely matrixed. Members of a Level 5 organization acknowledge, embrace, and foster community business analytics. Their content strategy is closely linked to the corporation's product strategy. They provide leadership in information architecture not only within the organization but within the broader information-development community. They are seen as a destination hiring organization throughout the field. They often receive awards for their innovative activities.	Level 5 organizations are very hard to sustain. The forces pulling any organization back to Level 1 are strong. Reorganizations in which major leadership is lost and staff members are dispersed among different business units and work for managers unequipped to promote innovation can cause any organization from Level 2 through 5 to collapse back to Level 1.

Maturity Levels of the IPMM

The five maturity levels of the IPMM are closely related to the ability of your organization to innovate in information design and increase your productivity and efficiency. Process maturity relates especially well to evaluating the ability of an organization to move to a content-management environment for planning, developing, managing, and tracking their information-development activities.

Level 0: Oblivious

Although there are five primary levels in the IPMM, I find it useful to describe a sixth level that I call Level 0. Level 0 describes an organization that produces technical information but does not recognize the importance of employing people who are experts in information development. The technical information is usually produced by technical experts in the context of their work in product, process, or service development. If a Level 0 organization employs anyone to produce information products, the work is primarily dedicated to formatting final deliverables and correcting basic grammar and spelling.

Not Yet Ready for Prime Time

The head of customer service and support at Hyperactive Development is concerned that the maintenance engineers cannot easily find the information they need to repair customer equipment. They have access to a file server that holds documents and technical bulletins prepared by the design and development engineers and the people in manufacturing that assembled the machines. Customers have complained that repair work is often slow, extending equipment downtime and affecting their productivity. If the problem gets much worse, Hyperactive faces fines for violating service level agreements.

The department head attended a conference on content management that included sessions espousing the advantages of moving documents into a content-management system and using XML markup to enable reuse[2] and repurposing.[3] He announces to his staff that the company needs to "get into" content management right away for all its marketing and customer documents, especially the product manuals.

Unfortunately, the organization has no one who understands content management and how information should be developed to meet the needs of the field staff. They decide to bring in The Center for Information-Development Management (CIDM) to conduct a process assessment and provide recommendations for improving their information delivery.

At the first meeting with the support team, the assessors ask who creates the information and what processes are used to develop, review, and publish the maintenance information.

The service manager explains that many people in the company create the documents. The research and development engineers write their own product specifications, and a few of the engineers and programmers put together the user manuals for the products. Manufacturing engineers write technical bulletins as they assemble

[2] *Reuse refers to the process of using chunks of content in more than one context or document.*
[3] *Repurposing refers to delivery the same information in more than one medium print, HTML, help, and so on.*

the products in the factory. The field installation and maintenance team in customer support creates install guides and maintenance manuals for their own use. Customer telephone support develops FAQs for the company's website and creates ad-hoc documents that they send to customers in response to queries.

In brief, documents are created everywhere by everyone. They each develop the documents any way they like, with no common look and feel. Company officials have vehemently opposed hiring technical communicators in the past. They feel that the engineers know the products best and should be able to write about them.

When it comes to publishing the final versions, once again, staff members use a wide variety of processes, each developed and implemented independently. They put documents on a company internal website, which does not have a search system. They also develop CDs with PDFs of the documents but not a search system either.

In the Information Process Maturity Model (IPMM), Hyperactive Development is at Level 0: Oblivious. In an oblivious organization, management has not yet recognized the need for standard information produced following a quality assurance process.

Despite the service department's interest in content management, Hyperactive Development is not yet ready for prime time. They need first to pursue several important and potentially difficult steps toward standardization before content management makes sense:

- ✔ Review who creates documents and what documents they are creating (an inventory of existing materials).

- ✔ Understand the processes being used throughout each document-development life cycle.

- ✔ Query customers, both internal and external, about the successes and failures of the current documentation set.

- ✔ Create a standard set of documents related to the products being delivered and the users' needs for information.

- ✔ Adopt a standard document design that takes into account relevant differences among the document types. Create a template for each document type with common style names across the set.

- ✔ Train staff members on using the standard.

- ✔ Institute a quality check process to ensure that the outgoing documents follow the standard design and template.

Most importantly, to move to Level 1 or higher level of process maturity, Hyperactive needs to hire experienced information developers to participate in the information-development process.

Rest assured, these steps will not be easy to pursue. Dealing with an essentially ad-hoc organization is always difficult. In an oblivious organization (not yet ad hoc), people don't see the point of standards or of using a regular information-development process. They value their independence in creating anything they want, rather than the need to deliver usable, accessible, and effective information to customers and staff. Impressing everyone about the need for standards is especially difficult when no experienced technical communicators are part of the picture. Occasional information developers have little interest in standards or patience with a defined process. Because document development is not part of their regular job description, it's low on their priority lists.

Even if you are working with experienced technical communicators, you are likely to encounter resistance from staff who have long been independent. We all know, of course, that even under the best of circumstances, communicators find it difficult to compromise on style preferences. Everyone thinks his or her way is best.

Quite clearly, an oblivious organization is not ready for content management. I find that if they invest in a content-management system, they will use it as an expensive file server with version control. Version control alone provides little calculable return on investment. The only path available to an oblivious organization wanting to take advantage of content management and reuse is through standards and a professional information-development team.

Perhaps the best means of achieving some success is to ensure the support of a champion in senior management. Champions among senior management are most likely to emerge in response to customer complaints or out-of-control costs. Customers unhappy with the current confused state of affairs often become quite vocal about their problems. Staff members who cannot find information that supports their work voice complaints as well.

Oblivious organizations are on the slow path toward information-development solutions. It often takes two years to move an organization one level in the IPMM.

Level 1: Ad hoc

Level 1: Ad hoc is the first recognized level of process maturity. In most cases, a Level 0: Oblivious organization has no mechanism for recognizing the need for a more mature process, that is, unless someone in senior management decides that problems with customer or staff dissatisfaction have reached critical levels.

Level 1 organizations often appear when information developers are first hired, although at early stages, the information developers may have no experience in the profession and little understanding of standards for product or process. In other cases, people with experience in the profession are hired by individual product or process development teams to work exclusively on their information development. Information developers report to people managing the technical teams, rather than to experienced information-development managers. The individuals may be skilled and experienced, producing excellent work and instituting their own good processes, or they may be completely inexperienced with little idea of what is required to deliver quality information in a cost-effective way.

Independence Rules in Ad-Hoc Organizations

The CIDM has been asked to conduct an IPMM assessment for Checko Systems, a huge manufacturing company with a wide range of products. When it was a fledgling start-up, Checko had a central technical publications organization.

Today, however, Checko has a number of autonomous information developers, each associated with a set of products. The information developers are co-located with the engineering teams and are distributed around the world at the company's many development locations. Some of the information developers work in small teams and have set up their own standards and processes. Others work alone. Overall in the company, no standard approach to information development exists.

The IPMM assessors' initial investigation centers on structure standards:

- ✔ How is information created by the diverse, autonomous departments and individuals?

- ✔ Have any of the teams set up standards for document design? Is anyone using a style guide?

- ✔ Are there any common processes in place among the teams? Are the staff willing or able to cooperate with one another?

After several meetings with individual contributors and small teams, the CIDM assessors conclude that they are dealing with a Level 1: Ad-hoc organization. Processes, information structures, and even basic design standards differ markedly from team to team. Although all the documents are delivered to customers on a corporate website and a single CD-ROM, simple searches reveal that each document is a complete surprise. Terminology is not standard; tables of contents for similar products are different; the users cannot easily tell which product or release they are reading about when they use full-text search.

From the 30,000-foot view, a Level 1 in process maturity appears much like Level 0. The difference is that in a Level 1 organization, professional technical information developers are, for the most part, responsible for information development. Some Level 1 organizations may have a few small departments with experienced communicators. Other Level 1 organizations, closer to Level 0, depend upon technical managers to supervise information developers. Still others have a mixed approach.

With so many diverse approaches to information design and development processes, Checko needs to consolidate and standardize before taking on any new projects, especially content management. Individuals and small teams will have to pool their resources, creating a unified approach to information development and moving to Level 2: Rudimentary.

Individual initiative in Level 1 organizations

In observing many Level 1 organizations, I have always noted the strength of individual initiatives that foster innovative approaches to delivering content. But it is this individual behavior that limits Level 1 organizations to ad-hoc approaches that succeed only when they are limited in scale. For an organization to work toward a uniform, standardized solution involving more than a handful of people, staff members will have to work together. Ad-hoc organizations, by definition, do not have the degree of collaboration or cohesiveness necessary.

Once staff members in a Level 1 organization decide that they will benefit by working together and agreeing on standards in process, design, and technology, they are already on their way to becoming a Level 2.

Centralizing production

Level 1: Ad-hoc organizations often begin to standardize by developing a central production team that is responsible for final deliverables. Typically, such centralization occurs when it becomes cost-effective for an organization to publish technical documents jointly. The organization decides to issue a single CD-ROM with all their documents or place the documents on a centralized website.

Typically, the work of a central production team is repurposing. They take the documents produced by the information developers in different parts of the corporation and issue them as a unit. Costs are saved because only a few people are responsible for production activities (producing PDFs, creating CD-ROMs, posting to a website) rather than everyone working independently or in small teams.

Smoothing the production process, automating many activities, and using non-proprietary tools result in significant returns on investment. By reducing production costs and time, documents are not only less expensive and less time-consuming to produce, but also more time is left before production to ensure accurate content.

For a Level 1 organization to move to Level 2 of process maturity, they may consider several strategies:

- ✔ To achieve consistency, individuals or small groups must form a coalition to develop information-design standards and implement common tools and templates for all the documents produced throughout the company.

- ✔ Once standards are in place, individual departments will need to train staff members to use the standards effectively.

- ✔ By working together, the small groups need to adopt uniform processes for information development, starting with creating information plans and collecting data about existing projects. Those data form a base for future project estimating.

- ✔ The central production unit may continue to repurpose existing documents to PDF for electronic delivery with no changes. However, without agreement on a corporate-wide information design and the development of standards for deliverables and content, customers will continue to be frustrated by inconsistency in the documents they receive.

The most critical development that must take place to move from Level 1 to Level 2 is leadership. Although it is possible for an ad-hoc organization to develop standard processes, without a manager to direct the organization, grass-roots improvements are difficult to initiate or to sustain. In most cases, a Level 1 organization remains at Level 1 until a management change is made. A new manager is hired or promoted with instructions to improve quality, standards, and processes, often by consolidating the work of distributed individuals or small teams into a central organization.

Standards in document design must be accompanied by developing unified processes. The same processes must be used in each sub-organization to develop documents so that documents are consistent, not only at the level of formatting but of content and style. For information to be shared in the future among different product areas, the level of detail and coverage, as well as the writing styles, must be the same.

An organization can move from a Level 1 to a Level 2 in the IPMM by implementing rudimentary standards in three areas: process management, information design, and technology. Level 2: Rudimentary describes the state that occurs when an organization begins to establish uniform practices and consistent designs. At Level 1, in which uniformity does not exist, creating a comprehensive and unified information-development solution is usually not possible.

Level 2: Rudimentary

Level 2 organizations are on a journey from Level 1 to Level 3. As a result, they are in a very uncomfortable position with a lot of changes to make. The more changes, the more instability staff members have to deal with.

Moving to Level 2

Checko Systems is determined to reorganize its information-development staff into a more coordinated group to take advantage of the cost savings and benefits of centralization.

Checko still has a number of autonomous individuals and small teams, but they are rapidly restructuring them into a central organization. The company has hired a professional information-development manager with more than 10 years of experience in the field and 5 years managing information-development in a competitor organization. They have management support for the change, but they expect it to take 2 to 3 years to complete. Two of the largest teams have already joined forces.

Following through on the IPMM recommendations, the new Checko central information-development organization is working on its Information Model. They are

- ✔ beginning a preliminary user study to discover how customers use information, how they judge the existing information, and what would provide them with increased value

- ✔ working to define standard information types, such as procedures, concepts, and reference materials, that all the information developers will be able to use

- ✔ deciding which information types are appropriate for each set of deliverables to a diverse user community

- ✔ deciding on a new information-development process with five distinct phases from planning and design through development, production, and delivery

Nonetheless, just because Checko has a plan to move to Level 3 of process maturity doesn't mean it's going to be easy. In fact, the level of complaining has already gone up significantly. The staff members don't necessarily agree on which information types are needed or how they should be defined. They don't have much information about their users and don't know how to begin their user study.

Preparing a Level 2 organization for content management

For an organization to move successfully into Level 2 with plans for Level 3, it needs a centralization strategy:

✔ To achieve consistency, individual departments merge to form a centralized organization with links to other independent publications groups within the company.

✔ The new, merged group begins to develop information-design standards and implement common tools and templates for all the information produced.

✔ The merged group provides training on the new standards and tools for their own staff and the staff of the remaining independent teams.

✔ They begin to develop a set of processes to govern their activities that can be tested and then passed on to others.

Level 2: Rudimentary describes the state that occurs when an organization begins to establish uniform practices and a consistent information architecture. As a Level 2 organization consolidates and unifies, it becomes increasingly ready to implement a content-management solution. It is important to point out, however, that the changes recommended here take time and concerted effort. If the company as a whole is unstable or the information-development organization encounters resistance, the process will be slowed and may be derailed.

Level 2 organizations, I find, are particularly unstable. There is great pressure to slip back into the ad-hoc world. But even if you don't achieve full Level 3 maturity at first, by experiencing what it will take to manage the change, the next effort should go more smoothly.

Making content-management decisions

Level 2 organizations are actually doing the work needed to prepare for more mature process and management.

As the staff members work through aspects of their Information Model, they should identify a potential pilot project to institute the new design ideas.

The pilot project is characteristic of all information-design planning, but it is particularly important for Level 2 organizations because they are not used to working collaboratively.

The pilot project must be a collaborative effort, involving representatives involved in all aspects of the new process. Stakeholders may include production, localization, training, and technical support, as well as customer representatives. If information is to change for the better and be brought under control, the potential internal users of that information should be involved in its redesign.

The pilot project not only tests new design ideas but also principally tests the new process. Those involved in the pilot must understand the goals of the new process in managing costs and assuring quality. If the goals are not well understood and defended, it is too easy for team members to revert to their old ways of working. At Level 2, most members of the organization are experienced at working independently and often believe they achieve better results than when they work with team members. The leader of the pilot project has a major role to play in defending the new process and ensuring that it doesn't get bypassed as soon as problems occur.

Level 2 organizations have an opportunity to create content that can be reused and repurposed. It might even be possible to introduce modular, topic-based authoring in a pilot project. By developing modules and standardizing the design of the modules with information types, information developers can develop content that is meant for reuse rather than for individual documents. New deliverables can be created out of modules that work effectively together because they are designed with the standards in place.

Level 2 organizations are capable of building a path to comprehensive, development-level content management and reuse, rather than simple repurposing of text into variations.

Centralizing production

In a Level 1 organization, production teams generally concentrate on delivering books or PDFs of books to customers in print and electronically. As the Level 2 organization moves toward a more modular approach to information design, the production team will have to rethink its processes as well.

Modular content is, almost by definition, more difficult to control than whole books. Smaller chunks of content need to be correctly assembled into appropriate contexts either for delivery as books or into the networked relationships of a website. I believe that production teams have a considerable challenge in defining the processes they will use to prepare content for delivery, both as static content organized into deliverables during production and as dynamic content that is updated on a regular basis.

Timing the effort

Generally, I have found that for large organizations of 20 or so members, the effort of moving from Level 1 to Level 3 takes at least two years. Some staff make the transition quickly; others wait to see what happens, and the laggards never change but spend a lot of time criticizing everyone else. As the manager of the new organization, you will have to convince other parts of the company that you are focused on change.

Making the changes to your information design is often your own affair in information development with some pushback from marketing. However, you may also have to convince product managers and engineering directors that the process you use to develop information and the information that crosses product lines must be standardized. That convincing will take time.

Many organizations find that cost savings are the only motivation for change, while others are convinced only if the change can be directly linked to overall customer satisfaction. As the new manager leading the move to Level 3, you need to develop a comprehensive vision. Clearly define where costs will be saved through process redesign and how the information redesign and delivery method will address customer needs. Deliver this message and sell your vision as often as possible to anyone who could affect the outcome of your project.

Level 3: Organized and repeatable

In the software-development process maturity model, an organization must be at Level 3 before it will be considered for government contracts. At Level 3, the organization has the stability in process that ensures that it will deliver a quality product on schedule and budget. The same is true in information development. An IPMM Level 3 ensures that your organization has in place repeatable processes that help guarantee results. If the processes are followed, the quality of the output should be assured.

As a mature organization, a Level 3 team has made a strong commitment to following its process. The most common quote at this level: "That's just the way we do things around here." Team members would not consider skipping an important step in its information-development process. Team members rely upon the process to "save the day" when an emergency occurs, rather than abandoning the process as soon as something goes wrong.

A Level 3 team has documented its process, following an ISO requirement. The primary phases of the process are well defined in terms of inputs and outputs. The details of the process in each phase is documented. People have been trained on the new process and shown that it results in quality work with less time taken up by low-value activities. At Level 3, the activities that are part of the process have been well established and are taken seriously by all the team members.

Because a Level 3 team has a well-defined process in place, they begin to have time to devote to activities that add more value to customer deliverables. They can add quality assurance steps that may have been neglected, including developmental editing, in which team members work together to plan deliverables and review each other's work, asking a senior editor to work with new or weaker team members. They can add time to track the activities in the process and eliminate or reduce the time devoted to those that don't directly contribute to customer value.

At Level 3, an organization can effectively plan and implement a content-management system that advances their goals of reusing information in multiple deliverables and for diverse audiences. Prior to Level 3, the collaboration and coordination of team members required for successful content management is impossible to achieve.

Settling on Mature Processes

Sonoita Technologies spent three years working out effective processes and practices to become a Level 3 organization in the IPMM, following their initial assessment as a Level 1. It was quite a struggle at times, particularly when the information-development manager faced a potential revolt from some of the information developers who preferred working on their own. In a few cases (a very few, fortunately), the staff members decided to move on.

Since the first of the year, the manager has felt that her staff has "made it."

They completed their first comprehensive user study. They discovered, to everyone's surprise, that the first-level technicians at their telecom customers were not given access to the product documentation. Instead, they received on-the-job training and a field notebook of the procedures they were to follow.

Team members in information development had succeeded in developing common standards with their sister organization on the east coast. Agreeing on standards had been difficult because both groups thought their way of doing things was best. Finally, they were able to reach a compromise that included the best of both.

The new process standard was still proving to be a challenge. Too many staff members still felt it was easier to dive into a new project rather than developing a content plan and estimating project hours. The planned project management training should at least ensure that everyone is using the same language to talk about the planning activities.

Despite the clear progress they have made, the management team is alert to the possibility of backsliding. Until they all have experience working and cooperating in a more mature business environment, they will struggle with change. The senior manager hopes that her recent promotion to director of technical publications signals senior management's support for their efforts.

When a Level 3 organization begins to consider a new process, an effective starting point is to assemble a small team interested in knowing more and leading a possible implementation project. The team members will work on several issues:

✔ becoming expert in the newly defined process standards, especially the rationale for their implementation

✔ understanding the requirements of their customers for specific content, especially the planning and surveillance engineers they have interviewed and observed during the site studies

✔ relating the customers' need for information to a set of clearly defined information types

✔ standardizing the information types so that they contained a standard set of content units

Once they have defined their information types and related them to customer needs, they will have the rudiments of a new Information Model.

To move their new process and design initiative beyond the information-development organization, the team must invite to the planning process representatives of other departments, including

✔ training

✔ customer support

✔ marketing and marketing communications

✔ contracts and proposals

Each of the stakeholders may prove to be interested in joining the initiative. Without their contributions, information development may not be able to support the full cost of new technologies that will help them decrease development time and costs. If other departments participate, they will each strengthen the business case needed to obtain approval and funding for new technology from the senior management.

One of the first assignments for the stakeholders is to create a vision of how they can work together to develop and contribute to the enterprise knowledge base. They must envision what the new information-development environment will look like if they are successful in reaching their goals.

It's important for the team and the managers to recognize that the consolidation of Level 3 activities is not going to happen overnight. You need to plan for 12 to 18 months for planning and implementation. Remember that team members have their regular deadlines to meet while they are working on new process and design projects. It will take careful planning to ensure that they have time for a reasonable level of involvement in the projects.

Starting small

You know that many changes in organization and process are ahead. It isn't enough to buy a technology; your team has to make changes in the way it handles its work. For the first project, your team might decide to start with a small set of reasonably well-structured information and work through the entire new process.

Level 4: Managed and sustainable

Information developers in a Level 4 organization typically have a standard process and information architecture fully developed. Everyone is committed to following the process as they develop their information. In a Level 4 organization, information-development managers have budgetary control over the organization's activities. That includes gathering considerable data about the cost of individual projects. With any new initiative, the managers will look for new ways to reduce development costs while improving the quality of deliverables.

A Level 4 organization has a strong base from which to work but typically the team needs to solidify the previous gains and introduce more innovative approaches. Without innovation, it is all too likely that a Level 4 organization can develop a complacent attitude and become increasingly bureaucratic. Many Level 3 organizations turn into dead-end bureaucracies because what was once a new process and a new design worked so well. Unfortunately, no one seems to have the initiative to question the status quo. If an organization fails to pursue continuous innovation and improvement in processes and products, it ceases to be an interesting place to work. If management has attracted a first-rate staff, they must remain motivated.

The characteristics required to sustain a Level 4 organization are as follows:

✔ Strong development processes to support the information life cycle, and complete dedication among staff members in following the processes.

✔ A desire to continue to innovate in information design and process re-engineering because the goals of the organization continue to change, just as the business goals with which they align also change in response to changing customer and market environments.

✔ A new prominence for customer studies. Although they may have conducted several customer studies in the past three years, the entire team is dedicated to knowing customers better.

✔ A leadership role in developing a tools and technology strategy for the organization, including active involvement in tools development, often through advisory roles with tools vendors.

✔ Dedication to acquiring and analyzing business data. Business analytics help to promote a closer aligning with corporate strategy.

✔ If the larger organization is acquiring new information-development teams or adding new projects, the Level 4 organization takes its outreach role seriously. They provide the learning opportunities so that team members are available to lead other groups as they form. They develop leaders who can seed new, emerging teams.

✔ A strong collaborative and cross-functional business model in partnership with stakeholders throughout the larger organization.

✔ A sustainable organizational structure that includes succession planning and attention to organization health.

✔ Considerable opportunities for growth among staff members.

✔ Industry-level awareness of the profession. The company provides examples for other information developers; wins recognition for business processes. They are beginning to be seen as a destination organization for new people coming into the profession, especially from prestigious graduate and undergraduate university programs.

✔ Willing to promote and fund research in the field, including the ongoing study of customer requirements.

Feeling Comfortable with Mature Processes

ODS continues to surprise the CIDM assessment team. They have followed their progress for several years, ever since the software-engineering area of the company earned a Level 4 in process maturity from the Software Engineering Institute (SEI). Implementing Level 4 processes had enabled ODS to turn things around from being a money-losing to being a money-making company. The careful application of systems thinking helped ODS management recognize that they needed to abandon some parts of their older technology and focus on customer-driven changes to their new products.

A long-time senior manager in the group has been promoted to director of information-development after the previous long-time manager retired. She is aware of the progress the earlier manager had made in stabilizing processes within the organization, but she also knows there is much to do for the information developers to equal the accomplishments of the software and hardware engineers. The earlier manager walked into a stable Level 3 organization; the new senior manager's job is to move the team to Level 4.

The manager works closely with assessment team to define the characteristics they must have in place to be recognized as a Level 4 in the IPMM (Information Process Maturity Model).

Innovation in design

For a Level 4 organization, maintaining an innovative approach to process and information architecture is essential for continued growth. Modular design, continuous publishing, and customer value—all are part of the innovations that Level 4 is prepared to pursue. They are closely aligned with corporate strategy, with a balanced scorecard in place to evaluate results.

Although there are skeptics in their midst, the very nature of a Level 4 information-development team means that most staff members are willing to give a new idea a try. They have all been intimately involved in studying their customers. As a result, they recognize the need for innovation in information delivery to better meet customer needs. The entire department is user-focused, which means that managers and staff work well together in developing a vision of a new user experience once the information transformation is complete.

Level 5: Optimizing

In discovering a Level 5 organization, an astute observer might be surprised by its external similarities to a Level 1: Ad-hoc organization. Everyone appears to be acting independently, but their independent behaviors have a context.

Information developers in a Level 5 organization are self-actualizing. Although they work closely together on information development, each team member is charged with the responsibility for developing content that is most useful to customers.

Communicators take personal responsibility for understanding their customers' needs. They are involved in user groups, call on customers regularly for insights into requirements, and have even begun to look at the customers of their customers for new ideas.

Communicators are accustomed to following standards because they know intimately how important standards are, not only in making their own jobs easier by taking time out of mundane processes but also by invoking a consistency of presentation that reassures readers that they have found what they are looking for.

No one in a Level 5 organization needs prodding to follow templates or maintain best practices in the information-development life cycle. They know that a lot of hard work has gone into designing effective information process and standards and that the best practices the department has instituted actually save everyone from boring work. In fact, one of the hallmarks of a Level 5 organization is the enthusiasm of the staff for best practices and new ideas and the esteem with which they are held by the rest of the organization. A Level 5 organization pursues innovation and continuous improvement. They are often engaged in customer research, benchmarking, and competitive analysis. They have the time, resources, and skills to incubate new ideas.

Going beyond the Ordinary

"Does a Level 5 of process maturity exist in the information-development world?" The Software Engineering Institute, from time to time, reports on a Level 5 organization, based on their analysis of software development. Most often, however, Level 5 activities are confined to particular projects that are run exceptionally well. Whole organizations that operate at Level 5 are few and far between.

The CIDM assessors are concerned about the apparent lack of Level 5 information development. They helped ODS define their Level 4 processes earlier in the year and have watched them begin a transformation to a greater focus on customers. However, Level 5 organizations are in the thick of customer studies, relying on customer input to guide innovation. In these tough economic times, a Level 5 organization might emerge in a leadership position.

An excellent Level 5 candidate I encountered a few years ago saw its organization dissolve in the face of layoffs, budget cuts, and the somewhat reluctant retirement of their inspirational leader. Under better conditions, this leader had surpassed almost every other group he worked with. The department members were recognized for their expertise in the user experience at the highest level of management. The senior manager held a respected position in the senior leadership group. He had gained for his team a significant role not only in innovating in publications and training but also in assuming responsibility for the design of user-friendly software interfaces for the products.

Unfortunately, as their industry lost economic ground, the underpinnings of this department's support were weakened. Even after winning an award of excellence from the corporation at large, they could not withstand the pressures to retreat.

Defining an optimizing organization

In the absence of a clear role model for Level 5 of the Information Process Maturity Model (IPMM), I believe it necessary to create a comprehensive picture of an optimizing organization, reflecting upon Clayton M. Christensen's management standard, *The Innovator's Dilemma* (Harvard Business School Press 1997). In it, Christensen argues that older companies in the high-tech industry are likely to be challenged by the innovations of newcomers. The newcomers are able to react more quickly and are hungry enough to take chances on new ideas.

Many of the organizations operating at Level 4 are part of mainstream high-tech companies. In fact, most of the information-development managers running these organizations had been in the field for 20 to 25 years. They have had considerable experience putting excellent processes in place and have been successful in introducing innovations. However, they have also experienced significant internal pressure to cut costs. They are often able to invest in content-management systems (CMS) to save costs but find themselves stymied when they try to change the way information is delivered to customers.

These observations have helped me to outline the requirements for design and development that might characterize a Level 5 information-development organization.

Characteristics of a Level 5 organization

My vision of a Level 5 organization centers on a customer focus and innovation. Rather than be satisfied at following industry trends in areas like electronic delivery, content management, and minimalist design, a Level 5 organization learns lessons about information development directly from its customers.

In a Level 5 organization, information developers are on the front line with the customers. It's too easy to take direction about customers from marketing, sales, or even engineering. But these organizations don't pay much attention to how customers use information to learn and maintain products. As a result, a Level 5 organization assumes several new roles:

- ✔ They never write an installation manual in the office. They send information developers out with engineers on the first few installations of a new product, watching what happens and taking lots of notes. After the installations, the information designer works with engineering, production, and maintenance to learn how the problems will be solved in the next installation. As the problems are solved and best practices are worked out, the installation procedures emerge.

- ✔ They give information developers direct lines to key customers. These customers are interested in communicating when they run into usability problems or are confused by instructions. Rather than be frustrated by a lack of appropriate information, these privileged few can directly communicate with an information developer who will find a solution. Together, the customer and the information developer reinvent the instructions so that they work better.

- ✔ A Level 5 organization has an advisory council of customers interested enough in information development that they'll spend time reviewing ideas and giving advice. They use advisory council members as entry points into customer site visits. They use site visits to conduct analyses of customer goals and tasks that will become a basis for improved content.

Spending time with customers is recognized as a core activity in a Level 5 organization, not something you do only after everything else is done first (which usually means never). What role, then, does technology play in a customer-focused organization? Doesn't the pursuit of continuous innovation preclude the use of technologies that tend to reinforce standards? Aren't standards stultifying? How can an organization be innovative when everything has to be written in the same way?

A Level 5 technology solution

In a Level 5 organization, technology like content management is simply part of the infrastructure, a little like email. Once you have it, you can't figure out how you existed without it. In a Level 5 organization, technology must be able to accommodate the special needs of a highly innovative organization. At Level 5, information developers help guide the development of new technologies that they need to keep abreast of efficiency and cost-reduction requirements as they find new ways to provide value to customers:

- ✔ Information designers must be able to update existing information types and create new ones in response to customer needs.

- ✔ Metadata needs to be designed to accommodate the needs of users rather than only accommodating the needs of information developers. With dynamic metadata, information can be directed to specific user needs through search (pull) and updating (push).

✔ A solution must have the capacity to grow with expanding needs. In a Level 5 organization, more opportunities are quickly discovered to manage information creatively for the benefit of customers.

Changes to the IPMM

Since I first introduced the IPMM in 1994, the essential characteristics of successful, well-run departments have not changed significantly. At the same time, some of the challenges have increased in magnitude and difficulty. The new millennium has added complexity to the responsibilities of information-development managers.

Mergers and acquisitions

Mergers and acquisitions have increased dramatically in the past decade, often accompanied by a reluctance to change the cultures of any of the parties, at least not initially. Information developers from a multitude of different organizations find that, at least at first, no real connection with the "corporate" brand is being asked of them. However, after a year or two, someone notices that the company has no information identity and encourages a blending of the information design.

The arm's length relationship can certainly continue. Sometimes, managers find that they have peers in the merged organizations. In other cases, managers discover a plethora of lone information developers who have no real sense of a corporate identity. The task of creating a single standard for information development is difficult even within a single department. Working across departments and across geographies and cultures can seem daunting, especially if the level of distrust is high.

However, the IPMM suggests that customers are best served when information developers collaborate and pursue best practices. That requires the unification of disparate entities into a single department or a confederation of like-minded departments, each led by an experienced manager. With such confederations, I often find a corporate-level group that serves as the coordinator, helping to set and maintain standards in information architecture, tools, production, delivery, packaging, training, and localization and translation. Whatever activities you can select that will benefit from economies of scale can be viewed as contributing to a higher level of process maturity.

Offshore information development

Another change that accompanies mergers and acquisitions is an increased reliance on offshore information development. Large, multinational corporations have always had information developers in many countries, usually because those countries also have product-development activities. More recently, however, companies are looking to low-cost labor in countries like India, China, Chile, Argentina, and those in Eastern Europe to provide product development services at a lower cost than in North America or Western Europe. Accompanying this move has been the addition of information developers, sometimes working along with product developers and sometimes working independently.

The promised savings in salaries is at least 80%. However, we find that information development is much more challenging to outsource to Third World countries than software programming. Because of language and cultural differences and the absence of a tradition of technical communication, the results of offshoring information development are often very disappointing.

I suggest to embattled managers trying to make a go of outsourcing and offshoring that they use the IPMM as a point of departure. It can be extremely useful to conduct a process maturity assessment with an outsource vendor or an offshore department to gain an in-depth understanding of the changes needed for the venture to be successful.

A Level 3 or 4 information-development organization suddenly having to rely on a Level 1 or 2 group of individuals can be challenging. An organization that is itself not Level 3 should not consider outsourcing or offshoring at all. Such an organization does not have the processes in place to communicate effectively with an outside organization or a completely new and untrained group of employees.

Demands for increased productivity and reductions in force

The third challenge that I believe has been exacerbated by changes in the corporate climate over the past 10 years is productivity. Reductions in force or the lack of new hiring to accommodate increased workloads has forced information developers to explore opportunities to reduce the work of developing technical information. I have seen an increased interest in task analysis, minimalism, and content management as groups look for ways to manage the demands with the same or fewer staff members.

Managers who have planned well and have developed a Level 3 organization find that they are better equipped to discover opportunities to improve productivity than those who rely on individual contributors and brute force to keep up with a never-ending workload. Using technology to improve productivity, at the same time that you pursue ways of innovating information development and delivery techniques, promises a chance for a solution.

Key Characteristics of the IPMM

During an IPMM assessment, an organization is evaluated according to eight key characteristics. These characteristics help to describe how a successful information-development organization functions. The focus, of course, is on structure, process, and best practices. I firmly believe that there is a close correspondence between the behaviors outlined by the eight key characteristics and the ability of the staff to produce excellent information products in a cost-effective manner.

However, it is possible, but highly unlikely, that an organization has all the behaviors it needs to be successful and still produces defective products. You know, for example, that highly bureaucratic organizations, such as you tend to see in the government, have in place all the rules and processes that one could think of. Yet, because they are reluctant to change old information-design models, their products rarely change. Such organizations are typically Level 3 and not higher, simply because the customer-knowledge characteristics of

Levels 4 and 5 are typically accompanied by innovative, customer-centric designs. It's hard to spend much time with customers and continue to ignore their needs for more effective information content and delivery.

Nonetheless, I need to remind managers and staff that the measure of quality for information products that really counts is customer satisfaction and performance. If customers cannot find what they need nor use the information to reach their goals quickly and effectively, then the information products are not successful.

No outsiders, no matter how experienced in information design, can tell you if the information you produce is excellent. Only your customers can be the judges.

Table 2-2 provides a brief outline of the eight key characteristics. A complete IPMM assessment includes many more distinctions about the nature of activities within each level.

 ## Best Practice—Organizational structure

Organizational structure is the first and most pivotal of the eight key characteristics of the IPMM. Without a sound organizational structure that supports the development of standard processes and architecture, all of the other activities associated with process maturity will not follow. Quality assurance, planning, estimating, cost controls, customer management, and hiring and training are all dependent on the leadership provided to the information developers. If an organization is able to implement quality practices, it typically has some organized leadership interested in the success of information development as a whole.

One of the first issues addressed during an Information Process Maturity assessment focuses on the structure of the organization:

- ✔ Is the information-development function centralized?

- ✔ Do all the information developers report to one manager?

- ✔ If there are several information-development teams, do they each report to a manager skilled in information development?

- ✔ If there are several information-development teams, do the managers of each team have a mechanism to support collaborative work on standards and practices?

- ✔ Are any additional information-development activities centralized, such as production and localization?

- ✔ Does the senior manager of the information developers have experience in the profession, and is he or she concerned about the efficiency and effectiveness of the staff and their work?

In a Level 1 organization, information developers report to various product-development groups and have little opportunity and less encouragement, except through personal initiative, to work toward common goals and objectives. Although excellent work might be done by individual contributors at Level 1, it is equally likely that much work is below par.

Table 2-2: Key Characteristics of the IPMM

Characteristic	Level 1	Level 2	Level 3	Level 4	Level 5
Organizational Structure An organizational structure that enables information developers to produce consistently high-quality work.	Information developers work for technical managers. Information developers usually work alone or in small groups.	A centralized information-development organization is in place. The organization manager is knowledgeable about information development.	A senior manager designates leads for individual projects. Specialized job functions have been developed.	Information developers are in a matrixed organization, reporting to a central group but working closely with cross-functional project teams.	Information developers have leadership roles on cross-functional project teams and with peer organizations.
Quality Assurance A series of activities specifically designed to promote uniform high standards of quality, including copyediting, developmental editing, peer reviews, and technical reviews of draft information products. Includes usability testing and customer studies to ensure that the quality achieved meets customer needs.	Information developers are responsible for their own quality assurance. Few or no corporate-wide standards and best practices are in place.	Standards are in place and designated individuals have begun to be responsible for maintaining the standards.	Designated individuals (editors) are responsible for maintaining standards. Developmental editing is in place to assist in developing consistent information design and architecture.	Usability assessments are a standard part of the information-development process.	The outcomes of quality assurance activities are measured as part of a continuous improvement process.
Planning Activities to ensure that every information product meets customer needs as well as the demands of schedule and budget. Includes the development of adequate resources and budget to ensure that required quality standards are met.	Individuals sometimes create Information Plans.	A standard Information Plan is in place and followed for many projects.	All projects begin with Information Plans. A standard information-development process is followed by staff.	Plans are regularly reviewed to encourage innovation and cost control.	The planning process is measured to ensure that productivity and performance goals are achieved.

Characteristic	Level 1	Level 2	Level 3	Level 4	Level 5
Estimating and Scheduling Activities to ensure that the information-development process is being followed to meet schedule and budget requirements. Includes project tracking to assess and accommodate the impact of project changes and changes to customer requirements through the course of the project. Establishes project histories to better inform planning for future projects.	Assignments are made without knowing if they can be accomplished by the deadline while maintaining quality.	Information developers apply guesses to determine if they can complete projects by the deadline while maintaining quality.	Projects are carefully estimated according to data on previous projects. Projects are carefully tracked to ensure they will be successful.	Projects are estimated and tracked so that adjustments can be made to resources, schedules, and scope of work in response to requirements changes.	Complete development projects are scheduled and tracked, and they include the requirements to meet quality goals in information development.
Hiring and Training Information developers are hired by knowledgeable professionals in the field, and hiring is based on a wide range of clearly defined professional requirements. Once hired, information developers are provided with internal and external opportunities for continuing training so that best practices in the field are understood and maintained.	Information developers are hired by technical and other managers. They are typically hired for technical and tools expertise rather than information-development skills and training. No regular training is provided.	Information developers are hired by knowledgeable managers and peers for technical and tools skills and sometimes for expertise in information development. Training is provided occasionally by request.	Information developers are hired for their expertise in specific specializations. Training is considered a required part of each person's professional development.	The skills of senior information developers are leveraged through hiring of entry-level staff. Training and mentoring are provided internally, and external opportunities for growth are regularly provided in specialized areas.	Information-development managers are provided with management training and development opportunities to increase their understanding of business objectives.

Continued

Table 2-2: Key Characteristics of the IPMM (Continued)

Characteristic	Level 1	Level 2	Level 3	Level 4	Level 5
Publications Design Activities to ensure that the organization is following the best practices in the industry. Design innovations are regularly introduced based upon research in the field, usability testing, customer studies, and practices learned through exposure to the work and ideas of industry leaders.	Information developers may design the publications they produce. However, the designs are often heavily influenced by others in the organization, including nonexperts in engineering, programming, and marketing. Few or no information design standards are in place.	Information developers are fully responsible for the design of their publications, although outside influence may still be a factor. Standards are being developed with incomplete compliance. Some specialization in design and publishing functions may be in place.	Information developers are fully responsible for the design of publications, following departmental or corporate standards they have established. Compliance with standards is complete. Specialized functions for design, graphics, editing, production, and others are in place.	Information developers, working with teams of specialists, are actively pursuing design innovations and testing these with users. They are aware of industry standards and best practices and compare their work with best-in-class designs.	Information developers actively contribute to the design of product interfaces. Information developers are actively engaged in sharing their design expertise with others in the industry and developing and disseminating industry best practices.
Cost Control The publications organization has budget authority for its activities and carefully tracks the costs of its development projects. Costs are well understood and regularly evaluated in terms of return on investment and value added. Budgets are defined by the need to achieve a stated level of quality in information products.	Costs are determined by headcount assigned. Total costs may include printing, distribution, and localization and translation.	Publications organizations have assigned headcount. Departmental budget allocations for training, printing, and localization and translation are beginning to be the responsibility of the manager.	The publications organization has a budget controlled by the manager who submits budget requests. The organization is active in cost-reduction activities and reports on these activities to senior management.	Senior management is well aware of the quality cost associated with publications, through the communication efforts of publication management. Efforts to reduce costs and increase productivity are well received by senior management.	Publications managers have instituted a continuous improvement process to reduce costs while maintaining or improving customer quality.

Characteristic	Level 1	Level 2	Level 3	Level 4	Level 5
Quality Management A series of activities directed toward complete and well-informed definitions of quality, including regular studies of customers' needs, regular usability assessments, regular assessment of customer satisfaction with products, regular assessment of the impact of poor quality on training, support, sales, and others. Strong communication of goals and strategies to senior management and peer managers. Recognition by the larger organization of the value added by technical communication activities.	No mechanism exists to measure quality of output. Quality is often equated with making deadlines.	The publications manager and staff are beginning to investigate ways to measure quality besides meeting deadlines. Customer complaints are addressed.	The organization is active in defining, measuring, and managing customer-driven quality. Customers are regularly polled and their issues addressed. Benchmark studies are pursued for the first time. Competitors' information is evaluated.	All aspects of customer-driven quality are regularly assessed, including satisfaction with information, calls to support, and complaints. Benchmarking is a regular part of the process.	Staff members have acknowledged expertise in the field at defining quality in publications. The organization is actively engaged in developing quality standards in the larger organization. An understanding has been established between the quality of information and the success and profitability of products and services.

Those managing the information-development activities are rarely knowledgeable about the practices and standards in the profession and are often completely uninformed about the work that information developers perform or the value they add to the product and services of the company. Given the diversity of skills and abilities in a Level 1 organization, I give the organization the label ad hoc. Each individual does the best job he or she is capable of doing, usually without supervision and frequently with few opportunities for learning.

In most cases, some event precipitates a change to a Level 1 organization. Of necessity, that event forces a move toward consolidation of activities and people, often with the introduction of a manager who unifies the individual contributors into a central organization. Some organizations make do with several small groups with someone in a managing role for each. Others create a single department to which all the information developers report. The stage is thus set for progress to a Level 2: Rudimentary organization.

Why, then, do I argue that growth toward more mature processes can only occur with consolidation into a centralized organization with professionally skilled management? Introducing mature processes is a complicated affair. The staff needs to learn to work together and agree upon standards for writing style, information typing, structured writing, and audience definition in addition to basic processes for information planning, estimating, scheduling, and tracking. Their efforts to develop standards should be guided by a manager or group of managers who have a vision for the future growth of the staff and improvement of the quality and cost-effectiveness of the work.

I believe the transition to an organization with a plan and goals cannot take place without leadership. Although some ad-hoc groups may occasionally be able to work together to develop a style guide or institute some common processes, such grass-roots efforts are difficult to sustain and rarely successful over a long term and across disparate teams. As soon as someone new is hired or someone decides not to cooperate, the effort is undermined. With progressive leadership, the efforts at standards and processes have the possibility of success although the road ahead may be quite rocky.

Is it possible for an information-development organization to be centralized and still operate in an ad-hoc fashion? Yes. I've visited several organizations over the years that have an information-development manager who is disengaged. Generally, this individual functions as a hiring manager, adding new people to the department and conducting performance reviews. The work of developing information is still entirely managed by individual contributors. Usually the disengaged manager knows little about their projects except for the deadlines.

Based on experience with disengaged managers, I find that a centralized management structure is necessary but not sufficient for the development of mature and consistent practices that are accepted and executed by all the staff. The necessary step is to have an organizational structure that is centralized, one in which everyone is working collaboratively toward the same business goals. If such a centralized organization is to be successful, the leadership must be fully engaged in the development of professional practices in information development. The leaders must understand the activities essential to the information-development life cycle, including information planning, information architecture, project estimating and tracking, editing and other quality assurance activities, and all the other best practices associated with the IPMM. The leaders need to commit the organization to innovations in service and to the needs of the customer.

Best Practice—Information planning

"We really don't plan. It takes too much time, and the product keeps changing."

Despite arguments to the contrary, you learn from effective managers who develop detailed information and content plans that the efforts pays for itself. Content planning, when done in the context of user requirements, decreases information-development time and alters the distribution of activities during the information-development life cycle.

Information plans that include details about the project, the process, and the design and deliverables are an essential ingredient of a Level 3 organization. In Levels 1 and 2, plans are not consistently developed, are cursory lists of deliverables with little detail, and rarely include estimates of project time and resources based on data from previous projects. One clear indicator of a managed and repeatable process is the development of thorough information plans.

Too often, information developers feel compelled to wait until products are stable before beginning to outline the topics they need to develop for users. In groups that focus on documenting the user interface rather than on documenting user tasks and concepts, planning tends to be neglected because so much of the plan depends upon the product. In contrast, groups that focus on user tasks and concepts can begin planning long before product interface details are established. Far fewer changes occur with respect to user goals for a product release than with interface details.

Rather then spending time rewriting detailed instructional topics in the midst of rapidly occurring product changes, during the information planning phase of a project, information developers should focus their attention on the users' goals. They can define use cases and scenarios that describe how users will interact with the product. They can focus on what users need to know to be successful and what they must learn how to do.

By concentrating your planning time on the users' need for information to support their successful use of your product, you can postpone including the user interface details until the engineering changes settle down. Once product details are better defined, at least in terms of the user interface and the ways in which users will need to perform tasks, these details can be added quickly and easily to your predefined set of topics.

Early input from product marketing, participation in product definition discussions, and input from the user community all provide sufficient resources to enable thorough, efficient, and productive planning early in the project life cycle. You find that user goals rarely change with engineering changes, although the details of task performance may be altered.

Topic-based architectures facilitate early planning

Too many organizations restrict their information planning to lists of final deliverables.

"We need one user guide, two command reference manuals, an administrative guide, and a help system."

Sounds like ordering from one of those '50s restaurant menus: you want one from column A and two from column B. Usually the list of deliverables for the current release is the same as the list for the previous release. You really don't know much about the scope of work required to complete the deliverables and meet the deadlines.

When your organization decides to pursue a topic-based architecture, you need to make a firm commitment to detailed planning. In fact, the level of planning required may come as a surprise because planning of this sort has been relegated to individual information developers rather than as a core activity of information architecture. In most book-centric information models, the details of a book's content is left to the individual contributor. The distribution of planning results in duplication of topic content among deliverables, even though the content may be written somewhat differently. It also results in missing content that no one has thought to include in individual books. Because topics are not planned across deliverables, fewer or no opportunities for reuse occur. Even if reuse were possible, the idiosyncratic nature of the content development precludes it.

In topic-based authoring, reuse is an important business and architectural assumption. Topics must be planned from the beginning so that their reuse potential is optimized. Without detailed topic-based planning, individual topics may continue to resist use in more than one context.

For a detailed description of information and content planning, see Chapter 16: Planning Your Information Development Project.

Remember that planning may seem an impediment to making a tight schedule. In fact, planning is essential to any schedule, tight or loose. As a best practice in information management, be certain to include careful planning in an early phase of your projects.

Best Practice—Estimating and scheduling

Information planning that includes an estimate of the resources required to complete the project successfully flows from the Level 3 planning characteristic.

However, even in Level 3, estimates may not be based on actual data obtained from tracking the time and resources required to complete previous projects successfully. By Level 4, estimating has become a science and tracking is viewed as essential.

Despite the bad rap that time tracking may have, it is, nevertheless, an essential ingredient in the management of professional activities. Yet, publications managers tend to neglect time tracking because they don't recognize opportunities for using the information to estimate the resources required for future projects, support requests for new resources, build a business case to fund an innovation, or determine the value associated with a task in the information-development process flow.

Estimating future project resources

The most obvious reason for time tracking on current projects is to build a database of metrics to use for future projects. Accurate information about the time required to complete current and previous projects is essential for responsible estimating. You can track each project by determining hours worked on the project by project managers, information developers, editors, production specialists, and all other participants from your team. Hours worked include writing and editing time, as well as time spent data gathering, performing user and task analysis, planning, estimating, scheduling, tracking, managing the project, and generally performing all tasks associated with the project. The only time that typically doesn't get counted is the time spent by outside experts in providing and reviewing information.

Using these data to estimate future projects means not only applying the average time spent on projects but also taking into account the time required for more complicated and less complicated projects. The dependency calculator in Chapter 16: Planning Your Information Development Project shows you how to account for project dependencies such as access to developers, amount of change in the project, and other elements that influence your ability to estimate individual projects.

In addition to using time-tracking data to estimate new projects, you can use the data to estimate the effect of project changes on your ability to complete the project. For example, you may find that you've used half the hours allocated to a project, only to discover that you have completed only one quarter of the work. This circumstance occurs when projects become more difficult over time, often because of changes in the development process and schedule. You can use your time-tracking data to estimate how much additional time you will need to complete a project.

Requesting new resources

Requesting new resources is fraught with difficulty because of the pressure to contain costs and increase productivity. However, without data about existing projects, you have little ammunition to make your case. With good information about what it takes to complete projects, you are better able to present your case to management for additional resources. In addition, you can use your project data to demonstrate that, without adequate project resources, you must make cuts in the scope of the project by eliminating deliverables, reducing quality assurance activities, or generally making the project less time-consuming to perform.

Funding an innovation

Time-tracking data can also play a big part in helping you calculate a return on investment that justifies funding an innovation. Today, managers want to engage the staff in user studies, fund the development of structured development and content management, support educational opportunities for staff, and so on. Each of these innovations may require funding.

However, begging for funding rarely gets you very far. By calculating your current costs (based on dollars per hour of time), you can show how specific innovations will result in cost savings. For example, you know that information developers have demonstrated that single sourcing reduced development time for a set of related information deliverables. Instead of requiring two information developers to complete a suite of documents for related products, one organization discovered that they were able to complete their project with one writer because the writer built a repository of topics that could be used among the deliverables.

By knowing how much it takes to complete a project today, you can demonstrate cost savings realized by the application of new technology or innovative information design. You can also use tracking data to show that an innovation is in fact costing more than expected. One organization demonstrated to management that offshore outsourcing was actually increasing, rather than reducing, their costs. Because the outsourced information developers had no product experience, limited English writing skills, and no experience producing technical documentation, the home staff was required to spend extra time in communicating, managing, editing, and more to deliver reasonable content to customers.

By showing with data that extra costs were mounting up, the organization was able to restore resources that had been lost.

Performing a value analysis

A value analysis allows you to calculate the costs of individual tasks that you and your staff perform during the information-development life cycle. Senior management asks you to remove costs from the life cycle by ensuring that each task returns value in relationship to cost. You may discover, for example, that production tasks take a significant percentage of costs during the last quarter of a project. By automating production and eliminating detailed manual work, you can reduce production costs while continuing to produce high-quality deliverables.

Note that time tracking has a significant role to play in professional management activities at Level 3 and above. Not only does a database of projects provide you with the means to manage current projects and estimate new ones, but a database of project information enables you to build business cases for your innovations. Remember that when you are trying to communicate with people who depend on data to support their decisions, you need to speak the language they understand, a language of numbers. In a Level 3 or higher organization, you understand the data that senior management needs.

 Best Practice—Quality assurance

In a Level 3 or higher organization, quality assurance represents a series of actions that ensure that content is accurate, complete, and meets customer requirements. These actions include

- ✔ reviews by subject-matter experts

- ✔ testing of content against product

- ✔ editing for consistency, compliance with standards, and alignment with customers' needs

Each of these quality assurance actions is fraught with difficulties, especially when project schedules are seriously curtailed as companies push developers to release product more quickly. The pressure to meet the next deadline encourages organizations at Level 1 and 2 to forsake quality assurance, including basic testing of products to ensure they perform as specified and contain minimal bugs.

Yet one very important mark of a mature organization is the ability to assure quality within a high-pressure environment. In fact, lean and agile approaches to product development place quality assurance at the top rather than the bottom of the list of critical project components.

In addition, a topic-centered, domain-specific approach provides you with the environment you need to strengthen your quality assurance commitment and ensure that you deliver usable information to your customers.

You can find detailed information on quality assurance activities in Chapter 21: Managing Quality Assurance.

Content reviews

Information developers recognize that reviews of content by subject-matter experts is essential to ensuring accuracy and completeness in the information they intend to deliver. Of course, you expect the information developers themselves to develop product expertise, especially after some years of working on the same kinds of information. At the same time, you know that others in the organization involved in the development of new products and features have an understanding of how products are expected to function that must be taken into account.

You also recognize that reviews by subject-matter experts in the organization are often difficult to obtain. Everyone on the product-development team is busy completing assignments. In most cases, team members have schedules for completing their work that do not include time for communicating with information developers or reviewing draft content. Even though everyone recognizes the importance of this collaboration, it is often on the bottom of the priority list.

Topic-based, domain-specific authoring provides two possible solutions to the fight for available time.

Topic-based authoring means that information developers develop content in smaller, stand-alone modules that are later assembled into final deliverables. Because content plans are developed around lists of topics and information developers are directed to begin preparing drafts in areas where content is more likely available, information developers have an opportunity to collaborate with developers earlier in the product-development life cycle. Information developers can ask developers to review task lists related to each individual's area of expertise and then extend the task lists by linking conceptual and reference information to the essential tasks, making content reviews more productive than reviews of tables of content. Information developers can ask developers to review snippets of essential content as task instructions are put together, facilitating feedback on the conceptual and reference information that will support the tasks. Concepts and reference are easier to envision when they are purposefully linked to tasks.

Product developers and other information resources like marketing, support, and training are more likely to review small topics or collections of topics more quickly when those topics are specifically directed to their domain specialties.

In organizations where information developers develop domain-specific knowledge, the information developers themselves become part of the team of subject-matter experts. In this capacity, information developers are more likely to work closely with a small group of developers who are focusing on a particular part product feature. Since developers themselves are assigned by feature, this domain-focused arrangement provides a powerful commitment to quality.

Testing information against the product

If reviews are difficult to schedule, product testing seems even more difficult. Many product testing teams have seen their staff cut and schedules curtailed. At the same time, you clearly recognize that instructions must be tested if you are to ensure that they are accurate and produce the correct results.

In some information development teams, the information developers are responsible for this most basic level of testing, as long as they can get access to product test systems. In other organizations, test systems are not available until late in the product-development life cycle or not at all, ensuring that the basic task instructions are likely to contain errors.

The solution, in these cases, must include the testing team in the process. Generally, testers are provided with documentation to follow as they perform the tests. Remember, however, that professional testers are not end users. What may be clear and useful to them may be unusable.

That's why direct testing by members of the information-development team is essential. Information developers, especially interns or colleagues in other domain areas can act as surrogate users, depending upon the instruction rather than prior knowledge to perform a discrete task.

Information developers can, of course, test their own instructions against the products if they have timely access to test systems. However, testing your own instructions is never as effective as testing someone else's writing.

Editorial reviews

In recent years, we have seen a continual degradation of editing functions in information development. Today, there are fewer editors working in departments than there were 5 or 10 years ago. Some organizations retain editorial functions, and others find that editors are the first to be eliminated during reductions in force.

Clearly, editorial review is even more important than ever in a topic-based authoring environment. Information developers are working on discrete topics that will be assembled into manuals, websites, or help systems. The work of multiple authors is more likely to be intermixed in final deliverables in this case than when authors worked solely on their own books.

A topic-based, domain-specific collaborative authoring environment begs for the essential skills of a developmental editor. In most cases, such skills come in the form of a senior writer or project manager who has a long-term perspective on the product and information and has enough knowledge of the information architecture to look beyond individual topics to the whole approach to the customer.

The information architect is certainly part of this editorial review, with an overview of the topics to be created for the new release or new project. The architect/editor ensures that the topics fit together and provide a sensible and usable information environment for the customer.

At the level of an individual or small group of topics, the same small groupings that are sent to product-development reviewers, a developmental editor is able to check for consistency of message, integration and flow of the information among topics, the presence of adequate conceptual, background, and reference support for tasks, and compliance with the architectural standards for structured development.

In too many organizations, editing is relegated to final copyediting, including proofreading; compliance with regulations in the form of copyrights, trademarks, and others; and final format checks. Although all of these tasks play a role in quality assurance, they seem easy enough to eliminate during reductions in force. Automated production systems can help reduce the time needed for final preparation of deliverables, but proofreading and regulatory compliance still need some human intervention.

Only by elevating the role of quality assurance in a Level 3 organization and making it an essential part of the process can you avoid further inroads into editorial standards. It might even be wise to rename the job positions to avoid "editing" as a title because that seems to raise red flags among managers who think of editors as their third-grade teachers.

Quality assurance provides us with the means to ensure that customers get what they need. Quality assurance means that you need to review, test, and edit your writing with a continuous eye toward customer requirements. Senior information developers who are skilled in user and task analysis, sensitive to the minimalist agenda, and know much about the customers are among the most valuable members of a truly collaborative team. They provide support and guidance to all the information developers strewn across the world and with diverse backgrounds, languages, and writing skills.

Best Practice—Hiring and training

When I look at immature information-development organizations at Level 1, I often notice that hiring is in the hands of the product developers. Information developers are hired by individual development teams to supplement the activities of the engineers and programmers themselves, often with the comment that "we could write this ourselves, but we're too busy." When information developers are hired as developers, the hiring criteria are evaluated in roughly this order:

- ✔ Technical expertise in the developers' subject-matter area
- ✔ Tools knowledge, i.e., MS Word or desktop publishing
- ✔ Ability to write effectively and meet the needs of the audience

Because writing ability is difficult for non–information developers to evaluate, the ability to develop usable and readable content is often given short shrift. In fact, if you tell them you can write, you probably get the job as long as you know something about the technology.

This fixation with technical skills runs rampant in the hiring that occurs among many offshore outsourcing vendors. You hear horror stories about agencies hiring computer programmers as technical information developers with the promise that they'll move up to programming jobs as soon as an opening occurs. As a result, the turnover among the so-called information developers is rapid.

Only when hiring shifts into the hands of information-development professionals do you begin to see a focus on the ability to communicate effectively.

Hiring a jack-of-all-trades

At Level 2 in process maturity, a time when an organization is in transition between an ad-hoc, technology-dominated environment to a professional environment, I frequently find that managers prefer experienced information developers who can manage the entire process and themselves in the bargain.

I often receive job announcements asking for applicants who are expected to be completely independent in their work environment. Quite clearly, from these requirements, successful information developers must be able to manage their own projects, write and edit their own work, handle all interactions with other departments, and compose and publish their own deliverables.

When I review the practices of a Level 2 organization, I find the individual craftsman model in place. Each department member is expected to perform the entire job, often in complete isolation from co-workers or the manager. In such an organization, I notice that managers, although responsible for hiring and evaluation, often know little or nothing about the work being done by the staff members. Nor do co-workers know anything about the practices being used or the content being developed by their peers.

As a result, you certainly have great independence and self-sufficiency among the staff. You also have duplication of efforts, inconsistency of style and format, and gaps in the information provided to the customers. Staff members spend a lot of time performing mundane clerical tasks because there is no one else available to assist with the process.

Although at Level 2, the organization may have rudimentary style sheets and editorial guidelines in place, the independent staff members feel privileged to override the corporate styles whenever they perceive a need for a new approach. As a consequence, the product delivered to the customers is frequently inconsistent because everyone is doing "his or her own thing."

Hiring for specialization

In the move up the process maturity scale, managers come to recognize that hiring practices must shift from total independence to specialization and teamwork.

Specialization often occurs as a department grows or as it moves into content management. In a larger department, with 10 or more staff members, the need for specialization becomes obvious. Types of specialization that I first begin to see include

- ✔ editorial experts
- ✔ production specialists
- ✔ graphic artists

Management learns that technical communicators become increasingly inefficient and expensive when they are expected to edit their own work, see all the deliverables through to final production and distribution, or produce their own technical illustrations. Each of these jobs is better handled by experts than occasional practitioners.

Specialization of tasks also becomes essential when an organization begins to implement structured authoring, reuse, and content management. Not only are editors required to ensure a common look and feel among the authors, but individuals must take on responsibility for information architecture, repository management, tools management, and information output design. Although these roles may be assigned part-time to authors and editors, they have unique jobs that are best learned through specialization. If you ask everyone to take on all these new responsibilities, you are both misunderstanding the promise of increased productivity and neglecting the potential for automation to provide a return on investment.

Certainly, you may want to cultivate new skills among your existing staff. However, if you find opportunities for new or replacement hiring, the identification of specialized skills and clear descriptions of roles and responsibilities will aid in identifying the best candidates. As you increase the specialization, you look for people with qualities and interests that are miles away from the jack-of-all-trades.

If you have a small team moving into Level 3 and 4, specialization may involve individuals taking on diverse responsibilities, but they still discover that specialization speeds tasks and improves quality.

Hiring for collaboration

Beyond the basics of establishing specializations within the organization, an increasingly mature organization, especially in Levels 3 and 4, begins to recognize the importance of a collaborative environment. A single-source information architecture means that team members must work together to produce a body of knowledge from which information can be customized for delivery to users.

Collaboration seems to be mostly a state of mind, an ability to work with others toward a common goal and to abandon an ego-driven attachment to one's own control over everything. Much to the detriment of a collaborative environment, professionals have cultivated an increasingly unfortunate independence in the technical communication profession.

In a collaborative environment, responsibility shifts continuously among specialists. People with expertise in architecture help direct planning activities early in the life cycle. Those with expertise in information design take responsibility for final deliverables, developing the look and feel of electronic and print output. Throughout, however, staff meet and consult, agreeing on a single approach to content development. In such an environment, team members produce a well-conceived body of knowledge about a subject domain in the form of stand-alone topics. Topics, authored by specialists in the subject areas, are combined into myriad forms to best serve the needs of diverse customers.

The best collaboration doesn't exist in a vacuum of leadership. In a collaborative work environment, leaders make final decisions but also encourage consensus building. Leaders manage projects to ensure that all work is on track and that conflicting demands are reconciled and compromises reached that are both responsible and ethical. Leaders, in fact, know what is going on throughout the development process at the same time that they rely on the expertise of their specialists.

Hiring in a new global environment

Quite obviously, hiring for a specialized, collaborative environment is not easy. You look for people with special interests that can be fostered, and you look for people who thrive on teamwork. You need managers who want to be fully engaged but are anxious to take advantage of everyone's contribution.

Not so easy, you might say, to find and develop a new breed of information developers, but essential if you are to build mature organizations that can be efficient and effective in an increasingly competitive global environment.

 ## Best Practice—Information design

Most of the key characteristics presented thus far focus on issues of process management: planning projects, estimating and tracking effectively, ensuring quality, and managing costs. However, two of the characteristics allow us to examine the effectiveness of an organization in serving the needs of customers: information design and quality management.

Information design asks how innovative ideas are brought into the organization. When I look at an immature organization, I find that information design hardly exists, except in the hands of some energetic and talented individuals. To become a more mature organization, managers and team members must find ways to bring new ideas about information design into play.

What is information design?

Since the advent of desktop publishing, information design has been confused with page layout. When asked if they have innovated in information design, many technical communicators point to changes in the look of their publications. "We've changed the layout, written a new style guide, changed the fonts, and provided an HTML-based style to make reading easier on screen."

All of these design changes may in fact have benefited customers by making a publication more attractive and readable. Marshall McLuhan reminded us many years ago that the presentation medium is an important contributor to the message (if no longer regarded as the only thing).[4]

But information design is much more than layout. Information design is the process of developing content that meets the needs of the audience—all the needs of the audience. Excellent information design meets the needs of the audience in extraordinary ways. When I interview users of exemplary information design, they tell me that they ". . . were able to get started quickly, find the answers to their questions easily and achieve their goals with great ease."

If you compare information design with product design, you understand that good (or great) information design provides value to the customer. Customers with great information in hand use the products and services they buy from you with ease.

How does information design happen?

Information design in organizations occurs in two ways: in-depth understanding of customer information needs and exposure to innovative design ideas in the profession.

Excellent information design requires customer research

By understanding your customers and how they use information, you are prepared to design more effective and powerful communications. Take, for example, the technical professional who uses the product documentation to investigate and troubleshoot difficult problems. That individual is most likely to look for reference data (specifications and measurements) and background information containing the accumulated insights of other experts. The troubleshooting professional is unlikely to find much assistance in step-by-step procedures that explain how to complete screens and dialog boxes.

Team members who understand the needs of this customer may begin to focus their information-development activities on assembling and updating critical data and making it easily accessible online. They may also exert pressure on technical experts inside the company and from the user community to provide valuable conceptual information. In

[4] Marshall McLuhan and Lewis H. Lapham, *Understanding Media: The Extensions of Man*, Cambridge, MA: MIT Press, 1964.

one case, such a team worked with industry consultants to "ghost author" technical papers on the nuances of using the products.

But innovation does not occur only through customer studies. Knowing what the customers want to know does not always lead to the best solutions when the team has no resources other than its own experience. In fact, in such circumstances, the team brainstorms a solution that may have already been rejected by skilled and innovative designers in the field.

Excellent information design requires exposure to design innovations in the community

Innovations in information design occur regularly in the technical information community. The Society for Technical Communication (STC) awards innovations in its annual technical publications competition. The top winners in the STC competition are available to local chapters for display and discussion. Whenever we had the traveling exhibit in Denver, I made sure that my staff got a good look at the design innovations of the winners. One of the online information design winners, Autodesk, has served as a key example of structured writing. Autodesk's structured online documentation is an excellent example of using structured writing to build consistent information for the customer.

Earlier this year, Palmer Pearson, senior manager at Cadence Design Systems and member of the CIDM advisory council, started a Council for Innovation among companies in the north Boston area. Their group has expanded rapidly because members believe it incumbent upon them to learn about innovations in information design. By learning new ideas introduced in sister organizations, they are better prepared to innovate in their own organizations.

How, then, do you, as managers, ensure that your staff members are challenged to explore innovative design ideas? You must ensure that they are exposed to new ideas in the community by

✔ funding participation in CIDM, STC, and other industry conferences and insisting that attendees return with reports on innovations in design

✔ encouraging participation in local organizations devoted to design, including not only STC but also local chapters of the HFES (Human Factors and Ergonomics Society), UPA (Usability Professionals Association), IEEE PCS (the Professional Communications Society of the IEEE), and others

✔ investigating training opportunities locally and nationally and making them available to team members, including training in minimalism

✔ asking team members to subscribe to publications on design, such as the *Information Design Journal*, publications of various STC Special Interest Groups, and others and report regularly to team members on successful new ideas

✔ asking team members to read books on information design, such as Karen Schriver's *The Dynamics of Document Design* (Wiley 1996),[5] and report to the team

---

[5] Karen Schriver, *The Dynamics of Document Design,* New York: John Wiley & Sons, 1996.

Measuring the success of an innovative design idea

Exposure to innovations in information design helps to encourage team members to think creatively about their work and to take an active role in finding better ways to respond to customer needs. But innovation for its own sake is dangerous. You always want to introduce new designs carefully and measure the results. You must ensure that the innovation delivers value to the customer rather than only delivering kudos to the designer. You must ensure that the cost of the innovation is balanced with a measurement pointing to the design's effectiveness and the value proposition that it supports.

Take, for example, Beth Barrow's minimalism redesign efforts at Nortel Networks a few years ago. Beth employed a consulting team to study the customers' information needs. In response to the study's findings, the team decided to drastically reduce the volume of documentation and to focus on procedures rather than "filler" and background information. As a result of their minimalism innovations, they reduced the volume of documentation by 75%, representing hundreds of pages of information customers did not need.

Since the goal was to increase the usability of the documentation, the team members waited with "bated breath" for the results. Once the documentation was released, they were somewhat surprised to learn that the number of customer calls had increased significantly, rather than decreasing as they had expected. However, once they investigated the reasons behind the increase, they discovered that customers were calling to point out errors in the documentation. In fact, the errors had been in the documents for years. Only with the reduced number of words were the customers reading and finding the mistakes. The result of their innovation was impressive. Customers were using the documentation actively, apparently for the first time.

A return in customer satisfaction and engagement with documentation, in lieu of calls to customer support, is valuable, even when the innovation involves a degree of risk. To mitigate the potential risks, test a new design idea in the customer community. Early feedback will help you iterate your design ideas and arrive at better solutions. Recognize that innovation can disrupt traditional patterns of use among legacy customers. Introduce an innovation in small enough pieces that it is not too disruptive.

At the same time, recognize that all innovations are disruptive to some people some of the time. Realize that you must innovate to remain effective. Many managers tell me that they cannot improve the quality of their documentation because the translation costs would increase. They keep delivering information that is obviously flawed. Insist that no excuse is acceptable if it means delivering bad information. If your innovative ideas are sound and well tested, they will return value far exceeding the cost of implementation.

Introducing innovation to your team

To begin a design innovation program in your organization, expose staff members to innovative ideas about information design:

  ✔ Study your customers' information requirements and use the results to stimulate innovative thinking about design.

  ✔ Pursue a minimalist agenda. Minimalism is, in my view, the single most stimulating program for design.

✔ Assess the value of the innovation. If it does not deliver clear value to the customer, it's probably not a good idea.

✔ Weigh the cost of innovation against its potential value. But don't let initial costs bar the path to innovation.

Best Practice–Cost control

I should no longer be surprised by hearing information development managers tell me that they don't know how much their information products cost to develop. Too often, managers do not track the time it takes the staff to complete projects. They have nothing on which to base estimates of cost.

Equally problematic is the absence of budget lines for information development organizations, even in some Level 3 organizations. Managers are assigned a headcount, which may include both direct and contract staff, but not a departmental budget. In addition to personnel costs, a budget might include funding for equipment, infrastructure (office space, furniture, telephone, and basic supplies), travel, training, staff development, and other resources. In too many instances, the information-development manager must gain approval for every expenditure beyond the basic infrastructure. At the very least, expenditures below a threshold should be allowed without layers of approval.

In addition to departmental budgets, cost controls also affect individual projects. It is around projects that you should be controlling costs and not around the small expenditures that allow you to develop your staff. The projects are where the substantive costs can best be tracked and managed. Without knowing what projects cost, you are more likely to become a victim of second-guessing in which someone decides that your projects can be done less expensively in a country where personnel costs are lower.

Most information-development project costs are attributable to people time. Except for translation, print production, and sometimes distribution, the principal costs associated with your projects come from the work of your team members. You can use the fully burdened costs of your team members to calculate the project development costs—as long as you know which projects people are working on and how much of their time is devoted to each project.

If team members are tracking their time by project, you can use fully burdened costs to calculate project costs. However, if your projects include multiple deliverables, you may not know the cost of individual deliverables unless they are assigned to different individuals. To calculate the cost of each deliverable, you need to know how much time was devoted to each one. Consequently, you must ask team members to account for their project hours worked by deliverable.

Once you know the amount of time people have spent on each project and deliverable, it is quite simple to calculate the people costs.

For example, you can use a typical fully burdened cost for direct employees of $66 per hour, based on an average salary of $62,500 to which you have added all the overhead of taxes, benefits, office space, and so on. An hourly cost of $66 includes overhead equal to salary, or 100% overhead, typical of all but the most capital-intensive hardware development companies.

If you have spent 100 hours on a project, the total person cost of that project would be $6,600. If a staff member works for a year on a project, at an estimated 35 hours per week for 48 weeks (allowing for vacation, holidays, and personal leave), then the project costs 35 x 48 x 66 or $110,880. What that total implies is that each person in your organization costs a little over $110,000 for project work.

If you have contract staff, you can multiply by their hourly rate plus some percentage of overhead. Consult with your finance director to know what that percentage should be. If the contractor uses your facilities and equipment, you have a higher burdened cost than if the individual works at home or at an outside facility.

As you can see, estimating project or deliverable costs is not difficult. Doing so provides you with important information you can use to decrease costs and evaluate the feasibility of contracting or outsourcing.

With cost information in hand, you can implement processes to reduce costs, including minimalism, topic-based authoring, and topic reuse. Any topic that an information developer does not have to write reduces your development costs. You can also effectively evaluate the real savings from outsourcing or moving tasks to low-cost personnel. If you must continue to devote resources to manage or assure the quality of outsourced or offshore projects, you must include those costs in your total project costs. Without cost information, you have little evidence with which to argue that you are not experiencing the promised savings.

Not only do you benefit from understanding and controlling costs, but you can use cost information to communicate with your management. Senior managers often have no notion of the real costs of information development or the effects on costs of poorly managed, out-of-control product development projects. By maintaining data on each project and the deliverables, you can produce analyses that show which project costs the most and why.

Cost control is a powerful management tool, but many managers have little data to work with. Only when you understand and control costs can you achieve a high level in process maturity.

Best Practice—Quality management

Quality management is the process of ensuring that you understand your customers' information agendas and communicate your understanding to senior management.

Quality management is a hallmark of a mature information-development organization. In fact, quality management is a major determining factor in assessing an organization as a Level 4 or 5 in the IPMM. Without a program in place to ensure that customer needs are being evaluated and addressed, an organization cannot be judged as superior to most others in the field.

Beyond quality assurance and information design

Quality assurance encompasses activities such as developmental editing, copyediting, reviews by internal and outside experts, and even usability assessments of document design. All of these activities ensure that the information you develop meets your own standards. Quality assurance addresses such issues as accuracy, readability, core function-ality, and presentation.

Information design activities also contribute to the quality of the information delivered. Excellent information design requires that customer needs be taken into account from the beginning of a design process.

But quality assurance and information design activities are inadequate without a thorough understanding of how, when, and where your customers need and use information to meet their goals.

What is the essence of quality management?

Quality management, as a key characteristic of the IPMM, encompasses a variety of activities, all related to ensuring that customer needs are met and the organization provides value to customers.

Customer studies

Customer studies provide the core of quality management. Customer studies include

- ✔ regular assessments of quality, including surveys of customer satisfaction

- ✔ development of comprehensive user profiles that describe in some detail the characteristics of representative members of the user communities

- ✔ task analyses so that information developers know how users articulate the tasks they are trying to accomplish using the company's products and services

- ✔ customer site visits that focus on direct observations of users in the workplace to ascertain the users' agendas in using products, services, and information

Many information developers tell us that they have no time for customer studies, even though they know they are important. Others tell us that they are barred from customer contact by others in the organization, including senior managers who do not believe that information is important to the customer's experience. Still others have the opportunity to conduct customer studies but choose not to. They may believe that they already know what the customers want, or they may be reluctant to expose customers to their lack of knowledge of the product and the customer's environment.

In each case, an immature organization finds reasons for avoiding customers even though they admit that their information development suffers as a consequence. Some information developers prefer to interact with product developers, in part because these individuals may exert powerful influence on the information developer's career.

Mature organizations take the opposite position, finding ways to meet with customers even when there are barriers. They go out of their way to ensure that staff have regular customer contact.

Managing for quality includes knowing about the competition

I have generally been surprised to learn how little many information developers know about their competition. Understanding what the competition is producing is critical to quality assessments. That means acquiring samples of competitor documentation and asking questions:

- ✔ Are customers happier with the competitor's information than the information you produce?

- ✔ Are competitors conducting studies of their customers' information needs?

- ✔ Are competitors engaged in cost reduction activities, such as content management, topic-based authoring, and minimalism?

- ✔ Does competitor information appear more usable and accessible?

- ✔ Are competitors offering information on websites or providing other innovative ways for customers to find and use information?

Benchmark studies are one vehicle that mature organizations pursue in understanding competitors. In 1999, CIDM conducted its largest benchmark study, involving most of the large companies developing telecommunications hardware and software. The amount of information exchange among competitors was remarkable, resulting in a level of understanding of competitor information development that has been rarely equaled in other industries. However, few information developers in other industries appear willing to engage and invest in benchmark competitive studies. Product managers in the same industries are usually thoroughly familiar with competitive products; information developers need to study competitive information products as well.

Knowing what competitors are up to is a key competitive advantage in any field.

Managing for quality includes responding to customers directly

In 2004, the Microsoft Office Applications team reported on the continuous publishing program they had put into place to respond to customer concerns. The program, called Contact Watson, led to the first annual CIDM Rare Bird Award going to the Office Applications team. In the program, customers are asked to evaluate each topic they access through the web-based information system supporting Word, Excel, PowerPoint, Outlook, and other parts of the Office product line. Customers can rank the topic on a scale of one to five and provide comments about what they found to be useful and not useful in their search for information.

The customer comments and rankings are reported daily to the information developers, who decide which topics need to be updated or completely rewritten. Once a new or updated topic has been reviewed and approved, it is immediately made available to readers. Although the information developers don't respond to customer comments directly, they use the continuous flow of customer feedback to improve information quality.

Managing quality means responding to customers' concerns with information accuracy, completeness, readability, and usability. Organizations that have programs in place allowing them to respond, feel much closer to their user community. Consider programs in which information developers listen in or actually respond to customer inquiries with the assistance of the support organization. Consider processes in which every customer complaint is logged and explicitly addressed by the information-development team. Organizations that take an active role in addressing customer concerns will come out high in a process maturity assessment.

Partnering with customer support

Customer support is the perfect partner for information development. Both organizations are concerned with delivering quality information to a diverse community of users from novices to experts. Information-development managers should build alliances with support managers, especially since it is possible to show a clear relationship between quality of information and cost of support. Information developers who establish a partnership with support find that improved information content and delivery can have a dramatic effect on the cost of support, particularly by reducing calls for support.

In one study, I found that improved information quality reduced support calls by more than 60% and reduced the duration of an average call from 10 to 2 minutes. At a CIDM Best Practices conference, Angela McAlister described how her staff of information developers was responsible for ensuring the effectiveness of the customer-accessible knowledge base. Their work to improve the quality of information contributed to the steady growth in popularity of the support information website.

Promoting quality in your larger organization

Information developers in many organizations take a leadership role in customer quality. They review product interfaces to help ensure usability. They engage in direct customer studies that influence product development. And they engage their larger organizations in promoting quality.

Too often, you hear about information developers whose quality concerns are denigrated, often because they lack direct evidence of the importance of information in the customers' acceptance of and success with a product or service. You also find that when the evidence is there, in the form of direct customer data, quality concerns gain recognition. To earn a place in the discussion of product and information quality requires that you engage with the customers themselves. From direct customer visits to surveys and other data gathering, engagement with support, and through partnerships with customers themselves, you can elevate the discourse around quality in your larger organization.

Does quality count in these days of focus on the bottom line and building stockholder value? Only if it helps to improve sales and revenues. Does quality in information improve sales and revenues? It certainly does when it helps to increase customer satisfaction.

New Characteristics of the IPMM

Since its inception, the IPMM has looked at the eight key characteristics to evaluate an organization. Those eight key characteristics have been sufficient to distinguish immature from mature practices. However, more recently, I have evaluated the need for additional characteristics, among them collaboration and change management, to be added to the model.

Collaboration

Although the IPMM addresses collaboration as a goal of hiring, I believe that building a collaborative working environment must be addressed more specifically in our assessments in the future. The practice in which information developers work alone, responsible for defining and creating content based upon their interactions with subject-matter experts, is clearly a characteristic of an immature, Level 1 organization. A lack of collaboration is a key characteristic of an organization that develops information using ad-hoc processes. Given the need to share content among deliverables (single sourcing) and the need to ensure that customers receive consistent and complete information to guide their performance with a product or process, information developers must become increasingly collaborative in their working environment. It is no longer acceptable for writers to work in isolation on their proverbial mountain tops.

Chapter 20: Managing in a Collaborative Environment presents business best practices for establishing a collaborative environment at the project level. These best practices suggest that by working together to plan, design, and develop content, you both reduce development costs and increase quality.

In adding collaboration as a key characteristic of the IPMM, I suggest the descriptions in Table 2-3 at each of the five levels.

Table 2-3: Description of the Collaboration Key Characteristic

IPMM Level	Collaboration Key Characteristic
Level 1	Information developers work independently, designing and developing their content in isolation from other developers in their organization.
Level 2	Information developers occasionally coordinate their efforts to avoid producing the same content more than one time. They occasionally find opportunities to share content developed by other team members, typically through a cut-and-paste process.
Level 3	Information developers are encouraged to form teams to plan, design, and develop content regarding the same product or process. Opportunities for sharing content among deliverables increases because developers are more aware of the content being created by their colleagues. Developers frequently form self-organized teams to jointly produce a result.
Level 4	Information developers regularly engage in collaborative processes that include planning, design, development, and review. Team members trust and respect the work of colleagues, believing that together they can build superior products than they could individually. Project managers and team leads facilitate collaboration as a core business practice.
Level 5	Information developers regularly collaborate with colleagues from other parts of the organization, encouraging a free flow of information and frequent interactions. They are continually looking for new opportunities to collaborate. At the same time, they find ways to avoid constant meetings that threaten to bog down progress. As professional communicators, information developers help foster communication among colleagues who are not effective communicators. They work together to develop new ideas that are more than the ideas offered by any individual team members or domain experts.

Change management

Change management offers another opportunity to develop a new key characteristic for the IPMM. Because I expect mature organizations to be continually innovating, challenging the status quo, and embracing continuous process improvement, mature organizations must understand how to manage change effectively. Without change management in place, organizations risk having team members refuse to embrace innovations or deliberately avoid adopting an innovative practice in their own work. When people do not understand the business necessity behind a new practice or design and do not understand how the innovation will affect their personal work practices, they reject or undermine the change.

A mature information-development organization must be prepared to manage change effectively if they want the pursuit of innovation to become a hallmark of their work. I suggest the descriptions in Table 2-4 at each of the five levels of the IPMM.

Table 2-4: Description of the Change Management Key Characteristic

IPMM Level	Change Management Key Characteristic
Level 1	Information developers have no mechanism available to foster change in their practices or design. Only personal persuasion of interested parties may help changes to occur.
Level 2	With new consolidation of individual information developers into groups led by a professional manager, change becomes an integral part of achieving process maturity. In fact, change is at heart of a Level 2 organization. However, since change is new and everything is changing, managers at Level 2 may not have developed a protocol supporting best practices in change management.
Level 3	The information-development manager has introduced a change-management protocol to the organization in hopes of consolidating gains won by achieving Level 3 of process maturity and providing a mechanism for introducing additional change.
Level 4	Managers have integrated change management best practices into the organization so that all team members understand what they need to do to foster continuous change. Communication about change is a regular part of the organizational culture.
Level 5	All team members are able to manage change within the organization and to work with colleagues in other parts of the larger community to foster change.

As this discussion of possible new key characteristics reflects, the IPMM itself undergoes continuous change. You can find updates to the model published on the CIDM website at http://www.infomanagementcenter.com. CIDM is always open to suggestions for clarification or improvement of the model.

Summary

The IPMM delineates the key characteristics of a successful, innovative, customer-oriented, information-development organization. In the IPMM, I have developed critical success factors:

- ✔ A centralized management structure to which information developers report that has the ability to develop and enforce standards in process, information design, and publishing.

- ✔ An information-development process that has measurement at its core, including the ability to estimate projects as thoroughly and accurately as possible and to adjust estimates when workload or resources change.

- ✔ A business-oriented, strategic perspective on the value and role of information development to the success of the enterprise.

- ✔ The ability to hire and train a professional staff and build with them a vision of the future.

- ✔ A customer-oriented perspective that assumes ownership of the customer's learning experience and productive use of the enterprise products and services.

With these critical perspectives in place and the ability to assume responsibility for the strategic direction of your organization, you are more than likely to succeed as a senior manager and assume a place in the corporate hierarchy that allows you to align with corporate objectives and increase the effectiveness and future profitability of your organization's products and services.

If you are challenged to increase productivity, reduce costs, and manage remote teams of individuals or merged departments, consider beginning with a process maturity assessment. Study the maturity level of your operation, those of sister organizations, and outsource vendors. Use the results of the assessment to strengthen your position with senior management, gain support for your process-improvement initiatives, and resist hasty and ill-considered cost-reduction schemes.

The next two parts of this book are designed to help you become a Level 3 organization in process maturity. Part 2 describes the role of the information-development manager in aligning the organization with the corporate goals and objectives. Part 3 provides an outline of the processes needed to manage projects at Level 3. All the ingredients for a successful organization are in place, waiting for you to implement them.

Part 2

Portfolio Management

Chapter 12
Developing as an Effective Leader

Chapter 13
Promoting Innovation in Information Development

"We have to remember that a return requires investment."

Chapter 3

Introduction to Portfolio Management

> Strategy-Focused Organizations use the Balanced Scorecard to place strategy at the center of their management processes.
>
> —Robert S. Kaplan and David P. Norton,
> *The Strategy-Focused Organization*[1]

Strategic portfolio management is a central responsibility of business-savvy information-development managers who operate mature organizations and take responsibility for the efficiency and quality of their work. Managing the portfolio strategically means setting priorities on the projects that your organization undertakes. It means deciding what can be done well and what should not be done at all. It means ensuring that the work done by your staff is worth doing and adds value to the larger organization and to the customers. It means honing processes, so that they emphasize activities that produce value and eliminate activities that might be fun but just waste everyone's time and energy. It means having a staff that is well-trained and enthusiastic about doing the right work well, rather than doing the wrong work well.

For many information-development managers in less mature organizations, the work their information developers do is assigned by other parts of the organization. Assignments come in from product developers or service providers. The project content is defined by the "customers," who specify that they

---

[1] Robert S. Kaplan and David P. Norton, *The Strategy-Focused Organization,* Cambridge, MA: Harvard Business School Press, 2001.

want a particular set of manuals, a help system, or material on a website. The manager takes the assignments and divides the work among the staff members. The set of all projects becomes the portfolio for which the department provides resources. All the decisions about the priorities and the content to be developed are determined by people outside.

Consider this scenario typical of a Level 1: Ad-hoc organization in the Information Process Maturity Model (IPMM), described in detail in Chapter 2. The information developers are either assigned work directly by product developers or other project leads or they are assigned the same work through their departmental manager. In either case, responsibility for defining the portfolio is outside the control of the people doing the work. In this scenario, the information developers are essentially clerical, doing work assigned by others with little or no role in the decision making.

Then consider another scenario typical of a Level 4 or 5: Optimizing organization in the IPMM. The information-development manager with the aid of senior members of the organization negotiates the shape and scope of the project portfolio with senior management and other business managers. Some projects are given a high priority because of their value to the larger organization, representing key product- or service-development efforts that promise to increase revenues or customer satisfaction. Other projects have lower priorities, depending upon their importance to the business and the customers. Some projects move entirely off the list because they have minimal value.

However, not all of the projects in a Level 4 or 5 portfolio are the information-development projects that you might be managing to support products and services of the larger organization. Some of the projects are developed to support innovations within the organization, such as new technologies to support content management, process improvements to gain efficiencies in the workflow, and new information architecture to increase quality to the end users and decrease duplication of effort. These projects are also included in the portfolio because they are deemed essential to the future health of the organization.

Clearly, moving from a Level 1 to a Level 5 organization centers on portfolio management and making the best use of ordinarily scarce resources. It means that you take control of your projects and ensure that you are delivering meaningful value to the business, not just doing somebody else's busy work. It means that you take a strategic view of your obligations to the larger organization and ensure that the work of your staff meets the corporate objectives. It means that you are placing people, tools, and budget on the right activities and that you constantly monitor the work to ensure that the projects are on track and not going off in unproductive directions. It means deciding on when to shut down a project if it isn't producing the desired results, even if the project is the favorite of someone inside or outside your organization. It means telling a product manager that you aren't going to assign resources to his or her project because it doesn't come high enough on the priority list.

Why Portfolio Management Is Critical

As an information-development manager hoping to move to a higher level of process maturity, you must begin to take charge by actively managing your organization's portfolio, not waiting to be told what to do by others. Portfolio management is essential to your set of responsibilities, especially if you are accountable for the deadlines and the quality of the work produced. But taking hold of the responsibility, as well as the account-ability, is not easy in an organization where the business and the product managers are used to being in control. To gain control for yourself and your team, you must prove that you understand the corporate goals and align with those goals as you make strategic deci-sions about managing your portfolio.

Benefits of portfolio management

Stepping up to the responsibilities of portfolio management in your organization provides a host of benefits, especially if your efforts are supported by your senior management:

- ✔ You are able to optimize your resources by assigning them to the higher priority projects and minimize your risk of spending resources on activities with little ben-efit to the organization.

- ✔ You will improve your ability to plan your organization's schedule so that every-one works more efficiently and isn't required to jump from project to project.

- ✔ You reduce the number of projects that have little value and make them easier to eliminate entirely.

- ✔ You help to ensure that your staff is thinking in terms of value to the corporation and the customer rather than thinking only about their personal project values.

- ✔ You enable a better alignment with corporate goals and improve the level of pro-fessional communication with business colleagues in the rest of the corporation.

By asserting control of your organization's business portfolio, you place yourself in the ranks of responsible and accountable senior management. That's ultimately a stronger and more satisfying position than being thought of as something equivalent to a typing pool. In addition, the experience of senior information-development managers shows that strategically aligned portfolio management actually saves money. Because they can devote resources more effectively to innovative and high-priority work, they can decrease what is spent unproductively on projects that return little value commensurate with their investment.

Barriers to portfolio management

Despite the clear advantages in managing your own portfolio and allocating your resources effectively, you are likely to encounter opposition as you try to move from an organization that takes orders to one that manages its affairs in line with corporate goals. If people are used to giving you orders, they will want to continue to do so. If others insist that you follow their priorities for your organization without regard to the overall portfo-lio of projects you have before you, then you will meet resistance to any change.

Here are some of the barriers you are likely to encounter and suggestions for removing them:

✔ Portfolio management implies taking power away from some people and acquiring it for yourself and your organization.

People certainly like to think that their projects are the most important. They want to make and enforce personal decisions rather than working with the team. Your move toward portfolio management implies a more open and democratic decision-making process. If you have the support of the executives for your move, others will see the benefit of working together.

✔ Getting the information you need to make good decisions about the projects in your portfolio is never easy.

In many organizations, project information is not made readily available to information development. You don't know what projects are coming, so you can't plan effectively. Projects get added to the mix unexpectedly with new priorities. You also have difficulty identifying the scope of many of the projects, often because the product developers haven't estimated their own scope and resources required.

Once again, you need to gain the support of senior management for bringing some order to the chaos. If your organization has no way of prioritizing its development projects, it's likely to be very immature. As you realize, you will have difficulty trying to be more mature than your internal customers. Nonetheless, if you have the support of your executive sponsor for prioritizing projects and you provide a sound business case for your decisions, others can do little except protest your decisions. The more politically savvy you can be in developing your criteria for prioritizing, the more successful you're likely to be.

✔ Portfolio management is all about making tough decisions.

How important is promising more than one can deliver? Informing senior management about workload problems results only in negatives:

"That work is important. It must be done. I don't want to hear any excuses. Do what it takes, but get it done."

How many of you have heard such responses to a reasonable assessment of the workload?

Some managers take a different position, learning that some of the time they must "just say no." Prioritizing assignments is never easy, but introducing reality into your decision making is a sensible course to take. Without some agreement about the priority of activities, you are left with individual, ad-hoc decisions among your own staff. One writer decides to leave out quality checks; another chooses to reformat engineering documents without regard to their usability; yet another decides not to rewrite older information because redesign is too time-consuming.

Avoid trying to please everyone. Don't become the manager who always tries to fit everything in, knowing full well that the resources just are not there.

In a recent user and task analysis workshop, writers voiced their concern about the added level of effort that investigating user needs and tasks might entail.

"Where will we find the time?" people moaned. "We can hardly get our work done satisfactorily as it is."

Fortunately for this organization, information-development managers acknowledged the problem and decided to make the user study project a top priority. They understood that without additional information about users, they had no reasonable, customer-focused means of prioritizing the workload. By understanding more precisely what customers' need, they could omit what was not necessary from the documentation.

At the same time, managers need to take a hard look at older legacy documentation. Perhaps the best solution to the time crunch is to leave some legacy documents alone. If there is little to gain in terms of increased revenues and customer satisfaction or decreased support costs, that legacy is not worth redesigning or even maintaining. Better to put scarce resources where they will make the most difference.

Best Practices in Managing Your Strategic Portfolio

Do you need a strategic plan to manage your organization effectively? Although some information-development managers are serious about strategic planning, most consider it something nice that they never get around to doing. But without a well-considered strategic plan, you will find yourself blinded by the demands of the everyday work schedule. You will have no way to evaluate your team's activities so that you can decide if you are expending your resources effectively. Strategic planning is a responsibility of managers at every level of the organization, not simply the executives. For a corporate strategy to be executed effectively, it must be carried out at every level, including information development.

Consider the following best practices for managing your portfolio and demonstrating its alignment with corporate goals and objectives:

✔ Creating and managing your project portfolio

✔ The Balanced Scorecard: Translating strategy into action

The best practices in this chapter explain how to evaluate your portfolio and develop criteria for prioritizing. Then you learn to create a Balanced Scorecard for your organization that demonstrates how your activities support your customers more effectively and contribute to the financial goals of the corporation.

 ### Best Practice—Creating and managing your project portfolio

You're in the right place and time to take control of your project portfolio and increase the maturity level of your organization. You want to win the respect of senior management and convince them that your organization should be part of the strategic plan.

Consider the steps to take in creating and managing your organization's portfolio of projects:

- ✔ Conduct an inventory of your projects
- ✔ Identify the projects that are truly strategic
- ✔ Categorize the projects and set priorities[2]

Once you have your ranking system in place and use it to prioritize your projects, you are ready to place your results into a Balanced Scorecard so that you can track how well you are doing.

Conducting a project inventory

You may already have a good idea of the projects that your team is handling. If so, you need only transfer them to a table or spreadsheet so that you can look at them as a whole. However, if you don't know much about the projects, use the inventory as an opportunity to talk with your team leads and information developers to know exactly what they are working on and what resources they are devoting to each project. Many managers find that the staff is already subtly ok prioritizing projects on their own volition. Unfortunately, the priorities are done on an ad-hoc basis rather than through the decision-making process of the team.

With a list of all the projects in place, add the actual or predicted start and end dates, the information to be delivered, and the estimate of resources for each project. If your team members create project plans, as described in Part 3, ask for each of the plans so that you can better understand the rationale for each project.

Figure 3-1 shows a portfolio report developed by the information-development managers at a division of Hewlett-Packard.

Identifying the projects that are truly strategic

After you have identified all the projects, review them carefully with your team leaders. Study the rationale for each project, focusing on the audience analysis and the projected deliverables. Then, arrange for a meeting with your senior manager to discuss how the various projects fit into the corporate strategy. You want to consider several of the issues discussed in Chapter 5 on the Technology Adoption Life Cycle as one example of prioritizing. You may want to consider how well each project is funded as a whole as an indicator of its level of importance to the company. Learn as much as possible about what projects are candidates for cancellation or are approaching end of life. Work with your senior manager to understand how each project is viewed at the business or corporate management level. If you don't have a senior manager who is in the best position to help you evaluate the projects, find an informal manager who can give you advice quietly.

[2] Todd Datz, "Portfolio Management: How to do it right." *CIO Magazine,* May 1, 2003.

Project Rollup Month 200X

Product	Division	Program Changes	Slip Y/N	Current Start Date	Current Release Date	Q3 Planned Vendor Spending	Q3 Planned Perm Spending	Q3 Total Planned
TechDoc Projects								
Admin								
TechDoc Q3 Localization FY0X	1 TechDoc Projects	Scope -	N	5/1/05	7/31/05	$26,148.83	$0.00	$26,148.83
TechDoc Q3 Production FY0X	1 TechDoc Projects	Scope -	N	5/1/05	7/31/05	$80,525.83	$51,428.56	$131,954.39
TOTAL Admin	2					$106,674.66	$51,428.56	$158,103.22
Special Projects								
TechDoc Single Sourcing ~ Q3 FY0X ~ Q3 FY0X	1 TechDoc Projects	Scope -	N	5/1/05	7/31/05	$39,047.83	$52,701.77	$91,749.60
TechDoc Subteams ~ Q3 FY0X ~ Q3 FY0X	1 TechDoc Projects	Scope -	N	5/1/05	7/31/05	$35,232.50	$14,913.35	$50,145.85
TechDoc Warranty ~ 10 ~ TechDoc Warranty	1 TechDoc Projects	New	N	4/21/05		$0.00	$0.00	$0.00
TOTAL TechDoc Initiatives	3					$35,232.50	$14,913.35	$50,145.85
TOTAL TechDoc Projects	5					$141,907.16	$66,341.91	$208,249.07
Project A								
Project A								
Project A DG2.1.1	1 Project A	New	Y	3/1/05	5/15/05	$0.00	$0.00	$0.00
Project A DG2.1.2	1 Project A	New	N	5/16/05	7/15/05	$287.70	$2,874.82	$3,162.52
Project A JP321	1 Project A	New	Y	4/15/05	6/1/05	$119.88	$2,874.82	$2,994.70
Project A XSA 1.2	1 Project A	New	N	2/1/05	5/1/05	$0.00	$0.00	$0.00
Project A XSA 1.4	1 Project A	New	N	3/15/05	6/1/05	$287.70	$4,744.62	$5,032.32
Project A Release XAW 3.2	1 Project A	Scope -	Y	4/1/05	5/1/05	$2,538.22	$2,874.82	$5,413.04
Project A Systems 1.0	1 Project A	Scope -	N	5/15/05	11/15/05	$20,584.83	$0.00	$20,584.83
Project A Systems 1.1	1 Project A	New	N	11/8/04	5/1/05	$0.00	$0.00	$0.00
Project A Systems 1.2	1 Project A	Scope -	Y	1/1/05	6/1/05	$10,028.68	$4,312.22	$14,340.90
Project A Systems 1.3	1 Project A	Scope -	Y	11/1/04	5/1/05	$4,251.29	$5,749.63	$10,000.92
Project A Systems 1.4	1 Project A	Scope -	Y	7/1/05	10/20/05	$26,336.55	$2,874.82	$29,211.37
Project A Systems 1.5	1 Project A	Scope -	N	12/1/04	5/1/05	$7,273.50	$5,749.63	$13,023.13
Project A Systems 1.6	1 Project A	Scope -	N	5/15/05	9/1/05	$4,598.57	$5,749.63	$10,348.20
Project A Systems 1.7	1 Project A	Scope -	N	5/16/05	8/22/05	$18,650.85	$5,749.63	$24,400.48
Project A Systems 1.8	1 Project A	Scope -	Y	11/8/04	6/1/05	$7,972.29	$2,874.82	$10,847.11
Project A Systems 1.9	1 Project A	Scope -	N	5/15/05	9/1/05	$14,133.69	$2,874.82	$17,008.51
TOTAL Project A	16					$22,105.98	$5,749.63	$27,855.61
TOTAL Project A	16					$22,105.98	$5,749.63	$27,855.61
Project B								
Modell								
Moda c8873 1.x	1 Project B	New	N	5/1/05		$0.00	$0.00	$0.00
Moda c8873 1.0	1 Project B	New	N	3/1/05	8/15/05	$0.00	$0.00	$0.00
Moda c8873 1.1	1 Project B	Scope -	Y	7/1/05	10/10/05	$4,763.96	$15,550.14	$20,314.10
Moda c8873 1.2	1 Project B	New	Y	7/1/05	10/10/05	$0.00	$0.00	$0.00
Moda c8873 1.3	1 Project B	New	Y	5/1/05	7/18/05	$0.00	$0.00	$0.00
Moda c8873 1.4	1 Project B	Scope -	N	6/1/05	8/15/05	$5,783.96	$15,550.14	$21,334.10
TOTAL Modell	6					$5,783.96	$15,550.14	$21,334.10
Maxell								
MAX SXH 3232 1.0	1 Project B	New	N	4/1/05	7/19/05	$0.00	$0.00	$0.00
MAX SXH 3232 1.1	1 Project B	Scope +	N	2/1/05	5/25/05	$1,200.00	$2,613.47	$3,813.47
MAX SXH 3232 1.2	1 Project B	New	N	5/1/05	6/15/05	$0.00	$0.00	$0.00
MAX SXH 3232 1.3	1 Project B	Scope +	Y	2/5/05	7/26/05	$0.00	$2,874.82	$2,874.82
MAX SXH 3232 1.4	1 Project B	Scope -	N	1/1/05	5/15/05	$6,100.00	$0.00	$6,100.00
MAX SXH 3232 1.5	1 Project B	Scope -	N	11/1/04	7/31/05	$7,860.00	$0.00	$7,860.00
MAX SXH 3232 1.6	1 Project B	Scope -	N	5/1/05	6/1/05	$3,400.00	$0.00	$3,400.00
MAX SXH 3232 1.7	1 Project B	Scope -	Y	5/1/05	7/31/05	$7,200.00	$0.00	$7,200.00
MAX SXH 3232 1.8 -	1 Project B	New	N	6/15/05	1/15/06	$0.00	$0.00	$0.00
MAX SXH 3232 1.9	1 Project B	Scope -	N	11/1/04	7/31/05	$2,100.00	$0.00	$2,100.00
TOTAL Maxell	10					$27,860.00	$5,488.29	$33,348.29
TOTAL Project B	16					$33,643.96	$21,038.43	$54,682.39
Software								
Data Protection								
Data Protector ~ 5.6 ~ Junction	1 Software	Scope -	Y	11/1/04	3/3/06	$159,100.02	$0.00	$159,100.02
TOTAL Data Protection	1					159100.02	0	159100.02
ILM								
File System Extender ~ 4.x ~ tbd	1 Software	New	N	2/1/05	9/1/05	$0.00	$2,010.36	$2,010.36
ILM Database ~ 1.0 ~ Resell (Phase 1)	1 Software	New	N	3/14/05	6/1/05	$0.00	$1,005.18	$1,005.18
ILM Medical Imaging ~ 1.0? ~ Genesis	1 Software	New	N	5/1/05	6/27/05	$0.00	$2,010.36	$2,010.36
ILM Migration Agent ~ 2.0 ~ Information Migration Agent	1 Software	New	N	2/10/05	6/10/05	$30,000.00	$1,005.18	$31,005.18
Mendecino ~ 1.0? ~ Niagara Falls	1 Software	New	N	5/1/05	10/30/05	$0.00	$1,005.18	$1,005.18
Reference Information Storage System ~ 1.2 ~ Lancaster II	1 Software	New	Y	2/1/05	5/27/05	$0.00	$9,180.39	$9,180.39
Reference Information Storage System ~ 1.2 ~ XML Conversion	1 Software	New	Y	2/1/05	5/30/05	$0.00	$219.45	$219.45
Reference Information Storage System ~ 1.3 ~ RISS/RIM 1.3	1 Software	New	N	4/15/05	7/29/05	$55,500.00	$21,962.81	$77,462.81
Reference Information Storage System ~ 1.5 ~ Sheffield	1 Software	Scope +	Y	4/1/05	3/20/06	$0.00	$6,985.88	$6,985.88
Reference Information Storage System ~ Patch xxxx ~ CPE 2005.06 patch	1 Software	New	N	2/18/05	6/30/05	$0.00	$15,763.94	$15,763.94
RISS: Disk2Disk ~ 1 ~ Fuji	1 Software	Scope +	Y	12/8/04	3/20/06	$0.00	$4,112.80	$4,112.80
TOTAL ILM	11					85500	65261.53057	150761.5306
Software								
Server/Storage Management ~ TBD ~ Etna	1 Software	New	N	2/1/05	8/1/05	$0.00	$8,530.92	$8,530.92
Server/Storage Management ~ TBD ~ Quake	1 Software	Scope -	Y	11/1/04	7/22/05	$60,000.00	$26,519.28	$86,519.28
Storage Area Manager ~ 3.1/3.2 ~ Phoenix (DPIs)	1 Software	New	N			$0.00	$0.00	$0.00
Storage Area Manager ~ 3.2.2.x ~ OV SAM 3.2.2.x	1 Software	New	N	5/1/05	7/15/05	$0.00	$6,228.77	$6,228.77
Storage Area Manager ~ 3.2.x ~ Phoenix (DPIs) next	1 Software	New	N	5/1/05	7/15/05	$0.00	$9,103.58	$9,103.58
Storage Essentials ~ 4.0, SP1 ~ Storage Essentials	1 Software	New	N	4/11/05	5/15/05	$0.00	$603.11	$603.11
TOTAL Software	6					60000	50985.66445	110985.6644
TOTAL Software	18					$304,600.02	$116,247.20	$420,847.22
TOTAL ALL DIVISIONS	55					$502,257.12	$209,377.16	$711,634.28

Product	May	June	July	Total	Total Vendor and Perm Spend for Q3	localization cost forecast Q3		budget rgy	Q4 planned vendor/perm	Q4 planned localizaiton	% CMS deliverables
TechDoc Projects											
Admin											
TechDoc Q3 Localization FY0X	$0.00	$0.00	$0.00	$0.00	$0.00	$0.00	$0.00	$26,148.83	$0.00	$0.00	0.00
TechDoc Q3 Production FY0X	$0.00	$0.00	$0.00	$0.00	$0.00	$0.00	$0.00	$131,954.39	$0.00	$0.00	0.00
TOTAL Admin	$0.00	$0.00	$0.00	$0.00	$0.00	$0.00	$0.00	$158,103.22	$0.00	$0.00	$0.00
Special Projects											
TechDoc Single Sourcing ~ Q3 FY0X ~ Q3 FY0X	$0.00	$2,281.60	$2,909.04	$5,190.64	$5,190.64	$0.00	$0.00	$86,558.96	$9,867.92	$0.00	0.00
TechDoc Subteams ~ Q3 FY0X ~ Q3 FY0X	$0.00	$0.00	$285.20	$285.20	$285.20	$0.00	$0.00	$49,860.65	$285.20	$0.00	0.00
TechDoc Warranty ~ 10 ~ TechDoc Warranty	$0.00	$1,711.20	$1,140.80	$2,852.00	$2,852.00	$0.00	$0.00	-$2,852.00	$2,852.00	$0.00	0.00
TOTAL TechDoc Initiatives	$0.00	$1,711.20	$1,426.00	$3,137.20	$3,137.20	$0.00	$0.00	$47,008.65	$3,137.20	$0.00	$0.00
TOTAL TechDoc Projects	$0.00	$1,711.20	$1,426.00	$3,137.20	$3,137.20	$0.00	$0.00	$205,111.87	$3,137.20	$0.00	$0.00
Project A											
Project A											
Project A DG2.1.1	$0.00	$0.00	$0.00	$0.00	$0.00	$0.00	$0.00	$0.00	$0.00	$0.00	0.00
Project A DG2.1.2	$570.40	$855.60	$4,278.00	$5,704.00	$5,704.00	$0.00	$0.00	-$2,541.48	$5,704.00	$0.00	0.00
Project A JP321	$1,140.80	$0.00	$0.00	$1,140.80	$1,140.80	$0.00	$0.00	$1,853.90	$1,140.80	$0.00	0.00
Project A XSA 1.2	$0.00	$0.00	$0.00	$0.00	$0.00	$0.00	$0.00	$0.00	$3,273.00	$0.00	0.00
Project A XSA 1.4	$1,140.80	$1,426.00	$570.40	$3,137.20	$3,137.20	$0.00	$0.00	$1,895.12	$3,137.20	$0.00	0.00
Project A Release XAW 3.2	$1,711.20	$0.00	$0.00	$1,711.20	$3,826.20	$0.00	$0.00	$1,586.84	$14,674.60	$0.00	0.00
Project A Systems 1.0	$0.00	$570.40	$570.40	$1,140.80	$1,140.80	$0.00	$0.00	$19,444.03	$2,281.60	$0.00	0.00
Project A Systems 1.1	$570.40	$0.00	$0.00	$570.40	$570.40	$0.00	$0.00	-$570.40	$570.40	$0.00	0.00
Project A Systems 1.2	$2,281.60	$0.00	$0.00	$2,281.60	$6,511.60	$0.00	$0.00	$7,829.30	$22,750.80	$0.00	0.00
Project A Systems 1.3	$1,996.40	$0.00	$0.00	$1,996.40	$3,054.40	$0.00	$0.00	$6,946.52	$10,294.00	$0.00	0.00
Project A Systems 1.4	$0.00	$1,426.00	$1,426.00	$2,852.00	$2,852.00	$0.00	$0.00	$26,359.37	$5,704.00	$0.00	0.00
Project A Systems 1.5	$6,274.40	$0.00	$0.00	$6,274.40	$8,389.40	$0.00	$0.00	$4,633.73	$37,896.00	$0.00	0.00
Project A Systems 1.6	$0.00	$1,426.00	$1,426.00	$2,852.00	$2,852.00	$0.00	$0.00	$7,496.20	$5,704.00	$0.00	0.00
Project A Systems 1.7	$0.00	$285.20	$285.20	$570.40	$570.40	$0.00	$0.00	$23,830.08	$1,140.80	$0.00	0.00
Project A Systems 1.8	$1,711.20	$0.00	$0.00	$1,711.20	$9,643.20	$0.00	$0.00	$1,203.91	$31,548.40	$0.00	0.00
Project A Systems 1.9	$0.00	$1,140.80	$1,426.00	$2,566.80	$2,566.80	$0.00	$0.00	$14,441.71	$2,566.80	$0.00	0.00
TOTAL Project A	$1,711.20	$1,140.80	$1,426.00	$4,278.00	$12,210.00	$0.00	$0.00	$15,645.61	$34,115.20	$0.00	$0.00
TOTAL Project A	$1,711.20	$1,140.80	$1,426.00	$4,278.00	$12,210.00	$0.00	$0.00	$15,645.61	$34,115.20	$0.00	$0.00
Project B											
Modell											
Moda c8873 1.x	$0.00	$570.40	$0.00	$570.40	$570.40	$0.00	$0.00	-$570.40	$570.40	$0.00	0.00
Moda c8873 1.0	$570.40	$0.00	$0.00	$570.40	$570.40	$0.00	$0.00	-$570.40	$1,711.20	$0.00	0.00
Moda c8873 1.1	$1,140.80	$0.00	$0.00	$1,140.80	$1,140.80	$0.00	$0.00	$19,173.30	$3,707.60	$0.00	0.00
Moda c8873 1.2	$570.40	$285.20	$570.40	$1,426.00	$1,426.00	$0.00	$0.00	-$1,426.00	$10,267.20	$0.00	0.00
Moda c8873 1.3	$3,707.60	$2,852.00	$1,140.80	$7,700.40	$7,700.40	$0.00	$0.00	-$7,700.40	$9,982.00	$0.00	0.00
Moda c8873 1.4	$1,140.80	$1,996.40	$2,852.00	$5,989.20	$5,989.20	$0.00	$0.00	$15,344.90	$13,119.20	$0.00	0.00
TOTAL Modell	$4,848.40	$4,848.40	$3,992.80	$13,689.60	$13,689.60	$0.00	$0.00	$7,644.50	$23,101.20	$0.00	$0.00
Maxell											
MAX SXH 3232 1.0	$0.00	$570.40	$0.00	$570.40	$570.40	$0.00	$0.00	-$570.40	$570.40	$0.00	0.00
MAX SXH 3232 1.1	$1,711.20	$1,825.28	$0.00	$3,536.48	$4,736.48	$0.00	$0.00	-$923.01	$15,232.88	$0.00	0.00
MAX SXH 3232 1.2	$0.00	$0.00	$0.00	$0.00	$0.00	$0.00	$0.00	$0.00	$0.00	$0.00	0.00
MAX SXH 3232 1.3	$741.52	$1,311.92	$1,882.32	$3,935.76	$3,935.76	$0.00	$0.00	-$1,060.94	$7,814.48	$0.00	0.00
MAX SXH 3232 1.4	$0.00	$0.00	$0.00	$0.00	$4,600.00	$0.00	$0.00	$1,500.00	$13,200.00	$0.00	0.00
MAX SXH 3232 1.5	$0.00	$0.00	$0.00	$0.00	$5,400.00	$0.00	$0.00	$2,460.00	$13,500.00	$0.00	0.00
MAX SXH 3232 1.6	$0.00	$0.00	$0.00	$0.00	$0.00	$0.00	$0.00	$3,400.00	$0.00	$0.00	0.00
MAX SXH 3232 1.7	$0.00	$0.00	$0.00	$0.00	$0.00	$0.00	$0.00	$7,200.00	$0.00	$0.00	0.00
MAX SXH 3232 1.8 -	$0.00	$0.00	$285.20	$285.20	$285.20	$0.00	$0.00	-$285.20	$285.20	$0.00	0.00
MAX SXH 3232 1.9	$0.00	$0.00	$0.00	$0.00	$600.00	$0.00	$0.00	$1,500.00	$7,020.00	$0.00	0.00
TOTAL Maxell	$2,452.72	$3,707.60	$2,167.52	$8,327.84	$20,127.84	$0.00	$0.00	$13,220.45	$57,622.96	$0.00	$0.00
TOTAL Project B	$7,301.12	$8,556.00	$6,160.32	$22,017.44	$33,817.44	$0.00	$0.00	$20,864.95	$80,724.16	$0.00	$0.00
Software											
Data Protection											
Data Protector ~ 5.6 ~ Junction	$0.00	$0.00	$0.00	$0.00	$110,559.00	$0.00	$0.00	$48,541.02	$390,168.00	$0.00	0.00
TOTAL Data Protection	0	0	0	0	110559	0	0	48541.02	390168	0	0
ILM											
File System Extender ~ 4.x ~ tbd	$0.00	$570.40	$570.40	$1,140.80	$1,140.80	$0.00	$0.00	$869.56	$1,140.80	$0.00	0.00
ILM Database ~ 1.0 ~ Resell (Phase 1)	$0.00	$285.20	$285.20	$570.40	$570.40	$0.00	$0.00	$434.78	$570.40	$0.00	0.00
ILM Medical Imaging ~ 1.0? ~ Genesis	$0.00	$0.00	$285.20	$285.20	$285.20	$0.00	$0.00	$1,725.16	$285.20	$0.00	0.00
ILM Migration Agent ~ 2.0 ~ Information Migration Agent	$570.40	$285.20	$285.20	$1,140.80	$1,140.80	$0.00	$0.00	$29,864.36	$1,140.80	$0.00	0.00
Mendecino ~ 1.0? ~ Niagara Falls	$0.00	$0.00	$171.12	$171.12	$171.12	$0.00	$0.00	$834.06	$171.12	$0.00	0.00
Reference Information Storage System ~ 1.2 ~ Lancaster II	$0.00	$0.00	$0.00	$0.00	$0.00	$0.00	$0.00	$9,180.39	$25,486.20	$0.00	0.00
Reference Information Storage System ~ 1.2 ~ XML Conversion	$570.40	$0.00	$0.00	$570.40	$2,962.40	$0.00	$0.00	-$2,742.95	$5,122.40	$0.00	37.50
Reference Information Storage System ~ 1.3 ~ RISS/RIM 1.3	$285.20	$285.20	$570.40	$1,140.80	$1,140.80	$0.00	$0.00	$76,322.01	$1,140.80	$0.00	0.00
Reference Information Storage System ~ 1.5 ~ Sheffield	$855.60	$570.40	$570.40	$1,996.40	$38,206.40	$0.00	$0.00	-$31,220.52	$38,776.80	$0.00	84.62
Reference Information Storage System ~ Patch xxxx ~ CPE 2005.06 patch	$1,140.80	$1,140.80	$0.00	$2,281.60	$2,281.60	$0.00	$0.00	$13,482.34	$2,281.60	$0.00	0.00
RISS: Disk2Disk ~ 1 ~ Fuji	$3,707.60	$4,335.04	$3,992.80	$12,035.44	$48,245.44	$0.00	$0.00	-$44,132.64	$54,805.04	$0.00	0.00
TOTAL ILM	7130	7472.24	6730.72	21332.96	96144.96	0	0	54616.57057	130921.16	0	122.1153846
Software											
Server/Storage Management ~ TBD ~ Etna	$1,140.80	$1,711.20	$2,852.00	$5,704.00	$5,704.00	$0.00	$0.00	$2,826.92	$7,415.20	$0.00	100.00
Server/Storage Management ~ TBD ~ Quake	$7,187.04	$6,616.64	$5,989.20	$19,792.88	$57,067.88	$0.00	$0.00	$29,451.40	$135,814.88	$0.00	100.00
Storage Area Manager ~ 3.1/3.2 ~ Phoenix (DPIs)	$0.00	$0.00	$0.00	$0.00	$0.00	$0.00	$0.00	$0.00	$1,711.20	$0.00	0.00
Storage Area Manager ~ 3.2.2.x ~ OV SAM 3.2.2.x	$0.00	$0.00	$0.00	$0.00	$0.00	$0.00	$0.00	$6,228.77	$0.00	$0.00	0.00
Storage Area Manager ~ 3.2.x ~ Phoenix (DPIs) next	$0.00	$0.00	$0.00	$0.00	$0.00	$0.00	$0.00	$9,103.58	$0.00	$0.00	0.00
Storage Essentials ~ 4.0, SP1 ~ Storage Essentials	$0.00	$0.00	$0.00	$0.00	$0.00	$0.00	$0.00	$603.11	$0.00	$0.00	0.00
TOTAL Software	8327.84	8327.84	8841.2	25496.88	62771.88	0	0	48213.78445	144941.28	0	200
TOTAL Software	$15,457.84	$15,800.08	$15,571.92	$46,829.84	$269,475.84	$0.00	$0.00	$151,371.38	$666,030.44	$0.00	$322.12
TOTAL ALL DIVISIONS	$24,470.16	$27,208.08	$24,584.24	$76,262.48	$318,640.48	$0.00	$0.00	$392,993.80	$784,007.00	$0.00	$322.12

Figure 3-1: Portfolio inventory report

The dashboard report was provided by Charlotte Robidoux of Hewlett-Packard. The original data has been modified to protect its confidentiality. The dashboard report is created using NSight's project management software.

In your evaluation, you need to find the projects that are getting minimal funding. These are often projects that are "nice to do" when there is time. They are also candidates for cancellation if other demands take precedence. These are projects you may want to assign minimal support, or they may be projects that you should outsource because they give your department little strategic advantage.

Categorize the projects and set priorities

Categorizing the projects and setting priorities requires asking tough questions:

- ✔ Is this project producing increasing revenue for the company?
- ✔ Is the customer demand for this project's output increasing or decreasing?
- ✔ Are customers requesting better information or are they satisfied with what they have now?
- ✔ Does this project give you an opportunity to increase efficiency and decrease operational costs?
- ✔ If you work more effectively on this project, will an increase in customer satisfaction translate into more sales?
- ✔ Will an improvement in information accessibility for this project help to reduce support costs?
- ✔ Can you produce a better return on your investment in this project by providing more valuable information or delivering the information differently?
- ✔ Are you doing things because they are fun for the writers, not because they are valuable to customers?

Once you ask questions like these and try to find the best answers, you find it more and more difficult to justify activities that don't directly contribute to your success in the larger organization. By prioritizing work as an organization, with advice from peers and senior management, you have a better chance of focusing on what is really important to customers rather than what is important to individuals acting alone.

You will find it useful to add the answers to your questions to your project portfolio spreadsheet to create a standard method for ranking the projects. For example, you might consider the importance of the project to customer satisfaction. If a project will permit you to pursue an information innovation that you believe will increase customer satisfaction, it might rank higher than a project that is an update to information that is not heavily used by anyone.

You might identify a project on which you can apply a new technology to increase efficiency and reduce costs. That project might rank higher than one that is costly to produce and adds little value.

Look for projects that can become a new and effective way of producing information and that promise innovations that will increase customer satisfaction. Some projects are simply in maintenance mode and are prime candidates for limited resources or low-cost outsourcing. Others support the steady growth of your organization's products, needing regular and devoted maintenance to keep them strong. Still others require an entirely new and innovative approach that changes the way you do business.

Once you develop your ranking criteria with the support of your senior management, communicate it to your staff and ensure they understand the priorities. Because they have built their own allegiances to projects they may be reluctant to accept the priorities. By emphasizing the underlying business strategy and the acceptance of the plan by senior management, you help them understand and become enthusiastic supporters of the decisions.

Managing your portfolio

After you have applied the ranking and assigned priorities and resources to your project portfolio, you need to monitor the project to ensure that you're getting the value from them that you anticipated. I recommend setting up your projects in a Balanced Scorecard. The Balanced Scorecard helps you monitor the progress of a project or an innovation initiative through the four quadrants, as explained in the next best practice.

For example, consider this scenario: You identify a project that calls for the restructuring of a large volume of process documents that include detailed product specifications. This project is your highest priority, with full support from the CEO and your immediate vice president. The project is critical for achieving an improved level of customer satisfaction with your company's products and services.

You assign your best team to the project but you recognize that they are unfamiliar with the type of information that requires restructuring and rewriting and they don't know the audience well. You need to ensure that the team gets the training and direction they need to handle the content, and you need them to develop a new process to handle the restructuring and revisions of the content with the involvement of several subject-matter experts throughout the organization. You know that some of the experts do not support the changes that you want the team to make.

By creating a Balanced Scorecard, you can identify what you expect to happen on this project. Your scorecard might resemble the one in Figure 3-2.

CheckFree Corporation's Peak Performance program requires all divisions and business units to create a scorecard of goals and metrics in support of the corporate goals. The scorecard is designed to show how all three layers of goals relate to one another and to track the monthly progress of each goal.

The corporate goals are represented by the four shaded rows spanning the scorecard (Market Growth, Service Quality, Financial Performance, and Customer Satisfaction). The division goals are listed at the bottom of the scorecard and each is assigned a letter. These letters appear in the goal mapping table in the far-left column. The business unit goals and metrics are defined on the left side of the scorecard. These goals must be specific enough to easily translate into individual goals for each CheckFree associate.

To understand how the scorecard connects the three layers of goals and tracks results, look at the first goal in the list. The business unit goal is to support innovative development and implementation of products and features. The placement of this goal on the scorecard shows that it supports the corporate goal of Market Growth and division goals A, B, and C. There is no activity tracked for this goal in July or August. In September, 100% of the deliverables were on time and 80% of the deliverables went through a peer/editorial and technical review.

Fiscal Year
Peak Performance Goals & Metrics

Technical Communication Scorecard
Peak Performance Operational Results

GOAL MAPPING TABLE (A B C D E F G H I J K L)	PEAK PERFORMANCE GOALS	PEAK PERFORMANCE METRIC	METRIC GOAL	Jul	Aug	Sep	Oct	Jun	GAP
MARKET GROWTH									
	Support innovative development and implementation of products and features	Percentage of on-time deliverables for product releases	100%	-	-	100%			0%
		Percentage of deliverables going through peer/editorial and technical review	100%	-	-	80%			-20%
SERVICE QUALITY									
	Provide Best Practice Resource Center services for division	Number of selected gaps closed between current team practice and best practice	4	-	1	1			-3
		Percentage of usage of selected best practice methods in the division	50%	-	5%	10%			-40%
	Improve product release communications	Percentage of release communication content requirements met	100%	-	-	80%			-20%
		Percentage of release communication time requirements met	100%	-	-	50%			-50%
	Improve implementations speed and quality	Percentage reduction of time spent in the revision phase of the implementation documentation lifecycle	20%	-	5%	5%			-15%
		Percentage increase in client usage of doc as primary information source for implementation	30%	Research	Research	Design			-30%
		Percentage reduction of time to customize help during product implementations	50%	Research	Research	Research			-50%
		Percentage of specifications flowing through the update process	50%	-	7%	21%			-29%
FINANCIAL PERFORMANCE									
	Meet or exceed financial objectives	Percentage of variance of forecast to budget	0%	-4%	3%	0%			0%
CUSTOMER SATISFACTION									
	Evaluate usability of our information elements	Percentage of product usability tests that include scenarios about information elements	100%	-	100%	50%			-50%
		Access and analyze call log data to categorize questions by current/missing info elements	Two lists by 3/15/07	Research	Research	Research			-
		Address x number of call log issues coming from two lists by June release	x=from two lists, Q4 end	-	-	-			-
	Equip and enable each associate to achieve development plans	Percentage of completed development plans for each associate	100%	-	10%	93%			-7%
		Percentage of completed development objectives for each associate	100%	5%	5%	25%			-75%

A-Successfully develop and implement products/features
B-Provide resources in support of on-time product releases
C-Facilitate and assist division in achieving Sigma quality

D-Provide best practice services for division
E-Implement Release Management function
F-Implement Project Visibility and Tracking function

G-Improve implementation speed and quality
H-Meet or exceed financial objectives
I-Facilitate the "Six Sigma Shift" within division

J-Drive Peak Performance program
K-Implement initial customer satisfaction functions
L-Equip and enable each associate to achieve development plans

Division Goals

Figure 3-2: A representative Balanced Scorecard

This Balanced Scorecard report was provided by Ann Teasley of CheckFree Corporation. The original data has been modified to protect its confidentiality.

Best Practice—The Balanced Scorecard: Translating strategy into action

Robert Kaplan and David Norton, in *The Balanced Scorecard*,[3] argue that a business strategy is a set of if-then statements that attempt to predict how changes you may make to your organization and people will result in improved financial performance and customer satisfaction for your corporation. As you develop a business strategy and plan how you will manage your project portfolio, you can use a balanced scorecard to reveal the relationship between your activities in information-development and your company's success.

Consider this if-then sequence as applied to information development:

If you invest in training your employees to study your customers and their information needs, *then* they will be able to design and develop more effective information products. *If* the information products they design are successful, *then* customers will be able to lower their costs of doing business with your company and increase their staff productivity as they use your company's products. *If* the customers are more successful, *then* they will buy more products from your company thereby increasing its profitability and success.

Although the final measurements of success in this business strategy are financial, note that along the way you will be able to institute many other measurements of success:

✔ You can evaluate your customers' satisfaction with your information products.

✔ You can learn if you are effectively developing and implementing new processes that result in increased customer satisfaction.

✔ You can investigate the success of your innovation projects by measuring to what degree they have succeeded in transforming your organization and making it more efficient and effective.

The analysis you do on each of these measurements demonstrates how your activities to innovate and transform your organization lead directly to the corporation's customer satisfaction and financial performance.

Kaplan and Norton argue that the Balanced Scorecard approach works at many levels of the organization by accounting for a company's "intangible and intellectual assets" ("high-quality products and services, motivated and skilled employees, responsive and predictable internal processes, and satisfied and loyal customers"). They insist that, in a new information age, we cannot afford to fixate on bottom-line, expense-related, and often lagging measurement systems. We need to find new ways of predicting future success by implementing innovations today and evaluating their effectiveness. The authors' solution is the Balanced Scorecard.

[3] Robert S. Kaplan and David P. Norton, *The Balanced Scorecard: Translating Strategy into Action*, Cambridge, MA: Harvard Business School Press, 1996.

Traditional accounting measures tell us about what has occurred in the past, not what is likely to occur in the future. But what measures can we use to foretell the future?

The four perspectives of the Balanced Scorecard

To foretell the future, you need to approach carefully the four key perspectives of the Balanced Scorecard:

- ✔ financial perspective
- ✔ customer perspective
- ✔ internal business process perspective
- ✔ learning and growth perspective

The Balanced Scorecard's four key perspectives are derived top-down from your vision of the future and your strategy for getting there, as illustrated in Figure 3-3.

- ✔ **Financial perspective.** Are your actions contributing to the company's bottom line?
- ✔ **Customer perspective.** Are your actions contributing to satisfying customers?
- ✔ **Internal-business-process perspective.** Do you have the best processes in place to contribute to customer satisfaction and the bottom line? Are you creating entirely new processes to create better information products and deliver better service in the future?
- ✔ **Learning and growth perspective.** Are your actions with regard to people, systems, and organizational procedures creating long-term potential for improving and growing the business?

If you develop your business strategy to include all four perspectives and to make measurements everywhere, Kaplan and Norton believe you can achieve a proper balance between outcome measures (customers and finances) and future measures (processes and abilities). Such a balance includes both objective and subjective measures, going beyond bottom-line calculations. At the same time, they ask you to connect your strategic plans to your corporation's or your department's financial goals. The authors point out that it is not enough to improve internal processes; all process improvement and learning investments need to focus on improving the results for customers and shareholders. In planning and implementing a Balanced Scorecard, you need to make sure that all the pieces are carefully linked.

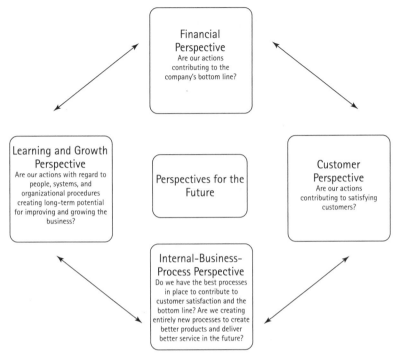

Figure 3-3: The four perspectives of the Balanced Scorecard

Why use a Balanced Scorecard approach?

I believe that information-development managers can use a Balanced Scorecard approach to both drive and measure the success of organizational change. With the Balanced Scorecard, you can align with your organization's strategic vision and turn this vision into explicit objectives and measures for your own department.

Kaplan and Norton like to point out that they agree with the maxim: "Measurement matters. If you can't measure it, you can't manage it." However, they insist that managers need more than standard financial measures to evaluate the success of a strategic business planning activity. As many business leaders have noted, a focus on financial measures among US companies has increased the importance of short-term gains (the latest stock market evaluation or the results from the last business quarter). Because financial measures are lagging indicators (we measure only after the damage has been done), they don't show what is being created or destroyed by current management actions.

The Balanced Scorecard provides managers with a more comprehensive framework to translate a company's vision and strategy into a coherent set of performance measures than financial success alone.

Remember that a strategy is a set of hypotheses about cause and effect. A good Balanced Scorecard shows how your planned actions are linked in a cause-and-effect relationship. You state: "If we do this, then this will happen." However, hypotheses are just that—hypotheses. You need a way to measure outcomes and drive performance improvements if your hypotheses are going to turn into facts.

In many cases, as you plan a business case, you know what outcomes you want. Customers who are positive about the information products they receive behave in desirable ways:

✔ They make fewer calls to customer service because they find answers to their questions in the documentation, the online help, or on the knowledge websites.

✔ They achieve a faster mean time to productivity because getting-started information is effective. Mean time to productivity refers to the average time a person or an organization takes to learn a new product or process.

✔ Customers experience less down time because information makes it possible to troubleshoot and solve problems more quickly.

As an information-development manager, you know that you would like to measure your customer's success with your information. However, these measures are still lagging indicators. Until the information is in the customers' hands, you cannot measure its effectiveness.

What about the leading indicators? What do you have to measure about the performance of your staff to ensure that you get the outcomes you want?

It's important to point out a corporate fact-of-life that Kaplan and Norton mention. Quality improvements don't always translate into financial success. Many process-improvement campaigns lead to improved capacity to perform the work (you're more efficient; you've eliminated unnecessary steps), which may mean that you now have resources that are really not employed as effectively as they might be. Process improvements leave you with unused capacity, or the ability to do more work with the same people, but you may not have realized actual reductions in spending.

By taking a Balanced Scorecard approach, you attempt to link process improvements directly to customer satisfaction and financial gains. You also measure all along the way to ensure that you are spending your process and employee training resources effectively.

Financial perspectives

The starting point of any Balanced Scorecard analysis is financial. Kaplan and Norton point out that you need to understand the goals of your corporation, business unit, or individual product line. In researching the goals, you are likely to find all three financial strategies: growth, sustain, and harvest.

✔ Growth. Grow revenues, grow sales, invest in new products, infrastructure, distribution, customer relations, and so on.

✔ Sustain. Enhance profitability. Increase the return-on-capital-employed. Increase the return-on-capital-invested.

✔ Harvest. Make no significant new investments. Instead, maximize cash flow from existing products.

In a high-growth environment, a company may not really be interested in controlling the costs of its resources. Rather, it may seek to build people, process, and technology resources to ensure that they can respond quickly to changing needs.

As information-development managers in high-growth environments, you need to figure out how to use your assigned resources effectively to ensure that revenues and sales increase. Your businesses' financial objectives may be reflected in developing new products as quickly as possible or finding new applications for existing products, new customers for existing products, new solutions in which more of a product line is sold, or in shifting the product market (for example, moving a product to less skilled customers and providing more services).

In a high-growth environment, productivity improvements are likely to be viewed as more strategic than cost reductions. Your high-growth company's management does not want you not to save money but to increase the revenue generated by each employee. They want you to do more work with the same number of people.

For sustaining businesses, the situation is just the opposite. Because such businesses or product lines are not growing, management looks more favorably upon strategies designed to reduce unit costs. The questions asked may include:

✔ Where are you spending money?

✔ Can you produce more efficiently (fewer people, more output, greater efficiencies)?

✔ Can you reduce the cost of a transaction (for example, by moving information online for customer self-service rather than inviting customers to call a customer-service representative)?

Your strategies must be designed to lower the cost of doing business. If you're in a sustaining business, it would be wise to investigate the operating expenses of the competition. It will be important to reduce expenses. But any reduction in expenses must be balanced on the scorecard with the quality of customer service and the implementation of efficient processes. Many sustaining companies see operating expenses as a burden that must be contained or eliminated. However, that viewpoint can be counterproductive when quality is reduced beyond recovery. In fact, Kaplan and Norton argue that managers should not focus on decreasing spending but on increasing efficiency. You need to closely examine the benefits being produced by your cost reductions and not endanger customer relations by cutting in the wrong places. I have long argued, for example, that bypassing a customer needs analysis is the wrong place to cut costs.

In a sustaining environment, you need to look closely at asset utilization and investment strategies. By measuring and reducing time to market, for example, you make better use of your people and capacity assets by decreasing the time before revenues are realized. You may also successfully argue for a centralized publications organization so that you can increase the leverage of a capital investment (equipment, facilities, production) by sharing the investment across more of the company. By leveraging a capital investment, you choose not to replicate physical and intellectual assets any more than absolutely necessary.

In both sustaining and harvest lines of business, you need to practice strong risk management around your resource costs. Risk management might include better forecasting of expenses and better project estimating. Your companies want fewer surprises that increase costs unexpectedly.

In a harvest environment, the best approach may be to eliminate costs altogether. Organizations that have decided to orphan (stop making any changes to) harvest products have usually made the right decision by drastically cutting costs.

Thus, your first step in constructing a Balanced Scorecard is to determine the financial goals of your organization and determine how to measure your success in contributing to those financial goals.

Customer perspective

If you are to meet your financial goals, you need also to ensure that your customers are well served by your products and services. You must measure your success by helping to acquire new customers and retain existing ones. Your company and your organization should assess customer satisfaction and market share, in addition to measuring the profit that you accrue from each customer segment. You know, however, in information development that mere measurements are not enough. You need to be proactive in ensuring that you provide your customers with value in return for their investment.

In information development, you attempt to add value by providing information that will assist your customers in using your products quickly and effectively. You help them to get started through accurate and usable installation and configuration instructions, and you help them perform tasks easily and efficiently by providing the right information at the right time. You can also use innovative methods of delivering information to your customers electronically so that you can update it as needed. You may even continue to support traditional paper-based publications. You embed information in the products themselves, provide built-in training opportunities, and even work to make the products more intuitive.

Kaplan and Norton once again ask you to consider how to measure your contributions to meeting customer needs. They ask you to focus on timeliness, quality, and price. In information-development terms, timeliness equals finding ways of delivering information at the time it is needed and quality equals accuracy, accessibility, and usefulness. Price should be measured, not by the price per page of document development but by the cost-savings that come with assisting customers in reducing the cost of implementation and maintenance of your products in their environment.

Information-development managers actively seek out customer measurements of performance, but you would be well advised to consider studying the cost of implementation and use. If a product comes with excellent information resources, customers get started more quickly, require less training, make fewer mistakes, and experience less downtime. Each of these measures leads to a lower cost per unit and increased customer profitability. Real measures like these can be extracted through customer research more easily than we can demonstrate that excellent information products generate additional sales.

Internal business process perspective

Perhaps the most exciting aspect of the Balanced Scorecard is the emphasis Kaplan and Norton place on innovating new processes to meet identified financial and customer needs rather than improving existing processes. If you begin by understanding what customers value about your information products, you have an opportunity to find new ways to provide information solutions. Although you are under pressure to reduce cycle time and spending on development efforts, you need to balance these pressures with the importance of developing innovative information products that effectively respond to customer needs. Too often in information development, managers and staff argue that because they have to reduce time to market, they cannot afford to innovate. However, that conclusion means that you spend time making poor processes more efficient rather than finding better processes that might result in delivering innovative products to customers.

To account for a process perspective, you need to identify the critical processes that you must perform very well if you are to meet the objectives for financial growth and customer satisfaction that you have already identified on your Balanced Scorecard. Once you have identified the critical processes, you need to find ways to measure their success. For example, if your customers value accurate installation instructions, do you have a way to measure whether you are producing them? If you are not creating accurate instructions, do you have a way to measure the effect on your customers? Can you find innovative ways to improve accuracy without increasing costs dramatically?

Learning and growth perspective

If you want to implement innovative processes and better meet customer information needs, you need the right mix of employee skills. To achieve the right mix implies that you provide employees with opportunities to learn and grow within the context of the organization's strategic vision. Traditionally, you may have measured success by evaluating employee retention, satisfaction, and productivity. You have learned that employee satisfaction drives retention and may drive productivity. However, you must be careful to ensure that the productivity is going in the right direction. As you know, employees can be very productive producing the wrong product.

The Balanced Scorecard asks that you evaluate how effective your employees are in directing their learning and growth toward organizational goals. For example, if you want to develop information products that better meet customer needs, you need to put in place new processes, like customer-needs analyses. To do so means that you need employees who can perform customer-needs analyses and use the results to create new and improved information products. Cadence Design implemented just such a program a few years ago and measured its success by tracking the number of employees who had not only successfully completed training in needs analysis but had also actually completed a customer site visit.

Too often, learning opportunities (seminars and conferences) are simply viewed as perks without a real connection to results. If you institute a Balanced Scorecard, you are obligated to demonstrate how learning has resulted in growth, how growth has led to new processes, and how new processes have led to improved customer measures of satisfaction with information products.

Outcome measures are not enough

As you plan your Balanced Scorecard, remember that measuring outcomes is not sufficient to evaluate the success of your new business strategy. Outcomes are often too late. As lagging indicators, they tell us only what went right or wrong. They give us no tools to judge current performance or ways to predict the future. Every Balanced Scorecard must include, indeed focus on, the leading indicators that drive performance. You must take time to decide what change will look like as it happens. Will you see more staff members engaged in innovative information design? Will a significant percentage of your staff visit customer sites and learn about their real information-use patterns? Will you find more staff anxious to change what they write and how they deliver information? Such evidence is the proof you need to show that your attempts to institute new processes and engage in strategic learning have paid off.

The Balanced Scorecard is a great tool for information-development managers. By implementing a Balanced Scorecard, you demonstrate how closely your organization is aligned to the goals of your corporation and how the actions of your team make the company more successful and profitable.

Example of a Balanced Scorecard

I prepared a Balanced Scorecard for The Center for Information-Development Management (CIDM). The purpose was to show how improvements in one quadrant lead to results in the other quadrants:

✔ Investing in employee learning leads to service quality improvements

✔ Better service quality leads to high customer satisfaction

✔ Higher customer satisfaction leads to increased customer loyalty

✔ Increased customer loyalty generates increased revenues and profit margins

As managers recognize, tangible assets are easy to value. They know the cost of buildings, furniture, and tools. Financial results are also somewhat easy to measure. Managers calculate profit by knowing the difference between revenues and costs. Intangible assets, such as investments in experienced, well-trained staff and excellent processes, are harder to measure, although they can provide significant long-term competitive advantages.

In creating the Balanced Scorecard for the CIDM, I began with a vision statement:

> The CIDM exists to increase the actual and perceived value of our members and their organizations to the larger organizations in which they operate.

Then I evaluated the four perspectives with regards to this vision. What follows are the four perspectives as they relate to the CIDM.

- ✔ Financial perspective. We need sufficient revenues to support our current activities and add new activities so that we can increase the value of the CIDM to our members.

- ✔ Customer perspective. If members are satisfied with the value provided by the CIDM, they will maintain their memberships, recommend membership to new members, and sponsor CIDM research and reporting activities.

- ✔ Process perspective. We must develop processes to deliver value to members in a cost-effective manner, including research into best practices, communication of business-critical information, education for superior performance, and community development.

- ✔ Learning and growth perspective. Our staff and associates must exhibit expertise in CIDM processes and understand how their actions in response to member needs directly influence customer satisfaction and financial success.

Based on this analysis of the four perspectives on the organizational goals, I created a series of targets for CIDM to meet in the coming year:

Perspective	Objective	Measure
Financial	Increase the revenue from memberships Increase the number of members Increase participation in benchmark studies and sponsorship	% revenue increase % retention of members % membership increase # of new sponsored benchmark projects
Customer	Build strong level of member satisfaction with services Increase member participation in all programs Strong expression of CIDM value to non-members	# of members satisfied or highly satisfied # of members attending the BP conference # of endorsements and testimonials
Internal Business Process	Work to involve CIDM members in CIDM activities Increase member contact Invite nonmembers to participate in CIDM activities as potential members	# of projects # of reports # of member contacts # of potential member contacts % of members contributing
Learning and Growth	Develop a training program Implement new tools and technologies Evaluation of staff concern with members and potential members	# of staff completing training # of new tools researched # of new technologies learned Employee survey of awareness of member needs

The measurements ensure that we have a method of regularly reviewing our objectives throughout the year to ensure that we are making the progress we expect.

Summary

In evaluating your strategic portfolio, you need to make decisions about the viability of your projects and their alignment with corporate strategies. Then you can use your decisions to create a Balanced Scorecard to evaluate the ongoing success of your initiatives. Remember that financial measures are only part of the picture. You need to include measures of increased customer satisfaction, improved processes, and enhanced opportunities for learning and growth among your staff.

By focusing on strategy and managing your portfolio of projects effectively, you will increase the value of your operations and the value of your information products to the customer. As you plan your strategy, consider the following:

- ✔ Account for all the projects in your portfolio in business terms.

- ✔ Evaluate the projects to ensure that you understand their business value and determine how your team can best support that business value.

- ✔ Prioritize each project according to the number and type of resources you should devote to that project. High-priority topics get the highest investment of resources and the most appropriate resource to handle the design and development of information products.

- ✔ Develop your strategic plan, with a vision of the future and an outline of the activities you will pursue to achieve the vision.

- ✔ Turn your strategic plan into a Balanced Scorecard, which illustrates the relationship of people and process to customers and finances.

- ✔ Determine a set of measurements you will use to determine the success of your strategic plan and to maintain your focus on its achievement.

No doubt, you will encounter resistance to strategic management of your portfolio. You will be tempted to give in to the demands of the most argumentative business leaders that their project rise to the top of the priority list. Ensure that you have a strong executive sponsor at your back to help you drive your business case for portfolio management. Work closely with your staff so that they understand the vision and the day-to-day decisions that will help your organization remain focused on achieving your goals.

"Our resources are limited. What gets the top priority?"

Chapter 4

Managing an Information-Development Budget

> In an empowered organization, people are free to make mistakes and equally free to fix them. Managers have side discretion in making decisions; as a result, they can obtain resources more quickly . . . without having to document [them] quite so elaborately.

> —Jeremy Hope and Robin Fraser[1]

One of the key characteristics that is researched in an Information Process Maturity (IPMM) assessment is budget. The more managers know about their budgets, the more likely they are to be at higher levels on the process maturity scale. In the IPMM assessment, I am interested in learning how an organization is funded and how well the managers understand the budgeting process in their companies. The questions in the IPMM survey focus on the degree of responsibility and control that the information-development manager has over the organization's budget.

[1] Jeremy Hope and Robin Fraser, "Who Needs Budgets?" *Harvard Business Review*, February 2003.

The questionnaire asks the following questions about the budget:

✔ Do you have control over the budget for your information-development organization?

✔ Do you prepare your budget estimates and proposals based on an estimated workload for the coming year?

✔ Is your information-development budget controlled elsewhere in the organization?

✔ Are you assigned a headcount (number of staff) but have no information about your organization's budget?

✔ Are you able to successfully request additions to the budget if the workload increases during the year?

✔ Do you estimate the cost of each project as part of your budget request?

✔ Do you track project costs so that you know accurately the actual cost of each project?

✔ Do you include external costs in your project budget, such as printing, localization and translation, graphics, multiple media, and so on?

✔ Does your budget include funding for training and professional development for the staff?

✔ Can your budget include funding for capital equipment purchases such as new content management tools?

In response to these and other questions, I find that organizations have widely differing approaches to budgets and budgetary controls, all with different effects on the management environment. Each funding method presents challenges. Each has a positive and a negative side. As a manager, you may have inherited a funding method from a previous management or the funding may be the standard approach used in your company. You may find that your current approach works well and provides you with the resources you need to do the job. Or, you may decide that your current approach makes you accountable for getting the work done but removes any means you might have to influence how the work gets done.

Funding methods are only one part of the financial picture. In the Balanced Scorecard, you must establish links between the work done by your organization and the overall revenue and profit objectives of the larger organization. Defining how your work is related to the company's financial goals may be difficult, especially if you do not directly earn revenue for your activities.

If you are operating a profit center, you are quite clearly contributing to revenue generation. Managers who have responsibility for delivering training to customers generate training revenue to offset the cost of delivery (trainers, facilities, expenses) and, in some cases, the cost of training development. If the training provides value to the customer for the price, the revenues should offset the costs and produce a percentage of profit. In rare cases, information developers manage a profit center by selling copies of product documentation directly to customers or through distributors. In one case, a department generated revenue for information development by selling its database of product information and the content management system they had developed. Customers purchased the content management tools and the source content. They could use the tools to customize the source content for their own needs and add their own information to the repository. The department generated about $8 million in revenue in its first year by selling information and training clients to manage it.

In most cases, however, managers of information development operate a cost center that is budgeted as part of product development, marketing, or services. A cost center does not generate revenue. As a cost center, information development is usually viewed as part of product development, marketing, or customer assistance. The costs of developing content are factored into the cost of developing the total product or the cost of selling, maintaining, and supporting the product through its life cycle. If your information-development organization supports internal policies and procedures development, produces business communications, or supports other internal information needs, your costs may be factored into the operations budgets of those business areas.

As a cost center, your responsibility to the financial well-being of your organization comes by increasing the efficiency and productivity of your staff. Instead of representing the direct revenue-generating activities of information-product sales or training delivery, your work is factored into the cost of doing business or the cost of goods sold. As such, your responsibility is to keep those costs as low as possible without jeopardizing customer satisfaction.

In Chapter 6: Developing Relationships with Customers and Stakeholders, you learn about ways of understanding customer needs so that you deliver what is needed rather than what is not needed. In Chapter 7: Developing User Scenarios, you consider ways to optimize operations and increase the productivity of your organization so that it is increasingly cost-effective. In this chapter, the focus is on understanding your costs thoroughly so that you can manage them more effectively and demonstrate the effect of your activities on the success of the larger organization.

Best Practices in Budget Management

As the products you support move through the technology adoption life cycle, you need to respond by managing your information-product portfolio in relationship to corporate business goals. As products succeed in crossing the chasm and moving into the early and late majority markets, you must change your approach from supporting product developers and early adopters to supporting customers who expect new products to be easy to adopt and implement. If the information developers have been content to describe product features and support convoluted product interfaces, you must change their behaviors or develop a staff that is increasingly customer focused. To manage information strategically, you need to understand why your organization is funded the way it is and how budget decisions are made. To respond to changing requirements from internal and external customers and from new technologies, you need to know how to establish a sound financial business case, calculate a realistic return on investment, and know how to make requests for additional funding or move monies among budget line items that are under your control. Without an understanding of where you stand in the corporate budgeting process, you are often working in the dark with little or no opportunity to influence the decisions being made for you.

Consider the following best practices for managing the cost of the services you provide:

✔ Understanding the costs of operating your organization

✔ Managing your organization's budget

The services you provide include the costs of developing information products and preparing them for delivery to customers. Those fundamental costs of the information-development life cycle usually begin with the internal costs of staff, equipment, and infrastructure and the external costs of localization and translation, printing, distribution, and other external budget items. Your services also include the cost of maintaining the competitive capabilities of your staff through training and professional development and investing in strategic projects that you believe will lead to productivity gains or added value to the products you deliver to customers. All of these services must be funded sufficiently to be viable but at a level that is competitive with other ways of producing the same outcomes, be those information outsourcing or moving information development into engineering, services, field operations, or elsewhere in the organization.

Best Practice—Understanding your operating costs

Unless you know what it costs to run your organization and to conduct your projects, you may find it difficult to ensure that your operations are cost-effective. Competition from lower-cost providers may threaten your team as the inhouse incumbents. You may be told to lower your costs without knowing what your costs are. Or, by understanding your operational costs, you may be in a position to demonstrate the wisdom of your operations decisions. For example, one manager faced pressure from senior management to outsource his printing operation. Only after he demonstrated that his operation was less expensive than the outsource vendors' was he able to keep his printing operation inhouse and control costs over the long term.

The starting point of understanding your operating costs is with your staff. Staff costs are most likely your highest cost resources. Base staff costs are generally a combination of salary and benefits (company tax contributions, healthcare, savings or pension programs, and others). You should be able to learn the costs of benefits as a percentage of salary from your human resources organization. In most companies, the benefit costs are approximately 30% of salary. If you don't know individual salaries, you can use an average salary plus 30% as a good estimation of people costs.

In some organizations, the total per-employee costs include the cost of the infrastructure, often referred to as fully burdened costs. Infrastructure costs may include such items as computer and other equipment, furniture, telephones, even the cost of the building, heat, air conditioning, and electricity may be factored into the total costs of running your organization. In most high-tech companies, these additional costs plus benefits are calculated to be two or two and a half times the cost of salary. In manufacturing companies, the fully burdened costs may be higher, accounting for the higher costs of operating manufacturing plants. To find out what your fully burdened multiplier is, consult with your finance organization.

For example, if your average salary costs per employee are $60,000 per year, the total cost of operations or the fully burdened employee costs are likely to be $120,000 to $150,000 per year (two or two and a half times the cost of salaries). Thus, a 10-person department will have an average budget between $1.2 and $1.5 million. To the cost of a fully burdened cost per employee, you must add what you pay to outside vendors and sometimes the costs of outside activities such as conference registrations, workshop fees, and travel.

These calculations added together constitute the full cost of your operations:

- ✔ Employee salaries and benefits

- ✔ Infrastructure costs of the department

- ✔ Costs contracted with outside vendors

- ✔ Costs of outside expenses for professional development

Adding these costs gives you the total operational costs of your department.

Comparing your costs with other parts of the organization

In some organizations, you may want to compare your estimate of total costs with the costs of other parts of the organization. Some information-development managers find that their allocations of budget are determined as a percentage of other budgets, particularly budget for product development. In benchmark studies of budget practices, I have found allocations that range from 2% to 25% of product-development budgets.

The percentages are often arbitrary, based upon a judgment of about the value of information development to the product. The percentage budget you are awarded may be based upon a real or imagined comparison with competitors. Some industries, such as telecommunications, may have a standard for information development that is widely accepted as adequate and competitive.

Your budget might also be expressed as a ratio of headcount in product development to headcount in information development. Some organizations decide that they should have one information developer for every three, seven, ten, or fifty product developers. In our benchmark studies of these ratios, I find that seven to one is a fairly common ratio in software-related organizations, and a higher ratio is common in hardware-related organizations.

In general, however, I find little logic in determining these percentages or ratios. They are most likely to be determined by the value those who control the budgets place on information development. If information is considered vital to the usability of the product, it is awarded a higher percentage of development costs or has a lower ratio of product to information developers than if it is considered a necessary evil.

Although you may not have direct knowledge of the cost of product-development in your organization, you should be able to calculate it by knowing the number of people involved and estimating a fully burdened cost per person. The supposed direct correlation between the costs of product and information development are demonstrated when product development costs are reduced by moving them to low-cost, developing economies around the world. When product development moves offshore, information development is pressured to follow, not because the quality of work is better offshore, but because information development is now perceived as too expensive because it has increased as a percentage of product-development costs.

Estimating project costs

Once you know or estimate your total organizational costs, you can estimate the cost of your individual projects if you have some idea of who worked on each project and how much time they spent. If your staff records the amount of time they spend on each project, you have a tool to estimate total project costs and you will know how much of your budget goes to overhead. Overhead costs include the time not associated with projects.

Organizations that bill back project time to the product teams calculate the project costs based on the number of people assigned and their total time multiplied by the average hourly rate (referred to as the run rate) of the staff. A fully burdened cost of $120,000 divided by approximately 2,000 hours per year equals $60 per hour. Using this run rate, a full-time project for one person for six months will cost $60,000. Any time not assigned to projects is considered overhead, representing time spent on company- or department-centric activities. You might include general meetings, special projects, and even vacation and personal time as part of the overhead of your organization. Because overhead increases in the run rate, staff are often under great pressure in some organizations to minimize the time not spent directly on project work. As a result, you often are asked to reduce the amount of time spent on innovation, professional growth, or special projects, often to the detriment of the long-term effectiveness of your organization.

You will find details on estimating project time and costs in Part 3.

Benchmarking project costs

When you know what your organization and the activities you support actually cost, you can benchmark with others in your industry. You can also research the costs of organizations you consider best in class.

Best Practice—Managing your operating budget

In working with information-development managers involved in IPMM assessments and other benchmark activities, we find a wide variety of budgeting mechanisms in place. Each method provides advantages to the organization as well as disadvantages.

The budget or funding alternatives range are usually associated with one of the following mechanisms:

- ✔ No budget
- ✔ Budgeting by project
- ✔ Budgeting as general administration
- ✔ Budgeting by taxation

Considering the "no budget" organization

At the most basic level, information developers are part of another organization's budget, typically product development, marketing, or customer service. They have no separate budget line and may have no information at all about the budget for their activities. In many cases, the information developers have no manager of their own but report directly to the product manager, project manager, or a more senior manager of the business unit.

If the "no budget" information developers have a manager, that individual provides leadership for the information developers, handles performance evaluations, manages the project portfolio, and manages any initiatives that the team members are able to pursue. In this case, the manager has no visibility into the budget for information development. Requests for funding that are not covered by the people costs of the standard headcount are controlled by a business unit or other senior manager. Requests for monies for training, professional development activities, conferences, equipment, or tools are all funneled through the people who do have budgetary control.

This basic approach to funding information development may prove quite satisfactory, as long as those who control the budget are supportive of the needs of information development. I know of managers who are happy with their level of support because the people to whom they report are willing to provide funds for training, conference attendance, publications, equipment, and new tools, as appropriate. Although they don't control the budget themselves, their requests for funding outside of basic headcount are respected and considered seriously.

In other situations, the basic approach is unsatisfactory, primarily because those who control the budgets do not believe that information development is a professional activity. They assign headcount but provide no other funding. They are often unresponsive to requests for additional headcount based upon projected workload. As a manager under such circumstances, your need for additional staff goes unheeded. You are told to simply make do with the people you have and try to get the work done anyway. You may attempt to prioritize the work, as discussed in Chapter 3 on the Balanced Scorecard, but more often you simply leave out many of the process steps that would help to ensure the usefulness and quality of your information products. It's not that the manager and information developers in these circumstances want to produce less effective information products, it's that they have no choice and do the best they can.

"No budget" situations are probably most suited to reasonably small organizations of 10 or fewer information developers. With larger organizations, the lack of predictable funding will become frustrating. If the management is seriously unsupportive, a larger organization will have trouble remaining intact for very long. It is more likely to operate at a Level 1: Ad-hoc of process maturity.

Budgeting by headcount

Budgeting by headcount alone may feel quite similar to the "no budget" organization. However, in this case, the information-development manager does estimate the work to be done in a fiscal year or quarter. The manager gathers information from the product or service groups that the team supports and catalogs the list of deliverables for the next version of the product or the policies and procedures. From the list of deliverables, the manager produces an estimate of the work to be done in the budget year. This estimate may be quantitative, based upon tracking actual projects, or it may be guesswork based on the assumption that the existing staff should be able to handle the predicted workload.

Increases in headcount for direct employees are often difficult to get approved. Management is reluctant to add to the number of people because adding people obligates them to additional costs for healthcare, savings plans, and various forms of insurance and other benefits for a longer term. Instead, management is more likely to support adding people as contractors employed through outsource agencies. The cost of contractors appears on a different part of the profit and loss statement from the cost of direct employees and often does not attract the attention from the stock analysts that adding additional permanent headcount might attract.

In most countries, adding and eliminating contractors is much simpler than adding and laying off employees. In countries with strong employment laws, companies are more likely to add additional budget for contractors than direct employees because they find it nearly impossible to layoff or fire an employee who is not performing adequately.

Managers are also able to protect their investment in long-term employees by adding contractors to handle peak periods of work. One manager I worked with always maintained her full-time staff at 80% of her work volume, using the 20% contract work as a buffer in the event that her budget was severely cut.

The same is true for adding contractors in offshore, low-cost areas. At present, North American and Western European companies are most likely to add contractor headcount outside of their primary locations because the costs are as low as 20% of their employment costs at home. If the staff is added through a contract agency, the costs are handled in the same way as domestic contracting costs.

Companies are also solidly advancing the cause of hiring direct employees in low-cost, offshore areas. Information-development managers inform me that they can significantly decrease their run rates (the overall cost per hour of work or the cost of goods produced) if they can employ part of their staff at a fraction of the cost of the employees in North America or Western Europe.

To meet their budget requirements, managers look for either less expensive employees, including offshore direct employees or contractors, domestic contract employees, or even interns hired locally. All help to reduce the average cost of the work done in the organization.

If you are managing and budgeting by headcount, finding ways to reduce the average cost of each "head" may keep your organization's expenses low and reduce the risk of losing your most experienced staff members.

Budgeting by projects

Budgeting by estimating the cost of each project is perhaps the most common method used to fund information development. The budget year begins with a list of potential projects and an estimate of their scope in terms of numbers and types of deliverables. The sources of this project list are usually the product or project managers in the larger organization. In well-organized, more mature organizations, the lists of projects at the beginning of the budget year may be reasonably accurate. The information-development manager has a fairly good idea of what projects will be on his or her plate. In less mature organizations, the list of projects is often inaccurate and likely to change frequently, making the budgeting process an exercise in futility.

As an information-development manager, you check with product and project managers to get their estimates of future work, discuss the estimates with your own senior manager, look at your budget from the previous year as a starting point, and inquire with other knowledgeable individuals in service, training, support, and marketing to fill in the gaps in the picture. Out of this research comes your list of expected work and enough information, albeit incomplete in most instances, of the scope of work as it affects your department.

Based on all the data, you and your staff produce detailed estimates of the project costs, including

- ✔ Hours to plan the details of the various projects
- ✔ Hours to complete the deliverables through the multiple drafts
- ✔ Costs of production of the final deliverables (print, CD-ROM, and so on)
- ✔ Costs of localization and translation in multiple languages

To the estimate of time, you apply your average "run rate," which is the hourly rate you will charge for people time in your organization. The run rate includes the cost of running your organization, including costs for general administration and benefits, infrastructure, training, professional development, and anything that you do to have a productive and responsive organization. Your estimates of the costs of the projects are likely to be a starting point, open to negotiation before the budget is complete. You will find complete information on estimating project costs in Part 3.

You will find, of course, that the departments that must pay the costs of your projects want to ensure that they are getting the value they expect. They may negotiate for less expensive deliverables or a lower cost per hour. They may claim to be open to decreasing the quality of the work delivered, although this claim is often abandoned once they get to review the more simple deliverables. One manager tells me that the product teams chose instructor-led training rather than the e-learning preferred by customers because developing instructor-led training is less expensive. However, she is also responsible for operating the training facilities and providing for the training logistics to conduct classes, all of which increases her hourly cost per employee. She works hard to inform the internal customers that choosing the less acceptable option may have repercussions in terms of customer satisfaction.

When your department budget is funded by an internal customer organization, you are open to competition from lower-cost providers. Product managers may argue that they can contract with outsource companies to do their work at lower cost than you can do it. The outsource companies do not have to support your higher staff and infrastructure costs, but they are unlikely to have the access to the developers or to the product, increasing the costs of managing their work. Your responsibility is to explain the real costs of using outsource suppliers, especially the cost of managing those suppliers by the product-development organization. You are also responsible for remaining as competitive as possible by exploring your own budget options. Implementing process and tools efficiencies, hiring contractors to supplement direct staff, developing offshore resources under your control, introducing efficiency-building technology, and pursuing minimalism, among other possibilities, can help you remain competitive.

Even more importantly, you must consider the value your stable staff provides in knowledge of the product and of the customers. The more your information developers become experts in the product and experts in the customer, the greater your competitive advantage in the larger organization.

If you are managing by project, you are, in effect, operating an internal contract and consulting agency. You are being paid for the work done, after negotiations with your internal customers. As a result, you are not required to prioritize your projects. That prioritization is under the control of your customers. If they choose to pay for low-priority work, you may not have an opportunity to argue that your resources would be put to better use elsewhere. However, the lack of prioritization encourages inefficiencies. You are often under less pressure to find better, more efficient ways of working than if you were budgeting by a fixed headcount or an overall departmental budget.

With project budgeting, the pressure to prioritize and look at the value being produced by increasingly limited resources is often on you rather than on the various business units who are your customers. As a central services organization, supplying products to these internal customers, you may find it advantageous to strongly recommend careful examination of priorities. In many companies, budgets for product development may be set independent of corporate goals and objectives. Projects are funded by politics rather than by sound analysis. As a result, you may obtain funding for a project that is unnecessary or overly expensive. The project wastes resources that could be better allocated elsewhere to meet corporate objectives. Unless you have a higher level of management that you can reason with to allow you to set priorities, you may become frustrated at the project-oriented budgeting process.

Product managers may ask for the least expensive solutions from information development, even when these solutions risk frustrating customers or increasing costs elsewhere. A poor information solution is likely, for example, to increase the cost of customer service or make maintenance and repair more difficult and time-consuming than it needs to be. As long as product managers are not responsible for the overall well-being of the company and its customers, you will be asked to develop information products that are less than optimal. The product or business unit managers who are your internal customers are more focused on their own budgets and performance bonuses than they are on the welfare of the corporation or the customer.

I believe that many information-development managers are actually very good at assessing the customer requirements or seeking alignment with corporate objectives. They often find themselves in the frustrating position of having to produce information products that customers find unusable or unsuitable to the tasks they have to perform because they are viewed as "less expensive" than a better solution.

If you experience this level of frustration, you may want to consider a move toward departmental funding rather than project funding, or some combination of the two. This more equitable solution has its own shortcomings but may provide you with opportunities for better decision making and control of the outcome.

Budgeting by department

At the other end of the budget control spectrum is the information-development manager who has full control of the budget and the support of senior management for a strategic approach to information development. It's likely that a manager with this degree of control, accountability, and responsibility has a Level 3 or higher organization on the process maturity scale.

If you have a budget for your organization independent of the projects you undertake, you are usually responsible for developing a budget proposal for the coming year, based at least in part on your estimates of the coming workload plus the additional costs you support and the activities you hope to pursue. Because it is often the case that you work for a larger organization that is more mature, you are likely to have a more predictable workload, based on better project estimates from the parts of the organization you work with (product development, customer service, training, and so on). Your budget proposal may include the cost of permanent and temporary staff, professional development, printing and distribution, localization and translation, and everything else under your control.

You may talk to product managers and others responsible for projects in the larger organization to paint a picture of the future, although your success in doing so depends upon their predictions. You may also have an allocated headcount for permanent staff plus a budget for contractors who are not counted as staff and receive fewer benefits, thus costing less.

In most cases, you will experience a great deal of negotiation before your final annual budget is fixed. Your management may concede that you should maintain the same budget levels as the previous year, especially if your workload will be much the same. You may receive a budget increase to cover increased staff costs from higher salaries or changes in benefits. Or, you may be asked to reduce your budget by a certain percentage, leaving you to work out the details.

Annual budgets usually are assigned specific monies by line item. The total budget is the sum of the line items. With an annual line-item budget, you may have complete control over the line items, enabling you to shift monies from one item to another, depending upon changes that occur during the year. You may, for example, be able to use unfilled permanent staff positions to hire more contractors. You may be able to shift training monies to tools acquisition or vice versa.

In other organizations, you may be required to stay within your line-item budgets and ask for approval before you are able to shift funds from one item to another. In either case, your control of a total budget allows you to set priorities and make decisions about what you can and cannot afford. You may want to invest in technology innovations, shifting funds from other budget areas to allow you to pursue an innovation that will reduce process costs or reduce the cost of external expenses like the cost of translation.

Prioritization may extend to the projects for which you are responsible. You may find that some projects demand a higher investment of staff and resources than others. You may have a project in your portfolio that is important to the future prospects of your company, such as a new high-potential product or a high-profile customer. You may decide to allocate more staff to this project than to others that are in maintenance. Your flexibility with regard to budget allocations allows you to make decisions that are aligned with corporate goals and objectives and benefit the larger organization rather than each project individually.

Responsibility for your organization's budget also aligns your position as a middle manager with other middle managers with budgetary control. Although it increases your level of responsibility, a budget also increases your accountability and makes your operation much more visible in the larger organization. If you're not carefully managing your money, you are increasingly at risk for outsourcing or other management decisions to bring your costs in line.

Business unit or project managers often prefer budgeting by project because they can clearly associate the products they are getting from your organization with their costs. They generally ask that certain individuals be assigned to their projects for the long term, helping to build product knowledge and personal relationships between information and product developers that make projects go more smoothly.

However, these same managers are always tempted to remove their projects from the general services group and establish their own internal information-development teams. In this way, they have complete control over the people working on their projects and their cost allocations show up on their profit and loss statements. Unfortunately, such fragmentation usually increases costs. In a central organization, you may be able to use your resources more efficiently by spreading their work over several projects. You are able to develop services such as graphics, production, and translation coordination by sharing costs across all your internal customers. Shared services are less expensive than if each department had to pay the total costs of these services themselves.

Department budgets by taxation

Budgets that are handled by taxation are often the least viable and short-lived in information-development organizations. In this case, each business unit or other organizations for which your team develops information products must pay an allocation to support your department. I have found that business unit managers often vigorously oppose this budget method because they believe they do not have enough control on how their monies are allocated. They are suspicious that you are using their money for someone else's work. They want to be certain that they get resources allocated to their work commensurate with the amount of money they pay for the upkeep of the information-development organization.

In many cases, the departments that are budgeted by taxation find themselves broken up, with staff resources moving to the customer departments. Of course, the cost of the staff resources is generally higher than the costs of maintaining a central organization which can spread costs across many common services.

Many managers tell me that they have great difficulty getting additional headcount allocations, or even getting additional contract help, if they have budgets by taxation. One large company decided on their own to break up their department into one for each major business unit because of headcount problems. The total number of information developers in the corporation increased fourfold after the large, central department was distributed among the business units. The business unit managers were willing to fund the additional resources needed to produce information products as long as they knew that the resources weren't going to another part of the organization.

Using budgets to encourage prioritization

One of your strongest assets in encouraging prioritization and focus on the customer is your department's budget. If you understand the market placement of each product you support, as discussed in Chapter 5 on the technology adoption life cycle, you will be able to develop a business case for proper prioritization. You may want to make highly skilled staff members responsible for highly visible and successful innovations. You may want to use a contract agency to handle maintenance on projects that are not changing very much. You may want to investigate adding a low-cost team for products being developed offshore.

As long as you are well-informed about customer needs for information, you can guide the decisions about where to best place limited resources. As a knowledgeable consultant to your internal customers, you are able to explain why resources are best allocated unevenly across projects. You have the ability to analyze each project that comes your way and decide if it is worth investing in.

As you no doubt recognize, many of the projects you are asked to do are not worth the investment in time, money, and people. They may be inappropriate for the customers, too costly to develop and maintain, or difficult to deliver effectively. As a professional manager, you want to be responsible for the work your team decides to take on, as well as accountable for meeting deadlines.

The rest of the chapters in Part 2 provide you with many possibilities for increasing your effectiveness by working more closely with customers, achieving operational efficiencies, developing metrics, and pursuing innovation. Each of the chapters is directly affected by your control over how your organization is funded.

Summary

Some management experts argue that budgets as we know them today should be abolished. They believe that the traditional methods used to allocate budgets and control performance around budgets is not only outdated but seriously detrimental to the health of the enterprise. Much time is wasted scrutinizing trivial budget details, like overspending this quarter's telephone allotment.

The result of corporate budgeting practices is manipulation to ensure that performance standards are met, often by unethical behavior, as we have seen by the rash of indictments of corporate leaders in recent years.

The alternatives to budgets are determined to make staff members more sensitive to customers and internal efficiencies. Departments are judged not by making their budgets or coming in under budget but by how effective they are compared to their peers in world-class organizations.

As you struggle with your own budget requirements or if you have an organization that operates without departmental budgets, you can still choose to follow the ethical high ground.

✔ You can make decisions that are aligned with corporate objectives and customer requirements.

✔ You can make the reasoning behind your decisions clear to your internal customers.

✔ You can pursue innovations with your staff that add value to your information products and increase the value you deliver to your external customers.

✔ You can establish performance objectives for your entire organization with the agreement of your management that ensures that you are both efficient and effective.

In the analysis of budget methods in this chapter, you should recognize that there are positives and negatives to each method. You may have inherited a method that fits your needs very well, with a senior management that recognizes your commitment to efficiencies and effectiveness. You may have inherited a budget method that makes it difficult for your team to do its best work, wasting time on activities that add little value and are controlled seemingly by whim and personal opinion.

Budget responsibility, being able to work with senior management to decide how to best use your resources, is a key characteristic of a mature organization and an astute manager. It doesn't matter if you budget by project, by headcount, or even if you have no budget at all. As long as you know how your resources are being allocated and you allocate them effectively, you can keep your organization attuned to the Balanced Scorecard you have established.

Budgets or other funding sources must first give you an opportunity to grow the professionalism of your staff. You need to pursue innovations with them, through training and professional development. With a skilled and effective staff, you can improve your operational efficiencies, reducing the pressure on your budget for activities that contribute little to your overall performance. You can use operational efficiencies to gain time to become expert not only in the product but also in the needs of your external customers. By promoting their ability to learn about your company's products and services quickly and easily and to not use less of the product functionality, you have the capability of increasing customer satisfaction and product sales. Eventually, by running your organization well and spending time with customers, you hope to influence the financial success of the company as well.

"I think our audience is changing."

Chapter 5

Understanding the Technology Adoption Life Cycle

> Innovators will ". . . forgive ghastly documentation, horrendously slow performance, ludicrous omissions in functionality, and bizarrely obtuse methods of invoking some needed function—all in the name of moving technology forward." Conservatives in the mature market want ". . . the gradual incorporation into the product of all the little aids that people develop, often on their own, to help them cope with its limitations".
>
> —Geoffrey A. Moore

In his 1991 book, *Crossing the Chasm*, Geoffrey A. Moore outlines the stages of the technology adoption life cycle as it applies to high-tech products. Using the life-cycle model originated by Everett M. Rogers in his work, *Diffusion of Innovations* (Free Press 1995), Moore demonstrates how customers change significantly through the life of a successful high-tech product.

Rogers demonstrated that customers for new, innovative products can be arranged on a standard bell curve, as shown in Figure 5-1. The earliest to become interested in a new idea are called innovators, representing only 2.5% of the potential customers. If a new product succeeds in meeting the needs of the small number of innovators, it may be embraced by the early adopters. Only later will more conservative customers in the early and late majority accept a new product, although they represent a significant part of the market. The fifth category, the laggards, never accepts the new ideas but can provide useful information during the life cycle if developers listen to their objections and criticisms.

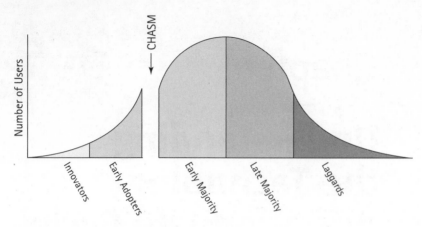

Figure 5-1: The technology adoption life cycle

Geoffrey Moore added to our understanding of Rogers' model by asserting that the move from innovators and early adopters to more mainstream customers is never simple and frequently is unsuccessful. Instead of a smooth transition, Moore argues for a chasm between the innovators and early adopters and the majority markets that are ultimately the most important source of revenue. The chasm exists because the characteristics of a product that appeal to early risk takers are entirely different than the characteristics of more conservative customers who will wait until the product begins to be accepted by their peers.

I recommend that information-development managers read *Crossing the Chasm* (consult the most recent edition). Moore provides insights into high-tech marketing strategies that can assist all of us in better understanding our information and training customers and in developing our own strategies for knowledge transfer. Moore helps us recognize the struggle our marketing colleagues are going through to position high-tech products effectively. The better you understand why product managers and marketers think the way they do and what they are trying to accomplish, the better you are able to contribute to the success of the products you support.

Moore's role as a consultant to high-tech businesses has been to demonstrate the flaws in the traditional approach to marketing high-tech products. Most companies expect, Moore explains, to move smoothly through the technology adoption life cycle, beginning with innovators and early adopters and moving to early and late majority purchasers. In fact, the needs and user profiles of these early adopters are distinctly different from the needs and user profiles of the conservative majority. Not only are their product needs different; but their needs for usability, information, training, and support are also distinctly different.

Information products have a significant role to play in all of the markets involved in the technology adoption life cycle. However, that role changes as a product moves from early adopters to the majority. Information-development managers must be prepared to evaluate the position of each product they support in the adoption life cycle, investigate the needs of the changing users of the product, and redefine the types of information products required.

Best Practices in Managing the Information–Development Life Cycle

As the products you support move through the technology adoption life cycle, you need to respond by managing your information-product portfolio in relationship to corporate business goals. As products succeed in crossing the chasm and moving into the early majority market, you must change your approach from supporting product developers and early adopters to supporting customers who expect new products to be easy to adopt. If the information developers were content to describe product features and support convoluted product interfaces in the past, they need to become customer-goals-focused. You need to pursue an understanding of the customer's as well as the end user's agenda in using the products. You need to concentrate on redesigning the information to support a task-oriented, minimalist agenda. And, you need to find ways to fund information development after the initial flush of new product funding is no longer available.

Consider the following best practices for managing the relationship of information development to the technology adoption life cycle:

✔ Positioning information development to match life-cycle needs

✔ Funding information development through the life cycle

The best practices in this chapter provide you with a method of analyzing products, defining information requirements, and presenting your business case for funding.

Best Practice–Positioning information development to match the needs of the technology adoption life cycle

The first step in correctly positioning information development to match the needs of products is to identify where in the technology adoption life cycle each product in your portfolio is positioned. The second step to understand how the buyers and the users are changing as a product move along the bell curve of the life cycle. Without understanding the changing user community, managers and information developers are more likely to continue producing the same information long after it is no longer appropriate.

Innovators and early adopters

At the earliest stages, innovators and early adopters are most interested in the newest advanced capabilities that innovative products offer them. In most cases, innovator and early adopter customers are integrally involved in helping a company develop the product in the first place. They serve as sources of initial use cases and help define the functionality that will best meet their own needs. They are usually the most technically astute customers, often requiring minimal support once a product has been installed.

If you consider these earliest of adopters in terms of their information requirements, you will often find that innovator and early adopter customers require a subset of information products. In fact, they may need very few information products at all. In many cases, they are best served directly by installation and service professionals who are intimately involved with the product under development, even to the extent that they may actually redesign the product during the first few field installations. These advisors, sent to guide early installations and support these initial customers, may perform many of the installation and configuration tasks that will later be performed by the customers themselves.

Early stage customers require a special set of information products, often quite different from the information you will eventually produce for more mainstream customers. They want and are often given access to product developers. Product developers see them as seasoned technical professionals like themselves. When the developers describe in detail how the product is designed, the innovators and early adopters use these descriptions to understand how they need to implement the product in their environment.

Managing Content for Early Adopters

A scientific equipment company had developed one of the first gas chromatographs, allowing customers to analyze the chemical contents of a sample. The first model was designed in close cooperation with key early adopters, most of whom were PhD scientists working in university or government laboratories. The equipment was designed to meet their specifications. The information they needed focused on the details of how the equipment functioned and how to configure it to test particular substances. Because the users of the equipment were highly trained professionals in the field, they rarely used the basic task instructions provided in the product documentation. They were more interested in the conceptual and background information that gave them insight into the machine's characteristics and performance. In more cases, they developed their own protocols for running tests.

The most effective sources of information for these early adopters were the product engineers and the experienced senior service personnel. Many of the developers had the same level of education as the customers and spoke the same technical language.

Many of the information developers hired by the company were themselves technical professionals, some with doctorates in chemistry or biology. They easily understood the information they gathered from the product developers and also knew the language of the customers. They were successful in developing a large volume of detailed conceptual information that the customers found valuable.

Unfortunately, the close relationship between product developers and innovators and early adopters can be counterproductive. Moore's contribution to our understanding of these early adopters was to point out that they generally fail to help a product move from the small niche market to a broader market with greater potential for revenue growth.

Early and late majority

The technology adoption life cycle model indicates that new technologies are first introduced to anxious innovators and early adopters. After they are on board, companies attempt to use recommendations from these customers to make the transition to the more hesitant, pragmatic, and conservative members of the early and late majorities markets.

The customers in these groups represent the greatest potential sales for the products because they encompass the largest customer base. It is among these majority groups that the greatest revenues and profits are to be realized, especially as the technologies mature and the effort expanded on the initial development efforts begins to pay off.

Unfortunately, as Moore demonstrates, the transition looks much easier than it is. Many technology start-ups fail to get past the early adopters. He calls the transition from visionaries among the early adopters to the pragmatists of the early majority, "crossing the chasm." Too often, companies with great products and exciting new technologies descend into this chasm, never to re-emerge. The limited niche market among early adopters is quickly exhausted, making new revenue scarce at the same time that the companies are expending considerable resources to improve the product, add functionality, and increase sales efforts to attract the early majority.

To overcome the resistance of the early majority to new technology, Moore recommends that companies find a special niche in the market and pursue it single-mindedly. To succeed, the company needs to present these potential customers with value that meets their special needs. To meet special needs, Moore advocates the development of a "whole product" message. Information-development managers can use Moore's perspective in *Crossing the Chasm* to look at the potential information needs of customers. In fact, with this understanding you may even be able to contribute to the difficult job of crossing the chasm.

Innovators are the classic techies who will try a new technology just because it is there. They have great, almost unanswerable, demands for information about the new technology. They do not need task-oriented information because they are likely to try to use the technology in new and innovative ways, and they are more than likely to learn through experimentation rather than through following standard procedures. They have enough experience with new technologies that they can often intuit how to make the technology work for them with only the interface and some notes from the developers.

It is important to note that innovators are more like developers than they are like mainstream customers, which is why developers are so comfortable with them. Because developers understand how innovators are motivated, they know a great deal about the kinds of information that will be both useful and exciting. As Moore explains, the technology enthusiasts will "forgive ghastly documentation, horrendously slow performance, ludicrous omissions in functionality . . . all in the name of moving technology forward."

Early adopters are less technically focused than the innovators, but they are patient with new technologies, willing to spend considerable effort understanding how to apply them to the complex business challenges. These visionaries need good information to support their goals, but they are willing to spend time in learning, even if the information is not complete. Both the innovators and the early adopters are willing to give your technology the benefit of the doubt; they want you to succeed.

Not so in the mainstream markets. The early majority are pragmatists; they want good service, and they want good information and training. Members of the late majority, who represent one-third of the potential market, are even more skeptical. They are believers in keeping things the way they have always been, not innovating. They find themselves trapped into using new technologies. They resent products that are difficult to use, and they hate it when they can't find the answers to their questions quickly and easily. They especially don't want to call customer service because they consider most of the people on the other end of the line to be ignorant about their needs. Members of the late majority want, even demand, all the job aids, wizards, and performance support aids that they can get to make their jobs easier.

Managing Content for Majority Customers

When the scientific equipment company decided to "cross the chasm" to a new, larger market for the gas chromatograph, they first decided to redesign the product by adding functionality important to chemical analysis labs outside of universities. They also had the foresight to hire my information-design firm to rethink the information products for the new customers.

The first activity of the information designers was to conduct a user study of the new customers. It soon became evident that the people who would be using the equipment were very different from the previous users. Most of the earlier users were PhD chemists and biologists and their graduate assistants. The people working in the industrial labs were technicians with minimal scientific education. They were trained to run samples with the goal of maximum throughput. They had no time and little interest in learning the intricacies of the product's design.

The designers recommended that the new information products follow a minimalist approach, focusing on helping the users run sample quickly and get reliable results. This approach meant minimizing much of the conceptual and background information found in the previous generations of the documentation and emphasizing basic task-oriented instructions, accompanied by lots of troubleshooting guidelines. They also suggested that the company produce a small booklet that described the basics of gas chromatography to help educate the new customers.

At first, the recommendations generated resistance from the product developers and from some of the information developers in the company. These individuals felt that a minimalist approach was insulting to the customers. They also were reluctant to relinquish their own special role as experts in the field. They didn't want to be seen catering to the needs of "dumb" users.

In the end, the minimalist recommendation prevailed and the company successfully crossed the chasm with the product, although more attention had to be given to a redesign of the user interface. Usability studies helped identify ways to make the tasks simpler to perform rather than relying on the documentation to lead minimally trained users through actions designed for experts.

The role of information in developing the whole product

Geoffrey Moore is not the first to introduce the concept of the whole product. He is able to use the whole product concept to reinforce his argument about the importance of adopting a new perspective on the product in order to cross the chasm. Moore points out that there is often a gap between the promise that marketing has made to the customer and what the actual product is able to do. If this gap is to be closed, the generic product must be enhanced by services and other products "to become the whole product." The whole product, Moore illustrates, must include effective documentation, training, and support. The services included with the whole product must enable the pragmatist and conservative customers to gain full value from the product by putting it to use completely in their organizations and realizing productivity and quality gains.

Information-development managers are all too familiar with the problems of getting their companies to understand the need for the whole product. Developers prefer listening to the innovators and early adopters, adding functionality, creating multiple paths to the same results, making functionality ever more complex to negotiate, and ignoring usability. Documentation and training often, under these conditions, emphasize what developers believe to be most important. Marketing is also paying close attention to the early adopters, who have lots of visionary ideas about what the product ought to be. They are not especially interested in building the whole product.

What marketing should do, according to Moore, is focus on the chasm and the problem of convincing the early majority that the product is safe, effective, and usable. At this point, a strategy-minded information manager should be participating in the dialog, even leading it. Information, training, and support hold the keys to success. If the information manager is not aware of the potential, it is highly likely that the documentation, training, and support will continue to be focused incorrectly on the early market. There will also be enormous pressure to reduce costs, especially by cutting documentation, because the product usually is not generating sufficient revenues to satisfy the venture capitalists or the stock market analysts.

Information-development managers must, however, be careful not to fall into their own information chasm. Communication and training professionals become quite adept at following the lead of the product developers and creating documentation that explains how the product works. However, the pragmatists and conservatives of the new mainstream market care a lot less about how the product works and much more about how they can reduce costs and gain efficiencies by using it effectively in their workplace. To meet the information needs of the mainstream market, information-development managers must insist upon going to work for the customer. That means visiting the customers' workplace, learning about their productivity issues, understanding their goals, and learning how they make use of information, training, and support. From this information base, information-development managers can focus their teams on building value for the majority customer.

Analyzing the products in the technology adoption life cycle

"We can't afford to spend millions of dollars developing information and training for a product that never sells beyond the first innovator client. Until a product produces adequate revenue, we will provide only a basic set of technical information in support." So argued an information-development manager who applied her considerable business acumen to her portfolio management. It took some convincing among senior managers, but the process of evaluating products helped to deploy resources effectively and reduce information-development costs.

As a portfolio manager, you need to understand the positioning of each product and service you support with respect to the technology adoption life cycle:

- ✔ Meet with each product manager and discuss the technology adoption life cycle. If the product manager is unfamiliar with Crossing the Chasm, prepare a brief slide presentation of the main ideas. Discuss your understanding of where the product is along the bell curve and ask the product manager to agree or disagree.

✔ Place each of the products your team supports along the bell curve. If possible, develop a presentation to senior management that describes your analysis and gain their agreement. You will find that individual product managers may dispute your positioning of their products, especially if they are in the innovator or early adopter stage.

✔ Determine the basic level of information support for each life-cycle phase. Consider, for example, that new products being designed for one innovator customer may require only the most basic information. It may be more valuable to work with product managers to develop informative white papers than to write task-oriented user manuals.

✔ If possible, develop a set of criteria to use in evaluating each product and gain support for the criteria among senior management. The checklist in Table 5-1 shows one manager's checklist for evaluating each product.

✔ Use the life-cycle placement of each product to distribute your resources. Your decision to provide minimal information for products in the innovator stage may result in assigning a senior information developer to attend product development meetings and work part-time on the development of white paper. A decision to support a product that is generating revenue from sales to long-time late majority customers but with few training requests and calls to support may result in minimal information updates and no redesign.

Table 5-1: Product Evaluation Checklist

Question	*Evaluation Criterion*
Is the product generating revenue? Usually new products being designed for an innovator/early adopter market generate minimal revenue initially. Highest revenue levels are often generated by mainstream products that have crossed the chasm.	• No revenue–Basic level of information support for an innovator/early adopter audience • Rapidly increasing revenue–Increased level of information support with full information set; increased support for training and customer support • Steady-state revenue–Minimal updates to full information set
Is the product generating requests for training? Training needs to grow as a product crosses the chasm. Training must address the needs of less skilled audiences as customers work to decrease costs.	• None or few training requests–Minimal updates to full training and information set • Rapidly increasing training requests–Increased level of training development focused on new analysis of customer training needs • Steady-state training revenue–Minimal updates of existing materials; move to e-learning as quickly as possible.

Question	Evaluation Criterion
Is the product generating calls to the support line? Calls to support always increase when new products or product versions are introduced. However, support calls also increase when the product has usability problems for majority customers. Such calls can be reduced by better information and training.	• Few if any support calls–Minimal updates of existing materials • Rapidly increasing number of support calls–Analysis of existing information products to initiate a redesign project to reduce support calls. More investment in information and training for the support staff. • Steady-state support calls–Minimal updates of existing materials. Increased analysis to find better ways to support customers through online knowledge base.

It is possible that you will be asked to make exceptions to the life-cycle checklist results. Your company may be developing a new product that has a high level of support with senior management. They may be willing to fund an extraordinary effort to develop information that will meet the needs of conservative customers even though the product is in the early adopter stage. You must be cautious, however, not to be pushed into devoting resources to new developments and neglecting products that need to be redesigned in response to a move across the chasm. As you will see, the need for resource funding for information, training, and support is likely to increase as a product moves into the majority markets.

Best Practice–Funding information development for the majority market

In the first flush of the development of a new, innovative, promising product, senior management understands the importance of funding the development effort, even though little if any revenue is forthcoming. When a promising product proceeds into new releases, introducing additional functionality to satisfy the needs of important, early adopter customers, development funding continues to meet demands. When a product reaches the steady state of a majority market, senior managers look for opportunities to reduce development costs, often moving the "cash-cow" products to offshore maintenance developers.

Funding for information development tends to parallel the funding for product development. As funding is increased, information developers have opportunities to staff-up for the new, exciting product. As funding is decreased, information developers find their teams outsourced or offshored with the development teams, unless the new development is being handled offshore.

Information-development managers must take advantage of increased funding when it is available early in the product-development life cycle. However, that often means developing information products without sufficient knowledge about the needs of the audience.

Innovator and early adopter audiences are often atypical, which means developing to meet their needs may not prepare the information or the product to cross the chasm to the majority customers.

Majority customers demand different information and increased usability in the product than do the early adopters. As a result, the funding necessary to change the approach to the product information and training is often insufficient.

Figure 5-2 shows the typical relationship of funding for product development to the funding for information development. Figure 5-3 shows how the funding should increase during the technology adoption life cycle for usability, information products, and training.

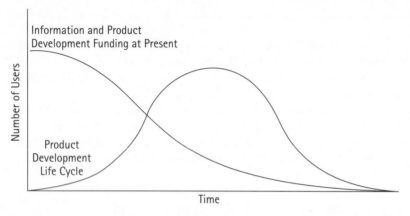

Figure 5-2: Relationship of product and information development funding during the technology adoption life cycle

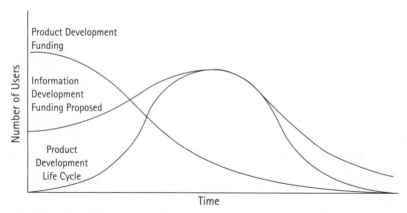

Figure 5-3: Proposed funding model for information development during the technology adoption life cycle

By analyzing the proposed funding model and finding support for increased funding by actively pursuing new customer requirements, you can develop a business case to support a move toward a more usable product that is better supported by information and training. At this time in the information-product development, you may find increased opportunities to merge the work of information and training development. Because information and training must now be redesigned to provide more targeted assistance, the opportunity for merging the development processes is expanded.

Taking Advantage of Merged Content for Information and Training

For one mature product, I worked closely with the information developer and the instructional designer/trainer to develop a new integrated introductory product course. The content in the introductory product manual paralleled the new course model. The content in the product manual could be used intact in the training, leaving the instructional designer with more time to develop new laboratory exercises. The same course, once the labs were fully tested, also had the potential to be transformed for an e-learning environment. Both participants in the project benefited from the dialog and synergies that emerged. The information developer gained insight into focusing the information in the product manual more explicitly for the majority user. The instructional designer found that a new emphasis on experiential learning with reliance on the background material provided in the manual attracted more customers to the training and enlivened the learner's experience. The majority learners were able to focus on hands-on experience with the product rather than long, detailed presentations on the product schema, for which they had little patience.

Finding funding for this experiment in redesign for a majority audience was not easy. Declining enrollments in the traditional courses and appeals by the customers for better information and training helped push the product manager toward a positive funding decision.

Summary

By understanding the technology adoption life cycle and the position of the products that your team supports on the technology adoption life cycle bell curve, you gain an important business perspective on the needs of the user communities you serve. You also gain the ability to make critical and sometimes unpopular business decisions about the level of support you can offer product managers. No information-development manager has a surfeit of resources to support every product at the same level. Many products require or deserve much less support than others. Using the insights developed through an analysis of each product's position in the life cycle, you can make better business decisions and support your decisions quantitatively.

- ✔ Understand the technology adoption life cycle so that you can discuss each product's position with product marketing and development managers, as well as with your senior management

✔ Develop a plan for evaluating each product and establishing a level of support with regards to information development

✔ Associate information-development resources with each product category by determining the appropriate level of support required by customers

✔ Develop a funding model that allows for support of mainstream customers with a renewed emphasis on accessibility and usability of support information

Developing a technology adoption life cycle-based approach to resourcing and funding depends on the support of upper management. If you can demonstrate that your support recommendations are based upon a business analysis of product needs and if you can align your recommendations with corporate business goals, you will enhance your standing in your corporate community as a business-savvy manager. Begin by having discussions with the product managers and marketing professionals. If they are already aware of Geoffrey Moore's recommendations for crossing the chasm, you will gain a level of understanding that promises to enhance your own management role.

"We want to work with you to make your
employees more successful."

Chapter 6

Developing Relationships with Customers and Stakeholders

> Customers want to "hire" a product to do a job, or, as legendary Harvard Business School marketing professor Theodore Levitt put it, "People don't want to buy a quarter-inch drill. They want a quarter-inch hole!"
>
> —Clayton M. Christensen

The greatest weakness of contemporary information-development organizations is the almost total absence of customer contact. Although information developers generally espouse the importance of knowing the customer, they are often content to speculate about the customer or to take the customer profiles provided by product marketing, product development, subject-matter experts, customer support, and others at face value. Unfortunately, the information coming from the so-called experts is rarely based on a systematic analysis of customers as users of information. Sometimes the customer profiles represent the best guesses of product marketing about who might be buying the product or service, sometimes they are based on actual market research into buyer characteristics. Sometimes customer profiles are built on a set of assumptions based on individual encounters with certain customers, usually those that are innovators and early adopters who are actively involved in defining requirements. Sometimes customer profiles are defined by those who call for assistance or are encountered by field personnel responsible for solving problems. Sometimes they are based on past experience at a customer location by someone in your organization. Sometimes they are defined by trainers who have customers in their classrooms.

Each of these information sources about the customer can be useful, of course, in providing you with insights about the customers' information needs. But each is likely to be filtered through a huge set of assumptions and biases. Customers who play the role of buyers are often not those who will be using the product or service. Even if they have both roles as buyer and user, their profiles as users will differ from their profiles as buyers. Customers who regularly interact with product developers are often more experienced and expert in their use of products or services than average customers. Customers who call into the support line are often self-selected and may be significantly different from customers who never call. Those who attend training classes are often specially selected by their companies as the most likely to benefit from the training. They are often responsible for training others on what they have learned.

Although you are wise to develop partnerships with internal stakeholders who have a perspective on the users of your information, you are also responsible for knowing more directly and personally the people for whom you and your staff develop information. Their role as information users is probably quite distinct from the roles they play as product users, participants in training, people phoning in with questions, or certainly as buyers.

Developing Content for Internal Stakeholders

In one user study I conducted for the US Veterans Administration hospitals, I was anxious to learn what information would benefit a host of people who used patient records on a regular basis. The users included physicians, nurses, technicians, and even administrative personnel. One of our first meetings was with a team that had been set up to shepherd the study through the organization. At the meeting, which included hospital administrators and physicians who were department heads, the participants began to tell us what they thought staff members needed to know. When I began asking questions about the ways in which they thought staff interacted with information daily, it became increasingly clear that they were defining the profiles of users that they did not fully understand, certainly not to the extent necessary to promote information usability. Only after the intent of our questions became clear did the participants acknowledge that they didn't know the answers. They recognized that the only way to answer the questions and make recommendations about information design was to meet directly with the intended users. In fact, the high-level participants expressed surprise when they finally realized that they couldn't speak for the needs of their own staff members. The result was a very successful user study that led to a superb interface and information design.

As a manager of information development, you must find whatever ways possible to learn about your users directly. Although surrogates who have their own experience with customers can be helpful, they are not sufficient. So many times when I make this same recommendation to a manager and staff, the answer is that "they won't let us." While senior management and others who control access to customers may erect substantial barriers, you should never take "no" for an answer.

This chapter covers one of the most powerful methods for understanding how users interact with information, the customer partnership. I also refer you to the valuable insights into customer studies that my colleague, Ginny Redish, and I included in *User and Task Analysis for Interface Design* (Wiley 1998). Study the methods presented in that book and here. Find ways to get you and your staff members into direct contact with information users. You will always be surprised at how much you will learn and how your experiences will challenge all your previous assumptions.

In addition, this chapter includes suggestions for establishing working partnerships with others internal to your organization who are involved in developing and delivering information to customers. Included among those internal stakeholders you will find trainers, training developers, and instructional designers, customer support personnel both in the field and on the phone, product marketing, product development, customer consultants, and a myriad of people who want to help customers be successful and gain in their understanding of products and procedures.

Best Practices in Customer and Stakeholder Relationships

Information directly through interaction with customers is essential to producing useful and usable information products. Without an in-depth understanding of customer information needs, you are probably as likely to produce too much information or information no one needs as to produce information that is vital to customer satisfaction and performance. Finding ways to partner with interested customers will enhance the quality of your information development in ways not available through any other means.

Not only should you gather information directly from customers about their information needs, but you should gather information about customers from others in your organization who interact with customers regularly. Many of those who interact with customers also develop information that they deliver directly to customers. They may find that a partnership with your organization is beneficial to them in reducing costs, allocating resources, and delivering superior information.

In this chapter, you learn about four best practices for including customers and other stakeholders in your information-development processes:

✔ Analyzing customer information requirements

✔ Partnering with customers to design and develop superior information products

✔ Analyzing others with a stake in providing customers with superior information

✔ Partnering with those stakeholders by identifying mutual business goals

The best practices here provide you specific programs for establishing partnership arrangements.

Best Practice–Analyzing customer information requirements

As information developers face the challenge of planning, designing, and developing for different customers into integrated online and printed information products, you must be dedicated to actively pursuing in-depth knowledge of your customers' information requirements. These analyses should tell you how much and what kinds of information your various customers need, as well as what media to use to deliver that information.

Several well-established techniques, including contextual inquiry, focus groups, and usability testing, provide valuable guidance about the information that information customers need. Contextual inquiry, which involves observing customers using products in their work environments, shows technical communicators how products and documentation are used in the workplace. In my work with Ginny Redish in user and task analysis, we found that direct observations of customers at work was essential for understanding requirements and knowing exactly how information fits into the customer's agenda. In many customer studies, we have learned that customers invoke a variety of methods to gain information they need to perform a function and use a technology product, starting with asking colleagues who are more experienced. They use information products when and where they are most appropriate and fit well into their workplace.

In addition to direct observations of customers, focus groups and usability testing can help you better understand how customers use information. Focus groups take users out of the workplace and into a discussion of their needs with their peers. Focus groups provide insight into users' attitudes and opinions about products but do not tell us how the participants actually use products or documentation at work. Usability testing, which allows you to watch participants actively using products and information, provides significant insight into problems with information access and usability but does not normally allow for long-term contact and interaction about the customer's needs.

Other sources include customer information gathered from support calls, customer surveys, telephone interviews, and attendance at user group meetings. In one complex information design project, we used contextual inquiry, focus groups, and usability testing together to assess users' information needs. Many of these methods of gathering customer information are described in detail in *User and Task Analysis for Interface Design*. All of these techniques provided valuable direction about the kinds of product support users needed; however, I found that none of them provided us with sufficient information to design an information library that met users' needs. To supplement these needs analysis techniques, we developed customer partnering.

Best Practice–Establishing a customer partnership

Customer partnering is a technique used to design information products by creating a long-term relationship between representative customers and information developers. Through a series of guided interactions, customers and developers investigate how current information products are used. They use this knowledge to specify the design of improved and new information products.

Like other forms of participative design, customer partnering uses aspects of contextual inquiry and focus groups, but it provides much more detail about how you can meet customers' needs, and it involves the customer much more deeply in the design process. Neither contextual inquiry nor focus groups provide the rich array of information needed. Contextual inquiry provides good information about how a customer uses a product in a particular setting, but it doesn't allow customers to help develop ideas and provide more thoughtful and focused feedback. Because contextual inquiry limits by the amount of time you can spend with a busy customer, you often are unable to see the entire spectrum of behaviors they exhibit with information products.

The focus group, on the other hand, often doesn't allow time to concentrate on design details because participants are together only briefly and can examine problems only superficially and out of context. Neither technique allows for repeated contact between you and your customers so you can collaborate in formulating ideas and opinions over time.

You need a combination of techniques that would allow you the time to learn in depth about your customers while still providing an atmosphere conducive to brainstorming and collaboration.

How to implement customer partnering

Customer partnering defines the essential relationship between a customer for a product or service and a developer who seeks to improve the quality of that product or service. I recommend that at the beginning of your implementation of customer partnering you employ an individual or group skilled in partnering techniques to serve as a facilitator of the customer partnering activities. In this way, implementation will not be biased by preconceived notions about what the customer wants or needs or by defense of existing information products. The *customer* represents the people who use your information products and volunteer to take part in the customer partnering effort. The *facilitator* is the person who designs the study, runs the information-gathering sessions, writes reports about the study, and makes recommendations about information delivery methods suggested by the study.

Facilitators should be skilled in the customer-partnering technique. For projects involving information design, the facilitators should also be experts in the information design process. This knowledge is needed in designing homework assignments, helping your staff understand the information gathered, and keeping the discussions focused on topics that will aid in information design. Just as with focus groups, it is often beneficial to have facilitators who are not part of your organization so that they can maintain a degree of objectivity during the sessions; however, customer partnering is not limited to facilitation by outside firms. Any group can use customer partnering internally to its own benefit so long as they develop expertise in the technique and engage an independent and skilled facilitator.

The facilitators in charge of a customer partnering program must first ensure that everyone understands and accepts the proposed method and its purpose. The commitment asked of customers is great, and any confusion between your organization and the facilitators about the goals of the study will disrupt the focus of the meetings and quickly discourage customer partners from contributing to the meetings or even attending them. Once the purpose and method are clearly established and understood, it is time to establish criteria for selecting members of the customer-partnering group and planning the partnering program.

Selecting customer partners

Customer partners should represent a broad spectrum of those who use your information products. Having participants in the study whose experience ranges from novice to experienced, who represent different sizes and types of companies, and whose use of the product varies ensures that customer requirements and diverse use models will be addressed during group discussions.

In addition to having the necessary experience in their field to be considered "typical" users and representing a broad range of customer company types, customer partners must also

- ✔ express interest in the project and willingness to commit to the time and schedule

- ✔ be knowledgeable about their own documentation needs

- ✔ reside close to the facility where the working sessions will be held, if travel budgets are not provided

- ✔ commit to attending five to six working sessions over a period of several months

- ✔ commit to completing "homework" assignments on their own time in between working sessions

Adhering to these requirements may, however, add some bias to a study. Unless the area in which the working sessions are held is a large metropolitan area with many potential customer companies, it is possible to have too limited a sample of users for the sessions.

It is also possible to introduce bias by the selection of volunteers. By their willingness to volunteer, the customers may be substantially different than the majority of users who don't volunteer. This bias, however, is endemic to all participative study techniques, including contextual inquiry and focus groups. The customer-partnering technique seeks to involve customers in product design who have a strong commitment to the product's success, who are interested in creating the best product possible. At the same time, they are selected because they represent mainstream product use; they are not outliers. If a list of potential volunteers is suggested by the sales force, those on the list may be "good" customers in the eyes of sales, that is, they may be friendly customers, but not truly represent the mainstream customer.

Planning the format and focus of each working session

Careful planning of customer partnering working sessions is essential to the success of the study. Because the customers' experiences often influence the direction of conversations during the sessions, the facilitator must have a clear idea of what is to be covered during each session to stay on schedule and to maintain the focus of the group. For example, you may want to focus initial meetings on discussions of current information products. You may want to reserve the final sessions for a discussion of prototypes and future information needs.

The key is to plan highly focused activities that provide specific feedback about customers' information needs, keep the sessions focused, and maintain participant interest.

Activities I have used in previous customer partnering sessions include asking participants to draw their workspaces with colored markers, work in groups to evaluate online documentation produced by our client and its competitors, and tape documentation topics to a wall-sized information organization chart. For example, to maintain the focus of a discussion of documentation delivery media, we asked participants to complete a delivery media preferences worksheet (see Figure 6-1). After participants indicated on the worksheet the documentation delivery medium they preferred for each of the tasks, there was a group discussion about delivery media. Asking participants to complete the worksheet before the discussion gave them time to consider their delivery media needs for a wide range of tasks and formulate their ideas before the discussion. In the analysis of the working session, I was able to refer to the participants' worksheets and their discussion comments.

Providing homework for group participants

Customer partners should be given assignments to complete between working sessions. Giving participants homework assignments enhances the working sessions in the following ways:

- ✔ Participants maintain their focus on documentation during the breaks between working sessions.

- ✔ Participants have time to think independently about the company's product and decide what their own needs are without peer pressure or time constraints.

- ✔ Participants are able to bring concrete examples and well-formulated ideas for product improvement to the working sessions.

- ✔ Facilitators gather valuable accounts about the experiences of the partners and their co-workers in using the company's products on the job.

Homework assignments may include polling co-workers, keeping logs of experiences with documentation, examining products or documentation produced by other companies, and designing documentation prototypes. At one session, in order to determine what type of online information presentation was most helpful to users, I asked the customers to bring examples of online documentation they particularly liked or disliked to the next working session. I found that the online documentation customers liked was task-oriented and organized into major task-oriented menu choices, such as installation, configuration, troubleshooting, and maintenance. Had I only asked customers to the company's existing online documentation, which was organized by product type and manual title, I might not have discovered customers' preference for task organization.

To ensure that homework assignments benefit the partnering venture, the facilitators must design assignments that are interesting, clear, relevant, can be accomplished in a reasonable amount of time, and that gives participants a good starting point for the next working session. They must also prepare all log forms and questionnaires necessary for each assignment. Customer partners must commit to completing all of the homework assignments and must come to working sessions ready to discuss the assignments.

Task	online delivery	paper delivery	online and paper delivery (explain)
selecting hardware			
selecting options			
researching specifications			
installing hardware			
installing options			
using SmartStart			
choosing a configuration			
installing an operating system			
tuning the system			
troubleshooting			
correcting hardware problems			
maintaining hardware			

Figure 6-1: Delivery Media Preferences Worksheet

Facilitating the working sessions

The relationship between customer partners and your organization develops during the series of working sessions. At the sessions, you are able to observe while your customer partners evaluate products, share their experiences, and talk about their needs. You can ask questions, present new product ideas, and solicit immediate feedback.

To guarantee productive working sessions, facilitators must prepare the necessary materials and discussion topics for each working session. Facilitators should not begin a working session until everything is in place, including equipment for running online, video, or audio documentation; documentation that customers will evaluate during the working session; and any paper or other materials needed for working session activities. Facilitators must also direct the working sessions, keep discussions focused, and prepare discussion questions and written surveys designed to elicit information on how customers currently use the products and on how they could use the products more effectively.

One main advantage of gathering information during working sessions is that the participants interact face to face over an extended period of time. Unlike focus groups, in which the participants never get to know one another, customer partnering provides enough time to allow the participants to form a coherent work group. This interaction contributes to information gathering in the following ways:

- ✔ Participants are able to play off each other's ideas and brainstorm solutions together.

- ✔ Facilitators can gauge customers' reactions to existing products and new prototypes by listening to their comments, watching their review of the products, and observing their body language.

- ✔ Your staff can see first hand how customers react to their products.

Successful facilitation of the sessions requires a firm focus on the objectives of the partnering and the ability to promote useful conversation. The facilitator must ensure that everyone has an opportunity to contribute and that no one is permitted to dominate the discussions. If facilitation is done correctly, a rapport is established among group members that allows candid and meaningful dialog that can lead to the design of improved information products.

Duration of the study

The duration of a customer-partnering study depends upon the complexity of the product being designed. However, a primary characteristic of customer partnering is that the interaction between customer and developer takes place over an extended period. Over time, especially several months, customers grow in their understanding of the design problems under consideration. In one study, for example, I noted that the server administrators began by being positive but vague in their comments on the documentation. They told us that the existing documentation was "fine." By the fourth and fifth sessions, their opinions had changed and their articulation of their needs and the problems with the existing information had become increasingly, even surprisingly, explicit.

For example, at an early session, the customers told us that the installation pamphlet was "helpful" and they would choose to use it during installation. By a later session, they became adamant that the same pamphlet was not useful and had to be redesigned. The growth in understanding comes in part from a degree of politeness. People dislike being critical in front of strangers, especially in the presence of the developers of the documentation. Unless they have very strong negative responses, they are unlikely to make an issue of minor concerns. In addition, however, users often have not formulated a clear analysis and criticism of an issue that they are asked about. They really have no opinion but will fill the gap in the dialog with vague commendations. Only after several sessions will they gain enough perspective in the issues at hand to discuss them cogently and have specific recommendations.

Customers also learn from their interactions with other participants to differentiate between a "pet peeve" and a more universal issue. They tend to temper their statements of need to take into account the circumstances and patterns of use articulated by other participants.

Customer-partnering programs should consist of at least five working sessions with two to three weeks between sessions. Five sessions allows the group to become an organic unit that moves well past the usefulness of a focus group. Beyond six sessions, the sessions may begin to lose effectiveness. As the customers become more expert in the topic of the study, they cease to represent typical users. You might be able to reconvene customer partnering groups, however, for follow-up activities (for example, for feedback on prototype documentation developed in response to the findings of the study); but for any future customer partnering programs, you should recruit a new group of customer partners.

Recognizing customer participation

To help encourage and recognize those customers who had participated in the program, you should plan to distribute small gifts, invite the participants to dinner at a nice restaurant to wrap up the study, and write letters of appreciation. This recognition is important because the customers receive no compensation for their participation, commit a significant amount of time, and show a willingness to contribute to the effort to improve your information products.

Participants are very interested in knowing what has resulted from their recommendations. They want to see the fruits of their labors in improved, more usable products.

I recommend that program facilitators and your organization maintain a relationship with customers after the final working session. A continuing relationship provides you with opportunities to ask customers for feedback on new product ideas and to modify your efforts accordingly during the development process. Customer partners might even participate in new product usability tests.

Developing product prototypes

The time between working sessions allows you to develop product prototypes for customer partners to review in later working sessions. Getting customer feedback on product prototypes helps with product redesign in the following ways:

✔ You receive immediate feedback on new ideas for product design and presentation.

✔ You have the opportunity to improve prototypes to meet customers' needs and preferences.

✔ Your ideas are supplemented by customers' ideas for improving the prototypes.

Program facilitators, your organization, and customer partners work together to create prototypes that become the framework for your organization's new information products. You are responsible for using information gathered during the working sessions to create prototypes of new documentation and for preparing written evaluation forms designed to elicit participants' detailed evaluation of the prototypes. Customer partners evaluate each prototype and offer feedback for improving it.

Reporting the results and recommendations of the working sessions

During the working sessions, the facilitator should ensure that customer responses are recorded. An observer should attend each session to log the sessions in much the same way a data logger functions during a usability test. You can videotape the sessions for immediate review, future analysis, and presentations.

After each working session, the facilitator should summarize the findings and recommendations that result from the working session. The summary should include information about what participants like, dislike, and need from your information products. It may also include samples of products or documentation that participants bring to the working sessions, sketches participants make of their workplaces, or prototype design ideas.

At the end of the program, the facilitator should create a final report presenting the overall findings of the study and addressing its original goals. To draw on the individual working session reports but avoid redundancy, I recommend organizing the final report around an evaluation of delivery media, user experience levels, or stages of product use. For example, while the working session reports focused on what customers liked and disliked about each of the client's documents or document prototypes, the final report discussed the differences between novice and advanced users' needs. The report also made recommendations for providing an information library that supports a broad range of users and uses the delivery media appropriate for each task they perform.

Benefits of customer partnering to you and your customers

The major benefit of customer partnering is the relationship that develops between your organization and your customer partners through repeated contact. Studies that involve one-time contact, such as interviews, surveys, focus groups, and contextual inquiry only gather participants' immediate reactions. Customer partnering, on the other hand, creates an environment for an in-depth investigation of customers' needs and preferences. As a result of the close relationship that develops between you and your customer partners, both parties benefit. The specific benefits of customer partnering follow:

✔ **Customers receive a product that is custom designed to their needs.** Customer partners leave the program knowing that product improvements will be made with their needs in mind. They eventually receive a product that is designed specifically to meet their needs. Insofar as the partners represent the entire customer base, the new information product is customized for all customers' needs. If the developers are prepared to do so, working prototypes of information products can also be delivered to the customers' sites between sessions for inhouse testing, giving the developer even more detailed feedback.

✔ **Customers feel ownership for changes to future products.** As they participate in the program, partners gain a feeling of empowerment: the opportunity to give feedback directly means that they have helped design products that directly address their needs.

✔ **Customers learn how the process works and provide better feedback and information in the later sessions.** During the weeks between working sessions, the customer partners reflect at length on the product they are evaluating, complete homework assignments that help them develop their ideas, and evaluate your and other companies' information. As a result, participants are able to bring to each session design recommendations, well-thought-out opinions, and examples of products they particularly like. This continuing effort on the part of the customer partners is the primary feature that makes customer partnering a richer, if different, information gathering technique than contextual inquiry or focus groups.

✔ **You get frank and valuable information about the design of products and documentation.** From the relationship that develops between you and customer partners, you gain insight into customers' working styles and product needs. Most importantly, you receive candid feedback from customers on aspects of products they actually use and how they use them. As you get to know one another and feel comfortable with each other, all participants feel free to contribute to each other's ideas and to express straightforward opinions about the products.

✔ **You get to educate your customers.** During the program, customer partners learn about the full range of products available to them, and they receive answers to specific, work-related questions. Your organization can even use this information to help develop marketing efforts to help others learn about these products.

✔ **Customers' views of your company may become more positive.** Customers who participate in partnering programs are impressed by your dedication to improving its products and involving customer input in the redesign process. They appreciate the attention and see you in a new, positive light. This can, however, provide a bias that keeps them from being forthright. The facilitator must be aware of this bias and probe for issues they feel might underlie the comments being made.

✔ **Customer partnering benefits all involved.** The benefits of customer partnering to all participants are enormous. Customers feel they have direct input to products they will use, feel they have your undivided attention for the duration of the program, and finally have products and documentation that more nearly meet their special needs.

✔ **Information developers no longer have to guess about customer requirements for information products.** Rather than just hearing the "horror" stories of products and processes gone bad, hearing only from customers who love the product, or hearing nothing at all, you interact directly with a spectrum of typical users who can help fill in the comprehensive picture needed for good information-product development.

Customer partnering offers information-development managers with a way to unravel the complexity of information and organize large bodies of information into information that is useful for its intended audience. By involving customers deeply in the information design process, you can meet customers' information needs as effectively as possible.

Best Practice—Analyzing internal stakeholders

Similarly to the analysis you perform with your external customers, you need to analyze the needs of your internal stakeholders, especially in relationship to their focus on serving the needs of external customers. You are likely to find many different internal stakeholders who are all involved in information development, including

- ✔ training development and delivery

- ✔ customer support, both phone and field

- ✔ sales and marketing

- ✔ product management

- ✔ product development

Depending on the nature of your organization and the type of products or services you are developing, you may find internal stakeholders in departments as widespread as legal, manufacturing, regulatory affairs, and more.

Your starting point is, of course, a stakeholder analysis. Unlike the stakeholder analysis you might do for an individual project in which you investigate who has a stake in the quantity and quality of the information being produced, your analysis must include all others in your organization who have a stake in what is delivered to customers through which channel. I discuss stakeholder analysis for individual projects in Chapter 14: An Introduction to Project Management.

In approaching your stakeholder analysis at the portfolio level, you have a broader range of possibilities and parties believe they have a significant role to play in information development and delivery. You are likely to encounter resistance to a partnership as often as you will encounter a positive view of a potential collaboration. In fact, your responsibility will be to change resistance into win-win results for both partners.

Develop a set of questions that represent the criteria that you believe will lead you to a solid win-win partnership with an internal stakeholder. Consider, for example, some of the following questions:

- ✔ Are the stakeholders developing content that they are delivering to customers?

- ✔ Do the stakeholders believe that content development is an integral to their role and responsibility in the organization?

- ✔ Do the stakeholders believe that their content is often superior to the content developed by other parts of the organization in meeting customer needs?

- ✔ Do the stakeholders have a close personal or business relationship with the customer receiving their content?

✔ Do the stakeholders receive revenue from the customers based, in part, on content delivery?

✔ Have the stakeholders developed tools and technologies to support content development and delivery?

✔ Do the stakeholders use any content that is developed by your organization?

✔ Do the stakeholders change the content you have developed to better meet their needs or the needs of their customers?

✔ Are the stakeholders interested in reducing their content-development costs through reusing existing content?

✔ Does superior content provide the stakeholders with a competitive advantage?

No doubt there are additional questions to investigate but these provide a start. Your goal in asking these questions is to rank the potential internal stakeholders with whom you may want to do business. The highest ranking might go to those stakeholders who will reap the greatest advantage from a partnership with information development. That partnership may help them reduce operational costs, reduce time to market for content, improve accuracy and consistency, or be able to focus their efforts on business areas that are most productive.

Information-development managers often believe that the training organization is a good potential partner in information-development because much of the information being developed by training is the same as the information that your staff is developing. Training and documentation are often brought together under the same senior management to take advantage of potential synergies in information development. Information developed for the user documentation can often be used intact in the training materials, freeing the instructional designer or the training developer from developing content that already exists.

At the same time, the training organization may have a conflict of interest in partnership with information development. Training is often a profit center, intent on filling seats in classes or registrants for e-learning. If the information they offer their training customers is not unique, the customers may object to be charged for the same content they could find in the manuals that they get for free. Trainers often find the information in the manuals to be inaccurate or too unfocused to help customers meet specific learning objectives.

Another potential candidate for a partnership is the support organization. Support staff are often responsible for writing content in response to customer queries. Many organizations invest in a support knowledge base that provides customers with online access to information that answers specific questions. However, the task of developing this information is often in the hands of people who are inexperienced at writing usable and readable text. The support knowledge bases themselves often become clogged with duplicate, near duplicate, and out-of-date responses, making customers searches frustrating and rife with potential mistakes.

The possibility exists for information developers to assist support staff in developing consistent content and managing the repository. Information developers are often asked to write and edit content going into the knowledge base or add content they develop to the knowledge base so that it becomes a central store of customer-facing information.

Support organizations may also be interested in call-avoidance strategies. If customers can easily find relevant information in print documents, on information websites, or in online help, they are less likely to call support with ordinary, easily answerable questions. Many support organizations will partner with information development to reduce the numbers of calls that might have been handled in the documentation, leaving them time to handle the calls that are unique or more difficult to handle.

Sales and marketing organizations develop considerable information that goes to customers that could benefit from the content that your staff has vetted for consistency, accuracy, and usability. They may want to put a "sales" flavor into the content, but they may save time and increase accuracy if they begin with a reliable technical source. Sales organizations, if they produce new business proposals for customers, may also be interested in a repository of accurate content that they can use to build responses to a customer's request for proposal. Sales organizations often develop their own content management systems to store standard responses to questions they often see in requests for proposal. They may welcome the assistance that information developers can provide to ensure that the content in the repository is up to date and accurate.

Every potential stakeholder that you investigate will provide you with a possible business case for collaboration and partnership. In your stakeholder analysis, you need to identify compelling business reasons for working together before you approach a peer manager with a proposal.

Best Practice—Establishing stakeholder partnerships

Once you have identified potential stakeholders with whom you would like to develop a mutually beneficial relationship, you need to approach them with your proposal. Once your proposal is accepted, you need to define the parameters of your partnership, including your measurable business goals and objectives. Include details of how the partnership will function, the resources to be made available, the responsibilities of each party, the activities each will perform, and a way to resolve conflicts.

Defining measurable business goals and objectives

Stakeholder partnerships generally have two important business goals: reducing costs and increasing customer value. Partners may want to develop information resources that can be used in common, reducing development time and increasing accuracy. Partners may want to share information they have learned from customers to increase the usability of the information. Partners may each have special expertise that they can apply to increasing the value of the content. By collaborating on information development, the partners may be better able to fill gaps in the content or correct errors.

Cadence Design Systems created a partnership between technical communications and customer support. The goal of the partnership was to improve the quality of information going to the customer and lower costs for customer support. If the product documentation provided answers to commonly asked questions and made it easy for customers to find the questions and answers, the partners expected that the number of customer calls would be reduced.

If possible, define how you will measure the success of the partnership in meeting the goals and objectives. For example, if customer support can reduce the number of calls that address commonly asked questions by 50%, they can save money by reducing the number of support personnel, they can reduce the number of additional support personnel who must be hired, or they can shift support personnel to answering more complex questions, increasing their value to the company and the customer. At Cadence, the new information developed through the partnership lowered the call volume for a particular product by 50%, a significant improvement.[1] You might establish a similar goal if you too have partnered with customer support.

A partnership with the training organization might establish a goal of reusing content developed by information developers and instructional designers in the deliverables of each organization, thus reducing development time and time to market. In such a partnership, information developers might document tasks required to use the product effectively, and instructional designers would use these tasks as the basis of their classroom tutorials. The instructional designers might create the business-oriented overviews and concepts, making them available for the information developers to add to their print deliverables or product information website.

A partnership with engineering might ask product engineers to develop reference information that they need for specifications in a form that can be directly used in production documentation. Parameter lists, prerequisites, and technical details of all types might be best maintained by the people who develop them. An information architect might develop a simply XML-based form for the engineers to complete that is readily included in documentation. Deutsche Post, the German post office, developed such a system, allowing them to transfer reference information smoothly from software developers to end users without that information having to be reformatted by writers. The software developers were happy to use XML as a writing tool since they were already using the language to develop their software systems.

Defining resources and responsibilities

Include in the partnership agreement the resources each party will provide and the responsibilities that they will assume. You might, for example, want to make your content repository or your content management system available to your partner. They might also have their own repository of content. You might each have customer profiles or task analyses that are useful to both. Customer support or other partners maintaining a website may have data about customer visits to the site that indicates which information may be most valuable. Application engineers who work directly on difficult customer problems may have a storehouse of technical case studies that will enhance training materials or information being developed for experts.

In one partnership, expert application engineers who were intimately familiar with the problems actually experienced at customer sites agreed to work with "ghost writers" from the publications organization. The information developers interviewed the application engineers, developed outlines for proposed white papers, and worked with the engineers

---

[1] Krista Guglielmetti, "Technical Communications and Customer Support: Partnering to Publish What Customers Want to Know," *Best Practices* of The Center for Information-Development Management, August, 2000 and the *47th Annual Conference Proceedings* of The Society for Technical Communication, 2000.

to understand the content. The writers created the drafts, and the engineers reviewed, added ideas, and corrected errors, enabling the teams to produce in record time white papers that were high on the customers' priority lists.

I once worked closely with a customer support engineer who had assembled a detailed spreadsheet of customers' questions about the software installation procedures. I conducted a careful analysis of these questions, recast the tasks and background information, and edited the draft content together. The result was a vastly improved installation instruction that reduced support calls on this topic by 60%.

The team at Cadence discovered that the application engineers were writing examples that they used themselves to answer customer questions. The customers wanted more examples in the product documentation. As a result of the partnership, the information developers were able to include the examples from the application engineers in the user manuals and the online help.

To prepare for an agreement about roles and responsibilities, both partners should describe their processes. You will find that by understanding how information is created, reviewed, produced, distributed, and maintained by your partner that you gain insights into information transfer that you did not know occurred in your organization. You could learn, as I did on a project, that the trainers rewrote the end-user documentation so that it corresponded to the customizations that the systems integrators had configured. The original documentation was never distributed. I suggested that the trainers be provided with access to the unassembled topics in the repository so that they could more easily customize them rather than having to type their own versions of the tasks.

Developing a program of activities for each partner

By developing a workflow of all the activities that will lead to the desired outcome, you can ensure that each partner understands what needs to be done and assigns the appropriate people to the activity. If one of your team members needs to analyze data collected, discuss how the data will be made available. If you want to interview members of the partner's team to gain insight into the customers and their needs, decide when the interviews will take place and in what form (in person, through email, through chat or instant messaging). If you are dividing up writing assignments among partners, define the authoring guidelines and experiment with developing the content through a small pilot project.

As you work together to establish mutually beneficial activities, create a schedule for the activities. The schedule should, of course, take into account major deadlines in which either partner is involved. You will not want to schedule a joint activity during the same week that your major deliverables come due. Neither will your partner. At the same time, both partners need to make a strong commitment to ensuring the work gets done. As long as you have established a strong business case, that commitment should follow. Unfortunately, in many situations, the managers establish the agreement, leaving the work to be done by staff members. The staff members may not feel the same level of commitment to the joint activity.

In one partnership activity, the information-development manager established a relationship with the sales organization to enable customer site visits. However, after the visits were scheduled, the manager delegated the activity to staff members who were not sure of the purpose of the activity or how much time they were supposed to devote to it, especially when they had other work to do. As a result, the visits were delayed or cancelled more often than they were held. The information-development manager did not partici-pate in the site visits and neither did his writers, except for the two who were dedicated to the success of the partnership activity. As a manager, you have made a commitment to your peer manager and to your own organization in setting up a stakeholder partnership. Both managers must make certain that all the people they ask to become involved under-stand what is expected of them and recognize the importance of the commitment to the joint activities.

Resolving conflicts

I recommend that a leadership team, including the managers of the partner departments, be established and continue to meet regularly during the course of the partnership activities. It is not enough to "dump" the project on overworked staff. Managers need to ensure that the project is successful. With a strong business case in place, you are the most likely to benefit from the joint project's success. If you have informed senior management of the partnership and its business goals, you are also the most likely to bear the brunt of a failure.

A leadership team is responsible for continuing to communicate the importance of the project to team members, ensuring that activities are given the appropriate priority level, and resolving any conflicts that occur during the project. Without this level of commit-ment, the project is likely to fail.

Summary

Customer knowledge and insight into the customers' real agenda are critical for the devel-opment of excellent information products. Without them, information developers are likely to create information for its own sake and to develop content dictated by product developers, marketing specialists, and others who have a stake in the information prod-ucts. Unfortunately, many stakeholders who want to control the information your organi-zation develops are serving their own needs rather than the needs of customers.

By becoming an expert in the customers' information and learning needs, your organi-zation takes a leading customer advocacy role, a role that emphasizes the customer as user rather than the customer as buyer of the product or service. If you are supporting information customers who are part of your larger organization, maintaining a customer advocacy role is also often unique.

- ✔ Have you persuaded your organization that you need to become experts in the user experience with information?

- ✔ Are you well versed in techniques for gathering customer information, including customer site visits, focus groups, interviews, and surveys?

- ✔ Have you considered customer partnering as a key tool in your customer repertoire?

- ✔ Have you built relationships with others in your organization who have responsi-bility for communicating information to customers?

✔ Have you identified ways to reduce costs, increase efficiency, and work together to provide more value in the information you deliver to customers from an enterprise publishing perspective?

Customer partnering methods provide you with a direct way to involve your information users in the creation of information that genuinely serves their needs. By using customer partnering along with customer site visits, interviews, observation, focus groups, and even surveys, you expand your knowledge base in a way that supports increasing the value of your information to your customers without increasing your costs for producing that information.

Just as you use customer partnering to involve customers in defining their information requirements, you can use stakeholder partnering to focus the efforts of all information developers in the larger organization on reducing development costs, increasing efficiency, and providing a more effective approach to supporting the users during the product support life cycle. By developing partnering relationships with training, customer support, field maintenance, and other content producers, you can enhance the users' experience with information resources.

"We want to know what information
you really need on the job."

Chapter 7

Developing User Scenarios

> Instead of designing products and services that dictate consumers' behavior, let the tasks people are trying to get done inform your design.
>
> —Clayton M. Christensen[1]

Understanding how people use your information is the starting point for defining an effective information architecture for your organization. With that understanding, you can approach your architectural framework knowing that the information is explicitly designed to meet the users' needs. Without that understanding, you are more likely to develop an architecture that meets the needs of your authors or your subject-matter experts rather than those who need to use your products and services to meet their goals and get their jobs done.

As you work to understand your users and their information-using agendas, you build a set of user profiles and user scenarios. User profiles describe the key characteristics of members of your user community. User scenarios describe how the members use your information to get their jobs done, from following step-by-step instructions to performing tasks to reading about how your products are designed to work, how to diagnose and fix problems they encounter, and where to find data that they need in support of their task performance.

---

[1] Clayton M. Christensen, Michael E. Raynor, and Scott D. Anthony, "Six Keys to Creating New-Growth Businesses," *Harvard Management Update,* January 2003.

A typical user profile may look like this one:

John is a technician working in the network operating center (NOC). He has only been at the NOC for two months. However, he has two years' experience as a field service technician, handling preventive maintenance on cell sites, installing and repairing equipment, and troubleshooting software problems. Like many of his co-workers on the first-line maintenance team, John does not have a technical education. He is a high-school graduate with on-the-job experience with telephone networks. He has taken inhouse training courses as part of his previous job, but he has not attended any training sessions yet at the new company.

John enjoys the troubleshooting challenge of the front-line position, and he's pleased when he solves a problem in record time. He isn't much of a reader, although he reviews the information that accompanies a system alarm. If he gets stuck, he first asks his colleagues if they've seen the problem before. When the problem becomes difficult to solve, he brings in experts at the NOC to help diagnose what is going on.

Based on observations of John at the NOC, a typical user scenario may look like the following:

As a front-line alarm-monitoring technician, John is responsible for monitoring and troubleshooting alarms on the network and correcting problems in response to alarms. If John and his colleagues cannot resolve an alarm after a set number of minutes, they escalate the alarm to a more advanced group of technicians.

When an alarm appears that the technicians are unfamiliar with, they want to locate information that tells them what may be happening in the network that has caused the alarm. Typically, the alarm may have more than one possible cause. They want to know how to investigate the problem in more depth so that they can decide which of the possible causes is the most likely. Then, they want instructions that tell them what to do to resolve the alarm and correct the network problem. If this course of action fails, they want alternative actions to pursue. And, they want all this information as quickly as possible with minimal text to read.

This scenario is typical of technicians who are responsible for maintaining networks or similar systems. As you analyze this and other scenarios associated with NOC technicians, you recognize several ways that you might organize your Information Model to respond to their needs. You know that John and the technicians will need several types of information, including descriptions of the alarms and the problems that may cause them, placed for quick access and immediate lookup. John will need task-based instructions that provide step-by-step procedures to diagnose and correct problems, especially problems that occur infrequently. The instructions might include decision-support flow charts if they don't become too complex to follow. John may also need reference information that lists parameter settings to be checked or signals to be analyzed. Added to these direct information types, your model, based on your analysis of John's needs, might include process flows that indicate how data moving through the network is supposed to behave and how it might behave if something is wrong.

One fairly ordinary troubleshooting scenario like this one leads you to many considerations about the details of your Information Model. You want to consider, for example, how to deliver the information that John and his colleagues need. Quite clearly, they need information as quickly as possible because they are under time constraints for resolving the network problems. Because the alarms appear on their computer monitors, you want to deliver information into the same environment as the alarms. Some basic information about the alarms should appear as an integral part of the alarm message that the technician views. Additional information about diagnosis and troubleshooting should be linked electronically to the message itself. If the alarm is straightforward, the links should go directly to task instructions for correcting the problem. If the alarm is more complex, the technicians should be able to link directly to well-labeled additional information in the form of tables of data or easily readable process or descriptive information.

If your Information Model includes training information, your analysis of John's information needs may lead to the creation of basic and advanced troubleshooting workshops. Your solution could include on-the-spot e-learning opportunities that John's employer could use as part of an on-the-job training program.

Some of the descriptive, background, and process information that doesn't change often might be supplied as longer articles for printing or reading online. You might decide to use an online alert system to notify the network technicians about new problems and their solutions or introduce an online forum in which network technicians could exchange troubleshooting information with colleagues and company experts.

All of these avenues are possible parts of the Information Model that you define as a result of developing and analyzing user profiles and scenarios. Profiles and scenarios are your most important and powerful design tools. The more scenarios available to you, the richer your Information Model will grow. The more accurate and detailed the scenarios, the better your design decisions will be.

To design an effective Information Model requires that you understand how your users interact with information in the course of their jobs. To obtain this information means that you must have access to your users in their working environments and access to others in your organization who interact with the information users, including trainers and those in support services.

Best Practices in Developing User Scenarios

Developing user profiles and scenarios is key to a successful Information Model. As an information-development manager, with the assistance of your information architect, you need to promote an in-depth understanding of users. Your staff needs to learn as much as you can about those who use the information produced by your organization. In the four best practices presented here, you learn to identify the basic roles played by the various people in your user community and how their activities in those roles cause them to access and use technical information. You learn the best ways to communicate the information gained to your staff and stakeholders. Finally, you learn to use the scenario-based knowledge to build your Information Model.

- ✔ Cataloging user roles and their information needs

- ✔ Understanding the users' information agendas

- ✔ Using user scenarios to develop your Information Model

- ✔ Communicating user profiles and scenarios to team members

The best practices in this chapter focus on the starting point for building a responsive and effective Information Model.

Best Practice—Cataloging user roles and their information needs

To create user profiles, you need first to develop a catalog of the roles that you believe make up your user community. Research among people in your organization who regularly interact with users often provides a starting point:

- ✔ Marketing may have already conducted studies on the people who purchase the products.

- ✔ Members of your sales team can brief you about the people they encounter during sales calls and purchase negotiations.

- ✔ Product managers and product development leads may have created their own descriptions of possible users and may have already developed use cases that support product design.

- ✔ Those who install and service products or provide consulting assistance directly to customers have insight into the working environment at a customer location.

- ✔ Trainers can help you understand the characteristics of people who attend instructor-led training classes.

- ✔ Telephone support personnel have information about the people who call with questions and problems.

Just remember that these internal sources have their own agendas and biases about customers, and they rarely have much insight into the users' information agendas. Marketing and sales are concerned with buyers, many of whom never use the products themselves. Installation and service personnel may interact with specialists who need some information resources but may not be regular users of the equipment, software, or service. Trainers may have experience only with customers who have been carefully selected to benefit most from training. Telephone support interacts only with those who call or are designated to place calls for others in the organization.

The responsibility of your team members lies in identifying those people who need information to support their use of the products and services wherever they are in the customer organization, rather than people who buy the products, call for support, or attend training. If the products you support are sold to consumers directly, your user scenarios may even include personal use at home as well as at work. In many organizations, no one knows how, when, or why users reference information sources to get their jobs done.

Survey your customers for your catalog of roles

After your staff has exhausted their internal resources as they compile a catalog of roles, ask them next to turn to those customers who are willing to work with directly. The process addressed in Chapter 6 on customer partnering provides one method of meeting with customers to define information needs. If you have email or postal mailing lists of key customers, you can construct a survey that addresses job roles in the organization. The survey should include questions like the following:

- ✔ Do you use product XYZ yourself?
- ✔ What do you do with the product?
- ✔ What is your job title?
- ✔ If there are others in your organization that use product XYZ, what are their job titles and what do they do with the product?

Surveys can be conducted through telephone interviews as well as through email.

You may learn that people with diverse job titles use the product to accomplish the same tasks. You may learn that people have different responsibilities with regard to the product and their tasks. In one study, I learned that users included those who used the products directly as a primary part of their job responsibilities, those who supervised the end users and used the products infrequently, those who trained the end users and used the products infrequently, and those who repaired the products and performed scheduled preventive maintenance. Another study found that, in addition to trainers and supervisors, the users of the equipment (in this case, railroad cars) had between 10 and 20 different job titles and performed specific tasks that no one else performed.

In some studies, you are likely to learn that although the tasks performed with your products are the same, the job titles differ greatly among customer organizations. A server administrator in one company may be called a network manager in another. Once you understand more about these users, you will be able to note the diversity of job titles for the same job.

You are also likely to learn that organizations divide jobs differently. In a small company, each individual employee wears many hats in relationship to your product, while in a larger company, several employees may divide the tasks among them. In one company, the basic administrative tasks on the inhouse telephone system were performed by minimally trained clerical staff. The more complex tasks were performed by network specialists. In many support organizations, responsibilities are dispersed among a group of specialists, beginning with those who handle ordinary, easily answered questions to two or three higher tiers who handle difficult and specialized problems. I found an organization in which a clerical employee with no technical education or background had been trained to perform the task of determining the location of network sites. In other organizations, this same task would be performed by experienced technicians.

Despite all the different roles that you will encounter, the tasks performed and the user agendas are likely to be very similar in reference to your product. However, the product will be used in numerous user scenarios, but the core practices are more often same. Information developers often argue that they cannot conduct user studies because their users are too diverse. However, users may have a variety of jobs and responsibilities, but, most of time, the use cases fall into a same set of similar or identical patterns based on the functionality provided by the product.

Review sample user-role analyses

In their initial Information Model, the information architects at Nortel Networks used the user analysis of job roles and descriptions conducted by instructional designers in the training organization as a starting point for identifying role-based information needs. Then, they developed a master list of job roles that took into account users related to the various products and services produced by the many Nortel business units. Although the job titles and division of responsibilities differed among customer businesses, the task-related roles remained the same.

Their list of users' role-related tasks for telecommunications network administration included

✔ planning the implementation of the product to meet requirements and specifications

✔ installing and removing hardware

✔ installing and removing software

✔ bringing the product online and verifying that it operates according to specifications

✔ configuring the hardware and software to specifications

✔ controlling access and administering the hardware and software

✔ managing the measurement of resource usage for the purpose of billing

✔ managing the detection, isolation, and correction of abnormal operation

✔ managing the protection of resources from unauthorized or detrimental access and use

With the addition of fundamental descriptive information about the technology and its functionality, the Nortel analysis of user activities described the base tasks performed by their user community. This analysis became the core of the new Information Model, shown in Figure 7-1, that was used to organize the information provided to support users in their task performance.

From this task-based role analysis, the information architects determined, for example, that needed conceptual, procedural, and reference information to help them determine their capacity and provisioning requirements for their systems.

In our analysis of the Red Cross staff responsible for blood collection, my team determined that trainers required information to help them construct instructor-led and e-learning courses to teach the technicians how to use the equipment properly, follow the safety protocols, interact effectively with donors, and respond to emergencies. The technicians required step-by-step procedures for ordinary and extraordinary tasks. They needed safety requirements and guidelines for interacting with donors, and they needed instructions for completing the required forms.

Category	Customer Support	Technology Fundamentals	About the Product	Plan and Engineer	Install Hardware	Install Software	Commission	Configure	Administer	Manage Performance	Manage Accounting	Manage Faults	Manage Security
Category Definition	Contains information that facilitates customer interaction with the company.	Contains information about telecommunications and computer fundamentals that are the foundation of the product technology.	Contains information about product-specific technology fundamentals that apply to the product.	Contains information about planning the implementation of the product to meet requirements and specifications.	Contains information about installing and removing hardware.	Contains information about installing and removing software.	Contains information about bringing the product online and verifying that it operates according to specification.	Contains information about setting up the hardware and software functionality.	Contains information about controlling access to and managing the hardware and software.	Contains information about managing resource usage.	Contains information about managing the measurement of resource usage for the purpose of billing.	Contains information about managing the detection, isolation, and correction of abnormal operation.	Contains information about managing the protection of resources from unauthorized or detrimental access and use.
What's New	Understanding what's new in customer support.		Understanding what's new in the product's release.	Understanding what's new in planning and engineering.	Understanding what's new in hardware installation.	Understanding what's new in software installation.	Understanding what's new in commissioning.	Understanding what's new in configuration.	Understanding what's new in administration.	Understanding what's new in performance management.	Understanding what's new in accounting management.	Understanding what's new in fault management.	Understanding what's new in security management.
Fundamentals		Understanding safety requirements. Understanding industry terminology. Understanding computer fundamentals. Understanding networking fundamentals. Understanding data communications fundamentals. Understanding telephony fundamentals.	Understanding the management system user interface. Understanding the product's basic capabilities and characteristics. Understanding product terminology. Understanding how to find and use product information. Understanding the product's architecture. Understanding how the project works.	Understanding planning and engineering fundamentals.	Understanding hardware installation fundamentals. Understanding safety requirements.	Understanding software installation fundamentals. Understanding the software installation user interface.	Understanding commissioning fundamentals.	Understanding configuration fundamentals. Understanding the configuration user interface.	Understanding administration fundamentals. Understanding the administration user interface. Understanding the file structure and database.	Understanding performance management fundamentals. Understanding the performance management user interface. Understanding performance data.	Understanding accounting management fundamentals. Understanding the accounting management user interface. Understanding accounting data.	Understanding fault management fundamentals. Understanding the fault management user interface. Understanding fault data.	Understanding security management fundamentals. Understanding the security management user interface. Understanding security data.

Figure 7-1: Sample Information Model

Study the information needs of each role

The catalog your team created based on internal and customer research is a foundation for an in-depth study of the information needs of each role. You may find it possible to conduct surveys and telephone interviews to learn how individuals in each role use information. Even better is direct observation of users in their working environment. Users can tell much about how they use or don't use information, but you will observe practices that they are unaware themselves.

In our study of railroad maintenance workers, my team found no one among the mechanics who used written information to perform their tasks, except for basic reference data that was easier to look up than remember. For example, an electrician would look up the voltages in a particular circuit to troubleshoot an electrical problem. The procedures for performing tasks were used by the trainers to develop their inhouse training programs. Trainers read the manufacturer's documentation and extracted information they thought was relevant for the mechanics.

In a study Comtech conducted of users of software supporting real estate transactions, we interviewed clerical workers who were taught how to use the software on the job and needed no additional procedural information to perform their tasks. The task-oriented product manual was used only by the office supervisor to understand the tasks and resolve problems encountered by the staff. The supervisors never used the software themselves, except for one individual in each office who was responsible for software installation and developing new reports.

In interviews with telecommunications and network engineers, we found careful reference to procedural documentation because of the liability involved in performing a task incorrectly. If they didn't follow the procedures, they might void their warranty. In the medical equipment field, users are also careful to follow procedures to avoid injury to themselves or patients. Among medical equipment maintenance personnel, there was considerable interest in reading detailed theory of operations topics to become more proficient in troubleshooting and repair. People in industries that are regulated by some government agency are often diligent in following written instructions to perform tasks.

Consumers who buy their own products and use them independently may use some information, such as installation instructions, if it is convenient and necessary for them to succeed. If no one is available to answer questions, consumers may follow instructions to install equipment, perform infrequent tasks, or fix a problem. The willingness to use written information among end-user consumers is highly idiosyncratic. Some people read everything first, some read manuals occasionally, and others never look at them at all.

Conduct a professional user and task analysis

The best way to gather data on your users' information needs is by directly observing them at the workplace or home in situations when they are doing their jobs and might be using information. Although you can gather some information indirectly through survey and telephone interviews, there really is not any substitute for direct observation. Unfortunately, many companies make direct observation difficult, if not impossible, requiring you to use less direct methods and find data wherever you can. Despite opposition, you may find that persistence pays off to eventually win agreement for direct customer studies.

Conducting a user site study takes careful planning both before you visit and during the observations themselves. For detailed information on how to develop a site visit plan and conduct the visit, consult *User and Task Analysis for Interface Design* by Ginny Redish and me.[2] You will find sample site visit plans and lots of advice for conducting the visit professionally so that you don't bias your results.

Direct observations during site visits provide you with insights into the users' information agendas that cannot be achieved in any other way. I watched users at a Federal Express office consult their policies and procedures manual to review the procedure they should follow to research an incomplete or incorrect address on a package. They were not aware that I was observing. I listened as hotel clerks tried to perform a procedure with their new reservation software before consulting their manager for help. Although they had an online help system available, they never considered using it, even after the manager admitted that he didn't know what to do. In another site visit, Comtech researchers learned that the clerks had never used the help system even though they were performing key procedures incorrectly in archiving critical government documents. Their colleagues and management complained that no one could ever find the archived documents, but they assumed the fault was the clerks' idiosyncratic archiving practices. No one realized that the clerks were entering the keywords incorrectly.

When asked about the software manuals, the clerks told us that they had never received manuals and had only limited product training years earlier. The manuals were stored in the bookcase of the IT system administrator, who had also never looked at them.

Sometimes users are quite clear in explaining the problems they have with information access. They show you the difficulty they have using a search system or navigating a website; they ask why standard sections are missing from some manuals but are present in others; they wonder aloud why reference material is incomplete or badly organized for their needs; they point out help screens with titles but no content. The information design and delivery problems become obvious to the observers, and the new Information Model begins to take form.

Best Practice—Understanding the users' information agendas

Armed with user profiles and information from your site observations, interviews, and surveys, you translate your findings into user profiles and scenarios. A profile is a description of the user's characteristics with respect to the tasks performed and information needed to support them. The scenario describes how the user interacts with information to perform tasks. Together the profiles and scenarios provide the foundation for your Information Model and its implementation in individual information-development projects.

A typical user study begins with an information problem. Feedback from users, usually through calls to telephone support, reports to sales representatives, and direct feedback through websites, indicates problems in using information resources. You learn that users have difficulty finding the information they need to answer questions or perform tasks.

---

[2] JoAnn T. Hackos and Janice C. Redish, *User and Task Analysis for Interface Design,* Hoboken, NJ: Wiley, 1998.

When they find information, it does not meet their needs because it is too simple or too difficult or organized in ways that do not match their mental models.

In the 1992 study of the policies and procedures (P&P) manuals at Federal Express, the information developers learned that ground operations personnel, including couriers, service agents, and managers had difficulty locating and using the information in their manuals. The old Information Model, in which information was organized according to functional areas, was no longer working.

As Comtech researchers designed the study to assess the problems with the existing P&P manuals and develop a new Information Model, we included both site visits and usability testing. The study team felt it was essential to develop a rich picture of the challenges the users faced in the workplace. The team developed a list of representative users who represented ground operations in North America and around the world, visiting and interviewing 49 individuals in seven groups:

- ✔ Station managers
- ✔ Customer managers
- ✔ Trainers
- ✔ Couriers
- ✔ Station agents
- ✔ Ramp personnel
- ✔ Freight personnel

From the site visits, we learned that all the employees were willing to spend considerable time and effort studying P&P manuals, especially when they were learning about their jobs. Even after the initial training, they referred to manuals to answer specific questions, particularly about complex and infrequently performed tasks. Despite this willingness, the employees felt inundated by paper, often resorting to calling personal contacts or using email to find information. The manuals were frustratingly difficult to use; indexes and tables of contents did not use words and headings that users readily understood. Once they located information, it often was not organized for ease of reading, making it difficult to isolate step-by-step procedures or important guidelines from a mass of poorly organized information.

Based on the field studies, the team concluded that the P&P manuals would be more usable if

- ✔ the information were better organized
- ✔ the information were organized according to job roles
- ✔ the manuals contained more examples and more complete examples
- ✔ the content were presented in consistently structured manner

✔ tables of contents and indexes were improved

✔ consistent patterns were maintained across manuals

The usability tests corroborated the results of the site visits. In the test, Comtech wanted to measure the time required to find the answers to questions in the P&P manuals. The test participants had difficult finding information in the manuals. They were much more successful using quick reference guides. We learned from the participants that they generally avoided using the large P&P manuals, preferring to look up information in the quick reference guides, services guides, and newsletters. They also tended to trust information that was updated more regularly. The P&P manuals were perceived to be out of date.

We concluded, as we had learned from the site visits, that employees found the P&P manuals difficult to use, although they could find answers to their questions given sufficient time and persistence. We also learned that employees used multiple resources to find answers, suggesting that information was extensively duplicated among the publications and, in some cases, contradictory.

Typical user scenario at Federal Express

Based on the study results, we were able to develop user scenarios to describe how the ground operations employees used information to support their day-to-day tasks. A user scenario of a station manager might read as follows:

> The station manager is responsible for providing employees with accurate and up-to-date information regarding personnel policies. For example, an employee notifies the manager that his grandmother has died, and he requests personnel leave to attend the funeral. The manager knows that the company provides for personnel leave for funerals but is unsure of the number of days the employee can take with pay. He decides to look at the personnel policy and procedure manual for managers. He quickly scans the index, looking for a personal leave policy for a death in the family. He looks under several keywords, finding nothing that suggests a policy for funerals. He also scans the table of contents and flips through a few pages without result.
>
> In frustration, he calls a station manager he knows personally in another city. The colleague does not know the policy for funerals and cannot find one in his manual. The manager calls a human resources contact and after waiting for many minutes learns that the information is indeed in his manual, under the heading "Bereavement Policy." The manager remarks that he didn't think of that word and wouldn't consider looking for the information under that heading. He decides that the manuals are not especially useful in supporting his information needs. He's spent several hours trying to track down information that was in his bookcase.

If you consider the profile of this manager, you will find it typical of a person who is busy, handles a number of widely different questions during a workday, and is not an expert at information access. Given that the information is in paper, rather than electronic, he cannot search it easily. Even if the information were online in book form, he and his colleagues might still be frustrated in finding a specific piece of reference information among multiple manuals and hundreds of pages. Because the information is not labeled in ways that match the users' mental model of the task, it is nearly invisible.

Best Practice—Using user scenarios to develop your Information Model

The Information Model your information architect develops reflects the understanding developed of your users and the specific scenarios derived from the user investigations. At this level in the model development, the model describes the overall approach you want to take. At the next level, you will use project-specific scenarios to implement your Information Model on a project basis.

The overall Information Model includes both how you intend to deliver content to your customers and how you plan to build the content to facilitate the deliverables. As you apply the Information Model to specific projects, your project teams decide which deliverables they will support and how they will develop the content needed.

The user study and the scenarios that emerged from the Federal Express study described how ground operations staff used information in their working environment. The study led to the design of a new Information Model that better met the user needs. The model was then applied to the design of specific sets of information to support the activities of couriers, station agents, and managers.

Rather than organizing the information according to functional areas, the information architecture team at Federal Express, with our help, decided to organize the information according to job roles. The titles of the manuals were rewritten to make them more obviously related to the jobs. Policy information for managers was separated from procedural information for staff. New design patterns were developed for each collection of information so that categories of information were easier to find. Similar topics were grouped together, and the groups of topics were arranged so that they followed a roughly chronological sequence. Some reference information was organized alphabetically for ease of retrieval.

In addition to the overall restructuring of the manuals being delivered to staff, the information architecture team developed a new design for the procedural information, which included a standard sequence of content elements for each procedure, a new format for the content elements, and editorial guidelines emphasizing readability. The team also began to develop a controlled vocabulary to make the English standard procedures more easily understandable by nonnative readers.

Although the initial changes were made to the print P&P manuals, the architecture team began plans for electronic delivery of information. In designing a documentation database, the team decided to de-emphasize the original book paradigm and manage units of information as topics in the database. The new design resulted in the elimination of duplicate information because the database of topics could be used in multiple outputs.

In presenting their plans for a new information architecture to senior management, the team learned that the study results, especially the calculations of time and money saved by ease of access and more understandable information, could be used to develop their business case. Explaining to senior management that you can "save $3 million dollars per year and increase user satisfaction given the new architecture" is more convincing than arguing for more graphics or some sort of controlled language.

Understand the context of information use

Susan Harkus, as an independent information architect, points out in the following vignette that "the profiling of an audience only provides a framework for the real analysis. For each task, their context—their objectives, their performance environment and their agenda of priorities, assumptions and expectations—accounts for their behaviors and outcomes." Unless you understand what the users do with information to meet their personal and job-related goals, you cannot develop an Information Model that adequately supports them. To be most effective, your Information Model must be derived from knowledge of the users' agendas and how information might support those agendas, delivered in the right way and at the right time.

Ignore Context and Risk Your Project[3]

When a product fails, we know that somewhere, someone "got it wrong," sometimes several someones and sometimes at several points in the development project—but WHAT we got wrong isn't always obvious.

In 2000, I joined a team that was developing a strategic web-based system. The system's marketing advantage was to provide salary benefits to employees without incurring unacceptable administrative overheads.

Employees would self-manage their benefits package and carry most of the administrative load. However, the pivotal user of the system was the administrator.

Employees would submit a request for a pre-tax payment, for example, to purchase a notebook computer. The administrator would review and grant or reject the request. At each pay run, appropriate amounts would be deducted automatically from the employee's pre-tax salary.

The process was straightforward but the administrator's context was not. Legitimate benefits would be tax-free. Illegitimate benefits and benefits that exceeded specified thresholds would incur taxes at the highest tax rate.

Administrators would want to deny claims that incurred tax penalties for the employee or the company. For example, purchasing a notebook with pre-tax salary was only legitimate if the notebook was purchased for business use and if only one was purchased within the financial year.

Administrators would also be under constant pressure from employees to grant requests. They needed to make *informed* decisions.

When I was writing Help for the grant benefit request form, I realized that the administrator needed to access the benefit history of the employee to decide whether or not to grant the request.

The benefit history was not accessible from the form, but I'd seen the history use case in the project documentation. I suggested that a history query link be added to the form.

A history lookup was not possible for Phase 1. Why? Because when time and cost pressures had caused the project to be downsized, project management had deferred the history use case to Phase 2—in the words of the System Architect, "The use case is TECHNICALLY INDEPENDENT, and deferring implementation won't affect the Phase 1 deliverable."

continued

---

[3] Susan Harkus, "Ignore Context and Risk Your Project," *Information Management,* the e-newsletter of The Center for Information-Development Management, April 2004.

Ignore Context and Risk Your Project *continued*

Put the mathematics of the decision in a wider context. Abandoning one little use case saved several hours of programming, integrating, and testing, but without a history lookup, the administrator's task, making an *informed* decision about a benefit request, was almost impossible.

Task is an interesting word. In this project, as in many other situations, technical and user interface designers took *task* to mean what people do. Once analysis had identified and documented what users would do and what dependencies existed between what users would do, all decisions were driven by the definition of the set of do's (use cases).

The French existentialist, Jean-Paul Sartre, also focused on doing. He claimed that people are what they do, not what they think they are or what they want to be. Philosophically, Sartre may have a point but what people do goes beyond visible actions, as Malcolm Gladwell constantly argues in his book, *The Tipping Point*. (Malcolm Gladwell, 2000. *The Tipping Point: How Little Things Can Make a Big Difference*. London, Abacus.)

For Gladwell, context is the powerful determinant of what people do, and he emphasizes "the power of context" in the social events and research he describes.

For example, he recounts the behavioral study by Princeton University psychologists, Darley and Batson. Seminarians were invited to give a talk on campus. Some were asked to speak about religious vocation and some about the parable of the Good Samaritan. On the way to their talk, they were confronted by a man in agony and distress.

As each seminarian left to go to their talk, some were told they were late and some that they had plenty of time. Of the group who thought they were late, only 10 percent stopped to help the man in distress. Of the group who were told they had plenty of time, 63 percent stopped to help.

So many elements were at play in the engagement context: motivation for joining the priesthood; the compelling parable of the Good Samaritan who put himself in danger to help someone in distress; and the emotive sight of another human being in agony; yet the overriding determinant was whether or not seminarians thought they were late.

Two contexts were at play: the got-to-hurry-I'm-late context and the plenty-of-time-I'll-arrive-early context. The contexts were the primary determinants of what happened—not demographics, not vocational motivation, not principles.

How does the power of context relate to a project that failed to deliver a successful, user-driven solution for managing salary benefits?

In two ways.

First, in their traditional requirements gathering, the team explored their audience and their audience's goals, in much the same way as is currently promoted in goal-oriented, persona-based design methodologies.

The mistake they made was to analyze and document their users apart from their tasks. What Malcolm Gladwell emphasizes again and again is that actors, their context, and their acts cannot be separated.

Second, because the project followed a rigorous, use-case driven methodology, the non-existing, real-world association with context meant that when implementation plans were revised, technical management had no understanding of the importance of the history lookup.

The profiling of an audience only provides a framework for the real analysis. For each task, their context—their objectives, their performance environment and their agenda of priorities, assumptions and expectations—accounts for their behaviors and outcomes.

When we reviewed the lessons learned from Phase 1, I suggested that, in Phase 2, contexts be fully defined for all real-world user tasks and that each use case be linked to its real-world contexts. The real-world contexts could then be used as the determining reference for project decisions.

Develop user scenarios to create a new information architecture

User scenarios are narrative depictions of the users' agendas in interacting with products and information. They provide the most concrete foundation on which you can develop a sound information architecture for your organization. The following examples demonstrate just a few of the situations of use you might encounter as you study your users.

Insurance company office workers

The Comtech team learned that the office workers rarely, if ever, consulted the long, detailed, and out-of-date policy and procedures manuals. Many of the everyday tasks they performed were learned on the job from supervisors, trainers, and colleagues. When they encountered new problems or had to perform an unusual task, they first asked others in the office for advice. If no one knew how to perform the task, they would use the clues provided in the user interface of their computer software. In most cases, they could figure out what to do, especially for those applications for which the user interface had been newly redesigned and improved. For the older and more difficult-to-use applications, the office workers preferred to use their quick reference booklets because they contained just enough reference information to help them complete the poorly labeled fields in the application. The policy and procedures manuals were generally used as doorstops or plant stands.

As a result of the user study, the information architects decided to eliminate the large manuals and focus on embedding more relevant information into the user interface and developing more quick reference booklets.

Healthcare and administrative workers at blood-collection sites

Based on site visits and focus groups, the information architecture team and I decided to develop four distinct information types to replace long procedures manuals that were rarely updated. The four tiers of information developed began with policy statements, followed by process descriptions, step-by-step procedures, and supporting reference information. Each information type was defined with a strict sequence of content units and thorough writing guidelines. By creating individual topics rather than long chapters of longer books, the information developers could update the content as needed, rather than waiting, sometimes for years, before entire manuals could be updated.

As a result of the topic-based, structured architecture, small topics of a few pages each were delivered electronically and in print guides that were consistent, well labeled, and more easily accessible.

Telephone support personnel

Following site visits with staff members answering customer questions, the information architecture team changed the focus of its information from end-user customers to the support staff. Although minimal guides continued to be available for end users, most of the critical information needed to use the application software effectively was targeted for the support personnel. The company's policy of customer care encouraged end users to call for answers to their questions rather than look them up in long, complex manuals.

Software programmers using debugging software

Based on the findings of an industry benchmark study of information use by software programmers, the information architecture team concluded that procedural manuals were not as effective as annotated code examples. The programmers typically scoured the print and electronic information for examples that appeared to produce the outcomes they were trying to achieve. If they found an example that looked promising, they would carefully review the code examples and try to understand how the designer of the example had structured the code. If the information developers provided annotations that explained how the code examples worked line by line, the users found the information more helpful than procedures that did not lead them to the results they wanted.

One information architecture team, responding to the same user agenda, created an online tool that allowed users to execute the code examples, directly observe the processing steps, read the annotations, and copy the code for their own applications.

Consumer users of inkjet printers

A study of printer use in people's homes led the information architects to conclude that consumers wanted information related to their goals. For example, many consumers purchased printers for home use so that they could create picture albums of weddings, vacation trips, or other special events. Information organized to show them how to create and print attractive albums directly met their information needs, rather than general step-by-step procedures covering generic tasks.

User analysis leads to creative solutions to users' information agendas. When customers find information that matches their goals and helps them achieve those goals as easily and quickly as possible, they use information directly rather than rely on other methods. Many customers, especially those just beginning to use a product or whose use of a product is restricted to a small number of tasks, ask for information that is primarily task-based. They are action-oriented, hoping to accomplish what they want easily.

More dedicated users, especially those who make a career out of using a particular product, may want background information that supports their successful performance of tasks. They want to troubleshoot problems on their own, make better decisions, configure and manage products more efficiently, and generally increase their level of knowledge and expertise. For such users, you may discover aspects of an information architecture that result in more effective conceptual and reference information.

Develop patterns to define a set of deliverables

Based on your analysis of the user information agendas, your information architect and other team members must define the set of deliverables most appropriate for each agenda and related job role. The set of deliverables may include manuals, help systems, topic-oriented websites, personal digital assistants, and other media. The entire set of deliverables may not apply to every user role but provides a standard set that you use in planning the deliverables for a specific product or services.

For example, your information architect may conclude that a software-development audience is best served by online access to command reference information. The command reference information may be organized by product, function, or other metadata that helps the developers assemble the command information that is most relevant to their tasks. The architect may further decide that end users of a hand-held device should receive setup information in a printed pamphlet and task information embedded as help in the device. The hardware installers might need a manual that they can print from a PDF on the website to take into the field. If they carry laptops, the same installation manuals might be viewed on screen. As these examples illustrate, a standard set of deliverables supplies a portfolio of choices for your project teams.

Each information collection that the architecture team defines must be further defined by a design pattern. A design pattern formally defines a solution to a specific user problem. For example, you might have a design pattern that defines how an installation how-to instruction should be organized. The illustration in Figure 7-2 shows a typical structure for an installation design pattern.

Because the people who install hardware for a living are focused primarily on performing tasks, the design pattern also focuses on tasks both following the regular installation steps and troubleshooting problems with the installation. The conceptual information to support the installation is provided through introductory conceptual information that explains how the installation process works and notes decisions to be made prior to beginning the tasks. Within each procedure, additional but brief conceptual information helps the user orient the specific task to the installation process as a whole and reiterates configuration decisions needed at this point in the process.

Each installation manual your team produces should follow exactly the same structure, down to the names and characteristic content of each introductory, procedures, or troubleshooting sections, so that users benefit from a reliable and consistent presentation of the subject matter. As your set of standard design patterns is completed, your authors will know exactly how they should structure the information they are designing. In using design patterns for each deliverable, your authors provide users with a framework for organizing content that makes information access easier and improves their overall experience with your content.

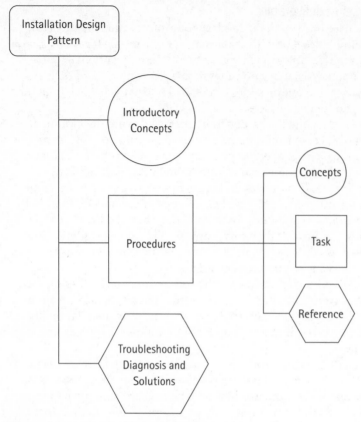

Figure 7-2: Installation manual design pattern

Extend the design patterns to define your standard set of information types

Technical and procedural content can be usefully organized into a series of distinct information types. Each information type defines the content of a standalone topic that represents a complete piece of information for the users. A typical set of base information types for technical content includes

- ✔ Concepts
- ✔ Tasks
- ✔ Reference information

A typical set of base information types for procedural content includes

- ✔ Policies
- ✔ Processes

✔ Tasks

✔ Reference information and job aids

A design pattern for each information type defines the standards for content units and the hierarchy and sequence in which they must be presented. For example, you may define a policy as shown in Figure 7-3.

Policy Title and Number

Policy Date

Policy Approved by

What this policy is about
This section explains the reason for the policy and states the subject area to which the policy applies. Do not detail organizational values in this section.

To whom this policy applies
This statement clarifies who is affected by this policy.

Underlying principles
This section refers to the organizational values that underlie the policy. It should not be a restatement of the "what this policy is about" section.

The policy
This section clearly and succinctly states what the policy is. It establishes the rule or guideline to be followed in conducting activities that are the subject of the policy.

Explanation
This section explains the context and reasons for establishing this policy. This section is optional.

Associated documents
This section lists documents that are integral to making the decision about this policy, for example, a regulation, an issue paper, a statement in an official document, or another corporate policy statement. This section is optional.

Figure 7-3: An example of a standard policy template

Each section is given a standard label and the content to be included is defined. A policy standard like this one would likely be accompanied by instructions for its implementation.

Policy writing guidelines
The guidelines for writing policies are as follows:

✔ Policy statements are simple statements of the rules that the organization follows to conduct its day-to-day business activities. Policy statements, because they represent the basic values and beliefs of the organization, rarely need to change— Once a policy statement exists, it should apply for many years.

✔ Policy statements are statements of fact. As such, they are written in direct language without qualifications—For example, a policy might state: It is the policy of the organization to hire people without regard to gender, race, religion, or national origin. That's a short, sweet, succinct statement of what is.

✔ Policy statements exist to tell why something is done. As such, policy statements contain no procedural information—For example, the hiring policy does not state that the origination accepts applications only from people who have completed the hiring application form. Completing a form is a procedure, not a policy. Policy statements contain no job aids or requirements because those documents represent what to do and how to do it.

✔ Policy statements, written effectively, help staff members understand the reasons behind the procedures they are asked to follow diligently.

✔ Policy statements are written in the universal present tense, as exemplified by this statement. Policy statements never use "will" or "shall," because these words imply a future rather than a present state. If you are planning to do something in the future, it is certainly not your policy now.

✔ Policy statements sometimes contain the verb "must," but many statements containing this verb are not policy statements—The statement, "directors must sign all ABC forms," is not a policy statement but might be a step in a process directive or a work instruction.

✔ Policy statements are written in plain language—Several federal government leaders in past administrations have endorsed a plain language policy. The organization does the same. The goal of such a policy is to ensure that every staff member and volunteer can read and understand what the policy intends for them and their day-to-day business activities.

✔ Policy statements are positive—We try to make policy statements positive rather than negative. Negative statements are always more difficult to comprehend than positive statements. We avoid statements such as "policy statements are not negative," because the double negative in this statement makes it almost incomprehensible.

✔ Policy statements should be complete enough to be understood on their own but not so detailed as to require frequent revision.

Guidelines of this type are developed for each information type and should be accompanied by examples to further clarify the intent of the guidelines.

A concept information type for an XML-based Information Model using the Darwin Information Typing Architecture (DITA) that is an OASIS standard might be defined in your Information Model, as shown in Figure 7-4.

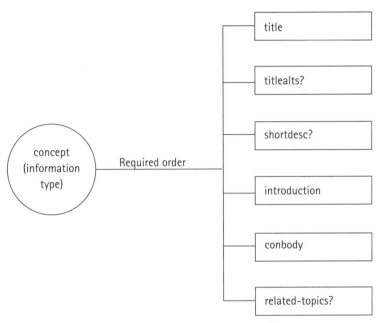

Figure 7-4: A complete description of the concept information type

Develop user scenarios to define content for specific projects

After you have implemented your overall Information Model, you must develop specific user scenarios to plan the content for each project. At the project level, user scenarios describe how people are most likely to use products. If you are planning a hardware product implementation, for example, you want to develop scenarios around installation, configuration, and basic and advanced use of the product functions. If you are supporting a software product, you consider how your users are likely to interact with the software functions. If you are planning a series of procedures to support operations, you must consider how the users perceive the tasks that they must perform.

Consider the user scenarios developed to plan the information needed by new users of a database marketing software system. The intent of the software was to help small business owners conduct more effective marketing campaigns. The user analysis targeted business owners who were usually computer novices and distrusted analyzing their customers with software. Their agenda was to increase revenues by getting paying customers into their stores. The Comtech staff had to convince them that the software could help.

Here is one user scenario that emerged from the analysis:

The manager of a small retail store believed her customers came primarily from the older part of town. They wanted to replace stained glass or add to existing glass displays. She ran her ads in a small local newsletter. After getting her customer information into the software database, she analyzed the customers according to zip codes. The analysis revealed that her best customers and the highest percentage of her revenue came from the new subdivisions. Customers wanted stained glass for their new entryways, dining areas, and other areas that needed an improved design.

On the basis of scenarios like this one, the information developers created a list of tasks that the business owner needed to perform to be successful. They also decided that the owners needed brief basic conceptual information to understand the benefits of database marketing. Based on the user analysis, the developers were convinced that the managers were not willing to read much about the product. They wanted to get immediate results and improve their business results.

Project-specific user profiles and scenarios lead to decisions about the content to produce and the manner in which it will be delivered to a variety of users.

Best Practice—Communicating user profiles and scenarios to team members

Communicating the products of your research and analysis to your team members and stakeholders is essential to garner support for the Information Model to come. Your plan for the new model may include developing a standard set of information types that information developers will follow to create consistent topics. The model may result in new ways of delivering content to customers that will require information developers to work in collaborative teams on information design and development. Your new model may require new skills in web or quick reference design in the team. You need their support to implement the Information Model. To gain their support, you need their understanding and appreciation of the user profiles and scenarios.

After NCR's check processing division conducted their study of the users of their check-processing equipment, they created large posters that were hung in the employee cafeteria so that everyone in the organization had an opportunity to learn from their user research. The information architecture team at Navision created a user profile website that included photographs of their "typical" users and detailed accounts of their jobs, responsibilities, product use, and need for supporting information. The website was made available to everyone in the company.

Intuit developed a set of posters with details about representative product users that provided a foundation for user scenarios that affected product, interface, and information design. Posters help to focus attention on user profiles and keep the profiles visible throughout the organization. They presented this information at an industry conference.

In organizations like Intuit, responsibility for developing user profiles and scenarios lies primarily with the usability and interface design teams. Unfortunately, these teams do not always include an analysis of information needs in their research. You may find that responsibility becomes yours as an information architect.

As you develop your understanding of the information users, give presentations to your team members so that they share your knowledge. If team members have contributed to the information gathering, include them in the presentations, especially if you have multiple product teams that need to take the user information into account in their information-development life cycle. In large organizations, you may want a person from each product group on your information-architecture team to ensure representation, learning, and communication of customer profiles and scenarios back to the team members.

Plan your communications carefully and follow through. Your team members may range from enthusiastic and receptive to mildly disinterested in user information. Especially among those information developers who are reluctant to change the way they develop information, you could have everything from a negative and skeptical response (I don't think customers really want that) to a pronouncement that the information about customers is irrelevant to their work.

Communications with stakeholders outside your immediate team are equally important. If product managers, marketing managers, product developers, and others are accustomed to dictating the type of information produced and delivered with products, the Information Model will be disruptive. If they understand that your model is based on data about the users, not upon guesswork, you will be more successful in gaining their support or at least their reluctant acquiescence. Many product developers believe that users are like them in their information needs. When you hear people state that users don't need information to support their use of the product, you're often hearing their own personal perspective on information. Bringing data to bear on your decisions about the Information Model will improve your chances of getting it accepted.

Summary

Developing user profiles and scenarios is the essential first step to building a user-centered Information Model. Without an understanding of your user community, their personal agendas, and the way they interact with information resources, you are more likely to develop a model that satisfies authors and product developers but that customers find inadequate. Primed with a thorough understanding of how your users are to be supported with information, you can create a model that fosters customer success and satisfaction.

To develop the information, you need to do the following:

- ✔ First determine the most likely jobs and roles that you believe users will perform.

- ✔ Confirm your preconceptions with real data by surveying, interviewing, and visiting users in their real-world locations.

- ✔ Based on the results of your user investigations, develop user profiles that accurately describe your users as real people with real personalities.

- ✔ Armed with sound user profiles, create overall scenarios that support your user's agendas with tasks, concepts, and reference information.

✔ Develop delivery solutions that will help users access the information they want quickly and easily.

✔ Communicate your Information Model to information developers and other authors and stakeholders so that they understand how the model was developed and why it will promote customer satisfaction and operational efficiency.

The best Information Models are products of sound user studies. With users in the forefront, you will design information that enhances customer satisfaction while keeping operational costs under control.

"If we keep measuring our information
quality, we will continuously improve."

Chapter **8**

Optimizing Your Organization's Efficiency and Effectiveness

> The New Six Sigma integrates best-practice processes with tools designed to help leaders in driving their business strategy for dramatic short-term business results while building sustained future capability.
>
> Matt Barney and Tom McCarty, *The New Six Sigma*

Senior management, concerned about the expense of supporting first-class information development activities, asks information-development managers to measure the productivity of their organizations and make a business case for the value of their information products to the customers. In response, you may be tempted to throw together a list of things you can count: number of pages produced per year, number of pages produced per writer, number of customer complaints about documentation and number of complaints resolved, number of typographical errors found in documentation, and so on. The result is an awkward set of countable items, many of which have nothing to do with productivity and provide little evidence that customers find value in the information they access.

Developing a comprehensive program to optimize the efficiency and effectiveness of your organization requires a much more thoughtful approach, one that ensures that you are measuring only what you need to measure to ensure that you are meeting your goals. By following a process of goal setting, instituting innovative practices to support your processes and products, and analyzing the successes or failures, you can create a useful improvement program and report measurable, even quantifiable outcomes to your management. The process you learn about in this chapter is related to the successful process called Six Sigma, originally introduced at Motorola and now part of most managers' tool kit. Six Sigma is based upon a program in which you make small, incremental changes in pursuit of a performance and productivity goal. As each small change is implemented, you measure the success of the change and decide if you are going in the right direction. If the results are negative or not as robust as you had expected, you can use the incremental change process to adjust the implementation or pursue another course.

Information-development managers often begin to take process improvement seriously when they are questioned about the efficiency and effectiveness of their operations by senior management. I usually receive calls from managers who are surprised to be asked to measure performance. They are often at a loss how to proceed and usually fall back on some aspect of operations that is easy to count. The danger of course is that you get what you measure. If you decide that you can easily count typographical or stylistic errors in documentation, you are likely to get documents that are error free but provide little information of value to the users.

The questions that most managers should ask themselves are quite different from finding something that is easy to count. And, managers should never wait to be forced to begin a program to optimize efficiency and effectiveness. As a business-savvy manager, optimizing performance should be one of your most important objectives. Consider the questions about performance that you might ask:

- Is the staff as productive as it should be?

- Are we more productive this year than last?

- How long does it take to produce our information products?

- How much does it cost us to produce each deliverable or library of information products?

- If we add new tools and technology, will we reduce costs and increase efficiency?

- If we change the way we architect our information, will customers be more or less satisfied?

- How do we measure the effectiveness of the information we produce?

- Are we providing customers with the value they expect from our information?

- Are we decreasing or increasing costs in other parts of the organization?

- Is the cost of our operation commensurate with the value produced for customers?

By pursuing answers to questions like these, you are on your way to running your operations in a manner commensurate with the demands typically put onto operations departments in most organizations.

Best Practices in Optimizing Efficiency and Effectiveness

Running an efficient organization is particularly satisfying to a manager who wants to be viewed as aligned with the goals of the larger organization. You will likely discover that your cost-reduction projects get the attention of senior management, at least until the next round of cost reductions begins. More importantly, however, is that developing and instituting efficiencies should give you time to be more effective in meeting the needs of your customers. If you can eliminate redundant or low-value activities, you will find more time for high-value activities that win customer support.

In this chapter, you learn about six best practices for optimizing the efficiency and effectiveness of your enterprise:

✔ Defining your goals for efficient and effective performance

✔ Analyzing your processes with respect to your goals

✔ Investigating industry best practices (IPMM)

✔ Developing methods for measuring efficiency and effectiveness

✔ Improving processes and measuring results

✔ Pursuing techniques for measuring effectiveness

The best practices here provide you a step-by-step process for making and measuring improvements in your processes.

Best Practice–Defining your goals for efficient and effective performance

The most critical aspect of developing your process optimization program is setting goals. Your goals should address the issues of greatest concern to your management. Those concerns are likely to be different, depending upon the nature of the business you are in and the position of your business with respect to the competition.

When one individual was a director of publications, her company was the industry leader with more than 80% of the market share. Given that leadership position, senior management was concerned with maintenance rather than advancement. Operations managers were informed that they should work to produce acceptable but not industry-leading information products. Clearly, the company was more interested in keeping costs in line than bolstering quality.

At the same time, the leading competitor in the industry was investing heavily in quality improvements to its information development. They were actively engaged in process improvements, adding new tools and technologies to their activities, and encouraging staff to pursue new approaches to information design. The publications manager actively pursued information about the position of products in the technology adoption life cycle so as to optimize the resources for the fastest-growing products.

Other organizations I have worked with found themselves forced to abandon industry leadership positions as their products and the information moved to commodity status. As commodities, their products, even the newest ones, were achieving minimal profitability. As a result, the major corporate goal was to drastically reduce operational costs, including moving much of the work to offshore locations in low-cost economies. The goals imposed upon the information-development manager were to make the low-cost operations as productive and efficient as possible. They actively pursued technology improvements that would reduce costs and development time.

At the other extreme, I see high-flying organizations in which enhancing customer value through superior usability and information is the focus. In such organizations, the information-development managers are growing their departments, pursuing innovative ideas, investing in the latest technologies, and ensuring that staff members obtain the latest training and exposure to the most progressive sister organizations.

Clearly, your goals for optimizing the efficiency and effectiveness of your operations must align with the position of your company and the products you support in the market. If you have not yet considered the position of each of your products in the Technology Adoption Life Cycle, be certain to review Chapter 5: Understanding the Technology Adoption Life Cycle.

Once you understand the overall corporate goals and the goals associated with each product, you are prepared to set your own organizational goals. One way of identifying goals is through an examination of your organization's strengths, weaknesses, opportunities, and threats, popularly known as a SWOT analysis. Note that SWOT stands for Strengths, Weaknesses, Opportunities, and Threats. In conducting a SWOT analysis, assemble either your entire staff or the key leaders in your organization. If you have contributors at dispersed locations, consider a joint meeting, face to face if possible, to conduct the SWOT analysis and use the results to focus on your performance goals. Table 8-1 shows the results of a SWOT analysis by one team anxious to improve the quality of its information deliverables while keeping costs in line. This particular team was less anxious to reduce costs than some others, although they saw opportunities in streamlining their processes and reducing the redundancy of the information they produced.

Table 8-1: Result of an Information-Development SWOT Analysis

STRENGTHS	WEAKNESSES
Skilled, professional staff members	Lack of automated process steps
We really know the subject area.	People are not used to collaborating.
We have pretty good contact with the customers.	Difficulty overcoming resistance to change
We've already started to analyze our information.	Used to working in product silos with little collaboration
We have experience in reusing content among deliverables.	Very time consuming final production process
We've already conducted some new projects that have been successful.	Duplication of content

OPPORTUNITIES

Increase our focus on the customers

Recognized as experts in the subject area and the customer needs

- Ability to move into continuous publishing with regular updating
- Chance to reduce development and translation costs
- Better process to produce online help at less cost
- Reduce in support costs due to documentation shortcomings
- New learning among team members; new tools; new skills

THREATS

- They'll reorganize us before we accomplish anything real.
- We need to make our deadlines.
- We don't have the resources to devote to learning a new way of working.
- We need quick results and this will take too long.
- We're already much too busy; how will we fit in anything new?
- We keep trying new things that don't work out anyway.
- We're much too cynical for this to succeed.

As a result of your SWOT analysis, you may develop goals like the ones on this organization's list:

✔ Reduce the duplication of content in the documentation of different versions of the same equipment. Write once and reuse in multiple deliverables.

✔ Reduce the overall volume of content delivered to both internal and external users to that necessary to support their job performance.

✔ Reduce the cost of maintaining the information among multiple deliverables and across multiple versions of the products.

✔ Reduce the time required to prepare multiple deliverables for publication.

✔ Increase the number of languages being supported while reducing the cost of translating the content into each language.

✔ Increase the consistency of structure in the content so that users can more easily find and understand the information.

✔ Reduce the number of calls to customer support by increasing the online accessibility of information.

✔ Increase customer satisfaction with the technical information they receive.

✔ Reduce the duplication of effort between information- and training-development.

Note that while each of these goals might lead to process improvements, none of the goals requires the implementation of new tools and technology. In that, the goals differ from organizations that frame their goals in terms of obtaining a content management system or authoring in an XML editor. Those activities provide the means for reaching the goals, but they are not the goals themselves.

Best Practice—Analyzing your processes with respect to your goals

With your goals identified, your next step is to analyze the processes that you use to reach your goals today and to develop any processes that do not yet exist. For example, if one of your goals is to reduce the content that is duplicated among deliverables, you need to understand why that content is being duplicated in the first place. You may find that each deliverable for a suite of products is assigned to a different information developer who works independently. As a result, concepts that support all the products are written about multiple times, often with key information presented differently. Tasks that are performed for all the products are written separately even though most of them are identical or have minor differences in procedures. Reference information such as installations parameters, equipment lists, and specifications tables are produced uniquely for every version of the product even though the information is exactly or nearly the same.

Obviously, to achieve the goal of reducing duplication of work and content, you will have to change the way assignments are made and information developers work. In this case, a project manager or team lead and the information developers involved need to develop a new process in which they plan the content collaboratively, parcel out work assignments to the appropriate developer, account for the required differences in similar content using conditional text or its equivalent, and create a single source of topics that can be assembled into the appropriate deliverables required for each product version.

Consider another set of goals—that of increasing the number of languages into which the information is translated at the same time decreasing the cost of translating in each language. A number of processes may be affected in trying to reach these goals. Clearly, the localization process needs to be streamlined to remove costs from the system. That streamlining might include a change in authoring as well as a change in the process of moving content to the localization service provider:

- ✔ If only the topics that have changed from the previous release of the product are sent to the localization service provider, the cost of reviewing unchanged topics will completely disappear.

- ✔ If it were possible to mark only those sections of an existing topic that had changed from the previous release, the number of words to be translated might be further reduced.

- ✔ If the topics were sent to the service provider when groups of them were reviewed and approved rather than waiting for an entire deliverable to be complete, the translations could be completed in a more timely manner.

- ✔ If the organization instituted more automated publishing processes, it might be possible to produce final deliverables in multiple languages without expensive desktop publishing at the last minute.

Not only would these steps require changes to existing processes, they might also require new tools and technology to support automating the final production work.

Once your goals are established, you can examine existing processes and identify potential areas for improvement. However, you will find it tempting to try to change everything at once. I recommend that you approach the changes incrementally, starting small and tracking the change to ensure that it is delivering the expected improvement. For example, you may decide that you can fairly quickly automate your process of assembling content into multiple deliverables before you can change the process by moving to topic-based authoring. By instituting an incremental change, you can measure the outcome and decide if that change is helping you to reach your overall goal.

You might be able to reduce the duplication of effort by sharing easily reused content among deliverables before you tackle the problem of creating master topics that include all the differences among the products. In any optimization effort, you will be considered more successful if you can deliver improvements quickly rather than take years to put every change into place. Quick wins are a way to demonstrate to your senior management that your process improvements are bearing fruit. Quick wins enable you to garner support and funding for additional projects that may require more time and cost more to implement. One team, for example, decided to introduce structured authoring using their existing desktop publishing system before moving to a new XML-based authoring environment. As a result, they could concentrate on redesigning and improving the content and pursuing a minimalist agenda before asking everyone to learn new tools and a new way of creating content.

Your objective should be to identify incremental changes in process that you can easily measure to discover if the change is producing the effect you want.

Best Practice—Investigating industry best practices (IPMM)

It is possible for you and your team members to pursue innovations without ever learning from any other organization, but you will find it more efficient and productive to learn about the best practices adopted by other industry professionals, especially if those best practices have yielded measurable improvements.

I recommend that every information-development manager engage in benchmarking by comparing processes with successful managers. The benchmark projects I have managed have been overwhelmingly successful. In 1999, for example, Comtech Services developed a benchmark project among information developers in the telecommunications industry. Thirty-five independent departments in ten organizations took part in the study. Each department manager and staff members responded to detailed questionnaires and took part in on-site interviews. During the interviews, they identified practices in their organizations that had proven to be successful. They also revealed areas in which they knew they should improve, especially in terms of customer knowledge. The resulting benchmark report detailed the state of each department and described a large number of documented best practices. Each department received a detailed set of custom recommendations and a listing of the best practices that had been derived from their work. At the benchmark conference attended by more than 100 participants, department managers and team leaders reported on their best practice and exchanged ideas for process improvements. It was most interesting to watch the exchange among departments from companies that were direct competitors. After some initial trepidation, the managers discovered that they could all contribute to the value delivered to what were often mutual customers.

Other benchmark studies are not so wide-ranging and often include not direct competitors but organizations with some characteristics in common. One benchmark study focused on companies that are accountable for maintaining a high level of service to their corporate customers through defined service level agreements. Another benchmark looked at organizations that were pursuing incremental process improvements and were concerned with measuring the results. A recent benchmark study included organizations trying to establish information-development teams in China.

Managers and staff members who attend national or regional conferences, take part in industry training opportunities, and maintain contact with like-minded managers through listservs and user groups can learn about innovations pursued by others in the industry. Learning about the practices of organizations that have won awards in their professional areas provides an informal benchmarking opportunity. Comtech Services ran a benchmark study that looked at the practices of companies that had won national awards for the quality of their customer service, a study that resulted in the customer partnering best practice described in Chapter 6: Developing Relationships with Customers and Stakeholders. The only caution I offer about informal benchmarking is to recognize that you should not blithely accept the word of a manager and organization that you know nothing about. Managers ask questions on chats and listservs that attract responses from people who are otherwise unknown to them. The responses might be interesting and thought-provoking, but in such an informal setting, they are difficult to validate. In a formal benchmark, you can carefully select the participants because they represent organizations that have been identified independently as well run or successful in reducing costs and increasing value.

The Information Process Maturity assessment described in Chapter 2: The Information Process Maturity Model is itself a benchmarking opportunity. Because the IPMM was developed through an assessment of best-in-class information-development organizations and continues to be updated based upon a wide knowledge of well-run departments and effective managers, it provides a baseline from which to judge your own efforts. For each of the nine key characteristics in the original model, the IPMM establishes a set of best practices that represent the work done by increasingly sophisticated and effective organizations. During an IPMM assessment, you can compare your own practices to those of organizations that have achieved a Level 3, 4, or 5 in the IPMM.

In pursuing industry best practices, follow a few simple steps:

✔ Determine what characteristics of benchmark partners are most important to your comparisons. The characteristics might include industry (competitors), organizational structure and size, use of tools and technologies, depth of customer relationships, award-winning publications, and others that are similar to the issues you have been chosen to examine.

✔ Identify a group of potential benchmark partners that you understand to have the characteristics you want as points of comparison. Use your networking or work with a benchmark consultant that knows the industry players well to find the best possibilities.

✔ Develop a benchmark plan, including the goals you would like to achieve, and share the plan with the managers of the partner organizations. Solicit their interest in pursuing the benchmark study. Recognize that it is always more difficult to develop a benchmark study with your competition.

✔ Develop a questionnaire to collect basic information from each partner.

✔ Develop a set of questions to address in visits to each partner's location or through telephone interviews. Personal visits always yield the richest information about the partner's best practices and everyday processes.

✔ Gather your notes and create a report of the study results that you will share with all the partners. In the report, you may have permission to reveal the identities of the partners. In some cases, you will have to disguise the identities, especially when the partners are in the same industry.

✔ Send the report to the partners for review. Make factual corrections to the report after it has been reviewed by all the partners.

✔ Arrange for a conference call among the partners after they have all had a chance to review the report. Consider continuing the relationship over a longer period of time with periodic conference calls.

✔ If possible, arrange for a meeting of the participants in which everyone shares ideas and discusses setting up ongoing activities.

A benchmark study of this sort does take time and energy to set up and conduct. However, the results are worth the effort. The knowledge you can gain about successful practices in other organizations provides the basis for your own process improvement activities.

Best Practice—Developing methods for measuring efficiency and effectiveness

During many benchmark studies, I find it frustrating to learn that there is no quantifiable data to measure the effectiveness of an interesting process improvement. The manager and team members have moved to topic-based authoring, introduced XML editing, added a content-management system, developed a new minimalist approach to their information, created specialists in quality assurance and information architecture, all without one shred of factual evidence that the innovation is successful. In most cases, they have informal, anecdotal accounts of improved productivity, more efficient processes, or better quality to the customers. But without measurable outcomes, they often have nothing that persuades senior managers, especially finance-oriented senior managers, that the innovations have been worth the effort.

Measurements are not difficult to establish when you examine a new process, but they do require that you develop baseline data on your existing processes before you institute the incremental change. In Figure 8-1, the flow chart shows the recommended process for developing your measurements methods in direct relationship to your business goals. The steps illustrated in the flow chart are provided in more detail here:

- ✔ Define your initial goals.

- ✔ Establish the required measurements for each of the goals. You may have multiple measurements related to a single goal.

- ✔ Set up a process for collecting the data and educate staff members (a change management process).

- ✔ Collect data and use it to establish baselines for each measurement and goal.

- ✔ Analyze the data for the identified projects during the six months for the project.

- ✔ If the goals are not met, identify issues and plan improvements to achieve the goals.

- ✔ Educate staff members who are charged with making the improvements.

- ✔ Reiterate the improvement process until the goals are achieved.

- ✔ Train the staff to handle the metrics in the future to provide ongoing information about the efficiency and effectiveness of your processes.

Establishing your business goals is the first step in the process. Note one of the goals identified by the SWOT analysis in Table 8-1. The manager and team members want to reduce duplication of effort by sharing topics among deliverables in writing content for those topics once rather than multiple times.

The second step is to define the metrics you will use to evaluate your progress toward meeting the goal. What measurements would be interesting to gather that would show that the goal of reducing duplication of effort has been accomplished?

- ✔ By examining the legacy content in the deliverables for a suite of products, determine how many individual topics or sections of content were produced by all the information developers who have been producing their content independently. Count the number of existing topics and the number of words or pages of text they represent.

- ✔ Institute the new process of developing topics collaboratively and assembling them into the required suite of deliverables. Count the number of topics produced using the new process and the number of words or pages they represent.

- ✔ Determine if the new process has resulted in a net decrease in the total number of topics now being produced.

- ✔ Calculate the amount of time required to produce a new release of the information given the legacy content. You will likely know how many developers worked for how long to update the existing content.

- ✔ Calculate the amount of time required to update the information by producing a reduced set of topics.

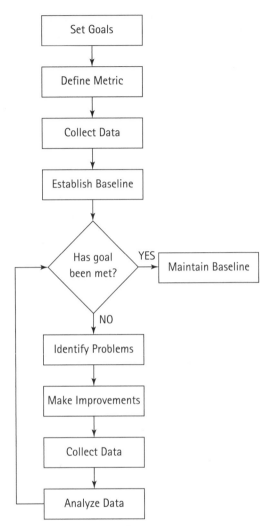

Figure 8-1: Information Development Efficiency Improvement Process

Daphne Walmer, director of publications at the CRM division of Medtronic, conducted a study similar to the one described here as a pilot project following the implementation of their new content management process. She reported a 60% reduction in writing time to produce a single set of topics from which more than one deliverable could be assembled.

Another publications manager discovered that the new release of the information, based on the reuse of topics in multiple deliverables, reduced the time required for information development by 50%. One information developer was able to complete a project that had traditionally required two people full-time.

Sometimes the results of a change can be evaluated by examining a representative sample of the new output. For example, one of the goals I reviewed was to improve the consistency in the structure of like information. In one organization, the information developers followed the Darwin Information Typing Architecture (DITA) to produce concept, task, and reference topics. The goal was to ensure that the topics followed the authoring guidelines and were uniform in their presentation of information to the user. The first measurement was applied to the legacy content.

A selected sample of topics could be compared to discover how consistently structured they were. Let's say that in an examination of 100 topics of each type, we found that 25% followed a very similar structure, 50% showed minor deviations from the standard structure, and another 25% were structured in a decidedly different way. After the new authoring guidelines were instituted, existing topics were restructured and new topics produced following the standards. An examination of a new set of 100 topics revealed the following results. Sixty percent of the topics were structured according to the standards with no variations; 20% of the topics showed minor deviations from the standard; the remaining 20% were still different enough to be considered problems. As a result of the before and after comparisons, you might conclude that a marked improvement in consistency was achieved but additional work was needed to bring the remaining content into compliance with the standard.

Best Practice–Improving processes and measuring results

Once you have developed your goals and defined the measurements you will use, your next step is to develop a method for instituting the measurement in your organization. It is time to introduce the method to the staff members who will be responsible for collecting the data, typically your information-development teams and the team leaders. In many cases, you will find that your teams are reluctant to collect data about their projects. They feel threatened by the suggestion that their work will be measured quantitatively. They believe the data will be used to evaluate individual performance and result in a staff reduction. They argue that they are too busy doing their primary task of creating content to waste time on data collection.

Your key to success when you are confronted by these objections is change management. You will find it best to begin with your business goal, stated clearly and forcefully to emphasize the importance senior management places on measuring results. In change management, managers often find it important to create a "burning platform," a scenario designed to impress upon people the necessity of the change. For goals and measurements, you can emphasize that if you do not become more efficient in your processes, senior management may be encouraged to look for other opportunities to reduce the costs of your operation, including offshore outsourcing to low-cost economies. Once your team understands the importance of goals and measurements and the necessity of process improvements to meet business goals, they are most likely to become strong partners in the process change rather than cynical observers.

When you institute new measurements to your teams, be certain that you find an efficient way to gather data. If the data is available without people's involvement in its collection, find ways to collect what you need without them. For example, if you have a content management system that provides reporting on the total number of topics in the repository, you can extract the data directly from the system before reviewing it with your team leads and project managers.

Be extremely careful to limit the data you collect, especially if you expect individuals to collect it. Nothing is more demotivating than collecting project data that is never used to draw conclusions and spur new actions. If your staff finds that you ignore the data you have asked them to collect, it is increasingly difficult to achieve their compliance and learn anything about your operational efficiency.

Collecting baseline data establishes the starting point for your process improvement activity. You cannot measure improvements unless you know where you are starting. Many managers are reluctant to measurement improvements because they have no baseline data. They argue that they don't know exactly how much time it takes to produce individual topics or final deliverables. They don't know how much content they have in terms of topics. They don't have records of how much content has been reused in the past, if only by cutting and pasting.

Even if you have little specific data on past activities, don't use the lack of data as an excuse. By going back into your content or reviewing the approximate amount of time devoted to various information-development activities in your organization, you can often make a reasonable approximation of the baseline data. You can always refine your approximation when you review the new measurements.

The following example shows how you might specify the relationship between your goals, your metrics, and your baseline measurements.

Efficiency Measurement Process

Template for the development of goals and metrics for process improvement.

Goal

Reuse content to minimize the number of topics in the repository. Lower the ratio of repository content to produced content from 1 (no reuse) to 0.8.

Metrics

1   Number of topics of produced content (PROCON)

2   Number of topics of repository content (REPCON)

Baseline

Determine a base line for PROCON for topics of content in the previous release. Determine a base line for REPCON for topics of content in the repository at the time of the previous release. If the count is the same for both metrics, the content reuse is zero. The ratio of REPCON/PROCON will vary from a small fraction to one.

Analysis

A ratio of one represents no reuse. The optimal ratio is something less than one and depends on how many times topics can be reused. Our goal is to bring the ratio down to 0.8. This does not imply a percent reuse of 30%. If two identical documents were produced using the same content in the repository the ratio would be 0.5 while the percent reuse would be 100%. If three identical documents were produced using the same repository content, the repository-content to produced-content ratio would be 0.33 while the percent reuse would still be 100%. The repository-content to produced-content ratio is a better measure of reuse than percent reuse.

continued

Efficiency Measurement Process *continued*

While developing the next release of the documentation, measure Metric 1 and Metric 2 and calculate the ratio of Metric 2 to Metric 1.

Improvement

Look for ways to reuse content in delivered documents. Reuse can be maximized by designing topics for reuse before content is added to them. Normalize your content by designing topics to be reused and topics that are specific to one instance. Use variables and references to independently maintain content within your topics and to handle specific content wherever possible. The goal is reached when the ratio of repository content to produced content is 0.8.

If the goal is not met, make design changes to topics in the next release until the ratio falls to 0.8. If the goal is met, maintain the ratio for future iterations or set a new goal.

For each of the goals you have identified, use this template to identify your process improvement and the measurement you will use to evaluate success. You may have more than one metric associated with a goal.

You may find it difficult to decide how to create an effective metric. Remember that you need something that you can count: number of words written, topics created, topics written accordingly to authoring guidelines, topics shared with the training organization. Each of these metrics looks at the efficiency of your organization in performing its work. In general, you can use efficiency metrics to decrease the amount of time and effort it takes to produce your information. Productivity metrics are evaluated for entire projects or for the processes in your organization as a whole.

Although you can use productivity metrics to evaluate the work of individual information-development team are rarely equal:

- ✔ Certain information developers are assigned work that is more difficult or easier to accomplish.

- ✔ Certain subject-matter experts are more or less cooperative, knowledgeable, and communicative.

- ✔ Subject matter is difficult and ill-defined, making it difficult to explain it in a useful way.

- ✔ Users are poorly defined so that it is difficult for information developers to decide what they do and do not need to know.

- ✔ Products undergo significant changes in direction during development, contributing to delays and rewriting.

- ✔ Regulatory changes require policies and procedures to be reworked multiple times.

✔ Stakeholders who have not been involved in planning the work suddenly want their points of view taken into account.

✔ Products don't follow the specifications. When writers finally get to see the product, it doesn't work the way it was supposed to.

For many reasons, some projects and assignments take longer to complete than others. The skills that your information developers bring to a project may also differ in ways that reflect on their productivity:

✔ New hires need to learn how to maneuver in your organization.

✔ New information developers need to learn what is expected of them in your organization and with information development in general.

✔ Even experienced information developers may be challenged by new technology to write about, new tools, or a new type of information development. For example, someone moving from books to topics may find his or her productivity temporarily reduced.

Nonetheless, you still encounter people and situations that don't produce at a level you expect. One caution: if you are yourself an old hand in the organization, having grown up with the company and the subject matter, don't expect everyone you hire to work at your pace.

Measuring individual productivity is possible without depending upon counting numbers of words, pages, or topics. One helpful metric is percentage of editing time. Editors are often fully aware of writers whose work requires considerable intervention to make acceptable. The time required of the editor in reviewing and managing changes to content and writing style can be an effective measure. For example, imagine an information developer with two to three years of experience in the field produced 100 pages of information. Next, an experienced editor spends between 25 and 30 hours reviewing, revising, and asking questions about that 100 pages of information. At 25% to 30% of the information developer's time, that high a percentage of editing time may reveal a serious problem with the information developers work habits and productivity. You might tolerate a high percentage of editing time for new information developers or for someone who is changing process, subject matter, or a delivery type. But the same percentage may be viewed as excessive for someone with experience. Other measures may provide you with the data you need to require that improvements be made in a timely manner.

Best Practice—Techniques for measuring effectiveness

Many of the examples thus far in this chapter focus on efficiency goals, ways to improve departmental performance and productivity and reduce the cost of operations. Despite the focus on cost savings and reductions in headcount that dominate corporate goals, the effectiveness of the information you produce in meeting customer needs is equally if not more important.

The only method most organizations use to measure customer satisfaction with a company, product, service, or sometimes information is a customer satisfaction survey. At the most, such surveys will have one or two questions that mention information that the customer receives to support use of the product. The survey questions are often quite high level, asking if indeed the customer is happy (yes or no) with the documentation received. Although a customer satisfaction rating can provide a baseline from which to measure improvements, the information gathered by a simple survey is rarely useful in developing innovative improvements. A simple, single-item question often does not reveal exactly what information the customer is thinking of in the answer. They do not even ensure that the customer responding to the survey actually uses any of the information provided. I have known of surveys answered by individuals in the purchasing office or on the shipping dock, rather than those who might actually find the information valuable. I have witnessed surveys answered by high-level executives who have no knowledge of the information provided to those who actually use the products. As a result, simple customer satisfaction surveys are notoriously unreliable indicators of value.

To gather reliable data about the customers' satisfaction with information requires a more creative approach. You should focus not on asking if customers are satisfied but if they receive value from the publications in line with their expectations. In many cases, the customer who receives the primary value from the information is not the end user but is an internal customer who serves the end users as well. Customer support personnel, trainers, installers, consultants, and others who interact with the customers may find value in the information your organization produces.

An indirect but potentially more effective way of measuring customer satisfaction with information requires cooperation from customer support. Effective, easy-to-use, and accessible information can have positive effect on reducing customer support calls. Your goal, metrics, and improvement program might look like the following example.

Efficiency Measurement Process

Template for the development of goals and metrics for process improvement.

Goal

Reduce by 50% in a one-year period the number of calls to customer service that might be avoided by improved user information.

Metrics

1   Number of calls to customer service attributable to shortcomings in the user information (PRECALL)

2   Number of calls to customer service attributable to shortcomings in the user information after the improvements are instituted (POSTCALL)

Baseline

Determine a base line for PRECALL numbers based on consultation with telephone support management. Review records of existing calls and count the number that might have been avoided had the user information been more complete, accurate, or easily accessible. Determine the POSTCALL number over a set period of time (six months, one year, etc.).

Analysis

A ratio of one represents no change in customer behavior. The optimal ratio is something 2 to 1 or less, depending on how many calls are avoided by improvements to the documents. Our goal is to bring the ratio down to 2 to 1 by reducing the call volume by 50%. If the PRECALL number were 1000, the POSTCALL number must be 500 or less to mean we are reaching our goal.

Improvement

Review call logs, talk with phone support personnel about the kinds of questions asked most frequently, review possible additions, corrections, or changes in the architecture of the user information.

Interview users to understand the ability to access information easily. You may discover that users cannot navigate your information website or the CD-ROM they receive because the search mechanisms don't work or the topics lack appropriate metadata to help narrow a search. You may find that the end users don't even receive up-to-date copies of the information at all, as we found in one customer study.

Develop a plan for improving the information content and accessibility as needed. Make the changes, deliver the improved content, and measure the results in number of calls received. In one project, we were able to reduce the number of calls by 60% and the average call duration from 10 minutes to 2 minutes, an enormous savings in customer support costs for a small cost of change in the information.

In finding ways to optimize the efficiency of your operations, you also want to avoid reducing customer value by also decreasing the effectiveness and value provided by your information. If you can do both at the same time, you have a winning solution to use in your Balanced Scorecard report.

Summary

Managers are expected to measure the efficiency of their operations and work to improve staff productivity and reduce operational costs. At the same time, they are expected to maintain or increase customer satisfaction with the value provided by the work they perform.

As you develop your process for optimizing the efficiency and effectiveness of your operations and measuring customer satisfaction, consider the following:

✔ Determine the goals you want to achieve for your organization by analyzing your strengths, weaknesses, opportunities, and threats.

✔ Consider the improvements projects you need to put in place to reach your goals and improve efficiency and effectiveness of your operations.

✔ Determine the measures you want to use to evaluate your progress toward meeting your goals.

✔ Make incremental changes in your processes and measure the results.

✔ Determine if you have made the progress you want. If not, make another set of incremental changes and measure again.

✔ Continue to make changes and measure the results until you have reached your goals.

✔ When you reach your goals, consider resetting the goals or finding new ones to continue your progress toward an efficient and effective organization.

Once you start using metrics to promote the productivity and effectiveness of your organization, you will find that you have a common language with which to communicate with senior managers, especially those most interested in reducing costs. Without the language of metrics, information-development managers are often viewed as out of touch with the mainstream goals of the business and unable to communicate about their work in terms that senior managers understand. As you build your business acumen and become an effective middle manager, your knowledge of business metrics will be your most important and visible communication tool.

"Which tools best meet our requirements?"

Chapter

9

Supporting Process Improvements with Effective Tools

> If the only tool you have is a hammer, you tend to see every problem as a nail.
>
> —Abraham Maslow

Process improvement initiatives in your organization should be designed with specific business goals in mind. You might decide to improve a process to increase the productivity of your staff by decreasing the time devoted to activities that add little value and increasing the time for value-added work. You might further decide in specific terms, as one group of information developers did, that too much time is taken up by meetings. Their process improvement was to restrict meeting time and redesign the way they managed meetings. Their goal was to increase the amount of time that people devoted to "content work." Process improvements to achieve business goals can obviously take many forms, including improvements that help foster collaboration among team members and cooperation among process stakeholders.

None of the process improvements mentioned thus far require changes in tools. It's important to keep in mind that many improvements you might pursue have nothing to do with tools. It's too easy for technically oriented people to disassociate improving processes from acquiring new technology toys.

Tools are rarely the only answer, or even the best answer, to process improvement questions. However, despite all the many opportunities to institute process improvements without new tools, many innovations and improvements throughout the information-development life cycle are supported by the acquisition and implementation of tools that promise to make some work easier and faster.

Consider a brief history of tools in information development. Thirty years ago, most information developers used typewriters. Word processors were introduced in the 1970s but were often reserved for pools of word-processing specialists who entered your hand-written documents into their systems. Only with the introduction of personal computers into the workplace did most information developers use word processing regularly. Even then, most documents looked like they had been typed. Only the largest companies like IBM, Hewlett-Packard, Unisys, and a few others, used word-processing systems that allowed you to transform tagged text into attractive print results.

The most significant tools change came in the mid-1980s with the advent of desktop publishing, first on the Apple Macintosh and then on the PC. Information developers were overjoyed at the ability to create typographic fonts and attractive page layouts with-out the expense of typesetting or the imposition of tags. The tagging systems largely disappeared, replaced by comprehensive desktop publishing tools with WYSIWYG layout capabilities.

When help systems first became available, they too required awkward coding of content. Once again, tools became available to make help topic development easier and more visual. The same process occurred with HTML development. By the mid-1990s, information developers were equipped with a suite of tools to guide the output of print, PDF, help, and web deliverables.

Document-management systems became part of information development in the 1980s, first to provide basic library functions like version control and check in/checkout security and automated workflow systems. Content-management systems were developed to include the management of structured, tagged content. The addition of automated work-flow systems to content-management systems supports the movement of content through the information-development life cycle. It has become easier to route documents through the stages of writing, editing, review, and approval.

More recently, content-management systems specifically designed to handle technical information have addressed the requirement to make smaller chunks of content easier to use in multiple contexts and multiple media. Added to these component-centric content-management systems have been tools to better support authoring in XML and implement-ing a topic-based information architecture. Systems like the Darwin Information Typing Architecture (DITA) support assembling topics into books, web pages, or help systems rather than authoring directly into the chapters and sections of books.

Enhancing the authoring process are increasingly sophisticated tools for editing and reviewing electronic copy, including systems that allow multiple reviewers to see each other's comments, avoiding duplications and reducing the number of contradictory suggestions.

An ever-increasing number of production and publishing tools are available to support developing outputs in multiple media quickly and easily. XML-based authoring is sup-ported by publishing tools that add formatting automatically at the end of the process, rather than requiring every author to be responsible for his or her own design and format-ting. XML tools also allow writing in the traditional WYSIWYG formats that many writers are used to, even though the final formatting may differ from the working version or may require multiple formats.

Added to the portfolio are the tools used to facilitate the localization of content and its translation into multiple languages. By using translation memory and translator's desktop systems, you use less time producing output in multiple languages and for diverse countries and cultures.

Most of the tools development today is in the hands of professional tools vendors, rather than inhouse teams that lose interest in supporting the tools or lack the funding to keep homegrown tools up to date and usable. With the introduction of international standards like DITA and DocBook, you can take advantage of the cooperative work done by vendors, consultants, and users to enhance the tools portfolio beyond the capabilities of any individual or company.

Despite all this interesting development of more versatile and comprehensive tools for information development, information developers themselves have tended to be conservative. They tend to accept what the tools will do as inevitable, rather than insist that the tools reflect sound design and usability principles. For example, when Microsoft's help system was first introduced and vendors developed tools to facilitate help topic development, most help systems tended to look exactly like Microsoft's because that was what the tools were designed to build.

Instead of blithely following the lead of the tools vendors, as information-development managers, you should first defer to the needs of your customers and staff both in selecting a tools set and configuring the way the tools will be used. Next, you should consider the role a tool set plays in affecting the productivity and business value of your operations and your information products. Ask yourself and your team members the following questions:

- ✔ Do the tools we use today enhance our productivity?

- ✔ Would new or different tools increase our productivity?

- ✔ Do our tools allow us to work better and faster, rather than harder and slower?

- ✔ Do the tools we use introduce errors into the process rather than reducing the possibility of error?

- ✔ Is using the latest tools more important to our team than developing the best information products for our customers?

- ✔ Is using the latest tools important to retaining employees who want to keep their skills up to date?

- ✔ Is the work we use the tools to perform encouraging us to make the best use of our time and energies?

- ✔ Do the tools encourage us to perform activities that deliver value to our customers, or are the tools a distraction that occupies our attention to the detriment of delivering customer value?

You should find it obvious from these questions that tools can be a support or an unfortunate hindrance. Many information developers (and hiring managers unfortunately) judge their value by their knowledge of and skill with tools, rather than their understanding of the user community or the subject matter. Such a focus, unfortunately, can decrease the perceived value of your organization among senior managers who believe that information developers are highly paid formatters and tools jockeys rather than valuable contributors to customer satisfaction and the bottom line.

In the analysis of the source of competitive advantage, illustrated in Figure 9-1, the value of tools expertise as part of work practices is both easier to develop and easier to copy than becoming expert in the technology you are writing about (core technology), understanding your customers (customer knowledge), or developing your business acumen and alignment with corporate strategies to increase your value to your larger organization (strategic alignment).

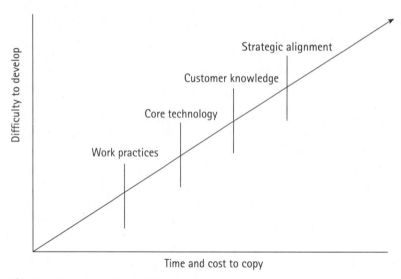

Figure 9-1: Sources of competitive advantage

Certainly, you recognize that having the right tools for the job is important. But first you must understand the job you are being hired to do.

Best Practices in Implementing Effective Tools for Process Improvement

As you begin to develop a tools strategy, you need to take stock of the tools already in your portfolio. If you have begun using process improvement metrics, based on the best practices in Chapter 8: Optimizing Your Organization's Efficiency and Effectiveness, you may already know in what areas your tools may be limiting or enhancing productivity. If you have engaged in partnering with your customers, based on the best practices in Chapter 6: Developing Relationships with Customers and Stakeholders, you may have identified areas in which the tools you are now using provide an effective set of deliverables to the customers or where they fail to do so.

In this chapter, you learn about five best practices for developing a sound tools strategy and investing in new tools that may be required to increase the effectiveness of your information-development process:

- ✔ Developing a tools strategy

- ✔ Developing requirements for tools

- ✔ Researching and acquiring tools (deployment)

- ✔ Managing tools (training, configuration)

- ✔ Developing a tools strategy for the larger organization (Levels 4 and 5)

These best practices provide you a systematic method for developing a strategy and carefully introducing new tools into the work environment.

Best Practice—Developing a tools strategy

A tools strategy begins with a well-defined relationship between your business goals and your tools portfolio. One organization wanted to ensure that customers could find the information they needed on the documentation CD-ROM they received four times a year. They asked Comtech to investigate the customers' satisfaction with the information on the CD. During the customer site visits, we learned that most of the end users could not use the search and retrieval tool that was packaged with the CDs. They found the interface confusing, the search embarrassingly slow, and the search returns of almost no value. Referencing a CD full of huge PDFs of documents with 500 pages or more, the search system returned only the document titles. To find a specific topic, the end users had to laboriously open each PDF before using the PDF search routine. Even then, the search simply highlighted all the terms in the document that matched the words in the search string. Most end users simply avoided using the CDs completely. At some sites, the one individual who had mastered the art of finding useful information conducted searches for everyone else, not a particularly good use of his time.

Comtech reported that we found it impossible to evaluate the usefulness of the CD content, because most of the users never succeeded in finding any content at all. Without improving the search tool or packaging the content differently, the company would never achieve its goal. They needed a new search tool, and they needed to develop content that could be more easily and effectively searched.

Your analysis of your own tools portfolio with respect to your business goals might take the following form:

- ✔ What is your business goal?

- ✔ What tools do you use in the process of achieving the goal today?

✔ Do the tools now in use facilitate the process or make it more difficult?

✔ What do we require in a new tool or technology that would better support the process and help us achieve the business goal?

A typical analysis might look like what you see in Table 9-1.

Table 9-1: A Tools Portfolio Analysis

Question	*Response*
What is your business goal?	To improve end-user access to time-critical information
What tools do you use today?	PDFs of entire documents are accessible through dialup to the corporate file server.
Do the tools facilitate the goal?	No, slow dialup makes the file system difficult to access and the PDFs hours to download.
What is the requirement?	Need a way to reduce the amount of time to access the content Must search and retrieve the topic needed rather than an entire document Once found, downloading the relevant information must take less than five minutes.

This analysis might lead you to find tools that support the development of topics rather than entire books. You might decide that you need a tool that supports adding metadata to each topic to categorize the content. You might also want to develop an information website that delivers HTML topics with metadata to facilitate search. Likely you'll need a better search engine. If you follow the system that Oracle developed for its database technical information, you will also develop a tool that lets end users assemble topics into their personal "books," with links back to the source so that they receive regular updates.

At Oracle, this innovative method of delivering information to end users was supported by a technology developed inhouse. The website to visit to see the results is http://www.oracle.com/pls/db102/ranked?word=web.

At Comtech, our investigation of a new tool for content production began with an analysis of internal costs to prepare for our conferences. In our review of the costs, we noted that we had spent an inordinately large amount of time and resources formatting the conference proceedings. In discussions with the staff, we learned that using a popular desktop publishing package resulted in extra time to number pages every time a speaker made changes to a slide set or the sessions were rearranged. Because slide changes are quite common as the proceedings are put together, we needed a system that would make changes less time-consuming and expensive.

We worked closely with our print vendor to evaluate other packages that not only would decrease the amount of effort in house but would deliver the proceedings in a format that would require less intervention by the printer. Such a move would decrease printing costs and reduce the opportunity for errors. We finally chose a package that supports robust template creation, handles PDFs of individual slide sets, and completely automates the page numbering. We expect to measure the change in time and cost for the next conference.

Developing a business case to acquire new tools

Your analysis of your current tool set must, as the previous examples demonstrate, include an analysis of the costs and the capabilities. Your analysis should address the following questions:

- ✔ Cost avoidance—What costs could be avoided if the current tool were replaced with one that was better suited to the needs of the organization?

- ✔ Increased productivity—Would the organization be more productive with new tools, decreasing the cost of all work produced?

- ✔ New business requirements—Would the organization be better able to support new business requirements such as content reuse, decreased costs of translation, decreased time to market with new tools?

- ✔ Customer satisfaction—Would customer requirements be met more easily and with better results if new tools were available?

- ✔ Support costs—Would the cost of supporting the tools be reduced if the tool set were consolidated?

- ✔ Delivery options—Would new tools allow the organization to produce and deliver content in ways unavailable with the current tool set?

- ✔ Return on investment—Would using the current tool set ultimately be more costly than acquiring and using a new one?

Acquiring new tools is rarely free, even if the new tool is "freeware." There are always associated costs for training, implementation, and maintenance, in addition to acquisition and customization. However, the return on investment can be considerable if you account for the current costs.

In preparing your business case, explain carefully the current costs of using the tool set you have in place. In one business case, I included the cost of supporting multiple products used in different departments, the cost of maintaining multiple servers used to store departmental content independently, the cost of searching for content unsuccessfully as reported by most staff members, the cost of not finding the information needed, and the cost of duplicating information that was stored but could not be found in a timely manner. Then, we calculated the costs associated with acquiring the new web-content-management system, including acquisition, configuration, installation, and training. We used both costs to calculate the payback period, the point at which the savings of using the new system would exceed its cost, including the cost of maintenance. The payback period was approximately 2.5 years, far less than the payback period required for new technology acquisitions in the company.

The return on investment calculation also included qualitative as well as quantitative improvements, including helping departments develop web content without having to employ and train web experts. The departments could use their internal resources more productively by devoting them to the work of the department rather than to designing, implementing, and maintaining websites.

Note that return on investment (ROI) is a method of analyzing the cost of an initiative. If the initiative reduces costs or returns profits that exceed the cost of implementation, then it has a positive ROI. The initiative may not be worth doing, if, when you consider the total costs of implementation, the initiative costs more to implement than you can arguably recover through measures such as increased productivity, measurable changes in customer satisfaction, increased revenue, or increased profit. However, not all initiatives are approved because of a positive or negative return on investment in the short term. In many cases, the return will not be realized for many years. In other cases, senior management approves an investment because it promises to make an organization more competitive in the future.

Best Practice–Developing requirements for tools

After you have identified areas in which your current tool set is not meeting your needs and demonstrated that you have a business case for the acquisition of new tools, you need to develop formal requirements. As you proceed through your inventory of existing tools and your evaluation of how well they support your business goals, divide the inventory into categories:

- ✔ Authoring
- ✔ Controlled language
- ✔ Review and collaboration
- ✔ Metadata and taxonomy
- ✔ Production (print, help systems, web and wireless content, interactives)
- ✔ Graphics
- ✔ Content management (component, document, web content management)
- ✔ Translation
- ✔ Search and retrieval

An analysis of each class of tools and the characteristics required might look like Table 9-2.

Table 9-2: A Sample Analysis Table of Tools by Class

Class	Desired Characteristics
Authoring	Enables content creators to develop accurate and complete content in minimal time Includes word-processing and desktop publishing capabilities Facilitates structured authoring by subject-matter experts and professional information developers Supports implementing controlled language and standard terminology

Class	Desired Characteristics
Controlled language	Enforces terminology standards, grammar and spelling, and guidelines for minimalist and effective writing Is fully integrated with the authoring tools Is user-friendly and takes minimal time to apply to a text Can be configured to match specific business requirements
Review and collaboration	Enables electronic review of draft content Includes automatic notification of reviewers Allows reviewers to collaborate on their reviews by seeing each other's comments in real time Tracks all changes by reviewer and stores them for future access Provides a simple, time-effective method for accepting or rejecting comments and incorporating them into new drafts
Metadata and taxonomy	Provides a simple way to add metadata to content, providing a controlled vocabulary of attributes and values Automates the addition of standard metadata Supports controlled vocabularies Stores metadata in the online output files Makes metadata available to facilitate search and retrieval Provides a method for analyzing new content and assigning metadata automatically
Production (print, help systems, web content, interactives)	Facilitates the development of multiple outputs that are customized to meet the needs of different audiences, products, job roles, and media Allows production specialists to quickly and easily produce new content for multiple deliverables without costly and time-consuming layout manipulation Reduces the overall cost of production Supports the development of deliverables in multiple languages
Graphics	Enables the development and maintenance of graphics in multiple formats, including images, sound, and video formats Allows graphics to be used easily in multiple deliverables through reference Facilitates the translation of text in graphics without requiring the translators to manipulate the programs used to create the graphics (SVG graphics support)
Content management (component, document, document management, web content management)	Stores components based on metadata structures Facilitates the assembly of components into final deliverables (e.g., DITA maps) Provides version control, check in/checkout security Supports automated workflow Handles multiple file types Facilitates handling of components in multiple languages Supports the application of translation memory for pre-translation Supports hypertext links and cross-referencing between file types Supports multiple style sheets Delivers output automatically to web pages depending upon metadata Provides for multiple levels of archiving and restoration of archived content

Continued

Table 9-2: A Sample Analysis Table of Tools by Class (Continued)

Class	*Desired Characteristics*
Translation	Supports the application of translation memory Correctly interprets XML-based content Supports multiple language fonts Automates production and publishing processes Supports translation of text in graphics
Search and retrieval	Allows for faceted search based on metadata and taxonomies Prioritizes retrieval based on configurable rankings Supports Boolean search arguments Supports folksonomies (user-directed metadata categorization)

The characteristics listed in this table are not intended to be exhaustive. The characteristics that you identify for the tools you need should be carefully and thoroughly assembled based on input from your stakeholders and should be reviewed by those stakeholders for clarity and completeness. You will find it necessary to review your tools strategy annually to ensure that it is up to date. The tools environment changes rapidly, and many of your stakeholders may not communicate their problems or needs quickly or effectively.

Developing thorough requirements is absolutely necessary to acquiring the best tools. You may find it tempting to ignore developing requirements and simply ask tools vendors to demonstrate their products. You might have people on staff who prefer to review the products on their own and give you free-form feedback rather than following a strict review protocol. Developing requirements can take time and requires that you research the capabilities offered by new tools on the market. Without that research, you may omit from your requirements something that you did not know existed but would be extremely useful to meet your goals.

It's been my experience that groups that fail to develop formal requirements regret not having done so. They often discover that they have been swayed by the persuasive style of sales representatives into buying something that turns out not to meet their needs or requires expensive and time-consuming customizations.

Some organizations require that new tools follow corporate technology standards and are acceptable to those in information technology (IT) who have to maintain the tools. For that reason, you will need to consult with your IT professionals as you develop your requirements. You may find, for example, that IT has a database standard that will restrict the content-management systems you can consider. You may also need to work hard to help IT understand what you are trying to accomplish, especially if your needs run counter to general organization standards. For example, professional information developers may require a structured, XML-based text editor to achieve a good return on investment rather than using MS Word, even though MS Word is a corporate standard. You may find that you will need all your powers of persuasion to help IT understand and accept your business requirements even though those requirements mean changes to the corporate technology standards.

Once you have defined and reviewed your requirements to the best of your abilities, you need to develop a Request for Information, a Request for Proposal, or both. A Request for Information (RFI) allows you to ask a small set of vendors to provide their strategy for meeting your needs without having to compete on price. An RFI is the best way to gather information about a possible solution to your tools requirements without having to be too specific. If you need a solution for which the tools solution is not obvious, and if you really don't know exactly what your requirements should be, consider an RFI before you ask for specific prices. A vendor may present a solution that you would never have thought of. With an RFI, vendors feel less constrained by your solution prescription and can be more creative in recommending a good solution themselves. You can also use the RFI process to learn more about the vendors' capabilities.

Be certain to provide the same RFI to every vendor so that you can receive comparable information. If you keep changing your request, you will have great difficult in evaluating the responses.

Once you know more precisely what kind of system will meet your requirements, you can develop a Request for Proposal (RFP). At the RFP stage, you ask vendors to respond to a list of requirements by indicating if the requirement can be met with their standard product, can be met by customizing the product at a cost, can be met by configuring standard functionality in the product to meet your specific environment, or cannot be met at all. By providing a requirements checklist, you can ensure that the solution that is delivered does exactly what was documented in the proposal response. Without that checklist in hand, you may find yourself paying for a function that you thought was a standard part of the product.

One of the best approaches to an RFP is user scenarios. By including user scenarios, narrative accounts of how your authors and others expect to use the new tools, you can ask vendors to demonstrate how their products support the scenarios. The user scenarios become the content of product presentations by a small group of selected vendors. The scenarios also help your stakeholders understand how the products will meet their stated requirements.

RFPs also ask vendors to provide full pricing information. Unfortunately, the pricing often comes long after you had to anticipate the cost of the new tools in your presentation to management. If you find that the prices quoted differ drastically from your earlier estimates, you may need to revise your requirements to decrease your costs.

Best Practice–Researching and selecting vendors and tools

As you prepare an RFI or RFP, you should research tools and vendors that may meet your needs. I recommend that you send your RFI or RFP to a small list of vendors that you have already prequalified rather than sending the requests to many vendors that may not have appropriate solutions. The better your RFI or RFP is written, the more and better proposals you will receive. A small number of proposals fitting your needs is much better than lots of proposals not exactly directed to your goals.

You will find many sources of information about tools and vendors:

✔ Industry-wide reviews available on the web or through publications from organizations that provide comparative tools analyses. In most cases, you must purchase the reports provided by these organizations but they often contain detailed analyses not available elsewhere. The reports are especially valuable if the reviewers have tested the products rather than depended on information from the vendors themselves.

✔ Online user groups that provide information from individuals who are using the tools. Be careful about using the information provided only by advocates of a product. Try to uncover contrary opinions as well. Add your own questions to the user group listservs.

✔ Discussions with colleagues who have been through a recent selection process themselves. Finding people who have already completed their analyses based on criteria similar to yours will help you quickly eliminate inappropriate choices. Be certain the selection was recent because tools can change dramatically as new versions are introduced.

✔ Discussions with colleagues in professional organizations that have chosen similar tools for their organizations.

✔ Advice from consultants familiar with a wide variety of vendors and tools. Be certain that the consultant you select has no monetary ties to the vendors. Some consultants receive commissions for systems that they advocate.

✔ Visits to vendor booths at conferences. By attending an industry conference, you will have an opportunity to talk with many vendors at one location. You are also likely to meet customers and hear their presentations. Many conferences also provide opportunities to view vendor demonstrations without having to reveal details about your project.

✔ Information directly from the vendors' websites. Download white papers, read testimonials and case studies, print out specifications, and review online demonstrations. Sign up for webinars or demonstrations of the products. Attend presentations in your area. Download trial copies of the product if they are available and use them on your own content.

In general, you will find it much easier to research the vendors and tools when you understand your own requirements. Without requirements in hand, the candidates will all begin to sound the same. As you conduct your research, develop a rating sheet with questions from your requirements so that you can keep track of which vendor provides the functionality you are looking for.

When you are close to making your selections, contact the sales representatives and ask for preliminary presentations and demonstrations of the products. If you have them prepared, provide the vendors with your user scenarios to help them tailor the presentations to your needs. Once again, maintain rating sheets to help you remember what you saw and heard. The more demonstrations you see, the more difficult it will be to remember which vendor said what or had what feature. Invite your staff and your stakeholders to these preliminary presentations. They can add to your research and ensure that everyone understands the type of products you are looking for.

Evaluating alternatives

Many organizations find it difficult to fund new tools. As an alternative, you might consider open-source tools that are available at little or no direct cost. Many open-source tools have excellent functionality, although they can be more difficult to use because they are designed for early adopters rather than the late majority. If you have staff members who qualify as early adopters, which means they can handle programming logic or difficult-to-use interfaces, you may find a cost-effective solution in open source. Without technical expertise that you can depend upon for a reasonably long period of time, you will be better served by off-the-shelf products.

You may also have people in your organization who volunteer to create an inhouse solution that may precisely meet challenging and unique requirements. Many excellent inhouse solutions have been developed that rival commercial products. However, inhouse solutions are never free. You have to pay the salaries of the developers, and you have to have sufficient internal funding to maintain the solutions for many years. In my experience, inhouse solutions start with great expectations but often become difficult and costly to maintain, especially if the original developers leave the organization without documentation of their creations. You may also learn too late that migrating content successfully from a custom application into a commercial one is difficult, if not impossible.

Buying a commercial, off-the-shelf product means that you get the vendor's investment in continuing development. The vendor is responding to the needs of many clients, including you, developing new features and functions that will benefit your organization without the cost of funding the development. Of course, you need to be confident that the vendor will be around for the long term, something that should be part of your vendor-qualification process. If the solution you purchase is based upon industry standards, you also reap the benefits of a community of advocates who update the standard at no cost to you.

Making a final selection

After you have researched the options available, prepared your requirements, and received responses to your RFI or RFP, you are ready to make a final selection. Ask each vendor to prepare a presentation for your staff and stakeholders that includes a specific set of scenarios that you have provided. With the scenarios, provide the vendors with sample content so that everyone can see your information being used to produce content. Agree with the vendor on exactly what functions will be covered in the demonstration.

Provide an evaluation form for each person who attends the presentation. Get feedback from as many stakeholders as you can. Schedule meetings to review the evaluations with key stakeholders. Use a ranking system to evaluate the criteria in the requirements. Don't neglect, however, your general impression of the vendor and the presentation.

After you have narrowed your selection to one or two vendors, ask for references that have projects similar to yours. Either arrange conference calls with the references or arrange visits to their locations. Direct visits are usually more fruitful but can be expensive and time-consuming. Arrange for visits to the vendors' company headquarters. Your reception will indicate the importance of your project. Sometimes, the personal contact and your comfort level with the people who will be working on your project makes the differences between two vendors with similar products.

Be certain that you have a full accounting of all costs associated with the project, not just the hardware and software costs. Ongoing maintenance, customization, training, and ongoing support are all part of the long-term costs you will need to budget for. Ensure that you budget for the migration costs from your existing system to the new one. Even moving information from a file server will include some cost.

Arranging for a proof of concept

If, after all the presentations, conference calls, and visits, you are still unsure of your choice, you may want to arrange for a proof of concept project. In a proof of concept project, you ask the vendor to temporarily install the product at your location, incorporate a representative sample of your content, and support your staff in using the product directly to produce deliverables.

In most cases, vendors will ask you to fund the proof-of-concept project, unless you are an important and large client. Because the vendor must devote considerable personnel resources to the project, especially on-site at your location, they are entitled to compensation. In some cases, you will pay for the staff time at your location. In others, the vendors will offer to do the proof of concept as part of their sales process, as long as you are not asking for something that is unusual. The proof of concept is also an indication of a good fit between your requirements and the vendor's capabilities.

Also ensure that you have an internal team ready to participate in the proof of concept. Don't schedule the activity in the midst of critical deadlines. You want team members to exercise the product with their own content so that they can decide if the product does what they want. Include in the proof of concept time for staff training and support. You want people to have a positive experience with the products, not struggle with basic functionality because they are not well prepared.

Once again, ask all participants to complete evaluation forms so that you have as much feedback as you can gather. If you are one of the final decision makers, be certain to participate in the proof of concept yourself.

Before you make your final selection, ask the vendor for a best and final price quotation. Recognize that you may have added to the requirements during the proof of concept, resulting in additional costs. Be certain that you understand exactly what you will receive, including costs for professional services during installation and configuration and training for your staff. Understand the costs of ongoing product maintenance, often a percentage of the original product cost, and what is included in the maintenance contract. Be aware that you may have to pay the first year's maintenance up front.

It's critical to know what kind of service you are paying for. If you are running a system that cannot go down and you have a service contract that requires a call return within 24 hours, you may have a problem. Be certain the service level agreement is prepared as part of the initial contract, when the vendor is anxious to close the deal.

I assume that with any substantial purchase, you will have worked closely with your purchasing department. They may be responsible for the price negotiations. You just need to ensure that you are getting the product that you expected. Work with the purchasing agent to ensure that you are both working toward the same goal.

Best Practice—Introducing and managing tools

Once you have made a purchase decision, work closely with the vendor to determine a schedule of activities for installation, configuration, migration, and training. If you have someone who will be chiefly responsible for administering the new tools, arrange for administrator training. Develop a schedule for migrating content into the new tools and training your staff on their use. For more detailed information about implementing new tools, see my discussion in *Content Management for Dynamic Web Delivery*.[1]

Establishing a pilot project

For any tools acquisition that represents a change in process in your organization, you will benefit from establishing a pilot project. After the tool is installed, configured, and ready for use, have a team of enthusiastic early adopters use the tool for a real project. Your requirements team should develop a set of criteria for selecting a pilot. It should include real content with a schedule of deadlines that are not overly aggressive. It should exercise a substantial sample of product functionality. For example, if you have acquired a content-management system, you should be able to import legacy content, add new content, manage version control, set up a basic workflow, and use the content to create final deliverables. If you have acquired a new search and retrieval product, you should be able to optimize it for a sample of your content. You will need sufficient content to test the search results for completeness and accuracy. You will need to learn how to best apply metadata or keywords to your content so that the appropriate content is returned.

Ensure that the pilot team receives training that is timed for their schedule. Training that occurs weeks or months before anyone is ready has little value and will have to be completely redone. Training that occurs immediately before product use is the most valuable. However, the initial training should be supplemented with problem support and question and answer sessions once people have a chance to experiment.

As you develop your pilot project, decide how you will measure its successes and challenges. For example, your new search tool should deliver a high percentage of relevant results and a low percentage of irrelevant results if it is successful. Your new authoring tool should not significantly increase authoring time after staff are trained and comfortable using it. You may tolerate a small increase in time for the first project, but you should expect time reductions thereafter. If you don't achieve the promised reductions, you must develop a plan, based on a review of your original goals, to achieve the result you wanted. Research why you're not getting the expected results and use the outcome of your research to develop new project goals.

Managing tools for the long term

In a large organization, you may have a tools team that has been engaged with the requirements, selection, and implementation process. Along the way, they have gained knowledge and experienced in configuring the new tools. They are often actively involved in user groups or listservs and forums in which they learn from other users and exchange insights. Members of the tools team are involved for the long term in ensuring that the tools are well supported and maintained. In most cases, they track new releases of

---

[1] JoAnn T. Hackos, *Content Management for Dynamic Web Delivery.* Hoboken, NJ: John Wiley & Sons, 2002.

the products, determine when to acquire them, and how to introduce them to the user community. They also work closely with vendors to suggest new functionality that will make the implementation more effective and meet new goals for use and delivery.

In a small organization, you often have one or two individuals who are interested in the tools and will take responsibility for supporting others and learning the intricacies of the tools for effective use. These individuals should be supported and rewarded with advanced training opportunities and attendance at user group meetings and industry conferences. The danger, of course, is that your one tools guru will leave, putting your tools implementation at risk. You may want to investigate local support from an independent consultant until you can reestablish your internal skills. Or, you may want to ensure that you have at least two people on staff who know how to use the tools.

Having tools expertise inhouse is usually essential for the smooth operation of the tools and support of the processes you hoped to optimize. Vendors can help, of course, but rarely provide the hands-on insights that make everyone more productive. Be aware, however, that your tools team or person can become overly enamored of a tool and refuse to consider new opportunities. You want to foster an environment in which tools are always candidates for change. If your tools experts are regularly exposed to new entries in the tools environment through participation in user groups, listservs, local support groups, conferences, and other opportunities for learning, they will bring you new ideas to explore.

Best Practice—Developing a tools strategy for a global organization

Many senior managers find themselves responsible for a strategy that supports a global organization and often includes staff members and even entire organizations added through mergers and acquisitions. New groups of information developers are established for new products or at new development sites, often at distance locations. As a director or senior manager, you need to consider instituting an integrated tools strategy for the organization and determining how best to implement the strategy. Although some parts of the organization may be ready for the latest tools available, others are better served by a more conservative and gradual approach.

If you are responsible for a tools strategy but not for the overall management of the global organization, you are likely to need a strategy for encouraging independent teams to adopt a more global point of view. Providing them with an effective suite of tools, including training and support, makes it easier for more inexperienced or conservative groups to adopt the global solution.

An organization that is a Level 4 or 5 in process maturity has the ability to support a global tools strategy and present that strategy to parts of the organization that are more immature. It is important to acknowledge, however, that immature organizations may not be ready to adopt complex tools at one time. Through mergers and acquisitions, you may find yourself developing a tools strategy for a Level 2: Rudimentary or even a Level 1: Ad-hoc organization. They may first be asked simply to adopt the standard templates for the design of information products. They may need to begin, for example, with a content-management system that will manage their traditional chapter- and section-based content produced with desktop publishing tools. They may want to experiment with topic-based authoring and minimalism in a traditional desktop publishing environment, using techniques like text inserts to use the same content in more than one deliverable. Only after they are comfortable with a topic-based approach to authoring will they be ready to move to a structured authoring environment. Sometimes, the first solution is to use a structured version of an otherwise unstructured tool for an initial venture into structure. After structured authoring becomes commonplace, they may be ready to move into an XML editor or an otherwise different authoring environment.

More mature organizations with more early adopters on staff may be prepared to make a more rapid transition to a new tools environment that gives them a greater return on investment. However, even with early adopters ready to conduct pilot projects and implement new methods, you are likely to have confused and reluctant staff hiding behind "no comment." You will find it best to meet individually with those who demonstrate hostility toward the project, in private. That hostility usually hides fear about losing a job or losing hard-earned competence in the current environment. Individual meetings with assurances about continued responsibilities will help. Unfortunately, you may still have some staff who decide to leave for a more comfortable and familiar work environment. If you have instituted a change-management process, you have already planned to build a supportive team and analyzed all the internal stakeholders so that you have identified the people who are most likely to dissent.

Summary

In most organizations, information-development managers and tools and technology experts on your team should regularly evaluate the current tools environment. To be most fruitful, the evaluation should begin with business goals for productivity, time to market, operational costs, and quality of the information delivered to the customer. By evaluating tools in the context of key business goals, you can move beyond popularity and familiarity as the primary determinants of a tools portfolio. Remember that many staff members are heavily invested in the current tool set and are reluctant to change, even when change is in the organization's and often their own best interests.

A reevaluation of current tools is especially germane when you are trying to improve processes and reduce operational costs or when you are actively pursuing methods to improve the quality of information delivered to customers. Although tools and technology are not in themselves forces for change (or should not be considered in that way), they do provide underlying support to achieve a predicted return on investment. Tools simply automate what we otherwise design into our processes. Nearly everything we do today with tools in information development can and has been done manually.

As you work to develop a global tools strategy, consider the following process:

✔ Establish the business goals that you want to achieve

✔ Evaluate how well your current tool set supports or does not support your new business goals

✔ Look for instances in which work is made more difficult or you are unable to deliver content to your customers in ways that will increase customer satisfaction

✔ If you decide that new tools are likely to enhance operational performance and delivery of information to customers, begin the process of researching, qualifying, and acquiring new tools

✔ Develop a comprehensive set of requirements, involving all potential stakeholders in the development process

✔ Turn the requirements into an RFI or RFP

✔ Research the tools and potential vendors or open source options

✔ Consider inhouse development if that will be feasible and cost-effective over the long term

✔ Request proposals from a short list of prequalified vendors

✔ Evaluate the proposals carefully and supplement them with demonstrations and proofs of concept using your user scenarios and content

✔ Follow best practices in your vendor-selection process

✔ Establish a pilot project with skilled and enthusiastic participants once the selected tools are installed and configured for your needs

✔ Be certain the pilot team is well trained to use the tools effectively and productively

✔ Measure the results of the pilot against your original process goals

✔ Make corrections to the process and the tools and begin a global roll out to other parts of your organization

With this process in place, you are likely to develop a tools strategy and an acquisition process that will hold you in good stead over a long period of time and result in the support for process improvements that tools should enable.

"Go team!" "Gehen Mannschaft" "去队"

Chapter 10

Developing Effective Teams

> Teamwork is no accident. It is the by-product of good leadership.
> —John Adair

Developing effective teams is hard work and requires effective leadership. In conducting Information Process Maturity assessments for many years and working with numerous publications organizations on designing new processes, instituting project management best practices, and developing business metrics, I've had many opportunities to observe effective and ineffective organizations. So much of an organization's effectiveness is dependent on leadership that it's not surprising that organizations with strong, competent leadership are almost always more successful.

An effective leader, however, does not work in isolation. He or she develops an effective team that assumes responsibility for achieving business, organizational, and personal goals. The leader helps define the goals, especially by understanding the priorities of the larger organization, and communicates these goals to the team. The leader ensures that the people assume responsibility for meeting goals and understand the urgency with which they must be achieved. The leader helps to motivate the team both by supporting their activities and ensuring that those activities are going in the right direction. He or she ensures that team members don't get stuck in endless debate, especially when the matters debated become increasingly trivial.

As a manager, you know that you cannot afford to be complacent, happy with the status quo. Your senior management demands productivity improvements from your staff and expects you to reduce costs, not increase them. You have heard the oft-repeated mantra, "nobody reads the documentation anyway," and you know that it carries the threat of reorganizing your team out of existence or outsourcing the work to an independent agency or to another country. You have to respond by changing the way you do business. You have asked your entire team to participate in developing innovative and cost-effective information-development approaches. Your team must be fully engaged in any initiative that requires change. You need to focus everyone on business-savvy solutions that will gain the team acceptance and recognition among senior management. You know that won't happen by staying the same, and you know that it won't happen without everyone's commitment and active participation.

As an information-development manager, you are responsible for building a high-performance team that is ready to pursue innovative directions and increase the value of your organization and the information products you develop. Unfortunately, you may be faced with a team of people who are not collaborative or even cooperative with one another. Some of the resistance may be in your home department. Even more likely, resistance will come from globally dispersed team members, many of whom originally worked for different companies.

Perhaps the greatest challenge that effective information-development managers face is the increasing distribution and globalization of team membership. Managers find themselves responsible for bringing together managers and team members from newly acquired organizations, many with different structures, patterns of behavior, and degrees of talent. Managers find themselves managing team members who work in diverse environments, from large and small groups located together to single individuals working alone among product developers and individuals who work at home part or all of the time. You must learn to manage teams dispersed globally, through the acquisition of companies outside the home location, through outsourcing development work to other companies, and through establishing product teams in newly developing, low-cost economies. The workforce you now face as an information-development manager has never been more diverse.

Best Practices in Developing Effective Teams

Developing effective teams requires a variety of approaches, depending upon the nature of your organization. You may have responsibility for leading a local group with everyone in the same location or with team members dispersed geographically in the same country. You may be responsible for building a collaborative team among managers who lead independent departments added to your organization through acquisitions and mergers. Or, you may be charged with establishing a team that includes members dispersed globally, some of whom are direct employees of your company and others who work for outsource vendors. In this chapter, you learn about five best practices for developing strong and effective team resources in your organization:

✔ Developing collaborative teams

✔ Defining new roles and responsibilities

- ✔ Managing remote team members
- ✔ Working with global teams
- ✔ Outsourcing and offshoring

The best practices in this chapter focus on building your organization. Chapter 11: Managing Your Team Resources covers issues of personnel management focused on improving the effectiveness of your existing team members.

Best Practice–Developing collaborative teams

In the past, managers may have found it possible to assume a hands-off approach with information-development staff. Individual contributors were highly valued only when they were capable of working independently and handling the development of technical publications from research, design, and development through final production and even delivery. In a classic cottage industry, one individual performed all the tasks to develop a well-crafted final product. Information development as a cottage industry still exists in many organizations, especially in small two- or three-person departments or in cases in which an information developer works alone in the midst of a product-development team.

However, two forces are making it increasingly difficult to maintain the cottage point of view: single sourcing and globalization. Single sourcing requires that team members collaborate so that they share content development and avoid duplication of content and effort. For single sourcing to be successful, team members must plan what content should be developed and who should develop it. A lone information developer is able to single source content if he or she is responsible for multiple deliverables with content shared among them. But as soon as content must be created and shared among deliverables developed by a team rather than an individual, collaborative teamwork becomes a requirement. Everyone must know what everyone else is doing.

In addition, if content is to be truly reusable among a variety of deliverables in multiple media, the content must be produced according to established standards. That implies that team members can no longer develop information in any way they choose, personally crafting a book or a help system. Instead, each member of the team must follow standards and business rules to develop a consistent repository of topics that can be assembled into the final deliverables, often under the management of an information architect and a production team or team members who assume that responsibility during the design and production phases of a project.

Globalization, especially in a single-source environment, adds additional pressure to create collaborative teams. If the global team members are to produce interchangeable content or even content that is standardized and consistent, they need to work collaboratively. The collaboration requirements are exacerbated by distances, time zones, and cultures. Consequently, the information-development manager must become skilled at managing across diverse cultures and ensure that distance and time differences do not become barriers to achieving efficiency and customer quality.

Finally, consider what an effective team might look like in your organization. How would a team environment differ from the environment you now have? When I help organizations plan a move from a "book" to a "topic" information architecture, I find that they have a difficult time envisioning the changes that need to occur to make this new information architecture successful. The immediate assumption is that the information developers will continue working dutifully in their independent silos, producing topics that will go into their final books. Perhaps they will find a topic or two written by a colleague that they can use: copyright statements and warnings immediately come to mind. Perhaps they will use their topics to produce a help system or a web page. But, they will definitely not have to think differently about how they develop content. Unfortunately, if you have staff members who believe that nothing much will change except to learn a new tool set, you have a personnel dilemma to resolve.

If you do not change your work practices to accommodate a new approach to information development, you will have spent a great deal of time and money on tools with almost no return for your investment. You will continue to produce the same books you did before with an insignificant amount of content reuse, and sometimes at a higher cost per page. Almost the only savings that some organizations experience are through a reduction in production costs (less time spent tweaking final deliverables).

If you succeed in implementing a collaborative work environment at the project level and develop an integrated team pursuing corporate and departmental business goals, you will experience change that will carry you forward to continuing innovations.

Developing a team environment

Chapter 20: Managing in a Collaborative Environment, covers creating a collaborative working environment for a project. The approach I recommend is similar to the one you might use to create a team environment in your department. The key, of course, is communication, a meeting of the minds to ensure that team members pursue common goals and work cooperatively. Without communication built into the process, you will have individuals working in independent silos.

Communication, however, seems hard to establish in an atmosphere built around writing. Many people who pursue writing, even technical writing, as a career, are comfortable, and even prefer, working independently. The image of the writer on the mountain top composing a masterpiece is well established in technical writing.

As managers, however, you understand that your team works in a competitive business environment where the value of technical communication and information development is often questioned. By fostering a team environment focused on common business goals, you can help dispel the doubts and create an organization that is considered essential because it produces information products that are valued by the customer, and does so in a cost-effective way.

Many managers, pursuing a team-focused path, often find that their staff does not welcome the change. You will have to use your best powers of persuasion and a change management agenda to move reluctant staff to a new way of working.

The first step is to create a reason for the change, what in change management is often referred to as a "burning platform." A burning platform figuratively represents a degree of unease with the current working environment that is sufficiently critical that it persuades people to change how they think and how they work. People experience a sense of urgency that compels them to change.

In moving to a team environment and calling for increased collaboration in the work of the team, you must create a sense of urgency. You are likely to find many business circumstances that will be persuasive: the possibility of downsizing, outsourcing, or offshoring your operation; the breakup of the department and the dispersion of team members to various product-development teams; a decrease in funding that puts everyone at risk. One team was motivated to work more closely together based on reports that users were frustrated with their inability to find information they needed in the documentation on the website. Another team learned that customers were unable to use the CD-ROM they produced. The team that discovered that their primary information customers were the members of the support teams rather than end users at first balked at the news but then refocused their thinking and developed an effective strategy. An individual contributor who was reluctant to accept help in meeting a difficult deadline later gave a presentation at a national conference touting the effectiveness of working with her junior assistant.

Once you have succeeded in developing a burning platform and infusing a sense of urgency among your team members, you next must establish a communication plan that takes into account the styles and needs of each stakeholder. In Chapter 16: Planning Your Information Development Project, you learn how to conduct a stakeholder analysis when you are planning a project. When you are planning an organizational change of this magnitude, you must also conduct a stakeholder analysis, perhaps as a first step before discussing the urgency of the change. This stakeholder analysis initially focuses on your own staff members. Meet individually with each team member to evaluate his or her understanding of the new business requirements and acceptance of the need for change. Try to discover those who appear supportive but have grave doubts about the success of so pervasive and frightening a change. Statements of doubt are often expressed in terms like, "We've done this before, and it's never worked." "We don't have support from senior management, and the developers won't go along." "What about my job?" The underlying fear being expressed includes fear of losing one's job and fear of seeming to be less than competent in the new environment. People who are very good at the old tools and the old ways of working are afraid that they won't be able to handle the new demands. What if they're unsuccessful? They're afraid of looking incompetent and being judged harshly.

To focus your teams and make them more effective, you need to

- ✔ develop in the team members a sense of urgency in meeting the company's and the team's goals and performing as effectively as possible

- ✔ select team members who are skilled already or have the potential for learning new skills

- ✔ develop a vision of what the future will look like if the team is successful and develop the vision in your first team meeting

- ✔ develop a plan for communicating among team members and with other stakeholders

- ✔ remove the barriers to change

- ✔ set up opportunities for immediate wins so that the team has a feeling of accomplishment

✔ maintain the sense of urgency through the team's life and the life cycle of the projects

✔ make the changes stick through recognition and rewards that help to change the culture of the organization for the long term

In communicating with both enthusiastic and skeptical and fearful staff members, you need to paint a picture of what the new team environment might look like, at the same time admitting that you can't anticipate everything that might change. I've been reasonably successful with a set of cartoons, as illustrated in Figure 10-1, that vividly illustrate how the new team environment will look and feel. The cartoons show people working together on teams, consulting with specialists from other teams or colleagues in other parts of the organization, taking into account the needs of stakeholders outside the organization, bringing customer studies to bear on decision making, and finding innovative ways to respond. They communicate a vision of the future and set expectations about the new behaviors that are needed to support the vision.

Figure 10-1: Working in a new team environment

A team environment means that everyone participates to achieve business goals and find innovative solutions for improving processes and information design. Few team members end up defending the status quo.

Despite your best efforts in planning and communicating about the need for a new team environment, you are certain to encounter barriers to change. Reluctant team members will argue that the new methods really don't work and take up too much time. Outside stakeholders will be uncomfortable with the changes required in their work. For example, reviewers refused to review carefully and colorfully marked conditional text so that product variations could be built into master topics. They kept crossing out the seemingly similar content or procedures without noting that the small differences were deliberate. In another situation, the production manager was so fearful of change that he constantly harangued other team members and the outside consultants with his insistent message that the new tools were impossible to use.

If you are convinced that the change is necessary, you not only will have to communicate the message over and over again before it is heard, but you may also have to ask a negative team member who seems intent on preventing the change to move on.

Best Practice—Defining new roles and responsibilities

In many information-development organizations, everyone has the same job—information developer. Each information developer handles all tasks associated with producing a manual, including researching, planning (if done at all), writing, editing, revising, proofreading, graphics, layout, and final production. Some may even be responsible for shepherding a book through illustration, localization, and translation. Only in organizations that document hardware products are you likely to find dedicated technical illustrators, because they're often the only ones who can create acceptable product drawings.

As a team grows a bit larger, managers may ask someone to handle editing and proofreading. The task usually goes to an individual who shows interest and talent in those activities. Someone may eventually handle all the final production tasks, especially if the team expands to 10 or 15 members. In most cases, the specializations that develop should be based upon the strengths of individuals.

As your team moves to a topic-oriented, rather than a book-oriented development environment, you will need to add or grow additional expertise. Chapter 17: Implementing the Topic Architecture includes a table and descriptions of possible new roles, including information architect and information designer. An information architect develops the overall structure of information types and topics and designs the assemblies for final deliverables. The information designer is responsible for creating style sheets for all the media the team produces. The graphic designer develops illustrations for manuals, websites, and training. The localization coordinator works closely with an outside service provider. Each new role assigns an area of responsibility and accountability to an individual or small group.

Building on strengths

Adding people to your team or developing new roles and specializations for existing team members provides you with an opportunity to exploit the strengths of individual team members. Consider using the Strengths Test described in Marcus Buckingham and Donald Clifton's, *Now, Discover Your Strengths*.[1] They argue that people are happiest building on their strengths to develop new roles for themselves in an organization, rather than being constantly chided about their weaknesses. Information-development managers have used Buckingham and Clifton's strengths test successfully to help team members evaluate their own strengths and decide on new roles to explore that make the most of their potential.

[1] Marcus Buckingham and Donald O. Clifton, *Now, Discover Your Strengths*. New York: Free Press, 2001.

Based on a strengths analysis, you may want to ask people to volunteer for new roles in your organization. Then, you can actively support them with opportunities for training, required resources, and cooperation from other team members and the management team. Communicate about the new roles people have assumed to your own management and peer managers so that they are supportive rather than obstructive. If you note obstructive behaviors, find ways to understand the objections, clarify your purpose, and identify ways that you can improve communication.

Actively Supporting Change

One manager I worked with focused on encouraging his information developers to exert more control over the content, content that had been dictated in the past by product managers. The product managers became angry when information developers awkwardly informed them that text the manager had insisted be inserted in the documents was not needed by the users. The information-development manager had to meet with individual product managers to explain the goals and present evidence of the users' requirements, as well as meet with the senior development managers to garner their support for the change. He also needed his own senior manager to actively support the need for change at the highest levels of the organization.

By actively supporting his team and maintaining the vision of the future, the manager helped ensure that success of the new initiative. The next step, of course, is to reward the team members for their success and call attention outside the department to the benefits for the company and the customers.

Adding new members to the team

You may discover, of course, that no one in your organization shows a particular strength that you need to pursue innovation and achieve business goals. One manager identified a need to sell the services of his department to executives in the company's autonomous business units. Because his information developers were introverts with an aversion to "selling," he sought out a new team member from another department who had excellent people and persuasive skills, as well as the requisite technical knowledge. She became the department's promoter, finding new work for the team throughout the corporation. Officially, her role was to meet with executives, explain the department's capabilities, and capture the requirements of the new business opportunity.

With the move to a new tools environment for XML-based authoring and content-management systems, managers find that those with the technical skills to support the tools that are not always part of the skills mix in the department. The best solution is to find a person with the technical skills needed and who also has experience as an information developer. An individual who has experience in both areas is more likely to understand the working environment of the team than someone who is limited to technical tools skills alone. Early benchmark research indicates that organizations using enterprise tools for content management, XML authoring, and automated production need to devote between 5% and 10% of staff resources to tools maintenance. That percentage will increase if you decide to develop your own tools rather than use commercial off-the-shelf products and industry standards.

Creating specializations

The specialized skills required in a new team environment can often be found by recognizing the strengths already prevalent among your team members. By giving them opportunities to pursue new skills, they gain expertise and are often more satisfied with their jobs. If the skills you need don't exist in your team, you can define the requirements and pursue a staff increase, especially when you have executive support for funding new tools acquisitions.

In one organization I worked with, the executives wanted to know if existing staff could handle the new tools responsibilities. If not, they wanted to know what new positions were needed, if they could be outsourced or part-time, and exactly how much the new positions would cost as part of the request for funding approval. The organization argued successfully that part of the responsibility for the new environment could be handled by the two individuals on the existing tools team, but an additional full-time staff member was needed to manage the new content-management system. They identified the skills needed and hired a person with education and experience in information science. No one on the existing staff met the requirements for the new position.

Your first choice in developing the new specialized skills you need should be your existing team members. By identifying those individuals who are anxious to try out new skills and pursue innovative approaches, you foster growth within the team. Look for grass-roots activities that may already exist in your organization. One manager discovered that a small team had been formed to gather customer information and conduct on-site visits with customers. They had some training and a plan, but they were working undercover. By discovering their underground efforts and sanctioning them, the manager helped the customer team influence the rest of the department and foster improved customer relations.

In the absence of interest or expertise among your current team members, your second choice might be to hire new team members with the specialized skills in hand. The third choice might be outsourcing. Although many managers would prefer to have the skills they need inside the team, it's not always cost-effective to bring special skills inhouse. When new tools are introduced, outside consultants who are experts in their use can get your team up and running more quickly, perhaps gradually training and turning over responsibility to inhouse specialists. Outside specialists in graphics and technical illustration, service providers in the areas of localization and translation, experienced indexers, usability consultants, and others may be more easily added if they are not full-time employees or even full-time contractors.

Accounting for domain expertise

Many information-development organizations are structured around product domains. In many cases, the product development often takes place in independent business units, in new product development teams, or in entire product teams acquired through mergers and acquisitions. As a result, information developers tend to develop product domain expertise, which adds to the silo mentality and makes change to a team-oriented environment more difficult.

Domain expertise can, of course, be a decided advantage to your organization. You have team members who know the products and are viewed as valued members of the product teams. You've learned from unfortunate experience that many product developers are negative toward information developers who do not develop product expertise. They become frustrated when the information developers ask them what to write rather than researching the content on their own and asking intelligent questions.

On the other hand, team members may become overly attached to the product teams, identifying more closely with the product developers than with your organization. They may be reluctant to work with other team members and pursue common business goals, arguing that their business units and product managers have different agendas and requirements.

You may also encounter problems when your domain specialists would like to try something new or work on an innovative project. The product developers may be reluctant to let them go, demanding that they continue to work on their traditional projects. Your opportunity for cross-training others to work in a particular domain may be thwarted. One manager lost a valued staff member because he could never escape from documenting the same product he had worked on for years. The manager had suggested that another staff member be cross-trained in the subject area but the offer was rejected by the product developers. Of course, when the senior staff member quit, no one was prepared to assume responsibility for the information.

Topic-oriented information development strongly supports domain specialization, even to the extent of dividing responsibility among specialists more finely than when they authored books instead of topics. However, it is equally important to foster cross-training so that more than one person is able to handle content in the domain. You will also find it effective to ensure that members of your project teams become knowledgeable about all the content team members are creating so that they can understand the total product in more depth, identify opportunities where content can be used in more than one deliverable, reduce the number of content duplicates or near duplicates, and identify gaps in the existing content, especially conceptual or descriptive topics. Cross-functional teams in which learning from one another is encouraged means never having a team member who claims that he or she "has no idea what anyone else is working on."

Best Practice–Managing the managers

If you are the only manager in your organization, you must provide the leadership to create an effective team environment on your own, with the help, of course, of your most senior team members. If you are a senior manager or a director, you need to coordinate your efforts with other managers, especially if the other managers lead physically dispersed teams. As you look for ways to use resources more effectively and pursue your vision of a new organizational structure, you first need to enlist the support of your other managers. If those managers do not report to you directly, you must work carefully to win their support or your initiatives are unlikely to succeed.

If all the managers report to you directly, your starting point is to build a collaborative management team. Managers, like any stakeholders involved in a change, want to know what is going on and not be surprised by your actions or statements about organizational changes. When the managers report to you, invite them to an early discussion of the business goals and objectives that you have in mind. If they have been aware of the challenges in your business environment or some of the potential threats to the survival of the organization, they should be well prepared for an initial discussion of change. Even if some of the other managers don't report to you directly, take the initiative and invite them into the initial discussions. They may be interested in the opportunity to get a broader audience for their ideas, something that can happen most easily if all the managers work together.

If some of the other managers do not care to participate, you may want first to meet with them individually to discuss your plans and win their cooperation. If you have enough influence among senior managers, you may use a corporate-level sponsor to gain support for your initiative at higher levels. If individual managers find that you have garnered high-level support for the initiative, they may be more disposed to take part.

In either situation, you are obligated to explain how everyone is likely to win by participating actively in the organizational change. Together you have more resources at your disposal than independently. Together you are better able to approach senior management with a proposal and gain a more receptive audience. If the change is necessary for the long-term survival of information development, you are more likely to succeed by joining forces than working independently or presenting contradictory proposals.

You will have to use all your change-management skills in your meetings with the managers. Build a sense of urgency by communicating your understanding of the corporate goals. If you have heard rumors about outsourcing, offshoring, or reductions in force, communicate those to your management team. If you have a vision of the department's future and know why achieving it is essential to the team's survival, explain it thoroughly. Work hard to let the skeptics and the nay-sayers voice their concerns. You'll find it best to get the problems onto the table rather than allowing them to fester back in the cubicles. Find out what the concerns and objections are, but also ask the same skeptics how their objections might be overcome. Facilitate the discussions through which you establish a vision of the future, agree upon some goals and objectives, recognize strengths and weaknesses and opportunities and threats. Finally, after much discussion, ask for their commitment and active participation in the change.

One way of gaining commitment and active participation is to engage the managers in achieving short-term wins. Based on an assessment of their team members, the managers may more easily identify opportunities for making immediate changes in line with the vision of the future. Senior managers are appreciative of projects that deliver on the new vision, especially if they can be completed with minimal initial investment. One team, for example, decided to pursue a minimalist agenda and redesign the administrative manuals for which they were responsible before senior management approved the acquisition of a content-management system. As a result of their small redesign project, they were able to demonstrate the potential for reducing the volume of pages and provide a model for a new information architecture. They could show an immediate return on investment because the reduced page volume meant fewer words to translate and a smaller book to print.

The success of a short-term initiative allows the manager and the team to win recognition for their effort and praise for their accomplishments throughout the organization. By celebrating the small success, you gain a stronger commitment from the manager involved and all the other team managers. They begin to see the potential for a positive gain for themselves and their organization.

Team work makes managers more effective

When your managers meet regularly with their counterparts in your department or in the larger organization, the performance of all the managers improves:

- ✔ Managers make decisions that are better aligned with departmental and company goals and policies because they are sensitive to the impact of those decisions on the larger organizations.

- ✔ Managers develop professional standards for all the members to follow.

- ✔ The innovations that individual teams develop are open to the entire organization rather than hidden in individual initiatives.

- ✔ Because managers must work together, they learn to support one another. As a result, the managers become less dependent on their team members for support and more closely aligned with their peers.

- ✔ Decisions about hiring and promotion are made by the team of managers so that personnel decisions are more consistent.

- ✔ Training plans are developed globally so that all team members have an opportunity for growth.

- ✔ Decisions about compensation are also made more consistently across the entire organization, especially if compensation is linked to performance ranking.

- ✔ Your team of managers is less likely to engage in unproductive competition.

- ✔ You can ensure that your management team is fully involved in global initiatives to improve performance, increase productivity, or reduce costs.[2]

Barriers to team performance

Team success, whether the team members are managers or information developers, depends upon building trust. People often believe they are more successful on their own, especially in western cultures. They feel that team members slow them down or force them into less than optimal choices. Especially among introverts, team activities and collaborative work are uncomfortable.

Trust among team members and their managers begins with your advocacy and behaviors. If you demonstrate the behavior of trust that you want to instill in the managers, they are more likely to learn to trust one another and communicate trust-building behaviors to their team members. If you ask people to assume responsibility for team actions and be accountable for their activities, you encourage everyone to act responsibly and for the good of the organization.

Adding new members to any team can cause the team to become unsettled. You might find it useful to assign an experienced manager and team member to a new person to

---

[2] Thanks to Julie Bradbury for developing many of the ideas presented here on managing managers.

help them understand how the team works and what behaviors and outcomes are expected of them.

You may also be responsible for counseling uncooperative individuals who prove disruptive to others on the team. Sometimes training is effective for instituting a team spirit. In other cases, you may need uncooperative team members to look elsewhere for opportunities.

Best Practice—Working with remote team members

Many more managers today find themselves managing people who do not work in the same office space, building, city, or even in the same country. Some staff may be co-located in small groups of information developers. Other small groups of information developers may be co-located with development teams in various local or remote locations. Some may work as lone writers co-located with product-development teams but without professional colleagues. You may have team members who work at home full- or part-time in your immediate vicinity or others who have moved away and continue as part of the team. As a senior manager, you are likely to have teams that have joined through mergers and acquisitions with their own management, standards, and culture, as well as global teams located around the world, some outsourced and some directly employed by your organization.

Managing in all or even a few of these situations is always challenging and may ultimately be unsuccessful. Success requires that you develop strong communication skills and build a management team that works together toward the same goals.

As many publications managers recognize, managing remote teams takes much more involvement and care than any of us would have imagined, especially when the remote teams have a history of acting independently. In such cases, you encounter even more challenging issues when the remote teams are inexperienced in information development because they are located offshore in emerging economies.

The Nut Island Effect

In the March 2001 issue of the *Harvard Business Review*, Paul Levy describes the Nut Island disaster. For more than 30 years, Nut Island was the Boston area sewage treatment plant, located just far enough from Boston and city managers to become isolated. What seemed like a dream team of self-sufficient workers was responsible for the worst pollution disaster in Boston history, releasing billions of gallons of raw sewage into Boston Harbor.

The Nut Island disaster occurred because senior management paid little or no attention to the remote team at the sewage plant. They assumed that the staff knew what it was doing and would keep operations going, no matter what happened. The staff worked in almost complete isolation from the management, who were focused on more pressing and more politically relevant issues. As long as the Nut Island staff did its job and didn't complain, no one paid them much attention. Early on, staff and managers did request funding to maintain and upgrade failing equipment. But upgrading the sewage plant had few political payoffs for management, and so was ignored. The staff heard the message and became increasingly self-sufficient, even in the face of the impending disaster.

continued

The Nut Island Effect *continued*

Levy points out that the organizational pathology characteristic of Nut Island is not rare. Rather, many organizations will suffer from the Nut Island effect, as publications managers will no doubt quickly recognize. Not only should you guard against the Nut Island effect among remote teams of information developers, your own organizations, often ignored by senior management, are in danger of developing an "us against the world" mentality.

As Levy explains, the Nut Island effect goes through five fairly predictable stages, which sound frighteningly familiar.

In Stage 1, senior management decides that an important, although not critical task should be assigned to a manager and team. They give the manager and team almost complete autonomy to conduct business as they see fit, as long as deadlines are met and embarrassing disasters avoided. The team members usually have a strong work ethic and are often happiest working out of the limelight. They are very good at defining and managing their own activities, few of which are understood by senior management.

Stage 2 is often quite felicitous. Senior management is pleased that the team can function with little, if any supervision. When the team does ask for funding or responsibility for new functions (such as customer studies), they are often rebuffed. Why spend more money on a function that is holding its own? Over time, the team begins to resent senior managers who don't care about what they do.

During Stage 3, the team becomes increasingly isolated, taking on a strong "us against the world" point of view. They become increasingly skilled at disguising problems in front of outsiders. Think of the stories of remote teams who fail to mention major problems in file management, technical understanding, and editorial abilities until a deadline has been missed. Of course, until problems demand attention, senior management equates silence with good process control.

Stage 4 marks the extent of the isolation. Rarely exposed to outside ideas or industry best practices, the team makes up its own rules. The rules seem like good practices because there is no way to compare them. However, the house rules often mask serious deficiencies in process and standards. I've visited many departments who brag about the quality of their information design, only to find them dreadfully behind and oblivious to industry standards. Their stock of ideas is limited to what they can think up themselves.

By the final stage, the distorted view of reality has become difficult, if not impossible to correct. Think of a "legacy" staff, complacent about their processes, who are producing information products that no one uses and reflect practices considered out of date for 20 years or more. Eventually, some event breaks the stalemate in which management ignores the team and the team ignores management. In recent years, we've seen wholesale layoffs and outsourcing of expensive and ineffectual information-development organizations. The typical reaction is shock. "Haven't we been doing the job everyone asked of us?" "Haven't we worked extra long hours to meet deadlines?" "Why have we suddenly become obsolete?"

As Levy points out, "a team can easily lose sight of the big picture when it narrowly focuses on a demanding task. The task itself becomes the big picture, crowding other considerations out of the frame" (p. 58).[3]

---

[3] Paul F. Levy, "The Nut Island Effect: When Good Teams Go Wrong," *Harvard Business Review,* March 2001.

Despite its apparent ubiquitous presence in our organizations, the Nut Island effect can be avoided. Certainly, everyone wants team members who work hard and are dedicated to the team's success. But you must keep them from becoming isolated, unable to look at their work in a larger context. You must keep your own teams and especially the remote teams from becoming exclusively focused on deadlines and their own deliverables. Instead, they need to participate in a strategic vision that is aligned with larger corporate goals and customer needs:

1 Find ways to measure performance and give rewards that match larger corporate goals.

2 Keep senior management involved, especially through site visits and attendance at staff celebrations and other customer-critical events.

3 Integrate team members. Staff at the "home office" need to circulate among the remote teams. Remote team members need to work for a time at headquarters. All team members need to be actively involved with cross-departmental teams working on customer studies, requirements, product design, support activities, and so on.

4 Bring in outside experts to introduce new ideas and benchmark team actions against industry best practices. You need to involve everyone in the review and improvement process.

You also need, of course, to get out of the office yourself to learn directly from remote team members and managers and to interact with colleagues in your own field and associated fields. By building a community of professional colleagues and interacting in professional settings, you are continuously exposed to new thinking, decreasing your isolation, and providing you with challenges.

Paul Levy observes that putting good people into situations where they do the wrong things is not just avoidable, it is tragic. It's a waste of human potential and a threat to the corporation's survival. It's a good idea to seek out possible instances of the Nut Island effect and work very hard to prevent them from taking hold.

Helping remote team members succeed

Some of the ways managers successfully handle distributed teams include

✔ traveling regularly to every group location and getting to know staff members personally

✔ scheduling regular all-hands meetings and conference calls so that everyone hears the same message from management

✔ funding an occasional opportunity for people to travel to one location for a face-to-face meeting

✔ holding a department conference to foster the exchange of innovative ideas and successful projects

- ✔ making frequent use of conference calls and video conferences to encourage communication at all levels of the organization

- ✔ developing a departmental website and inviting team members to post pictures and bios

- ✔ developing a department newsletter or a news page on the website

- ✔ encouraging regular interpersonal communication among team members through telephone calls, emails, listservs, wikis, and instant messaging

- ✔ instituting a regular reward program to recognize staff accomplishments

- ✔ creating cross-location working groups involved in developing best practices for the entire organization

- ✔ holding team get-togethers in conjunction with regional, national, or international professional conferences

All of these activities are designed to keep lines of communication open and active. For example, I worked with a management team that scheduled a series of workshops for the staff, held at the larger locations. Lone writers or small groups are invited to the workshops, giving them an opportunity to meet and work directly with colleagues who they communicate with but have never met face to face. These events have been successful in forging relationships and opening lines of communication.

As part of your team-building efforts, you may have to set some basic rules that all team members are required to follow. Set up a working group whose responsibility it is to develop the communication ground rules. Basic rules the groups have instituted include

- ✔ responding to phone messages and emails promptly, perhaps with a worst-case time limit

- ✔ considering the tone of written communication to avoid unintended reactions

- ✔ following best practices for meetings and conference calls so that everyone has an opportunity to be heard

- ✔ taking responsibility for promised actions and letting others know when a scheduled delivery cannot be met, why it can't be met, and when one can promise delivery

Many organizations have people in human resources who can provide coaching on best practices for conference calls, meetings, emails, and instant messages. Training programs are available, often arranged through your human resources organization, that can help your entire team learn to adapt to working with people they seldom or ever see in person. With the right training, you can help your staff set their own goals for communicating with their counterparts in other parts of the organization. The goal is to keep the lines of communication open and avoid making people upset when behaviors differ.

Managing telecommuters and lone writers

In an effort to reduce costs, accommodate widely dispersed development teams, and accommodate employees who are reluctant to commute long distances, more organizations promote telecommuting and assign individual information developers to work alone, even though they are co-located with a development team. Although you will find some differences in the management of the two groups, they are remarkably similar.

Acquiring and retaining experienced, talented information developers is not easy, especially in more competitive markets around the world. In North America and Western Europe, staff members who are permitted to work from home are more likely to stay with the team. Experienced team members who want to move for personal and family reasons decide to continue working with your team if they can work from home.

After you have decided that a telecommuting team member is reliable, communicative, and a valued contributor worth keeping, you need to discuss together your assumptions about management and communication. Telecommuters should be able to manage their own project work and meet their deadlines without prodding from team leads. If your project managers are estimating their projects and assigning work in a reasonable way, the telecommuters will know what they are expected to produce on what timeline. If they are motivated to succeed, conscientious, and trustworthy, you should expect that work gets done as if the individuals were in the next cubicle. Because telecommuters are often free from the distractions of the office, you may even find that they are more productive than their in-office colleagues.

If you have a global team working in many different time zones, you may have a few telecommuters who agree to work in the evening or early in the morning to accommodate time differences. Because they can more easily adjust their schedules, they may help you enhance team cohesiveness.

Although the telecommuters find clear advantages to working at home, including avoiding long commutes, having more time with families, reducing costs, and working flexible schedules, they also find disadvantages in coordinating activities with other team members. As the manager, you must ensure that telecommuters are kept in the communication loop, invited to meetings, and included in the flow of information. For example, team members may have to learn how to accommodate remote team members during conference calls or ensure that they have access to all the information they need to work productively. Training on meeting etiquette should include limiting private conversations and asking speakers to identify themselves, as well as ensuring that everyone has the correct call-in numbers and meeting times.

Like telecommuters, lone writers working with product developers experience a feeling of isolation from colleagues. If you have set up a departmental website and instituted listservs, instant messaging, and other forms of electronic communication and team building, you can help bring these individuals into the team. Lone writers often need help asserting their independence from the people they work with every day. You need to ensure that they understand the department's vision so that they can effectively represent the department to their work group.

Best Practice—Working with global teams

Global organizations, by definition, have work centers worldwide, which often include both product and information developers. As a result, as a manager of a global team, you find yourself managing team members who speak different languages and make different cultural assumptions. Organizations become global through mergers and acquisitions, but they also find opportunities to reduce costs by creating work centers in emerging economies. In some cases, the information developers will have much the same education, training, and experience as those on your home staff. In other cases, you will find yourself with team members who are completely new to information development.

The strategies you use to develop an effective team in your own location will also work, with some modification, in a global environment. If you already have a set of goals for your original team, you will have to reexamine and possibly modify those goals for a global team. Your global team will, you hope, embrace a shared vision of the future in which everyone contributes to building efficient processes and increasing the value of information to the customer.

Achieving that common vision will likely require commitment and hard work by you and your management team. If you already have a management team, add the new managers to the team quickly. Asking existing team members to mentor newcomers will be especially valuable in bringing new managers into the fold. If the teams you have inherited lack a management structure, you may have to appoint a manager from among your existing team until you are able to hire a local manager. Once the local manager is hired, your existing manager should assume the mentoring role. Even if the new manager does not report to you directly, find ways to include that manager in your global team.

During mergers and acquisitions, many corporate executives promise newly acquired organizations considerable autonomy. They are allowed to function independently of the new organization, maintaining all their former practices and standards. That often creates a conflict with a senior manager who is responsible for developing a global perspective and ensuring that information goals are synchronized throughout the organization. In some instances, your overtures or collaboration may be completely ignored. In other instances, you may find that the independent teams and their managers are either willing to listen or are already enthusiastic about working together.

If the expectation was established from the beginning that the new manager and team members would become part of your organization, you need to be prepared to open the lines of communication quickly. You should immediately arrange to travel to the new location and meet personally with the manager and all the team members. Establishing a common understanding at the beginning is essential to the success of the integration. As soon as you can, ask members of the new team to serve on cross-functional teams together with veteran team members.

If the new teams are in countries and cultures that are new to your experience, you may want to enroll in a cultural awareness workshop before scheduling your first meeting. Even a basic understanding of the different behavioral expectations will improve communication. In many high-touch cultures, a personal visit from the manager is essential for establishing future working relationships.

Just as you would with any geographically distributed team members, you must find ways that team members can work together on special initiatives, even if they may not work on the same information-development projects. It will be most important to invite global team members to participate in existing initiatives. It will also be important to begin new initiatives that directly involve the new members as quickly as possible. For example, if you have a team of novice information developers, you may want to ask some of your experienced team members to develop a workshop in basic technical writing skills. These team members should work closely with the new members to develop the goals for the workshop and set the agenda. They might also train someone on the new team so that he or she is able to train new team members as they are added.

With your management team, develop a strategy for integrating the new global team members into the overall organizational direction. In the following vignette, JoCarol Gau of BMC Corporation explains how she developed a successful initiative to institute and develop a team in India.

Leading a Global Organization

By JoCarol Gau, BMC Corporation

After reading *Leading Change* by John P. Kotter, I decided to put Kotter's method to the test on a change effort in my department. The change effort is the offshoring of information development activities to India.

I found myself in a work crisis involving change and insecurity. Change was in the form of R&D jobs being offshored to India and long-time employees in the US being laid off. Our small operation in India had grown to about 400 employees in just two years.

What did this mean for the information-development organization? Many people thought that the offshoring of support, development, and quality assurance personnel would not cross over to technical writers. However, the Information Design and Development (ID&D) organization started researching the technical communication profession in India and the possibility of adding information-development jobs at our site in Pune, India.

Setting up even a small staff of information developers in India involved a huge mind shift, not only for the writing organization but also for the many extended team members in the R&D organization. In the US, questions were raised about the writing skills of non-native English speakers, about the availability of educational programs and training for technical communicators in India, and especially about the background and experience of software documentation professionals.

However, our research indicated clear advantages of having some writers in India: significant cost savings, the efficiencies of having writers working side-by-side with remote R&D teams, and the potential of offloading work for an overburdened workforce in the US. Also, we discovered that India indeed had a small but growing force of technical writers and an STC membership of about 400.

To take advantage of the offshore potential and, at the same time, to dispel the concerns of the home organization, the BMC managers decided to employ Kotter's eight-stage process. The process helped us to drive change, not only within the writing organization, but also with extended R&D teams. So far, the process has proven to be a practical and worthwhile approach to leading change efforts in the offshoring arena.

continued

Leading a Global Organization *continued*

Stage 1. Establishing a Sense of Urgency

The sense of urgency for us was created by an order from upper management to slash costs and improve operating margins. While other groups were lowering costs with cheaper Indian labor, ID&D found savings by moving writers at higher-cost commercial locations to offices at home, but these measures were not enough. Reducing head-count was not an option, as R&D teams were hiring abroad at a fast pace and we needed information developers to work with those teams. If ID&D management couldn't find cheaper ways of delivering documentation, writers in the US stood a great chance of being outsourced or laid off. The situation was urgent. We eventually came to the conclusion that by adding some lower-cost labor in India, we could actually save jobs in the US because this tactic lowered our overall expenses for the department.

Stage 2: Creating the Guiding Coalition

Our small team that is leading the offshoring effort includes the ID&D director and senior manager, a lead and senior writer, and an editor. The coalition has expanded to include our new Indian counterparts. This guiding coalition contains the key characteristics that Kotter suggests: ID&D management, with the support of the senior leadership team, has power to guide the change effort. The writers and editor provide expertise and leadership. Involving our Indian peers gives them a stake in the process and lends credibility that the new team will be successful. Along with the urgency established in stage one, having an authoritative, experienced, and credible guiding coalition has helped lower the resistance to change.

Stage 3: Developing a Vision and Strategy

I found Kotter's emphasis on vision to be one of the most enlightening parts of his work and have applied it judiciously to our offshoring effort. One of the first things the guiding coalition did was to "imagine a picture of the future" and write it down. Our guiding coalition knew exactly where we wanted to be in one year and developed a phased quarterly process (strategy) for getting us there. The phased process provides a realistic approach to building a foundation that will support an overseas operation.

Stage 4: Communicating the Change Vision

For our offshoring efforts, communication does not have a set beginning and end. Communication has to occur throughout the offshoring process with regular updates at all-hands meetings, project kick-off meetings, presentations by the guiding coalition at staff meetings, a global ID&D team website, and newsletter articles. These forums allow for two-way communication to receive feedback from employees about the offshoring process and to address concerns such as maintaining the quality of the documentation, mentoring less experienced writers overseas, and working with remote teams in opposite time zones. Ongoing communication continues to reinforce why we are offshoring and provides information about how we are doing it, who is involved, and what the plan is for success.

Stage 5: Empowering Employees for Broad-Based Action

Employees are empowered by serving as mentors to work with the new Indian writers. The mentors start new projects with remote Indian workers, determine the type of information that must be conveyed, and set expectations with development teams. In addition, the company provides intercultural training classes for both the US and India so that employees at all locations learn about the differences between working with low-context cultures in the US and high-context cultures in India. We also have training for "Virtual Teams" and "Leading from a Distance" to learn best practices of working in a virtual team environment. These actions remove barriers that might stall the change process. For instance, remote team members learn how to build relationships despite the limitations of electronic communications versus face-to-face interaction. As mentors, employees are accountable for ensuring a successful outcome. Working together toward common goals helps to remove the cynicism and doubts about moving some of the R&D operations to lower-cost "competency centers."

Stage 6: Generating Short-Term Wins

Some short-term wins are already happening as writers in India have become successful at completing documentation assignments. They are learning to publish independently without relying on US writers, a huge benefit to an already stretched workforce. Through surveys of US and Indian employees, we are finding out where improvements can be made.

Stage 7: Consolidating Gains and Producing More Change

I can't say we are at this stage yet, but I believe that eventually writers and development team members in all locations will see the benefits of working together in our new global environment. What we've seen so far is that most people enjoy working with different cultures and feel fortunate to have these opportunities. Relationships are forming across time zones and oceans. In the US, we've been forced out of our shells and grown because of it. That's not to say that we don't have some holdouts. Some of us still wonder about the long-term benefits of sending work abroad, but the number of us who do is growing smaller.

Stage 8: Anchoring New Approaches in the Culture

The future holds promising for enabling us to work together to reduce costs, maintain and improve quality, and stay competitive in a global environment. I believe that the global workforce is here to stay, and that in the near future, we'll accept it as the norm. I hope that our ability to compete opens up even more jobs for American workers and that we don't see degradation in our work life because of it. As my team approaches its one-year anniversary for offshoring, we must demonstrate continued leadership in the new global environment.[4]

In our studies of adding information developers in emerging economies like India and China, we have found that the efforts are often more successful when the new team members are direct employees of the company rather than employed by an outsource company. Direct employees are usually hired by your staff and have the skills or potential to meet your needs. They tend to be more loyal over the long term because they are brought into the organization and supported through training and career development opportunities. They begin to work as members of project teams with global colleagues and assume leadership responsibilities when they are ready.

---

[4] JoCarol Gau, "Putting Kotter's Ideas to the Test: Leading Change Through an Offshoring Effort," *Best Practices* of The Center for Information-Development Management, August 2005.

Beginning with a mature process

You are most likely to be successful working with global teams if your existing team has developed mature processes. In the Information Process Maturity Model (IPMM), a Level 3 organization has the following characteristics:

✔ A strong centralized management team

✔ Established processes for project management, including estimating resources required and tracking projects through the information-development life cycle

✔ Well-integrated quality assurance activities, including developmental as well as copyediting

✔ Policies for hiring and training new and existing team members

✔ Processes in place to foster innovations to improve efficiency and effectiveness

✔ A growing understanding of customer information needs

✔ Budgetary responsibilities, including responsibility for managing overall costs of information development

You can find more details about IPMM Level 3 and higher in Chapter 2: The Information Process Maturity Model.

If you do not have mature processes in place in your own organization, you will find it significantly more difficult to develop successful processes among new, often inexperienced global teams. The new team members, who may be more comfortable following established practices than creating their own, are likely to become frustrated at the lack of consistent direction. That implies that if you are asked to work with a new team in a new country, you need first to get your own practices into better shape, including writing process and style guides, establishing project management protocols, and setting objectives for quality and timeliness.

Having an adequate budget

You should already recognize that the promised cost savings of hiring information developers in low-cost economies are not likely to be achieved immediately. Your startup costs will be considerable in hiring, managing, training, and quality assurance. Be certain to track these costs, especially the time required by your existing staff to mentor the newcomers and the time your staff may need to spend editing and even rewriting work. Assume that you will not achieve the promised savings for two to three years after beginning, even if the salaries levels are much lower. One manager summed up her experience by comparing the new venture to hiring a team of summer interns with no prior experience and housing them 12,000 miles and 12 time zones away.

Developing an implementation plan

Planning should include startup and ongoing management, travel, relocation of managers or team members, hiring and training strategy, expected start-up and ongoing costs, communication, project cost tracking, calculations of return on investment, and more. It is much better to have some plan in place, even if it is subject to major changes, than no plan at all.

Begin your implementation with a carefully chosen pilot project. The pilot project will be most successful if it is carefully planned, involving well-understood existing documentation that requires maintenance and updating rather than information for an entirely new product. The more structure that can be provided initially, the more successful new information developers will be, especially when they have little or no training or experience in technical communication and no on-site experienced publications management. An experienced manager would give a new team such structured work even if the team were in the home location.

Involve as many of your local information developers with your global implementation as you can. Include individuals from local and global teams in the same project rather than relegating only legacy work to the new team members. They will learn faster if they are integrated with local information developers in projects, and the local information developers will be happier and more helpful to the global team members.

All of the successful implementations that Comtech has benchmarked were started with a reasonable number of information developers. The attempts with one, two, or even six information developers have failed. Many of the costs of developing global teams in emerging economies are fixed, but the savings depend on how much work can be done by new information developers compared with their American or European counterparts. The tendency has been for companies to inadequately support small implementations. Without a minimum number of people working together, it is difficult to develop the sense of community that is important to eastern cultures.

Many start-up costs and some ongoing costs are independent of the size of the group being established. These costs include travel, hiring, training, editing, and establishing process and management. Unless your labor savings are large enough to offset these costs, your implementation will not be successful. With a small number of information developers in the new global location, the absence of a workplace community results in high attrition levels.

Best Practice–Outsourcing and offshoring

Outsourcing information development often seems attractive to senior management because they can reduce headcount and obtain what they believe is an equal or better quality of service from a committed and responsible vendor. They hope to reduce internal costs, at least to the extent they show up on balance sheets by decreasing the number of direct headcount. As an information-development manager, you may also identify opportunities on your own for an outsourcing relationship. You may find it easier to hire contractors than direct employees. You may have overload work during project deadlines that could be done without increasing permanent staff. You may need the expertise of an outsource vendor's staff to work in an area that is new to your team members.

To begin an outsourcing effort, you need to define carefully and thoroughly the work to be done and your expectations for its quality. In fact, you need a statement of work that is detailed and explicit in defining your requirements. The best statements of work come when you already are experienced in defining internal projects and estimating and tracking their progress. If you have never done so internally, you have a considerable learning curve in defining a work effort for an outsource contract.

Perhaps the simplest type of outsourcing is to hire contract information developers through an outsourcing agency. Often, you need to define little more than the qualifications required and the length of the project. All the responsibility for training and managing the contract staff falls on you and your project managers. However, if you hope to have the work managed by the agency, you must define the scope and complexity of the project. You may find it valuable to estimate the cost of the project were it to be done internally with your staff as a point of comparison to evaluate proposals from potential vendors.

If you want the outsourcing vendor to handle the entire job, including employing and housing its own information developers, assuring the quality of the work, providing its own equipment, and managing the project, you are obligated to define your expectations in detail in a request for proposal. By knowing what it would cost to do the same work internally in your organization, you may even be able to make a business case to avoid outsourcing completely.

Select potential outsource vendors on the basis of the quality they can provide

You may have an opportunity to outsource to a local vendor whose staff may easily communicate with your team members and with subject-matter experts on the development and product marketing teams. You may be required to work with outsource vendors in low-cost economies, distant from your home location. Consider first the quality of work you expect and define this quality in a service level agreement that you include with the request for proposal.

A service level agreement should include definitions of the timeline for deliverables, including all interim deliverables for the project. Deliverables should include project plans, estimates of work, project-tracking reports, and periodic progress reports in addition to the final content deliverables. The agreement should state the requirements for the quality of the deliverables in terms of completeness and accuracy, as well as readability, terminology, standard English, and usability. You must define how each of the quality requirements will be measured. For example, you may define accuracy as passing a review by subject-matter experts or testing successfully with the product. Usability may be defined as successfully being reviewed in usability tests with typical customers. The more thoroughly you define your expectations, the less likely you will end up in a contract dispute.

In the service level agreement, define the qualifications you expect of the people hired to do the work. You will need to establish qualifications for each position, including years of experience in the information-development field, knowledge of the technical subject matter, education, and training. With the qualifications defined, you may have to accept anyone the vendor hires.

If you are a mature organization, you may want to require a defined level of process maturity from your outsource vendor, as measured by the five levels of process maturity in the IPMM. You are unlikely to find any organization with a process maturity level lower than Level 3 to be acceptable. If you do not know the IPMM level of the vendors, review the criteria that define a Level 3 organization and ask for evidence that the vendor has achieved this level in their work practices.

Many offshore outsource vendors claim a process maturity level of 5 according to the capability maturity model of the Software Engineering Institute. Of course, they have been measured according to their product-development skills, not their information-development abilities. However, as Level 5 organizations, they often have in place well-defined processes and best practices that ensure that projects will proceed smoothly from beginning to end. At the same time, as Level 5 organizations, they expect that you will be well-prepared to manage projects at the same level they do. If you do not have a Level 5 organization, as defined by the Software Engineering Institute, you may find yourself at a disadvantage in working with a Level 5 organization. You may be accustomed to relying on the individual initiative of your team members to understand users and research new products during the development life cycle. You may expect team members to work independently and meet final deadlines without tracking their progress or asking for estimates of the time they require to complete their tasks. In fact, you may not know much about the activities they are pursuing to get their work done. You may expect your team members to manage themselves. If so, you are most likely at a Level 1 or 2 of process maturity, at least in IPMM terms. As such, you may find it frustrating to work with an outsource vendor that has high expectations about the information you can provide about the projects. I do not recommend that you engage in outsourcing or offshoring until you reach an IPMM Level 3.

If your product developers are also operating at a low level of process maturity, you may have difficulty getting them to work effectively with a mature outsource vendor. The vendor will expect information to be provided in requirements and product specifications that are kept up to date. They will expect timely reviews of draft deliverables and the resolution of gaps in the information. They will expect formal notification of product changes that will lead to changes in scope and timing of deliverables. They will also expect to be paid for all their work, including extra work required to accommodate changes in the product that are outside the original project scope.

At the same time, you probably don't want to contract with an outsource vendor that is disorganized and unreliable in estimating projects and keeping to deadlines. You can't expect to obtain reliable and consistent work from an organization at Level 1 or 2. If you do not have a mature process in place in your organization, you probably should not be outsourcing at all.

Avoiding the low-priced vendor

Avoiding the low-priced vendor is often a best measure in outsourcing. However, you may be required to outsource only to low-cost economies where the only qualification is low prices. If at all possible, identify more than one potential vendor and ask for references. Talking with managers who have used the vendor's services will give you an indication of the quality and reliability of the work they have done in the past. As you talk with references, be careful to discuss requirements. You may have a different set of expectations than another manager. Ensure that the vendor does the type of work you are looking for. If you want contractors to work inhouse, be certain that the vendors on your list provide this type of service. If you want the information developers to work remotely and be fully managed by the vendor, don't select a contract agency that only sends people to your site.

As you develop your request for proposal, ask for a breakdown of services and their costs. You want to know what percentage of costs goes toward quality assurance or project management activities. If you are hiring a group of contractors, you want prices in terms of hourly rates for each skill level. If you are hiring a vendor to do the entire job for you, you should ask for a cost for the entire project. However, know that to obtain a useful total project cost, you must provide data about the scope of the project, even if you do not yet know exactly what the scope will be. The outsource vendor will need information about the nature and number of deliverables required in terms of pages or words, how much content is existing and how much will have to be revised or written new, the information required through the information-development life cycle, the number of drafts to be expected, the graphics required, and the amount of access to the product to review content and produce screen images. You will have to detail how the vendors will gain access to subject-matter experts, products in development, existing content, style sheets and templates, and anything else you normally use during your information development inhouse.

Remember that your outsource vendor is entitled to a reasonable profit as well. You will have a better relationship with a vendor that is successful than with one that is afraid of losing money on your project. Be prepared to anticipate and fairly evaluate requests for changes in scope. Don't expect outsource vendors to do everything you might ask of your inhouse information developers without additional compensation. One of the surprises that many organizations experience is the cost of overtime or additional hours that they are accustomed to getting for free from overworked employees.

Summary

Information-development managers are challenged today to develop effective teams across distributed organizations located around the world. The team members include those hired locally, those working alone or at home, those added through mergers and acquisitions, those set up in low-cost, emerging economies, and those who work for outsourced agencies both locally and globally. Few managers are equipped to handle all the diversity as individuals; for that reason, they must depend upon the skills of a management team that often includes individuals who were once independent managers in their own organizations.

To create an effective team environment among all the possible variables requires attention to change management and team building:

- ✔ Consider carefully the tactics you will need to develop a collegial and cooperative team environment.

- ✔ Establish a business strategy to support the need for a collaborative team, especially if you have not encouraged collaboration in your previous business model.

✔ Define the roles and responsibilities required of team members, focusing on developing the strengths of each individual and adding new staff to assume specializations required by a changing work environment.

✔ If you have a team of managers working for you, develop expectations for your management team and develop their best practices.

✔ Understand the challenges of managing people who are not in the same location you are, including remote team members in many different work environments.

✔ Learn how to work with managers and team members in different countries with different languages and cultures.

✔ If you decide to outsource some of your organization's work, learn to manage outsource projects effectively, whether your vendors are local or in low-cost, emerging economies.

Nothing is more satisfying to a manager than having established and maintained a smoothly operating and effective team.

"We all add our strengths to the team."

Chapter

11

Managing Your
Team Resources

Management is nothing more than motivating other people.
—Lee Iacocca

If you don't believe that good management contributes to the success of your company, consider the results of the study conducted in 2005 by McKinsey Company and the London School of Economics. The study surveyed 700 manufacturing companies in the US, the UK, France, and Germany and compared their responses to the financial performance of the companies, including market share, market growth, and shareholder value. They found that excellent managers produce excellent results, and mediocre managers lead to mediocre corporate performance. Those managers who follow best practices in defining standards, measuring results, and promoting the effectiveness and productivity of their teams, as well as corporate assets and capabilities achieve results that are measured in the bottom line.

The McKinsey study concludes that one of the best ways for managers to learn what works and what doesn't is through the sharing and publication of industry best practices. That, of course, is the principle on which CIDM is built. Good managers look for and rapidly adopt innovative practices among their competition and the best in class. Among the best practices that McKinsey studied were methods used by managers to decrease the cost of operations by adopting lean methods, to set goals and reward employees for achieving them and finding ways to find, attract, and retain talented employees.

One of the most interesting measurements that the study used to correlate management with company performance was Total Factor Productivity (TFP). TFP defines an efficiency measure that captures factors other than hours worked or capital investment. It includes management techniques, use of technology, the use of public infrastructure, and just plain luck. They found that one point on the management scale (1 to 5) of 18 characteristics correlated with 6 percentage points on the TFP scale. That percentage correlated statistically with a 35% gain in return-on-capital-employed (ROCE). That is quite amazing. It shows that good management produces best corporate results.

Better managed companies are generally more flexible and have workforces that are more likely to support change. They have more training and opportunities for professional development and they support more autonomy in decision making. Interestingly, they discovered that companies with more female managers tended to have more flexible and autonomous teams. Better managed companies provide for a better work-life balance.

Managing Information Developers

You certainly have heard the saying, "A team is only as good as its weakest member." Despite that truism, no one ever has, nor really wants, a team that has nothing but star performers. The egos alone would make the team difficult to manage. Any team is made up of people who have a variety of strengths and weaknesses. The team has people who are star performers in some situations and some of the time but not all of the time or in all situations.

As an information-development manager, you have the responsibility of encouraging all your team members to contribute to the best of their abilities. You must encourage individuals to take leadership roles when they are best suited to lead and to follow others when appropriate. You must find ways of motivating people to perform beyond their averages. You have to help people identify the objectives that they will work to achieve as a part of their short- or long-term performance. You have to intervene when problems occur to get an individual or the team back on track.

Many managers find personnel management their most difficult assignment. They would much prefer to manage deadlines and deliverables than people. However, unless the people you work with understand clearly what you expect from them, they will find their own direction, often in conflict with the larger organization, your department, or their own teammates.

Many resources are usually available in an organization to assist you in personnel management issues. You are responsible for hiring new staff, providing for training and professional development, conducting yearly performance evaluations, setting salary levels, rewarding exemplary contributions, and keeping people motivated to do their best work. You may be able to use resources that are provided by your human resources organization for your own training. Human resources may provide support when you have to handle difficult personnel situations or deliver bad news. Resources are also available in personnel management literature and through membership in groups like the Management Special Interest Group (SIG) of the Society for Technical Communication (STC) and the Center for Information-Development Management (CIDM).

Many of the resources deal with general personnel management issues. Few consider the special needs of information developers in an organization. Managers report that the special needs of their information developers are the result of personality types and the culture of technical communication:

✔ Information developers take pride in the quality of their work, especially their final deliverables.

✔ Information developers often enjoy learning to use new computer software that can add to the quality of their deliverables.

✔ Information developers are conscientious and hardworking, willing to put in extra time when necessary to meet deadlines and improve the quality of their deliverables.

✔ As conscientious workers, information developers often respect the traditions of the technical communication profession and of sound writing and editing in general.

✔ Information developers are often shy introverts, making them uncomfortable working in teams or with people they don't know personally.

✔ As shy introverts, information developers may be reluctant to meet with customers or speak out in meetings with people from other organizations.

✔ Information developers may have difficulty recognizing the relationship of their day-to-day work with larger corporate goals.

✔ As traditionalists, information developers may be reluctant to change the way they work.

✔ Information developers may not want to assert their own positions or those of their managers and colleagues in discussions with people they believe have more authority.

Clearly, these attributes are not shared by all people who choose a career in information development. However, to the extent they are shared by many individuals, they present special challenges for management.

Best Practices in Managing People Resources

A great deal of information is available on personnel management. In many companies, managers are offered training to help them manage people effectively and are supported by professionals in human resources when they have personnel problems. In this chapter, the emphasis is on managing information developers, taking into account their particular strengths and weaknesses.

You learn about six best practices to build an effective staff and ensure they are working in the most productive and quality-focused manner they are able:

✔ Developing a hiring strategy

✔ Investing in professional development

✔ Developing individual strengths

✔ Managing by objectives

✔ Delivering difficult messages

✔ Measuring productivity

In Chapter 10, the focus was on building a collaborative organization across different geographic and cultural environments. Here, I discuss issues of personnel management focused on improving the effectiveness of your existing team members.

Best Practice—Developing a hiring strategy

If you review the average job announcement for an information developer, you will most often note an emphasis on tools. Even though the announcement will include a reference to excellent writing abilities or years of experience in technical communication, the only specific qualification is the ability to use one or more of the standard desktop publishing or online help authoring tools. Knowing FrameMaker or RoboHelp is often a limiting characteristic of the job search.

At the same time, managers decry the lack of basic writing and editing skills among candidates. They are concerned that many applicants know little about information development, lacking an understanding of customer analysis, information design and planning, or topic-based authoring. To counter the emphasis on tools in hiring, managers need to carefully define the competencies required among their staff and find ways to identify these competencies or their potential in new applicants.

The changing skill set required of information developers today includes more than basic writing and editing skills, although they remain the foundation. Managers need team members who can

✔ plan a project and carry through on the plan throughout the information-development life cycle

✔ plan customer research and conduct customer studies, including visiting customers at their locations

✔ use customer knowledge to redesign information deliverables so that they better meet customer needs

✔ apply minimalist principles to information design, transforming too much useless content into information that provides measurable value to the user

✔ learn new tools and technology

✔ appreciate the growing need for topic-based authoring and be able to transform book-based content into consistent, reusable topics

✔ work effectively with product developers, instructional designers, support personnel, trainers, and others in the organization

✔ be a spokesperson for sound information design in the context of business requirements and constraints

✔ be open to and enthusiastic about changing from the traditional ways of working

This list is not meant to be complete. In your organization, you will have a somewhat different set of competencies. The key is to define what is needed, now and in the future. Not everyone will have all of these competencies, but without some of them strongly in place, you will have team members who are not fully able to contribute to the changing work environment.

Changing the personality mix of your team

Many managers are concerned that their team members are too reticent in their interactions with others in the organization, especially when they believe that those individuals are more powerful. Product or process managers may expect information developers to assume secretarial roles, writing what they are dictated to write by the subject-matter experts (SMEs). Managers expect the information developers to push back, explaining that they, not the SMEs, are experts in information design and delivery and demonstrating how information might be made more effective.

If you find yourself with a team that is more comfortable working behind the scenes, you should consider using your hiring opportunities to change the team's personality mix. In defining the type of personality you are looking for, recognize that individuals who are more extroverted and like the challenge of influencing others may seem radically different from your existing team members. They may be more talkative and outgoing, more comfortable meeting people and asserting their points of view. They are often more willing to interact with others directly, through phone calls and meetings rather than relying upon email to communicate.

You may also find that you already have team members who are more assertive. With a little training and a lot of support, they may learn to support the department's objectives in the face of ownership and opposition. You must be certain to support their initial efforts with new SMEs or those who want information "their way."

Whether you guide existing team members toward a more assertive role or bring in new people who are comfortable being assertive, you need to provide guidance. You will find it important to model the behaviors you want and ensure that everyone has solid training in the principles you want them to avow. When they encounter opposition, especially from powerful individuals, find ways to support them actively. One manager held lunch meetings at which team members presented their most difficult situations. They each presented a problem, which usually occurred when an influential individual insisted that something be included in a document that was not part of the standard or needed by the user. Other team members offered advice on handling the situation, presenting arguments and evidence that would demonstrate a better way to respond.

You must also concede that your team will not win in all cases. I worked with an engineer who insisted on adding a long dissertation on the theory of light to a manual for people who installed the company's product. Although we argued at length that the information was inappropriate for the users, he eventually won the argument. The information had no value for the users, it increased the size of the manual, and it was generally ignored by the users.

Hiring new graduates and entry-level employees

Many information-development managers focus on hiring only experienced information developers. As a result, they offer no opportunities for new graduates of technical writing programs. Yet, these new graduates often are a rich source of new ideas and technologies that can expand a department's skills portfolio.

A number of excellent degree programs produce graduates who have been exposed to new ideas and techniques, including usability testing, customer studies, topic-based authoring, and XML or HTML document development. What they lack in years of experience in information development, they make up for with enthusiasm and a fresh way of approaching traditional problems.

Entry-level employees, who have varied and interesting work backgrounds in other fields, can also add life to your teams. Individuals with experience or education in information sciences can help you define a new information architecture or help configure a content management system. Those with experience in support or training may bring a new customer perspective to the team. One manager hired an individual who had been a member of the user community for the company's products. He provided invaluable information about how the product was actually used in the field and how users interacted with information.

Entry-level employees may require more training in writing and editing or even the technology your team supports, but they bring innovative approaches that will challenge tradition and encourage others to consider new ways of developing and delivering information.

Testing for basic skills

Although you may be looking for a skill set in new employees that is different from the skill set of a traditional information developer, you may still want to ensure that new hires have basic communication skills. I have long advocated conducting writing and editing tests, even for administrative employees. I find that people fit better into an information-development environment if they can write.

Writing and editing tests have long been part of the hiring process in information development. In a few organizations, a human resources department might argue that tests are illegal. As long as everyone in the department, including existing employees, is able to pass the same test, you are fairly and uniformly administering it.

Basic skills tests do not have to be complicated. If you ask someone to write a basic procedure or rewrite a general concept that is badly organized or directed toward the wrong audience, you will be able to evaluate their writing abilities. Editing tests generally ask people to mark and correct a text that has both obvious and subtle errors of spelling, punctuation, grammar, and facts. Comtech keeps looking for the few who realize that April has only 30 days, not 31. Passing the editing test generally indicates that the new hire is sufficiently detail oriented to survive in our environment.

Whatever test you devise, be certain that you test everyone first and that the test reflects the baseline of skills you believe critical to success in your organization.

Involving team members in the hiring decisions

Giving tests on basic skills is a straightforward way to evaluate candidates. It is much more difficult to test for an entrepreneurial spirit, an assertive but not obnoxious personality, and the ability to enthusiastically promote the vision of information development in your organization. That combination of personality traits can be identified through interviews and discussions with previous employers and colleagues.

Most information-development managers involve team members in the evaluation of candidates and the hiring decision. Team members should have an opportunity to interview candidates individually, as long as they have had some training on interview techniques and the legality of certain questions. Once everyone has met the candidates individually, you can schedule a review session in which the candidates' qualifications are discussed and ranked. Be aware that if you are hoping to hire people who are different from your typical team member, you may encounter resistance. A group of shy introverts may find a gregarious extrovert uncomfortable. It's important that you establish the ground rules for the new hire and address people's concerns in advance. Otherwise, you are likely to hire only those who are very much like the existing team.

Best Practice—Investing in professional development

A core competency of a Level 3 or higher organization in the IPMM is its commitment to the professional development of staff. Professional development occurs in many ways, including opportunities for participation in training programs:

✔ inhouse training in the product technology or the service or process being developed and documented

✔ training and mentoring for new hires in the existing processes and best practices of the department, including tools and templates, style guides, authoring guidelines

✔ opportunities to attend external or internal training programs in areas of specialization such as usability, customer studies, task analysis, project management, minimalism, structured authoring

✔ training in new tools and technologies that are being introduced to the organization

✔ training for managers in project, personnel, and change management as well as strategic planning, metrics development, public speaking, and other topics designed specifically for managers

In addition to formal training programs offered inhouse or available externally, managers can foster professional development by ensuring that staff have opportunities to attend professional conferences. Attendance at general industry conferences, such as the annual conference of the STC, attendance at subject-matter-specific events, such as the Content Management Strategies or KM World conferences, or attendance in job-role-specific events, such as the Best Practices annual conference of the CIDM provide opportunities for learning. Through exposure to speakers and participants from other companies, participants in public events learn about innovations in the field or even standard practices that are simply unfamiliar to them. By reporting on what they heard, individuals who have attended industry conferences bring back new ideas to their entire teams.

Managers should also encourage staff to engage in personal learning projects. These may include reading about an innovative idea, investigating new tools, attending webinars, and more. The individuals should once again be prepared to bring their learning to the rest of the staff through brown-bag lunch sessions, internal conferences and webinars, or even through the development of electronic mailing lists and wikis to exchange information in an open forum. Department websites provide an opportunity for individuals to disseminate what they have learned throughout the team.

Funding professional development

Most information-development managers are active supporters of learning for the team members. Only rarely do I hear about a manager who will not support professional development because the subjects are just "common sense" that everyone should already know. However, many companies fail to fund learning activities adequately or at all. Managers have to work hard to get adequate budget lines for attendance at workshops and conferences, especially when travel restrictions are in place. Failing to recognize the need for learning among information developers often results from the assumption that writers are clerical workers. Clerical workers rarely receive outside training opportunities. Changing the perception of information developers in the organizations includes gaining support for training and professional development. I advise information-development managers to look at the opportunities provided in peer departments to help make a business case for more professional development funding.

Budgeting for training and professional development should provide sufficient funding to support each staff member. Some organizations establish training budgets at a minimum of 1% to 2% of the total department budget. If you have five people in your department and a basic budget of between $500,000 and $1,000,000 (based on fully burdened salary costs), you should have a minimum of $5,000 to $10,000 available for training and professional development.

If travel restrictions present barriers to training and professional development, managers should consider bringing professional development opportunities inhouse. Webinars are low cost but must be selected carefully. They are typically quite short, and participants may be overwhelmed by sales pitches. Inhouse workshops are available in almost every area of specialized learning, as are local programs sponsored by professional organizations.

Encouraging participation

Some managers have a training budget but they wait until a staff member requests training. Professional development should be built into the yearly objectives of every staff member. Otherwise a few people who are most interested will get most of the opportunities. Others who are uninterested or unwilling to speak up may be left out. Unfortunately, managers know that some staff are simply not willing to assert themselves and take responsibility for their own professional development. Such attitudes should be addressed during performance reviews. Individuals may do their day-to-day jobs adequately, but they offer little to the overall improvement of the team or to the challenge of increasing productivity and quality to customers.

Best Practice—Developing individual strengths

Many organizations spend considerable time and energy trying to correct perceived weaknesses in employee performance. Often, the effort results in unhappy employees who spend much time trying to do something well that is difficult and uncomfortable for them. That is not to say, however, that managers should not work with employees whose behavior causes problems for themselves, their colleagues, or the organization. However, a focus on correcting weaknesses is not the same as disciplining inappropriate behavior.

Instead of focusing on making people do better at something they're not especially good at, Marcus Buckingham and Donald O. Clifton, who developed their ideas in *Now, Discover Your Strengths*,[1] recommend that managers focus on enhancing people's strengths rather than eliminating their weaknesses. Several information-development managers have used the strengths model provided by Buckingham and Clifton to influence their staff-development plans. They find that individuals who are involved in tasks that align with their strengths are superior performers. They are people who everyone acknowledges to be expert at what they do because they perform so well. They are using their natural talents everyday. The results are significant because they are so different from our tendency to force people to concentrate on and correct their weaknesses.

Buckingham and Clifton provide an online survey based on their strengths book so that staff members can take the strengths test and compare results. You can graph the strengths for your entire team (or subteams) and see where strengths overlap or what strengths are unique for certain individuals. Figure 11-1 shows a strengths graph typical of a small team.

For example, the graph of your team's top five strengths may show that you have an individual who is especially strong in a characteristic called "context." According to Buckingham and Clifton, an individual with strength in context can be described in this way:

> If you are strong in context, you look to the past to understand the present. You don't trust the present because it appears unstable; you find your stability in the past. You want to find the original intentions of an idea or principle, and you make better decisions when you know the underlying structure. When you're asked to work with new people or in new situations, you may take a little time to get adjusted.

Someone strong in context respects tradition and looks to the "tried and true" answers to problems. They are reluctant to ask themselves or others to change without being sure that the traditions of the past are respected.

[1] Marcus Buckingham and Donald O. Clifton, *Now, Discover Your Strengths,* New York: Free Press, 2001.

Summary	16 Thinking	10 Relating	6 Striving	7 Impacting

JT–ENTJ	Maximizer	Strategic	Achiever	Input	Learner
RM–INTP	Learner	Maximizer	Strategic	Analytical	Ideation
SP–ESFP	Achiever	Relater	Adaptability	Positivity	Arranger
QR–ENTP	Arranger	Includer	Maximizer	WOO	Responsibility
MP–INTP	Relater	Strategic	Command	Analytical	Discipline
OS–INFP	Input	Maximizer	Individualization	Communication	Ideation
RT–ISTJ	Consistency	Responsibility	Discipline	Harmony	Deliberative
NF–INFJ	Responsibility	Empathy	Harmony	Learner	Adaptability

Figure 11-1: A strengths graph for a small team

As an information-development manager, you might immediately identify context as a strength you might want in your editors. Editors should know the foundations behind the principles of good grammar and writing style. At the same time, you may have to provide a person strong in context with a stretch objective to reexamine past practices to ensure that they don't enforce standards that are no longer relevant to your team or your user community.

Motivating through strengths

Managers have often tried to motivate employees by rewards and punishments. According to Frederick Herzberg, in his key article on motivation, most managers are "motivating with KITA" in which KITA is an acronym for what we might politely expand to "kick in the *pants*." While this phrase smacks of negative reinforcement, KITA can also be seen as rewarding desired behaviors. You can motivate a dog by pushing it from behind as well as by putting a biscuit in front of its nose. However, while you are motivated to move the dog forward, the dog is not. He is motivated either to avoid your foot or to get the biscuit.[2]

Managers commonly use a positive KITA to motivate employees. Spiraling wages, fringe benefits, sensitivity training, employee counseling, and telecommuting policies are among the many ways managers have begun to reward employees and provide incentives. However, according to Herzberg, strategies such as these have, at most, short-term

[2] Frederick Herzberg, "One More Time: How Do You Motivate Employees?" *Harvard Business Review,* January–February 1968.

benefits and don't really motivate employees to do their jobs any more than the kick motivates the dog.

To motivate people to be creative and innovative in their jobs, we must increase job satisfaction. Job satisfaction and job dissatisfaction are not opposites. The opposite of job satisfaction is no job satisfaction. The opposite of job dissatisfaction is no job dissatisfaction. The lack of job dissatisfaction is not sufficient to create job satisfaction. In fact, the factors involved in both are very different. In a study of 1,685 people at a large range of job levels, Herzberg found that factors related to job dissatisfaction tended to be external to the job itself. Those factors included company policy and administration, supervision, relationships with supervisors, work conditions, salary, relationships with peers, personal problems, relationships with subordinates, status, and security. Job satisfaction factors tended to be job-related and included achievement, recognition, the work itself, responsibility, advancement, and growth.

So what should you do as an information-development manager? By building on the strengths of your staff, you can redesign the work each person does to increase job satisfaction. In effect, you must enrich the job. Herzberg calls this task job loading. He defines horizontal and vertical job loading. Horizontal job loading entails adding more tasks or more varied tasks to increase job satisfaction. Vertical job loading increases the challenge of the job rather than enlarging the job. As examples, Herzberg suggests assigning individuals specific or specialized tasks, enabling them to become experts, and increasing the accountability of individuals for their own work.

Consider a 10-step process for creating job enrichment and building on each team member's strengths:

- ✔ Select jobs for job enrichment in which motivation will actually make a difference in performance.

- ✔ Approach these jobs with the conviction that they can be improved.

- ✔ Brainstorm a list of job-loading changes that may enrich the job.

- ✔ Screen the list to eliminate changes that are external to the job itself.

- ✔ Screen the list to remove generalities such as "give more responsibility," "growth," and "challenge." These terms have no substance.

- ✔ Screen the list to eliminate all changes that represent horizontal loading, which just gives an individual more of the same work to do.

- ✔ If you have a large enough organization, set up a control and experimental group to be able to directly compare the effects of the job changes.

- ✔ Be prepared for an initial drop in performance since the new job may lead to a temporary decrease in efficiency.

- ✔ Expect some anxiety and hostility from your first-line supervisors. They may feel that some of their responsibility is being handed over to their team members.

By working with your team members to increase their job satisfaction and play to their strengths, you build a team of individuals who are each performing optimally and enjoying what they do.

Best Practice—Managing by objectives

The practice of Managing by Objectives (referred to as MBOs) was first developed by management guru, Peter Drucker, in *The Practice of Management*.[3] MBOs have since become a common way of helping individuals align their performance goals with departmental and corporate objectives. As you work to build strengths and motivate your team members, consider how to ensure that the work they are doing aligns with your goals and the goals of your larger organization. Help each team member identify a small set of strategic objectives that he or she can meet in a year.

Your corporation may, for example, have a goal of reducing operational costs. You translate this goal to a department goal of writing your content in topics that can easily be used in several information deliverables, decreasing the cost of producing each deliverable. This department goal translates in specific objectives for your information developers. For each information developer, you work together to identify and carefully define a personal performance goal. For one individual, that goal might mean learning to follow the authoring guidelines and standards for each information type. For another individual, it might mean working collaboratively with team members to plan the topics needed for a suite of deliverables. Depending on individual strengths and objectives, team members define goals that they can achieve within a specific timeline, and together you define how you will measure everyone's progress toward reaching the goals.

The key to managing by objectives is understanding the corporate goals, developing goals for your organization, and translating those goals into specific objectives for individual team members. By creating MBOs, you avoid the trap of getting caught in day-to-day activities. You are constantly surrounded by the challenges of meeting deadlines and satisfying the demands of colleagues, peers, and senior managers for cost-effective, quality information. But, if you forget why you're doing these activities, you are in great danger of losing focus. Your team may, for example, successfully meet all its deadlines but produce information that is inaccurate, incomplete, and unusable by the customers. You've succeeded in the activity but failed to meet the goal.

Developing objectives

The objectives you develop with your team members should be aligned with corporate and departmental as well as individual goals. The objectives should follow the SMART guidelines:

✔ Specific—ensure that each objective is clearly and simply stated so that its meaning is clear to you, the team member, and your management. For example, an objective might state: "Develop a site visit plan and use it to plan and conduct a visit to a customer's site."

---

[3] Peter F. Drucker, *The Practice of Management,* reissued edition. New York: HarperCollins, 1993.

✔ Measurable—decide how each objective will be measured so that achieving the objective doesn't become subjective and open to argument. For example, a measurement might state: "Complete the site visit plan and conduct two user site visits by the end of the year."

✔ Achievable—be certain that the objective can be achieved by your team member. If your team member is faced with difficult deadlines throughout the year and has no experience or training in conducting customer site visits, the objective might be impossible to achieve. You need to ensure that the team member has the support, resources, and training or experience needed to achieve the objective in the time given.

✔ Relevant—every objective must be relevant to the team member and to the department and the company. A team member once asked me if the company would pay for her French class. She didn't need to write or speak French for her job, which meant that her objective of learning French was not relevant. A relevant objective supports reaching the organization's goals as well as benefiting the individual performer.

✔ Time-based—every objective must have a time stated for it to be accomplished. You may want to set both a final date and interim dates that will help the team member show you that he or she is making good progress. For example, a time-based objective might state: "Complete the site visit plan template by the end of the first quarter, complete the detailed plan by the end of the second quarter, and set up and complete the two customer site visits by the end of the year."

Obviously, developing objectives is a collaborative activity between you and each team member. Be certain that your goals for the team are clearly defined and effectively communicated to everyone. Effective communication of the goals helps you motivate the team and gets them to think about how they will help to achieve the goals. Ask everyone to examine their own strengths and identify areas in which they want to improve their performance or develop new skills. Ask them to identify at least one personal objective that will be a stretch for them to achieve, such as an area in which they want to develop skills but is outside their previous experience. Team members should know that you value objectives that enhance their creativity, their ability to innovate and initiate change, and their ability to show leadership.

After each team member has developed his or her own list of objectives, meet individually with each one to review the list and ensure that each objective is aligned with larger organizational goals. Be careful about lists of objectives that get too long, are difficult to measure, or have no well-defined timeline. Too many objectives may mean that the team member is not focused well enough on what he or she can really achieve, given all the day-to-day activities that will inevitably occur in the schedule. Be certain that every objective is measurable. If you don't agree on the measurement, you may have disputes at the end of the time period. Be careful about setting the timelines. A goal may not be achievable within the time period that the team member hopes to accomplish it. Try to be realistic but optimistic.

With objectives in place, you have the perfect vehicle to handle performance reviews. You and your team members should review the previous year's objectives, look at the data that reveals if the measurements and timelines were met. If someone has not achieved the objectives or met the measurement criteria, you can actively discuss what happened and how the objective can be met in the next review period. You may have to conclude that the objective cannot be met and set new ones instead.

If someone has successfully met an objective, you need to ensure that you reward the achievement appropriately. A set of rewards and recognition is invaluable in your organization to develop people strengths, show that their work is appreciated, and build their confidence and loyalty.

Defining the competencies needed in your organization

A good starting point for managing by objectives is a well-defined set of competencies that are required to be a successful contributor in your organization, depending, of course, on the job role and responsibilities of each team member. The competencies defined for a senior information developer may include the ability to plan a project and estimate the resources required to meet the project deadlines. An information architect may be required to develop authoring guidelines for each information type defined and specify the set of required and optional elements. All team members may have to organize their work so that they meet required project deadlines or be able to write a topic following authoring guidelines and requiring minimal editing and revision.

To define competencies will require the work of the entire team and consultation with others in the field. Reviewing the literature available on job descriptions and base competencies and finding examples of competencies developed by colleagues outside your organization will ensure that your set of competencies is aligned with industry standards and expectations.

After you have a set of competencies required in your organization, you and your team must define levels of achievement. The levels of achievement describe outstanding, average, and below average characteristics. Outstanding performance should prescribe the highest levels of achievements for that skill in your organization or in the field. Average achievement might be divided into very good and good levels, which means that the team member meets all the requirements of that skill. Below average performance means that the team member is not performing adequately and needs to improve or risk dismissal.

Some organizations require that managers rank all employees from overall highest to lowest performer, with the intent of dismissing the lowest performers if they do not improve within a specific time period. Many managers object to this ranking because they have worked over time to develop everyone on the team. They argue that they have no below average performers.

Developing criteria for ranking competency levels

Once you have established a set of competencies that you believe are necessary for high-quality performance in your organization and you have outlined a system of ranking those competencies, you need to define the criteria you use to define the highest level of achievement.

The following example shows the skills defined for a highly competent and expert information developer in one organization:

✔ Works on writing projects of any scope and technical difficulty. Develops documentation plans and product information suites. Works independently on difficult project development responsibilities. Acts as the first line of technical support and technical writing for other department members.

✔ Researches content and audience information. Uses specifications, background information, and interviews with customers and subject-matter experts (SMEs) to develop customer profiles and task analyses.

✔ Plans the scope, technical level, organization, and delivery medium of product suites to maximize their usefulness to the intended audience.

✔ Writes or revises descriptive and procedural documents for company products and tools. Delivers documents that meet corporate standards of style, writing quality, and format for online media or paper.

✔ Requires minimal editorial support to meet quality goals. Helps define or refine quality standards.

✔ Works with the electronic publishing department to ensure the timely publication of high-quality information.

✔ Contributes to product core teams, representing the information-development function on matters of product development, information quality, and project schedules.

✔ Acts as an early user of the product, providing developers with feedback on product design, usability, terminology, and implementation.

✔ Understands most technical information with some assistance. Offers suggestions to improve software product design and user interface.

✔ Can design practical examples.[4]

Encouraging peak performance

Why are some people excellent performers while others are not? It may have less to do with the individual players than with the working environment itself. You may find it helpful to identify six initiatives to ensure that your team members have the necessary tools to perform their jobs to the best of their ability:

✔ build an individual knowledge portfolio

✔ create mentoring and apprenticeship opportunities

✔ institute electronic conferencing systems

✔ develop an organizational knowledge repository

✔ create communities of practice

✔ establish a program of rewards and recognition

[4] Diane Davis, "Writing Performance Appraisals," *Best Practices* of The Center for Information-Development Management, June 2001.

An individual knowledge portfolio is a skills inventory that you and your project managers can refer to when you set up your project teams. Encourage each team member to develop and post a knowledge portfolio on the department website, listing projects they have worked on, people they have mentored, the type of work they enjoy most, and the strengths they bring to their work environment. Portfolios should grow, improve, and add value to the department and the individual. Once you have a good start on the portfolios, consider making them available to others in the larger organization to promote your team members and give them visibility.

Ask your more senior team members to mentor new hires and less experienced colleagues. When you acknowledge that an individual is an expert and able to teach others, you reward their experience. The colleagues they mentor also gain from the mentoring experience. Remember that when you ask a senior team member to take on a mentoring role, you have to credit the time they spend. When one manager instituted mentoring to train new team members in China, she asked senior mentors to devote 50% of their time to the activity and credited that time in their work assignments.

Most likely, your team members make liberal use of email to communicate with colleagues in the organization. If you have not already done so, consider other communication methods to enhance the ability to share information among your team members, including instant messaging, departmental websites, wikis, and listservs on special topics. Use these methods as substitutes for unproductive meetings or conference calls. You can ask questions or discuss pending decisions with the appropriate team members, and people can respond as soon as they have time.

Use your departmental website to create a repository of information that everyone in the department and outside can reference. You can include reading material that people have found informative, self-paced training material, and short articles or presentations created by team members.

Many more information-development organizations are creating communities of practices so that interested team members can discuss new ideas and best practices. People involved in such communities feel less isolated and more willing to contribute to a team effort. An informal community allows people to share, and build on, knowledge and work experiences.

Most managers I work with have a recognition program in place and know how important it is to reward innovations that contribute to the bottom line or succeed in adding customer value. Don't forget to recognize those who are mentoring new team members or managing new initiatives for change.

All the activities mentioned here are designed to motivate team members and encourage them to be innovative and collaborative. Motivated, enthusiastic team members make everyone on the team, including the managers, look good.

 ## Best Practice—Delivering difficult messages

Difficult messages come in two flavors in most organizations: First are messages that relate to the entire team, typically concerning reorganizations, outsourcing, changes in management, or layoffs. Second are messages that involve individuals or small groups. Individuals must be told about problems with their performance or behavior as quickly as possible, especially when they are being given a limited amount of time to correct the problem. Small groups need to know how they are perceived by others for the problems they are creating and given opportunities to improve.

Dealing with organization turmoil

Many organizations seem to be in a state of continuous reorganization. Information-development managers report that their departments are constantly being shifted to new reporting structures, often resulting in four to six new senior managers to whom they report in the course of five or fewer years. It seems that senior management in many companies does not know where to place an information-development organization. Sometimes information developers report up through engineering or software development organizations; in other cases, they report to marketing, user assistance, technical support, usability, even operations. Each business unit may have its own information-development team, some with a manager with a background in information development and others with a manager who is uninformed and disinterested. In many of the reporting structures, the senior manager has no idea what information development is supposed to do, and often doesn't care to learn.

As a result of all this turmoil in reporting structures, information-development managers often have to deliver difficult messages about still more reorganization, outsourcing, offshoring, and reductions in force. The objectives in the delivering the news should first be honesty and second calming fears and minimizing the disruption to the team.

Being honest and straightforward about bad news for your team is essential, within the bounds of your corporate obligations. Rumors are highly contagious and dangerous to the well-being of a team. As soon as you are able, communicate as much as you know and can tell your team. Explain how the decision was made and what the immediate affects are likely to be. Explain simply but quickly to keep the rumors at bay.

If you cannot explain everything that is going on, tell your team members that you are telling them everything that you can at the moment. They'll know that you're under some restrictions but will appreciate knowing what you can disclose about the reorganization, restructuring, outsourcing, or layoffs. Explain what you are doing to keep a positive point of view about the change and encourage them to do the same. Admit, however, that the situation is likely to cause disruptions and welcome people who want to talk to you individually. Arranging one-to-one meetings with each staff member may be time-consuming, but it does help dispel rumors and give people a feeling that their concerns and fears are being listened to.

Motivating your staff during turmoil

Constant change in an organization, brought about by downsizing and mergers, adds significantly to the stress experienced by staff members. This stress contributes to "low employee motivation, a lack of commitment to organizational goals, and reduced productivity," explains Mary T. Pellak in "Sustaining Motivation and Productivity During Significant Organizational Change."[5] Managers may face reduced productivity and a general feeling of

---

[5] Mary T. Pellak, "Sustaining Motivation and Productivity During Significant Organization Change," *Performance Improvement,* November–December 2001.

distress in their organizations. So, they need to know how to motivate staff members and sustain productivity in this time of change.

Managers play a vital role in sustaining motivation and maintaining productivity. Managers must decide to intervene in ways that both address the needs of staff and support evolving organizational goals. Managers in an organization might become aware of problems in motivation and productivity during periods of change when they encounter the following trends:

- ✔ staff anger, mistrust, or defensiveness

- ✔ more absenteeism

- ✔ increased customer complaints

- ✔ more errors or defects in the final deliverables

- ✔ diminished productivity from staff members

- ✔ staff reluctance to offer suggestions for improvement

- ✔ an increase in stress-related claims through staff assistance programs

You might refer to Abraham Maslow's Hierarchy of Needs[6] to understand how motivation is affected by unmet needs. When an individual's job is threatened because of an organizational change, the first level of need, physiological, is affected. At this level, an individual focuses on putting food on the table and having money to pay bills. Once these needs are met, an individual can focus on the next level of need, safety and security. Your team members may, at this point, seek positions with more security.

The need to belong is the third level of Maslow's Hierarchy in which staff members believe that they share common goals with their team members or the organization. Once the need to belong is satisfied, recognition or self-esteem becomes the motivator.

The fourth level is recognition, where an employee's self-esteem is satisfied from a sense of belonging. A manager needs to understand the need for recognition and acknowledge superior performance. Even a plaque or a letter of recognition presented in front of team members will be a significant motivator. A bonus or other monetary compensation is appreciated but not always necessary.

The ultimate goal for your organization is to foster an environment that supports motivation and self-actualization. A staff member who is self-actualized is motivated to be the best performer that he or she can be.

As an information-development manager, you need to employ skills that help you and your team face the challenge of organizational change: empathy, communication, and participation. You express *empathy* when you try to understand what employees are feeling and what might motivate them during organizational change. Support for training and professional development can help ease the transition. You must *communicate* with employees about the changing situation. By encouraging team members to *participate* in a change, you help everyone remain productive. Asking team members to help in decision-making processes or allowing them to provide input are opportunities for open participation.

[6] Abraham H. Maslow, *Maslow on Management,* revised edition, Hoboken, NJ: John Wiley & Sons, 1998.

You also need the support of your management and peers if you are to handle difficult changes effectively. Consider the positive actions you might take to

- ✔ involve staff in decisions about downsizing
- ✔ institute job-sharing or work-sharing arrangements to increase employment opportunities
- ✔ maintain clear two-way communication with your team members
- ✔ provide assistance and counseling to displaced team members

People need to know that they won't be punished if they participate in decisions made during a time of change. Whether the organizational change is due to downsizing, outsourcing, or mergers and acquisitions, people feel that their well-being is threatened, and they experience a loss of control. Organizations can help ease these feelings by informing staff about the reasons for the change and having leaders in the organization address the staff directly.

Departmental managers and senior management must work together to make the transition easier for employees. By providing leadership, open communication, employee participation, and management support, managers can succeed in a changing organization.

 ## Best Practice–Measuring productivity

Measuring productivity means figuring out how many goods and services are produced and how many people it takes to produce them. If the number of people working increases (full employment these days), so should the number of goods and services they produce. Some managers, however, not only exaggerate their productivity but base it on incorrect assumptions.

A False Productivity Measure

"I have the most productive technical publications group in the valley," James boasted. "We produced documents at a rate of one hour per page." "How so?" I asked. "That's way below industry average." "Well, don't tell anyone, but we counted all the pages we reproduced and shipped that came directly from the engineering group."

Not only was James cheating by not counting all the work done by the engineers, he was digging his department into a hole. He needed to improve the productivity of his information-development staff. They were well known in the company for producing masses of useless information, much of it full of errors. But any effort to improve the research and writing done by his department would make his productivity numbers get worse.

How do you bring this concept home for information-development professionals? Technical writers produce documents—lots of them. You could measure how much they produce by simply counting the documents. If Sam creates three new manuals in a year and Sally creates five that means their average output is four new manuals each. If Sam produces four new manuals next year but Sally only produces three, that might indicate that their average productivity has decreased from four manuals to three and a half.

You already recognize that simplistic measurements like counting the number of manuals produced gives you some information about your department's productivity but not much. So you start looking for more meaningful measurements.

Measurement One

The number of documents produced by your organization per year, including new and revised documents. The goal might be to produce more successful documents with the same staff.

You may have traditionally counted pages rather than documents, believing that by counting pages you even out the productivity measure. Sally may write five new manuals and Sam three in a year, but if Sally's manuals have 50 pages each and Sam's have 150 pages, Sam is clearly producing more pages and more words than Sally. That makes Sam more productive.

"Wait just a minute," argues Sally, "my pages are a lot more difficult to write than Sam's. It's not fair to count all pages as if they were the same." If you want to be fair to Sally, you have to add a complexity measure, giving her extra credit for producing difficult-to-write pages.

Measurement Two

The number of pages produced by your organization per year, new and revised, with a complexity-weighting factor added in. The goal might be to increase the number of weighted pages produced by the same number of people.

Software developers face the same dilemma when they count lines of code produced per person per year. A sloppy, disorganized programmer might produce more lines of code than a well-organized, effective programmer, even though they both produced the same number of functions. As a result, software development has begun to measure functions produced instead of lines of code. Functions refer to parts of the program that do something useful, such as sending text to a printer, drawing a graphic dynamically, performing a mathematical calculation, or searching a database.

In information development, what can you measure that might be similar to functions in software? You could measure how many user tasks are successfully explained by your writers. Counting user tasks in the content is easy; measuring success is likely to be more difficult.

You might try counting "errors." Errors are reasonably easy to detect if you verify the accuracy of the documentation by testing it against the product. Aside from outright mistakes, however, measurement becomes more difficult. You might use usability testing of a writer's work to see if typical users find the task-oriented information easy to follow. The more problems the users have interpreting the written instructions and background information, the less productive the writer has been.

"Well," I already hear you arguing, "what about the complexity of the information and the level of knowledge and experience the user brings to the task? And what about the product? What if it's so difficult to use the product that no one can explain it?" You're right! But if you spend a lot of time producing information that is not useful (or products that aren't easy to use), you certainly aren't going to get any rewards in the long run.

Measurement Three

The number of successful sets of functional information delivered to the users each year. The goal might be to deliver more successful functionality each year in proportion to the number of people on the staff.

You can also measure writer productivity by looking at the amount of extra work for others in the department that an individual writer generates. For example, Dan is an experienced and highly competent senior writer. He follows templates scrupulously and knows the department's style standards by heart. When his work goes to Mike in the editing group, Mike knows he can plan a quick review, handing minor comments back to Dan. When the production team gets Dan's books and help topics, they know they'll have only minimal cleanup to get them ready for printing and web publishing. Everyone in the internal support groups benefits from Dan's work habits. They would call him "very productive."

In contrast, when the editing team receives documents from Joan, they're almost afraid to open them. They know that Joan's work will be full of spelling mistakes (Joan can't be bothered to run the spell check program) and grammar errors. They also know that they'll have to "bleed all over" Joan's work because she has a difficult time organizing her writing and communicating ideas simply and straightforwardly. The production team also fears Joan because her documents are full of stylistic variations. She changes the standard styles constantly, which means that her files don't convert cleanly. Joan and Dan produce the same number of pages per year, but Joan is far less productive (and everyone but Joan's manager knows it).

Measurement Four

The amount of rework staff members generate for themselves and everyone else on the support team. An organizational goal might be to reduce the amount of rework done.

Rework could also apply to the relationship between the writers and the developers. Because you ask product developers to review draft documents for the accuracy of the information, the amount of rework they generate is another potential measure of your organization's productivity.

The list of productivity measurements quickly becomes long. You might measure, for example,

✔ the total number of hours it takes to move a document through the information-development life cycle

✔ the percentage of time spent on each phase of development activities (planning, organizing, writing, researching, editing, illustrating, and so on)

✔ the number of technical communicators compared with the number of product developers in an organization

Thus far, you have focused on measuring productivity by counting the number of documents produced, pages written, and projects completed. Now turn your attention to the information-development process itself and find ways to measure productivity before you have documents to count.

You might recall that our friend James boasted on the number of pages his staff turned out in record time. He was down somewhere around one hour per page. But he achieved this remarkable productivity by cheating. He counted work that his documentation team didn't do.

James would have been much better off trying to make his staff more productive as they produced their work rather than focusing only on the output. If he had, he might have found a number of potentially fatal flaws.

James's staff members love to use their desktop publishing tools. They spend many hours creating more and more elaborate page layout schemes. They add cute graphics to spice up the look of the pages and individually craft each two-page spread. In fact, if James looked at their overall time expenditures, he would discover that nearly 40% of their total project time is consumed by desktop publishing tasks.

To begin, however, James needs to measure the total time it takes someone in his department to complete an information-development project. Once he knows more about the total time (taking into account information type, project complexity, total work volume with new and revised pages), he can begin to break total time down into its components.

Measuring time on task

You can obviously measure productivity by looking at the output of your work effort. You can also measure the efficiency with which you perform tasks. If you can produce the same results with less effort or better results with the same effort, you have also increased your productivity.

Measurement Five

The time it takes to develop a document or complete a project from the beginning to the end of the information-development life cycle. Total project time gives you a starting point for other measurements that help you improve the time you spend on individual tasks. You begin with total time, and then you break that time into milestones and individual tasks.

In *Managing Your Documentation Projects*, I provide a recommendation for the percentage of total project time that goes into the phase milestones, as illustrated in Figure 11-2.

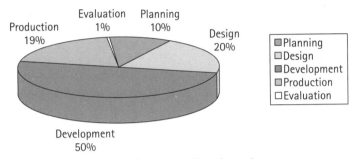

Figure 11-2: Information-development life cycle metrics

With these guidelines as a starting point, James and his managers began looking closely at the time spent reaching each milestone. Right away they realize that the time spent on Phase 4 is much too high. Forty percent of total time for production tasks means that little time is left for information design, content development, and validation testing.

Measurement Six

The percentage of total project time spent on each milestone. You may want to measure the actual percentage of time spent on each milestone compared to this (or your own) model. I have found, for example, that projects in which the writers spent less than 20% of total time on detailed design had problems later. Because detailed design (Phase 2) is so important to ensure well-planned, well-designed information, cutting the percentages meant design problems during implementation and testing. The overall project was less successful.

Because 40% of total development time in James's group is spent on Phase 4 tasks, as illustrated in Figure 11-3, his writers spend almost no time on planning. In fact, Helen is typical of the writers in the group. She claims that she has no time for planning. She just has to get started writing, even though her manager has asked her to turn in a content outline. But Helen feels that she can be more productive if she just gets started. She's been heard to argue that she can have the whole project done in the amount of time it would take her to create a plan.

Dan is perfectly willing to develop content plans; he just bases them entirely on the product specifications. He talks about getting out to user sites—the company even promotes doing so. He's just too busy. He can take the programmer's specs and whip out the documentation in record time. In fact, getting the content out of the way quickly means that he can spend more time with his real love—putting everything into a help system. He is an expert at playing with the help tools. He knows all about secondary windows, hypertext links, pop-ups, and so on. If anyone looked closely, they'd discover that Dan spends 25% of his time on content and 75% on converting the text into a help system.

274   Part 2: Portfolio Management

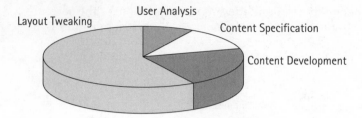

Figure 11-3: Problem with information-development metrics

Measurement Seven

The percentage of time spent on each key project task. You may also want to measure the percentage of time spent on key tasks. For example, if you have established a percentage for validation testing, you should ensure that this time is actually used for testing. If testing is cut, you are likely to have customer satisfaction problems later.

Figure 11-4 summarizes the series of productivity metrics listed in this discussion.

MEASUREMENT TYPES

1. Count the new and revised documents produced.

2. Count the number of new and revised pages, weighted by complexity.

3. Count the number of successful projects delivered.

4. Evaluate the percentage of total project time that is rework.

5. Measure the total amount of time to complete each project.

6. Measure the percentage of time spent on each critical milestone.

7. Measure the percentage of time spent on key project tasks.

Figure 11-4: Productivity metrics

Reclaiming Writing Time

An experienced senior manager at a CIDM member company instituted a best practice called "Reclaiming Writing Time." She asked every team to evaluate how much time they spent on activities that did not add value for the customer. The focus of the evaluation was to increase the percentage of time spent on developing content that customers needed. At the same time, they had to decrease the time spent on other activities because the total project time could not change.

The project goal was to reduce the time spent on non-value-added activities by 5% per quarter. Teams decreased the amount of time spent attending meetings, asked for and received faster equipment, reduced production time, and so on. They have found ways to increase the percentage of their time producing value for the customer.

In another case, the information-development team has conducted an extensive study of its process in an effort to find ways to streamline it. They were able to reduce the total number of steps in the information-development process by eliminating redundant and unnecessarily time-consuming activities.

Summary

Developing the people in your organization into effective team members is a high-level priority for every information-development manager. Without a great team, you can neither achieve the goals that you have set for the organization nor win the respect, if not the admiration, of senior and peer management. To create that team means working individually with team members to ensure that they are taking advantage of their strengths, developing as professionals, and are motivated to do excellent work. In working with team members, you must ensure that you

- ✔ hire the best people you can find for your team and use your hiring opportunities to enhance the team's capabilities

- ✔ provide adequate funding to support learning and professional development for every team member, ensuring that you bring innovative ideas into the organization

- ✔ understand each team member's strengths and give him or her the ability to contribute these strengths to the organization

- ✔ craft each individual's job to promote job satisfaction and real growth

- ✔ ensure that you have a good match between individual performance and growth objectives and your organizational goals, at both the department and corporate levels

- ✔ communicate carefully when there is bad news so that the team remains focused and motivated through the chaos

- ✔ measure the productivity of individuals, not to punish them but to ensure that you are providing all the support they need to be effective in their jobs

No doubt you will always have personnel challenges in managing a group of diverse individuals and building them into a collaborative team. The best practices in this chapter give you a starting point.

"A great leader will move us ahead."

Chapter 12

Developing as an Effective Leader

> In trying to tip the balance toward excellence, we try to identify great leaders' qualities and behaviors so we can develop them ourselves.
>
> —Robert E. Quinn, Professor of Business Administration, University of Michigan

Senior managers involved in The Center for Information-Development Management (CIDM) have written extensively over the past 10 years about leadership. The ideas they have espoused have come from their personal experience and from their reading in the literature of leadership. This chapter reviews a few of the leadership perspectives that I have learned from these managers.

In pursuing a leadership role for yourself in managing your organization, you can learn much from the best practices of skilled and effective leaders, both from those in your field and more broadly from leading theorists and practitioners about leadership and management. You can also learn from the practices of legendary leaders in very different professions. This chapter focuses on the experiences and leadership style of the Antarctic explorer, Ernest Shackleton, a superb leader who kept his team together under the most extreme circumstances. As you read about other leaders, you will also learn from their experiences and from the analyses done by management experts. I hope that the perspective here is only the beginning.

Remember that the management of an organization is not necessarily the leader. You can manage projects or departments by following the rules and enforcing them. The leader sets the tone for the organization and helps others achieve the organization's goals. You may find yourself most comfortable as a manager, relying on a team member to provide the leadership.

Best Practices in Effective Leadership

Effective leadership develops through experience, often by trial and error. You learn that a style that you fall into naturally is not motivating to the team you are leading or has a negative effect on morale. In this chapter, you learn about two best practices developing an effective leadership style for your organization:

✔ Developing your leadership style

✔ Learning from Shackleton on leadership

The best practices in this chapter focus on building your own style as an effective and responsible leader.

Best Practice–Developing your leadership style

Every manager has his or her own style of managing people and organizations. Most of the time, your leadership style is determined by how you feel most comfortable; it may develop without any active decision about an effective leadership style on your part. As a result, management is very uneven. No doubt, you have had experience with both competent and incompetent managers.

As you develop your own leadership style, you should consider exactly what you want to accomplish in working with your team members. You may decide to work on a style that is not natural but may be more productive in achieving your goals. Consider the following six basic leadership styles in your decision making:[1]

✔ coercive

✔ democratic

✔ affiliative

✔ pacesetting

✔ coaching

✔ authoritative

You are likely to find that one style does not work in every situation, but your predominant style establishes how your team responds.

---

[1] Daniel Goleman, "Leadership that Gets Results," *Harvard Business Review,* March-April 2000.

Coercive leadership

Have you worked for someone who demanded that you do exactly as you were told? The leader was probably a drill sergeant. Coercive leaders demand immediate compliance to their orders. Their message is, "Do what I tell you." These managers want to control exactly how people behave. Coercive leadership is also a style adopted by new managers who think that management consists of telling people what to do and punishing them if they fail.

The coercive style has its place in emergency situations, where quick action is necessary. However, it should be used sparingly in a professional organization because it tends to undermine self-confidence and initiative of team members.

Democratic leadership

Democratic leaders forge consensus through the active participation of team members. They ask, "What do you think?" Democratic leaders believe in collaboration, team leadership, and communication.

The democratic style builds respect for the leader by considering the opinions of all of the team members. This style works well when the team is highly motivated and competent. A drawback of a democratic style is that it may lead to endless meetings and conflict among members of the team. It will not work if the team members are inexperienced or incompetent.

Affiliative leadership

Affiliative leaders promote harmony and build emotional bonds among the team. The affiliative leader says, "People come first." Affiliative leaders try to heal rifts in a team and motivate people and help them overcome or endure stressful situations.

The affiliative style promotes team harmony, increases morale, and improves communication. However, this style can lead to poor performance, particularly if mediocrity is tolerated. Ultimately, tolerance for poor performance may lead to a general loss of morale, especially among the more competent members of the team.

Pacesetting leadership

Pacesetting leaders expect their teams to follow the same high standards for performance that they set for themselves. Their message is, "Do as I do." They expect the team to be conscientious, have the drive to achieve, and have the same initiative that they do.

The pacesetting style works best when team members are self-motivated and highly competent. The pacesetter frequently gives no feedback on how well team members are doing. Sometimes the expectations of pacesetting leaders are too high to be attainable by team members.

Coaching leadership

Coaching leaders develop people for the future. They try to develop team members, improve performance, and develop long-term strengths. Coaching leaders say, "Try this."

The coaching style focuses on personal development and works best for highly motivated employees. Because the focus is on development and not on immediate work-related tasks, this leadership style can be too slow to respond in some situations.

Authoritative leadership

Have you worked for someone who was able to get everyone to follow his or her lead? Authoritative leaders move to mobilize people toward a vision, usually their vision of the future. Their actions say, "Come with me." The authoritative manager wants to promote self-confidence and empathy among their team members.

The authoritative style motivates team members by making clear to them how their work fits into the larger goals of the organization. By communicating what the organization's expectations are, the authoritative leader maximizes commitment to the organization's goals. The authoritative style does not work well if the manager has little authority among the team. The manager may have little or no experience in the field or may not be as knowledgeable about processes and technology as some members of the team.

Each of the leadership styles has a place in an organization. The best leaders use a variety of styles, depending on the specific leadership needs of the moment. However, coercive and pacesetting leaders have negative effects on team morale. The most positive form of leadership in terms of morale may be authoritative.

Best Practice–Shackleton on leadership

Many of you may have heard of the incredible adventure of the Antarctic explorer, Ernest Shackleton. He is a classic authoritative leader. In January 1915, during their exploration of the Antarctic, his ship, *Endurance,* with 28 men on board was trapped in the Antarctic sea ice. He and his crew remained on board for nine months while the Endurance drifted with the pack ice. In October 1915, the ice crushed and sank the ship, stranding the entire crew on the ice with supplies and lifeboats they had removed from the ship. They drifted on the pack ice for another five and a half months before the northerly drift and summer temperatures melted the ice beneath them. In April 1916, they were forced to take to the lifeboats.[2]

After a week of sailing in heavy seas, they landed on uninhabited Elephant Island. Unfortunately, they had no hope of rescue because Elephant Island was far from the normal whaling routes. Shackleton took five of his crew and the largest lifeboat and sailed 800 miles in 17 days to the island of South Georgia, which had a whaling station. When they arrived, they were forced to land on the opposite side of the island from the whaling station and climb over a glaciated mountain range to reach the station. Once at the whaling station, Shackleton organized a rescue effort. In a few weeks, the entire crew was rescued.

The adventure itself is amazing. But what is even more amazing—no one in the crew was even seriously injured during the adventure, and many in the crew considered it the happiest time of their lives. Several volunteered for Shackleton's next expedition.

Upon his return to England, Shackleton authored *South,* a book about the expedition. Shackleton describes how he managed his crew both before and after his ship's misfortune.

Shackleton was interested in management and leadership. Many people over the years have studied his techniques, and some have tried to emulate him. In their book on

---

[2] Bill Hackos, "The Shackleton Way: Leadership Under Stress," *Best Practices* of The Center for Information-Development Management, October 2001.

leadership, *Shackleton's Way*, Margot Morrell and Stephanie Capparell describe the adventure and present the management and leadership techniques Shackleton used to save his crew.[3]

Learning about management from an Antarctic explorer

Besides being an entertaining story, Shackleton's adventure can teach us much about leadership and management. Unlike other Antarctic explorers, Shackleton did not come up through the military nor was he employed by any government. He came from the British merchant marine and funded his expeditions privately. He had to raise money for his expeditions, outfit himself, hire his staff, and make a profit, just like any modern high-tech entrepreneur. He suffered through the same problems that information-development management face every day—hiring, firing, morale, management, and profitability. (Once he reached the Antarctic, retention was not a problem!)

Hiring the best people

Nearly 5,000 men applied for the 27 posts on the *Endurance*. Shackleton personally interviewed all the applicants he thought had good promise. Although he obviously had to hire people who had sailing and scientific skills, he wanted people who he felt had the enthusiasm and optimism to carry them through the rigors of the expedition.

Personal interviews are a critical part of the hiring process, as many information-development managers would agree. In September 2001, the administration at the University of Denver decided to conduct personal interviews of all freshman applicants because they feared that relying only on test scores and grades eliminates many promising individuals. They believed that personal knowledge would improve the selection process.

Shackleton believed that personal trust was an essential part of assembling the best crew. He first hired a person he knew and trusted who would be second in command. He felt strongly that he needed at least one person whom he could trust and confide in. Because he had to demand absolute loyalty from all of his crew, he wanted to ensure that he hired men he could trust under any circumstances.

Technical skills were not uppermost in Shackleton's hiring practice. He was most interested in building a team, ensuring that the crew members could get along well under the most trying circumstances. He wasn't looking for special experience—he wanted people who were willing to learn anything they needed to learn to do the job.

Following Shackleton's method, information-development managers might want to reconsider the practice of hiring based on a set of existing job skills. Many managers prefer to hire people who have considerable previous experience. Then they assume that person produces good work without ever evaluating the work directly. Rather than searching for someone who knows FrameMaker, you might want to find the people most willing and able to learn. Then, you need to ensure that they get the training they need.

---

[3] Margot Morrell and Stephanie Capparell, *Shackleton's Way: Leadership Lessons from The Great Antarctic Explorer.* New York: Viking Press, 2001.
Ernest Shackleton, *South. A Memoir of the Endurance Voyage.* New York: Carroll & Graf, 1998.

Firing those who don't fit

Because Shackleton got to know each member of his crew personally, he was able to continuously evaluate each person's work. He knew that his choices for the crew were not always correct. He was convinced that if one of the crew did not fit in he should be removed quickly. In fact, Shackleton fired some of his crew on the intermediate stops he made along the way to the Antarctic. He was always fair to all of his men, respecting those he fired as well as those who stayed. He paid for passage back to England to those he fired during his stop in Buenos Aires.

Not only is it critical for you to make decisions quickly about keeping people who aren't working out, but you need to know a great deal about what your staff members are doing. Many managers are reluctant to fire people because firing anyone is uncomfortable. As a result, they retain negative and unproductive team members far too long.

Demonstrating concern for each team member's well-being

Shackleton demonstrated a sincere concern for each person in his crew. He would strike up a conversation with each one nearly every day and find out about any fears, concerns, and general well-being. He personally helped men when they became ill. In one case, he kept a man with a bad back in his own bed in his own cabin for weeks while he slept uncomfortably in a chair. Everyone in the crew was aware of Shackleton's concern for their welfare.

By knowing about an individual's work and the quality of his performance, he was able to make adjustments. He found one individual who was hoarding supplies, caching his own personal supply to the detriment of the crew. Rather than punishing him, Shackleton put this man in charge of supplies. His hoarding ability then became an asset rather than a liability.

Developing staff member skills

Shackleton was very concerned that his men were always productively occupied during the long voyage south. The men were placed on teams with both experienced and inexperienced individuals so that all of the crew could learn from the experienced. He assigned seaman tasks to his scientific crew and taught his seamen how to make scientific observations. He was involved in every aspect of the voyage, working right along with his crew. All of the crew felt that he took a personal interest in them.

An aloof management style is increasingly unproductive as we try to gain efficiencies in every part of our process. By following Shackleton's way, you should be involved with team members, assisting them and learning with them at every step. Even if someone is an expert in a complicated area or has a unique skill, you must take an interest even if you cannot be an expert yourself. It is especially important to know what the experts are doing and how they are making decisions. Technical experts often make decisions that may not be the most efficient or effective for the business and need guidance from those who are responsible for understanding the bigger picture and promoting the company's goals.

Focusing on teamwork

In my research into content management and the need for new processes to support a single-source strategy, I find that collaboration has become increasingly important. The old style of technical communication in which people work alone on their own books no longer meets the demands of cost-effective operation. In changing the style of the information-development workforce, you might be well served to follow Shackleton.

Shackleton dispensed with most of the hierarchical management style that was the norm on sailing ships. All of the crew received the same rations, ate together, played together, and did skilled and menial tasks together. He demonstrated to his men that all of their jobs were important to the success of the expedition.

He organized teams for every project and carefully selected who would take part. He mixed people's skills in staffing a project, thereby training the weaker people. He ensured that people had many opportunities to work on a variety of projects. People were not stuck for years with the same jobs.

Preventing cliques

Shackleton was concerned about the destructive nature of cliques that naturally develop and took action to prevent their development. He planned all kinds of activities that included the entire crew, from eating gourmet food together, to intense birthday parties, celebration of all holidays, skits, and practical jokes. These events broke the monotony of being stranded for over a year.

Eliminating negativism

When Shackleton found that one of the men was exerting a negative influence on others, he made a special effort to befriend that individual. Typically, the man was so impressed by the personal interest that he became a loyal supporter. Shackleton moved jobs and teams around to help anyone who was having trouble adjusting.

Committing to the leader

Shackleton worked hard to gain the commitment of everyone in the crew. The men's journals are full of statements testifying to their loyalty. The men adored Shackleton both before and after the destruction of the ship. They trusted that he would get them home safely. He was equally committed to the well-being of every man in the crew and was very conservative in taking risks with his men. Each time an accident occurred, he would personally find the cause and make sure that it didn't happen again.

Promoting optimism

Shackleton hired crew members based on their optimism. He himself showed his optimism to the men during the entire adventure. He kept all of his men informed about every problem involving them, informed them of their options, and detailed a plan for future action. He considered every setback a new challenge.

Moving leadership to a new age

The age of individual exploration is over. We might compare Shackleton's story with the near disaster of Apollo 13. But Apollo 13 was a heavily funded government effort with constant communication with Houston ground control. The crewmen followed the direction of the leadership on the ground.

However, entrepreneurs and managers in the modern high-tech world are managing efforts in an environment that seems to change as fast as the Antarctic weather. You often need to produce our company's technical publications with very limited resources. Not so different from the *Endurance* after all.

In the current economic environment, you cannot afford to continue wasteful management practices. Gone are the days when you had enough money to permit "hands-off" management. Gone are the days when you could tolerate the inefficiencies of unmanaged staff members. I recommend that you consider emulating Shackleton's leadership style by hiring carefully, quickly removing the weak producers, being aware of what your staff is doing, taking a personal interest in their needs, personally developing them as competent and optimistic professionals, and most of all, demanding loyalty.

Shackleton's crew members believed him to be the best leader that they had ever worked for. Following Shackleton's way might lead to the same result. What if your staff told others in your organization, "She is the best manager I've ever worked for, and I have complete confidence in her direction?"

Ten leadership steps for information-development managers

In his December 2005 article, "Making Project Management a Valiant Voyage," Mike Eleder outlined ten steps that information-development managers might use in following Shackleton's way.[4] He based his analysis on the book, *Leading at the Edge*, by Dennis Perkins.[5]

Shackleton's Way in Information Development

By Mike Eleder

1   *Keep the long-term goal in mind, while focusing immediate energy on achievable short-term objectives.* This is the vision thing! Know where the team is going. Communicate it on a regular basis. But focus on the small, achievable, top-priority steps that must be accomplished to get there. Focus on the results, not the actions. As things change, make adjustments. Many small course corrections are less disruptive than a few large course changes. Make decisions based on facts but make them quickly. Overcome uncertainty with structure. Clear roles and responsibilities give people purpose and divert attention away from negative thoughts and into productive energy. Vision and accomplishment work in concert to keep the energy flowing in the right direction.

2   *Lead by example, using visible, memorable symbols and observable behaviors.* Walk the talk! When under conditions of stress, the team needs to SEE the leader as a leader. Personal presence is a unique source of energy and power. The leader needs to sense when the energy of the team is ebbing and must mobilize them with an authentic, sincere message that will take the team forward. When things are the darkest, the team needs calm reassurance, straight talk, and unmistakable resolve. When mistakes are made, fix them and move on. Contrary to corporate culture, it is more important to fix the

---

[4] Mike Eleder, "Making Project Management a Valiant Voyage," *Best Practices* of The Center for Information-Development Management, December 2005.

[5] Dennis N. T. Perkins, *Leading at the Edge: Leadership Lessons from the Extraordinary Saga of Shackleton's Antarctic Expedition.* New York, New York: American Management Association, 2000.

problem than to affix blame. Use visible symbols to reinforce principles. People need to see leaders leading. The team needs to see someone on the bridge who knows the right course.

3 *Constantly show optimism and self-confidence, but remain grounded in the reality of the current situation.* Unflagging optimism is the hallmark of great leadership. If you don't believe it can be done, how will anyone else believe it can be done? The spirit of future possibility generates positive energy that is contagious and self-perpetuating. When presented with challenges, the optimist sees possibilities and focuses his or her energy on solutions grounded in reality.

4 *Maintain your own physical and mental well-being.* Success is often driven by high energy. To sustain a high-energy success engine, the leader and the team must have the stamina to cover the distances necessary to be successful. Leaders know they must push to excel, but they must also know when to stop and give people a rest. Leaders can show concern for others by monitoring their well-being. Sometimes this requires guidance from others around them to help choose what is best for the team.

5 *Constantly reinforce the team message that "We are one—we live or die together."* Establishing a shared identity reinforces team unity. The strength of the wolf is the pack. Helping each other resolve impediments strengthens unity. Frequent, honest communication also strengthens the bonds of unity. A unified team is one where each individual understands his or her role and the tasks required by that role. They feel a deep sense of personal responsibility for doing their part but understands how all the pieces fit together to achieve the group goals. Few teams succeed when vital information is hoarded or restricted to a few key decision makers. When making assignments, match individuals' skills to the task, but avoid any appearance of favoritism. The toughest assignment for the leader is dealing with poor performance. While individual feelings and team unity must be taken into account, poor performance must be addressed quickly and fairly. An unwillingness to deal with poor performance detracts from team unity. Performance actions must, however, avoid isolating an individual and provide a chance to recover and contribute.

6 *Minimize status differences, and insist on constant respect and courtesy.* Every member of the team, including the leader, must demonstrate core team values. Highlighting status differences and promoting special privileges undermines team unity. Avoiding unnecessary hierarchy promotes collaboration and teamwork for all tasks. An egalitarian spirit increases the resource pool for whatever needs to get accomplished and minimizes the anger and resentment that inevitably surfaces under conditions of stress when some are perceived as "more equal" than others. Always insist on common courtesy, no matter how stressful the environment.

7 *Actively manage conflict by dealing with anger in small doses. Engage dissidents and diffuse power struggles.* Many people fear conflict and approach its appearance with heightened anxiety. To prevent tension from rising to unmanageable levels, seek opportunities to deal with anger and conflict in small doses. Proactively seek out the opinions of others to recognize tension within the team. Many corporate cultures seek to drive conflict underground and replace constructive resolution with conflict avoidance under the guise of respect for people. Conflict is an early warning sign that problems are not being addressed. Bringing the conflict to the surface may be the only way to uncover the underlying problems. Worse, unresolved conflict will likely fuel tension that will surface in other, unproductive ways. Ignoring conflict increases the likelihood that the magnitude of the inevitable confrontation will be greater than if dealt with in small doses early on.

continued

Shackleton's Way in Information Development *continued*

8 *Constantly look for wins and celebrate. Look for humor and help people laugh.* Within stressful, time-constrained work environments, there seems to be little time for celebration. Yet, these are the times when celebrations can have the greatest impact on the well being of the team. Shackleton kept his team motivated by constantly looking for something new to celebrate. Small or large, celebration brings successful closure to a step or task and renews the team for starting on the next challenge. Celebration gives people a chance to feel good and enjoy a sense of accomplishment, and it builds confidence in expecting more successful outcomes in the future. During and between celebrations, look for humor and help people laugh. When things are the most serious is when laughter can be the most powerful elixir for rejuvenating one's spirit.

9 *Have the courage to take the Big Risk.* Successful leaders seldom take unnecessary chances, but when a risk is justified, they do not hesitate. Not making the key decision may do more to ensure failure than deciding to take a calculated risk that may be the only opportunity for success. Many corporate cultures promote risk-averse behavior. However, taking calculated risks of time, money, and resources for strategic opportunities can often yield wildly successful outcomes that would otherwise be missed.

10 *Never give up-there is always another move.* Leaders are relentlessly creative—they never give up. The vision of success they see in their mind's eye guides them toward the outcomes they believe are inevitable. Obstacles merely channel energy down another path to success. The game is never over. There is always another move, another path, and another opportunity-if only we can identify what it is. Tenacious creativity for finding alternative solutions is the hallmark of great leadership.

Summary

Becoming an effective and respected leader is not a simple process, as you may have discerned from the examples in this chapter. You need to work hard to encourage your staff members to become a productive team. A collection of individuals, as you may have noticed, does not make a team. A real team shares responsibilities and works effectively to achieve a goal. Its performance is measured by determining if the result of its work was successful, not that it was complete.

Everyone who comes to a management role has a personal style that influences how he or she chooses a leadership style. The examples in this chapter are designed to give you ideas to use to develop a personal style.

As a leader, you want to inspire your staff to become high-performing teams. Consider developing a pacesetting and coaching leadership style to motivate your staff, although you may need to employ other leadership styles at times.

"That's a great idea!"

Chapter 13

Promoting Innovation in Information Development

> Innovation distinguishes between a leader and a follower.
> —Steve Jobs

Information-development managers have a responsibility, both to the larger organization and to their staff members, to pursue innovations. When you are being pressed to reduce costs and increase productivity as well as deliver information that is valued by customers, maintaining the status quo is not one of the options. Among your team members, you are likely to have some that enthusiastically embrace innovations because they enjoy the challenge of thinking and doing something new and they believe that innovations can enhance their career development. Others on your team prefer to follow tradition because they are comfortable with the way they have been working and are reluctant, even afraid, to change. They may choose to quietly resist or they may become active naysayers, wreaking havoc behind your back.

You are obligated to find ways to promote innovation among your team no matter the resistance. As Clayton Christensen and Michael Raynor make clear in *The Innovator's Solution*,[1] failing to innovate dooms not only an entire company to be replaced by

---

[1] Clayton M. Christensen and Michael E. Raynor. *The Innovator's Solution: Creating and Sustaining Successful Growth.* Cambridge, MA: Harvard Business School Press, 2003.

competitors but will threaten an individual department. Your competitors are out there waiting for you to fail. They come in the form of product developers and others who believe they can develop technical information more quickly, accurately, and effectively on their own. They come in the form of outsource agencies that are knocking on your CEO's door to offer their low-cost services.

Only by pursuing innovation can you stave off the disruptive competitors and preserve the best your organization has to offer. If your team members fail to innovate and increase their value to the organization, they're unlikely to be around in five years.

For many information-development managers, new ideas are abundant. Team members want to redesign information deliverables, find better ways to meet customer needs, and make their own work more interesting. You may find that the most difficult challenge you face is to decide if the myriad new ideas presented by your team are worth pursuing. For an innovative new idea to be valuable, it must

- ✔ align with overall corporate goals and objectives and be politically acceptable
- ✔ add value to the customer by providing more useful and usable information resources
- ✔ reduce costs by making your team more productive
- ✔ reduce the time to market of information products

If the innovative new idea does not meet at least two of these criteria, it is probably not worth pursuing. If the new idea provides you with a more efficient way of working, adds customer value, and is aligned with corporate goals, it has a greater chance of succeeding.

Best Practices in Promoting Innovation in Information Development

Introducing and implementing sound innovations in your organization is not simple. It requires careful planning that begins with identifying an innovative idea that can be sold to senior management. It proceeds through managing the introduction of the innovation into your organization and ensuring its success against all obstacles. In this chapter, you learn about five best practices to help you promote innovations in your organization:

- ✔ Overcoming obstacles to change
- ✔ Understanding disruptive innovations
- ✔ Focusing on customer-centered innovations
- ✔ Instituting operational innovations
- ✔ Benchmarking with competitors and best-in-class colleagues

Introducing innovations into your organization and motivating staff members to accept new ways of working are always challenging. Consequently, I begin the best practices with some insights into overcoming the obstacles that make change difficult. As you work on innovation in information development, remember that change management is critical to your success. You may want to consult John P. Kotter's book, *The Heart of Change*, which provides a step-by-step plan for managing change.[2]

Best Practice—Overcoming obstacles to change

A perceptive study of leadership was conducted in 2002 by W. Chan Kim and Renee Mauborgne, management professors at Insead in Fontainebleau, France.[3]

Kim and Mauborgne conducted an in-depth study of William Bratton, who served as police commissioner in New York City in the 1990s and was instrumental in lowering the city's crime rate. They sought to understand how he accomplished the turnaround, especially in such a politically charged environment as New York. The tag line of their article reads,

"How can you catapult your organization to high performance when time and money are scarce? Police Chief Bill Bratton has pulled that off again and again."

Kim and Mauborgne outline four obstacles that managers face, distilled from Bratton's experience. In Table 13-1, I have applied them to the situations faced by information-development managers.

Table 13-1: Obstacles Faced by Information Development Managers

Cognitive hurdle	Our staff members respond best when they experience directly the problems our customers have in finding and using information to perform tasks or solve problems.
Resource hurdle	By targeting our resources on contributions that add value to the organization, we make best use of what we already have.
Motivational hurdle	Reforming everyone and everything at once may be appealing, but it's doomed to failure. We need to solve problems with key influencers who can help to spread the message.
Political hurdle	Powerful interests, such as product developers and marketing managers, often resist change. We must identify the naysayers and find ways to silence them.

Breaking through the cognitive hurdle[4]

Most people involved in decision making about a change don't respond to sensible arguments, extended examples, or even dollars and cents. Even the bean counters, who insist

---

[2] John P. Kotter and Dan S. Cohen, *The Heart of Change: Real-Life Stories of How People Change Their Organizations*, Cambridge, MA: Harvard Business School Press, 2002.

[3] W. Chan Kim and Renee A. Mauborgne, "Tipping Point Leadership," *Harvard Business Review*, April 2003.

[4] *The headings in this section are paraphrased from the Kim and Mauborgne article.*

that they are only interested in the numbers, won't act if they don't understand the problem or believe that any change is necessary.

Bratton's technique, at the core of tipping point leadership, is to help key decision makers actually experience the problem. Without a lesson in reality, the need for innovation never catches on. For example, Bratton decided that he had to convince his senior staff that the public's complaints about the subways were meaningful. They simply didn't think the problem was that serious. Besides—New Yorkers always carp about the subway. Big deal.

Convinced that making the subway safe was the highest priority in the turnaround, Bratton required that all his transit police officers commute by subway, himself included. Most of them had not ridden the subway in years. Not until they saw the graffiti, experienced the lawlessness of the gangs, and got mired in non-working equipment did they understand the public's point of view. And, only then did they agree that something needed to be done.

Bratton had gotten his officers over the cognitive hurdle. They had to gain a visceral and cerebral understanding of the problem before they were ready to respond to new ideas. A colleague of mine used a similar tactic a few years ago to convince his mechanical engineering staff that they had to redesign a critical piece of equipment used in treating gravely ill patients. He had feedback from users that the equipment was difficult and slow to adjust, resulting in unnecessarily long treatment times. They reported that patients, already anxious about their illnesses, were further stressed by the ill-adjusted and awkward equipment.

His problem—the engineers weren't especially concerned. So the technicians had a problem with the usability of the equipment. Couldn't they just be trained better? Our colleague used a drastic approach to move his engineers through the cognitive hurdle. He took them out to customer sites and had them experience the problem firsthand.

If that sounds somewhat like a customer site visit as part of a usability study, you're right. That's exactly what occurred. The engineers observed the users trying to adjust the equipment. They played the roles of patients, enduring the long adjustment periods flat on their backs with heavy lead weights on their bodies. It wasn't pretty. The experience changed their entire perception of the problem. They came back committed and energized about redesigning the equipment immediately and correcting all the adjustment problems.

Experiencing a problem firsthand can be a life-altering experience. I worked with a team of software programmers designing a medical-records system. It was clear from a first view of the prototype software that they were completely oblivious about the users' world. The prototype was completely unusable to anyone but the designers. I suggested that they visit some users, conveniently located across the street from the software-development department. They took me up on the suggestion.

Months later, meeting the same programmers over lunch in the cafeteria, they told me that the meetings with users and the direct observations of their work had "changed their lives." These were senior professionals, committed to doing good work. They simply needed to break through the cognitive hurdle.

I invite you to consider ways in which you can help decision makers in your organization overcome their reluctance to support innovation and change. Have they tried to find information on the corporate website? Do they know how it feels to use the products the company develops? Have they seen people at work having difficulty learning what to do? In the 1980s, both Hewlett-Packard and Sun invested in usability testing, hiring usability professionals and building usability labs, because their CEOs were unable to install their computers at home.

Information developers tend to be a group of introverts who sincerely believe that if they work hard and keep their noses to the grindstone, someone will notice good work and reward it. Despite such an optimistic belief, you need to sell the ideas you have for making things better, after you're certain, of course, that the ideas are better for the customers and the company, not just for you.

Sidestepping the resource hurdle

As soon as an information-development manager begins to research how to implement an innovation or a badly needed process improvement, the immediate reaction is to bemoan the lack of resources. The manager either has too few people or too little money or both. Bratton's police department certainly did not have unlimited resources or lots of new money. He succeeded in implementing his program by restructuring existing resources rather than seeking new ones.

After he had convinced his managers that a new strategy was needed by getting them to understand the problem, he had to find ways to implement the new strategy without an influx of cash or people. He had to find ways to make improvements that could be accomplished with existing funds and staff.

The greatest danger, after you overcome the cognitive hurdle, is to abandon your goals because you can't find a way to accomplish them with your existing resources. The answer is to find ways to achieve your goals without having to acquire new resources. Bratton looked first at how his resources were allocated to fighting crime. Although everyone acknowledged that a significant percentage of crime resulted from drug use, only 5% of his force was assigned to the narcotics division. Although everyone agreed that most drug-related crimes occurred on weekends, the narcotics division staff worked only Monday through Friday. Once everyone recognized the obvious problem, Bratton was able to institute a major reallocation of staff and resources.

When Bratton learned that police officers took an average of 16 hours to process an arrest, he found inexpensive ways to reduce the processing time to 1 hour. As a result, he gained back staff time that was being wasted on paperwork that everyone hated doing. When he found that one division had too many patrol cars but not enough office space, he located another division that had too much office space and not enough patrol cars. He then negotiated a trade.

As an organization manager, your role is to look for similar opportunities to achieve your goals without pleading for new resources from recalcitrant CFOs. Begin by carefully examining every step of the processes your staff uses to develop and deliver your information products. One manager, doing this analysis of the as-is procedures in the department, uncovered 27 pages of process steps. It was clear from the analysis that many of the steps were duplicated, unnecessary, or could be done more simply. By redesigning the process, the team members created a to-be procedure that had only nine pages of process steps, greatly reducing the time required in the information-development life cycle.

Another manager, asking team members to reduce unproductive time, found that the suggested process improvements enabled the team to increase their productivity without adding additional people. They cut out time from unproductive, required meetings. They eliminated writer tasks that could be handled better by their production experts. They looked for immediate opportunities to share content among deliverables.

A senior manager used a pilot project to demonstrate that the innovation was viable and cost-effective, which enabled her to get funding for the next stage of the project.

Many managers and teams become quite adept at finding opportunities to use resources more effectively. The minimalist agenda, for example, asks information developers to scrutinize information products for content that is rarely or ever used by customers. One team, after only a few hours examining existing manuals, decided to eliminate marketing information that was irrelevant to customers who had already purchased the product and was time-consuming for staff to keep up to date. They also found opportunities to reduce descriptions of features that had been added years before and were no longer needed by customers.

In moving toward minimalist documents, another team carefully analyzed the regulatory requirements that had to follow in developing documentation and included only the required content for end users. By eliminating redundant and unneeded content, they were able to reduce translation costs by 60% and speed time to market for global customers.

Costs saved in one part of the organizational process should be reallocated to other parts of the process that require bolstering. Pursuing a minimalist agenda and reducing the amount of time required to maintain and produce content might free time for customer studies. Time spent on customer studies might lead to further reductions in content that is not valued in the field.

The results of one organization's customer field studies showed that end users were not using the online help. In fact, the clerical end users learned their basic tasks on the job and rarely performed any new tasks. By eliminating the online help, the team could devote more time to redesigning the user interface, a task they had been asked to pursue but for which they lacked sufficient time. By removing a low-value task (creating online help), they could pursue a high-value task (making the interface more intuitive).

Jumping the motivational hurdle

As a people manager, you already know how difficult it is to motivate an entire staff to devote time and energy to instituting a change. The larger the staff, the more difficult they are to motivate because so many of the staff are physically or psychologically distant from the leadership.

The first step in tackling motivation is to recall the lesson learned from Malcolm Gladwell's *The Tipping Point*.[5] Gladwell demonstrates that change occurs not by a magical process in which everyone changes simultaneously, but from the influence of a small number of key individuals. Bratton knew that if he could motivate the key influencers in the NYPD, he could use them to influence change throughout the organization. He chose as his key influencers the precinct commanders, each of whom directed 200 to 400 police

[5] Malcolm Gladwell, *The Tipping Point: How Little Things Can Make a Big Difference,* Little, Brown and Company, 2000.

officers. If he could change the way each of these commanders handled crime, he had a chance of changing the entire force.

Bratton's tactic was to institute weekly meetings, bringing together all the precinct commanders and the top officials of the department, to scrutinize the crime statistics and the performance of the police precinct by precinct. Each week, one of the commanders was held accountable for his or her performance, explaining what was effective and where performance still lagged. The commanders became personally accountable for the success or failure of their operations. The commanders learned what was working or not working for their colleagues, encouraging them to take innovations back to their own teams.

In a recent project to develop an enterprise content-management system, Comtech advised the client to conduct a thorough stakeholder analysis. One of the outcomes was identifying the key influencers in the organization, people who everyone trusted to tell them what was a good idea and what was a waste of their time. Most of the key influencers were quite vocal about their likes and dislikes. If they didn't support a program, it was usually doomed. The liaison didn't like all of these people and would have preferred not having to work with them. However, with encouragement, he invited them to join one of the core teams working on the requirements and implementation of a new system. Their presence turned out to be critical because they were able to motivate others who were skeptical but could be influenced by someone whose opinions they trusted.

In another project, the lead rejected Comtech's advice, deciding to exclude an important key influencer from the information architecture team. As a result, the team's work was severely hampered because the entire group of subject-matter experts became aligned against the new information designs. Only after much work, and involvement of the key influencer, was the architecture team able to get back on track and make some progress in motivating the experts to become involved in the design process.

If your plan for innovation and change requires the cooperation of diverse team members, focus on the key influencers, those people whom everyone trusts. Involve them in the project from the beginning and provide a setting in which they are motivated to support the innovation and assume responsibility for its success. Their most important role is to motivate the others in the organization who will have to implement the changes.

It is important to frame your innovation in achievable chunks. Many of your team members will view a significant innovation as "pie in the sky," another unachievable idea foisted on them by out-of-touch managers. If your project seems too grandiose, staff members will not be motivated to take part. They'll just wait for it to fail, like so many others they have been through.

If you divide your project into parts that people believe they can achieve, they will more likely be willing to work with you. If they will take responsibility for their smaller parts of the project, even if they are skeptical that the long-term goal can be achieved, they will be motivated to participate.

Knocking over the political hurdle

Your innovation may be well under way and looking like a success when you face the barrier of political interests set up against you. Never underestimate the resistance people will show if they feel an innovation threatens their position or appears to be more successful than something they want to accomplish. If you've worked hard from the beginning of the innovation project to encounter and silence the naysayers, you may be ready for the opposition to be mounted against you.

If you have support among senior executives, you are more likely to be protected politically. Bratton anticipated opposition from the New York court system, which feared that his innovative project would increase the caseload. He sought and won the support of the mayor, Rudy Giuliani, to influence the legal system. He won the support of the *New York Times*, which helped enlist the public behind the changes.

You are likely to run into opposition from peer organizations. Product managers may believe that your innovations will take decision-making power away from them. Product development managers who are accustomed to controlling the content of technical documentation may feel threatened with losing their control. Marketing, support, training, or usability may insist that the customer is their property if your innovation is designed to improve customer satisfaction.

When the information-development managers at BMC decided to change the way they allocated staff to projects, they needed the support of senior management to overcome the resistance of middle managers on the product teams. The following vignette explains what happened:

BMC Software: A Case Study

In 2005, the BMC Software Distributed Systems Management (DSM) product documentation organization came to a crossroads. The company had been steadily moving to position its products as Business Service Management (BSM) solutions, which involves the use and integration of products from multiple lines of business, as well as products at different maturity levels. While the company was quickly moving to product integration and a business service model, the DSM writing organization continued to develop product documentation by line of business (LOB). The writing organization approached all products the same way: updating and enhancing all documentation deliverables for every product. There was no differentiation in the writing approach for new products, strategic products, or fully mature products. With an already reduced writing staff, flat headcount projections, and new projects on the way, the writing organization's management team realized that something had to change.

Over a six-week period, the DSM writing management team spent many hours in brainstorming meetings to

✔ innovate ways to fulfill customers' information needs by creating a customer-focused organization

- ensure coverage of all customer issues, which was done by creating four groups, spanning all Lines of Business (LOBs):

- Routes to Value (RTV). Writers would be responsible for working with marketing and program executives, as well as technical services field colleagues, to create technical documentation aimed at boosting sales by accentuating how products integrate to provide value-rich solutions.

- Customer Response (CR). Writers would be responsible for working with Customer Support and product teams to create answers quickly for customer issues and to provide workarounds and fixes. Deliverables include product-level white papers, technical bulletins, and high-priority flashes.

- Return on Investment (ROI). Writers would be responsible for working with product teams and product architects for strategic and new products to focus on installation, configuration, basic use, and best practices or implementation information to help customers see immediate return on their investment in the company's products.

- Classic. Writers would be responsible for working with product teams and product architects on mature products to highlight new and enhanced features, rather than updating all deliverables with all product changes in a large, stable documentation set.

✔ outline how each group would operate and the deliverables each would provide

✔ decide how to better align and design information deliverables to help boost sales

✔ determine the criteria for differentiating products by life cycle and maturity level

✔ centralize the editors to more effectively provide editing services across all writing groups

✔ assign writers to each group (based on skills, preferences, and so on)

✔ craft presentations to be given to upper management, product team management, and writers, to help in managing change

The DSM writing organization structure was implemented in May 2005. After 10 months, the DSM writing management team summarized their experience as follows:

✔ Despite some initial resistance from various product team members (most notably customer support and development), there is now acceptance and even recognition that the organization is not only innovative but is working.

✔ The team must continue working with customer support and product teams to find the best method for providing information about product issues: a bulletin, release notes, a readme file, or a resolution in the knowledge base.

✔ Writers were skeptical but are now enthusiastic proponents of the organization.

✔ Writing tasks are divided so that writers focused on documents specific to the next release of a product (ROI and Classic teams) do not have to interrupt their work to create white papers, technical bulletins, and so on.

✔ With more development groups moving to agile development environments using the Scrum methodology, the ROI and Classic writing groups are focused on developing manuals and help systems without interruption. With a sharper focus, writers on these teams can work more effectively in this environment.

✔ Marketing program executives and support colleagues are receiving more white paper, data sheet, and technical bulletin information than ever before because the RTV and CR teams are specifically tasked with these deliverables.

✔ With a cross-LOB approach, all writers are gaining more product exposure and expertise, increasing their value to customers and the company.

✔ Writers want to be loaned out to other teams to lend a hand, because they are supplementing their product knowledge and they now feel more comfortable with documenting a wider range of products.

Best Practice–Understanding disruptive innovations

Clayton Christensen outlined the forces of disruptive innovation in his book, *The Innovator's Dilemma*.[6] Disruptive innovations, Christensen argues, cause well-managed, well-established companies to fail, even when they are paying attention to their customers and investing in new technologies. These companies fail because their markets are increasingly dominated by small, innovative companies that introduce new technologies that are less expensive, often easier to use, and initially attractive to the low end of the market. The disruptive innovators chip away at the low end, gradually increasing their offerings to progressively higher-end customers, until they eventually force the once dominant company out of business, or largely so.

Remember when Sears was the dominant low-cost retailer in the US? Sears saw its market disrupted by aggressively less expensive competitors like Wal-Mart and Target that have themselves become the dominant players. Christensen shows how this pattern plays itself out over and over again, noting that the process is almost inevitable. Few companies have the ability to remain competitive in the face of disruptive forces.

You may, in fact, find yourself in a company that is being challenged by disruptive innovators. If you work for a longstanding company that is seeing its low-end customers moving to less expensive, more focused, and easier to use competitive products, you are witnessing a potential takeover. It's likely that your company is responding by moving further up the customer levels, choosing to remain the product-of-choice to its high-end customers and pretty much writing off the low-end to the competition. Unfortunately, as Christensen explains, that response, although it can be successful for some time, eventually fails. The disruptors produce better, more functional products, gradually absorbing more and more customers. At the same time, the high-end company offers more and more functions to its remaining lucrative customers until the product is so feature-dominated and difficult to use that the high-end customers won't pay the cost. They simply don't need much of the new functionality.

Christensen contrasts disruptive innovations with sustaining innovations. Sustaining innovations are the kinds everyone strives to make. Can we find new technologies or other ways to make our products better? Can we improve our processes of creating products to make them better or less expensive? Large, well-managed companies are masters of sustaining innovation. In the field of information development, we are constantly finding ways to improve the effectiveness and efficiency of our information product development.

In his *Best Practices* article, Bill Hackos explains the four principles of disruptive innovation described by Christensen.[7]

Principle #1: Companies depend on customers and investors for resources.

Most people think of companies and their management as being in control of their destinies. Companies succeed because management is brave and smart. Companies fail because of management mistakes. Christensen claims these notions are completely wrong.

[6] Clayton M. Christensen, *The Innovator's Dilemma: When New Technologies Cause Great Firms to Fail.* Cambridge, MA: Harvard Business School Press, 1997.

[7] Bill Hackos, "The Information Developer's Dilemma," *Best Practices* of The Center for Information-Development Management, October 2004.

Instead, companies are completely dependent on their customers and investors, even when management is doing all the right things. If a company doesn't satisfy customers, they don't bring in revenue that is crucial to company operations. Similarly, if companies don't satisfy stockholders' goals for growth, they don't get the capital they need for investment. No matter how hard companies try to violate this principle, all they really can do without customers or investors is go out of business. In fact, companies that produce products with features customers want and growth that investors demand will thrive. The best-performing companies market products or services with features that exactly match what customers want, and no more.

Principle #2: Small markets don't solve the growth needs of large companies.

Disruptive innovations result from the introduction of new technologies and are developed for small or not-yet-existing markets. Generally, the disruptive products or services are not as good for existing upscale customers, but they are cheap and convenient. They are usually developed by small entrant companies. Because of their small market and low cost (and small markup), they are not attractive to the large incumbent companies. After all, why should large companies fool with products that have small markets and low profitability. It is better for them to improve existing products for their high-markup customers. If large companies try to move into the disruptive products market niche, they are likely to slow rather than accelerate their growth. However, small entrant companies can thrive on revenue that is of little value to the established companies.

Principle #3: Markets that don't exist can't be analyzed.

Large, established incumbent companies are typically experts at market research. They are not able to throw resources into new, not-yet-existing markets because they have no way to do market research on those markets. Small companies can afford to aim toward not-yet-existing markets on the expectation that their disruptive product will create a market. Of course, many small companies aim wrong and go out of business before anyone knew they existed.

Principle #4: Technology supply may not equal market demand.

Most companies, large and small, make continuing improvements in products so that they can enter new, more upscale markets and take advantage of the greater markup in these markets to increase revenue as well as profit margin. Unfortunately, the large incumbent companies may improve their products to the point that the improvements are of no additional value to any customer group, a process often referred to as "feature creep." Christensen calls it "over-serving" the customer. Small companies take the same route of moving upscale, but their improvements enable their disruptive products to serve some of the same markets as the incumbent's products but at a lower cost. At this point, the products become commodities and the entrant company begins to erode the incumbent's market based on price. What happens next is often not pretty. Once this process begins, it may not be possible for the incumbent to recover.

Christensen goes through case study after case study to show how these four principles explain disruptive innovation in the real world. Here is a single hypothetical scenario of a disruptive innovation event.

Typical Innovator's Dilemma scenario

An incumbent company has achieved considerable success in a technology and maintains a large market share. The company continues to make improvements in its products based on customer and end-user research. Customers are happy and seemingly loyal. A new technology emerges, not necessarily high tech, but maybe a new process or business model. An entrant company seizes on the technology and creates a product that is at the low end of the market or even in a not-yet-existing market. Its low price, improved simplicity, or convenience creates a new market or may make inroads into the lowest-margin market of the incumbent.

Both the entrant company and the incumbent company are thrilled. The entrant is finally making some revenue. The incumbent is happy because it is losing its low-end market, which has the lowest profitability. Its average profitability has increased. "Good riddance!" The marketing department does a thorough market analysis and concludes it is not worth getting into a technology to compete with the entrant because the entrant's products will never be as good as the incumbent's, the markup potential is small, and the market is not large enough to bother with.

After a time, the entrant company starts using its technology to improve its disruptive product and begins eroding some of the high-quality market of the incumbent. At first, it's just pesky. The incumbent continues to improve its product by adding features that the customer likes but may not be willing to pay for. But the entrant continues to erode the market. Marketing is flabbergasted. "What is wrong with our customers?" "Why don't they understand that our quality is better than theirs?" Eventually, things get bad enough for the incumbent that management decides to act to protect its customers from this competitive onslaught. "We must produce that low-end product and compete directly!" But it's too late. Try as it may, the incumbent just can't compete with the entrant. It doesn't have the same expertise in the technology as the entrant. It is not as efficient, not willing to accept the lower markup, and has lost its brand advantage. It's convinced that its products are still much better. But why are their customers not loyal?

You probably can think of many well-known companies that are no longer with us or are only skeletons of their old selves. In fact, you may work for one of the incumbents.

What, then, is the Innovator's Dilemma? The dilemma occurs because incumbent companies react to the threat by pursuing sustaining innovations in their products that raise the price and may make the product more complex and less usable. The more market they lose, the more they try to compete by improving their product. It's a dilemma because they lose if they don't compete by improving their product and they lose if they compete by improving their product. The customer really wants a simpler, cheaper, easier-to-use, and more convenient product but the incumbent is unable to produce it.

Four principles for success

In her review of *The Innovator's Dilemma*, Ginny Redish summarizes Christensen's advice[8] by noting that successful companies can foster their disruptive innovators and reinvent themselves into the new world that the innovative technology will eventually bring.

---

[8] Janice (Ginny) Redish, "The Innovator's Dilemma: When New Technologies Cause Great Firms to Fail," *Best Practices* of The Center for Information-Development Management, June 2001.

Those that have succeeded have followed these four principles (paraphrased from Christensen, page 99 and Chapters 5 through 8):

✔ Put the project in an independent, autonomous unit that is allowed to have its own value system, its own customers, its own budget, its own profit margin—and that is often geographically away from the main organization.

✔ Put the project in an organization that is small enough to get excited about small opportunities and small wins.

✔ Recognize that the first attempts are most likely to fail. Assume that it will take an iterative process of trial and error to find the right product and the right market. Do not believe forecasts. Do not gear up for huge production until you have been through a few rounds of experimentation.

✔ Assume that the new technology will start out with a new market that values what makes this different from what your mainstream clients want now.

Best Practice—Focusing on customer-centered innovations

Christensen and colleague, Michael Raynor, expand upon the recommendations they made in *The Innovator's Dilemma*, in their 2003 study, *The Innovator's Solution*. In this work, they provide evidence for the efficacy of this advice in the 2003 book. They provide an empirically based plan that CEOs and other senior managers might pursue to achieve growth.

Now, unless you're the CEO, you're probably wondering what value *The Innovator's Solution* has for you as a manager in an information-development organization. How likely is it that an information developer will have an opportunity to pursue an innovation that is incrementally different from the company's previous business? In fact, how often can we pursue innovations that are recognized as being responsible for advancing our corporation's business in its traditional markets, which Christensen refers to as a "sustaining" rather than a "disruptive" innovation?

Christensen provides the solution for information managers to move into the mainstream of company innovation. The key is to understand the role information developers might play in disruptive innovation.

First, however, it is important to examine the differences between sustaining and disruptive innovations. Sustaining innovations are well known to most, because most information developers tend to work in mainstream companies that have been in business for a relatively long time. Sustaining innovations represent the progressive improvements our companies make to products in response to the demands of existing customers. The best customers tend to be those that buy the most expensive, state-of-the-art products with all the latest bells and whistles included. These mainstream customers help companies generate healthy profits by continuously moving up-market toward large customers willing to pay premium prices for better products—that is, until the products become too good.

Christensen explains that innovation along the sustaining model cannot last. Soon enough, customers get all the capabilities they are willing to pay for. More features and functions, as you well know, simply are not that attractive. The products pretty much do what they are needed to do as they are today. When you find that customers are not anxious to install the latest and greatest version of your product, you know you've probably reached the limit of the sustaining model. At that point, a company's mainstream product becomes a commodity in which the list of features and functions is duplicated by all the competitors. Eventually, cost-cutting efficiencies are the only ways left for our companies to manage profitability. You may have noticed in your own organization that commodity products are most likely to be outsourced for continued development and production, including, but not limited to, the outsourcing of technical publications.

Disruptive innovations work differently and are most often pursued by newcomers to the market, those companies hoping to win market share away from the big, dominant players. Disruptive innovations are those that successfully implement simple, more convenient products that cost less and appear to a new set of customers. In fact, many of the large companies that you may have worked for started life as disruptive innovators.

Christensen goes on to divide disruptive innovations into two categories: those that are low-end disrupters and those that are new-market disrupters. Low-end disrupters are familiar to all of us when Wal-Mart comes to town. Wal-Mart sells products as cheaply as possible, undercutting and disrupting many of the small businesses that had started happily in our communities. New-market disrupters look for the non-consumers, people who might use an innovative product if it were convenient and reasonably priced. For example, Christensen points to Canon and its desktop copier business as an example of a new-market disrupter. Canon made photocopying convenient by selling reasonably priced copiers that could be housed in individual departments or offices.

Here are the questions that Christensen suggests we ask about customers:

✔ Are there customers out there who don't have the skills, money, or equipment to do a particular job for themselves? Either they don't do it at all or they have to pay for experts to do the work for them.

✔ To use a product, do customers have to do something that is inherently inconvenient for them, like going to a central location to use the copy machine?

Think about desktop publishing in these terms. As Mark Baker noted in his article on authoring systems, information developers wholeheartedly adopted desktop publishing technology in the mid 80s.[9] Before desktop publishing, if you wanted a professional, type-set look for your publications, you had to use the services of an expensive and relatively inconvenient professional typesetter. Most of the time, the cost of the typesetter was high enough that you were forced to choose typewriter-style output for your technical publications. Even buying your own typesetting equipment and learning how to use it was outside the capabilities of most departments.

---

[9] Mark Baker, "What Makes an Authoring System Tip?" *Best Practices* of The Center for Information-Development Management, February 2004.

Desktop publishing became a new-market disrupter for typesetting. It was reasonably inexpensive and easy to learn. It allowed you to produce professional looking publications right in your own departments. Desktop publishing was so effective a disrupter that most typesetting businesses are gone.

Discovering products that will appeal to customers who are underserved by current products is not an easy task. However, information developers, if they are actively engaged with actual customers, may be in an excellent position to identify new-market disruptive innovations.

Christensen notes that over 60% of new-product development efforts fail. He insists, however, that the failures are not random and are both predictable and avoidable. He advocates abandoning traditional methods of segmenting markets that lead almost invariably in the wrong direction with regards to innovation.

Traditional market segments are identified by attributes of the product and the customers. Product attributes as market segmenters result in long lists of features and functions. The thinking goes that if you add more features to a product, it will attract more customers. See the addition of digital cameras to cell phones as a product-focused innovation. Customer attributes that serve as market segmenters result in the demographic categorization of customers. You are told, for example, that a particular set of young, male, upwardly mobile, MBA-type customers will want pocket-sized PCs that act as extensions of their desktop models.

Christensen argues that such representations of markets are inadequate to support disruptive innovations. He asks us to focus instead on circumstance-based categorizations that state explicitly what features, functions, and market positions will "cause customers to buy a product." Our ability to predict with confidence how a product will succeed is based upon our thorough, in-depth understanding of customer behaviors.

In keeping with the principles of user-centered design, Christensen argues that customers have jobs they need to get done, and they elect to hire our products or services to get the job done.

"Companies that target their products at the circumstances in which customers find themselves, rather than at the customers themselves, are those that can launch predictably successful products. Put another way, the critical unit of analysis is the circumstance and not the customer."

Disruptive products provide ways for people to hire the products to get a job done that has been impossible before. The example Christensen provides is the BlackBerry, the hand-held wireless email device manufactured by CIDM member, Research in Motion (RIM). RIM, Christensen argues, gained a "disruptive foothold [by] competing against nonconsumption by bringing the ability to receive and send e-mail to new contexts such as waiting lines, public transit, and conference rooms." By observing that customers were trying to fill in unproductive snippets of time, RIM found the circumstances, the job to be done, that would bring people to their product. Instead of producing yet another handless wireless device like the Palm Pilot (competing against product features) or developing a product for the business traveler (competing against customer demographics), the company took another path. By watching people use BlackBerries, Christensen notices them filling bits of otherwise unproductive time by accessing email. He argues that RIM competes, not against other electronic devices, but against the newspaper or the CNN Airport News or a boring meeting.

When information developers focus on the customer, especially when the customer is trying to learn what he or she may want to do with a product, you can learn not only what might make the information more useful but you might also recognize opportunities that are not being served by the current product. You may find, for example, many customers who choose not to use your products at all or who use only minimal functionality. Perhaps such customers would be better served by a low-end disruptive product at significantly less cost and with far fewer complicating features. You may also find customers who are in circumstances not served by your current products and who are looking for different solutions for different needs.

You might expect marketing professionals in your companies to be on the lookout for opportunities to create disruptive innovations. However, marketers frequently focus on the behaviors of buyers rather than the behaviors of users of the product. Information developers, in coordination with usability professionals when they are available in a company, are better placed to note the real circumstances and the real jobs that the customers are trying to do. Of course, part of the dilemma surrounding innovative and disruptive ideas is the difficulty of selling them in traditional organizations, especially when the ideas come from people who are not considered to be in the mainstream of idea generation.

How might information-development managers respond to Christensen's crusade to change the way existing companies pursue innovation? The best way possible is to pay close attention to those who are trying or trying not to use your products, including your information products.

Susan Harkus has written extensively on the importance of understanding and meeting the user's agenda in information products. In her article with Elizabeth Pek, Susan makes the argument that you must understand the user's real agenda before you can successfully design information products that meet their needs.[10]

Understanding the User's Agenda

By Susan Harkus and Elizabeth Pek

We know we have a good solution. We have done the research; we have analyzed our audience; and we have identified user tasks. We know that the solution meets the needs and goals of users. But, why are our users still dissatisfied with the offering? Is it the curse of the lurking user agenda?

As communicators, we want users to use our product. We want product manuals to be "dog-eared" from constant use; we want users to use the online help rather than call the support desk; and we want our Internet/intranet site's hits to increase.

By understanding and leveraging the user agenda, we can achieve the user and business outcomes that we are seeking. The user agenda is the missing link between the analysis and design that we do and the outcome we desire.

---

[10] Elizabeth Pek and Susan Harkus, "Building Success on the User Agenda," *Best Practices* of The Center for Information-Development Management, April 2002.

What is missing from our traditional methods is an understanding of where users are coming from at the point of engagement in the website or information. We need to delve deeper and understand beyond the needs and goals of our users; we need to understand and address our users' agenda at each point where they interact with our solution.

What Is the User Agenda?

The user agenda is complex and much more than just goals and objectives. It encompasses deeper subconscious needs such as priorities, expectations, issues, assumptions, and reservations.

For most one-off information or transaction interactions, users are unable to describe their agenda. Only where they have frequently performed a task and experienced the same frustration will you hear them exclaiming, "Why can't I . . ." or "Why don't they provide. . . ."

The agenda is only revealed at the point of engagement—when users read our document or interact with our web site. And even then, it is revealed progressively; often users do not know their agenda until they are performing activities with our solution.

Best Practice–Instituting operational innovations

In his April 2004 article in the *Harvard Business Review*, re-engineering guru Michael Hammer, takes on the unglamorous changes in everyday operations, revealing their potential to produce significant reductions in cost, increases in productivity, and improved customer satisfaction. "Deep Change: How Operational Innovation Can Transform Your Company" speaks to the operational innovations that are near and dear to the hearts of information-development managers.[11]

Hammer begins by telling the Progressive Insurance story, a company that has seen phenomenal growth in an otherwise staid auto insurance market. Progressive's growth came not through acquisitions or mergers, the stuff that puts CEOs on the front page of *The Wall Street Journal*, but through significant innovations in everyday operations.

You've probably seen Progressive's television ads inviting customers to use their rate-comparison website. Behind the scenes, they simply did their work better than their competitors. For example, if you have an auto accident, Progressive representatives are on hand 24 hours a day to take your call and schedule a claims adjuster. The claims adjuster works out of a mobile van, enabling a nine-hour turnaround rather than the industry-standard seven to ten days. The adjuster prepares an estimate on the spot and will, in most cases, write you a check immediately.

What provoked this innovation? At Progressive, you may notice a strong connection to the customer, the willingness to listen to customers' frustrations, and the common sense to act on them by changing the core of their business operations. As a result of customer feedback, they did not merely tweak the details of the claims adjustment process. They dramatically rewrote the process, resulting in significant cost savings for the company.

---

[11] Michael Hammer, "Deep Change: How Operational Innovation Can Transform Your Company," *Harvard Business Review,* April 2004.

More important, however, the hassle-free claims process keeps customers happy and loyal, reducing the significant burden of constantly replacing lapsed customers with new ones.

Hammer tells several stories of operational innovation, including Dell's direct sales model for computer sales and Wal-Mart's cross-docking practice, which moves goods from an inbound delivery truck to an outbound one going to a store, eliminating the need for expensive warehouse space. With every new story, you find substantial decreases in cycle time, development expenses, and operating expenses, at the same time that customer satisfaction is boosted.

Inventing operational innovations in publications

In information development, you have the potential to pursue innovations in two areas: development and operations. Although some of the work you do in information development is a part of product development, a high percentage of your costs are related to the operational tasks of producing and delivering information to customers. You package information for multiple products, produce graphics, facilitate localization and translation, handle final production activities, and supervise, if not control, delivery.

Your development innovations should focus on designing and developing unique new ways to affect customer productivity. Your operational innovations should involve finding new ways to do your work, including single sourcing and content management. By automating final production tasks, publications organizations have reduced cycle time by cutting days or even weeks off the end of the life cycle. By managing small chunks of information so that they can be translated earlier you can produce simultaneous product releases in non-English-speaking markets.

Despite the obvious productivity gains and cost reductions, information-development managers still bemoan the lack of support they get for their efforts from senior managers. Hammer sympathizes, noting the operational innovations are traditionally undervalued in corporate culture. "Operations simply aren't sexy," Hammer exclaims, in a business climate that rewards grand strategies and big deals. The everyday work of getting the product out the door is just too boring for executives to pay attention to.

As you know all too well, operations are often out of sight. How many times have you complained that none of the senior managers knows what you do? Only when the CFO notices the documentation department's budget line does anyone seem to pay attention. Hammer explains that you're not alone. Most operational innovations are grass-roots efforts begun by people who care about finding opportunities to improve processes.

Despite making the case for operational innovation, Hammer points to a shortcoming that strikes at the heart of information development—the deadline. Managers and staff must respond to the demands of the product release cycle deadlines. As a result, you often put aside opportunities for operational innovation. "We're just too busy getting the books out the door," you insist. "Where will we find the time to make changes in the way we work?"

Yet, the most spectacular operational innovations come under pressure. Offshore outsourcing, reductions in force, budget cuts—all these provide you with incentive to search for new ways of doing business.

Hammer lists four ways that help you discover new ways, not just better ways, of operating:

✔ Benchmark with companies outside your industry.

✔ Identify the assumptions that constrain how you think about your work.

✔ Turn special cases into everyday actions.

✔ Look closely at seven key dimensions of work.

Examining the seven key dimensions of work

The seven key dimensions of work represent a series of questions, all of which should be reexamined in the search for operational innovation.

✔ What results are required to be delivered?

✔ Who should do the work?

✔ Where should the work be done?

✔ When should the work be done?

✔ Does the work or all aspects of it need to be done at all?

✔ What information must be provided to the performers?

✔ How thoroughly does each aspect of the work need to be performed?

Consider just one of the questions: When should the work be done? In the past, information developers assumed that translation cannot take place until final documents are signed off. You considered the cost of making changes to translated text too high, and the translation vendors preferred to receive all the work at one time. Now, you have learned that you can reduce the translation cycle time by delivering modules for translation throughout the development process, without significantly increasing costs, if you're careful about what you consider ready to go.

As managers, you have also learned that you can reduce total development time for content if you spend more time in the planning stage.

Operational innovation means inventing and implementing news ways of doing work. It typically involves tough-minded decision making and unglamorous hard work. However, as Hammer concludes, " . . . it is the only lasting basis for superior performance. In an economy that has overdosed on hype and in which customers rule as they never have before, operational innovation offers a meaningful and sustainable way to get ahead—and stay ahead—of the pack" (p. 93).

Best Practice—Benchmarking with competitors and best-in-class colleagues

Benchmarking compares what you do and how you do it to what others do and how they do it. A well-designed benchmark study involves four basic steps:

- ✔ setting goals and recruiting partners
- ✔ collecting information from the partners
- ✔ analyzing the information and preparing a benchmark report
- ✔ acting on the benchmark results

Setting goals and recruiting partners

Depending on how the benchmark activity is designed, the partners may be selected before the goals are established or the sponsor may set the goals and recruit appropriate partners. In either case, the goals of the partnership are key to its success. Some benchmark studies have wide-ranging goals. The CIDM's 1998 Telecommunications Benchmark was designed to look broadly at the information-development departments in telecommunications companies to identify best practices in organizational structure, information architecture, customer studies, and more. The 2005 Information-Development Benchmark Study on China was designed to look more narrowly at the potential for instituting successful information-development departments in China.

After the goals are established, appropriate partners are identified. Partners may all be in the same industry, as was the case with the Telecommunications Benchmark. They may all be interested in a similar outcome, as was true in the China Benchmark. They may be interested in a particular innovation, which occurred in the Single Source Benchmark Study, which investigated the strategies that best-in-class organizations were pursuing to reduce information-development costs. In some studies, competitors are selected, which occurred in a benchmark study of organizations involved in blood collection. Whatever the criteria, the selection process must ensure that all partners can add value to the others involved in the benchmarking.

Funding for the benchmark study may come from contributions among all the partners or one partner may act as the sponsor. Funding allows for site visits and telephone interviews and the development of a comprehensive benchmark report. Unfunded, informal benchmark studies may produce lots of responses but they are difficult to evaluate. If you know nothing about the partners except for their responses to an email request for information, you do not know if you can compare their activities and outcomes to your situation. A formal benchmark study carefully establishes the framework of the study, sets the goals, and ensures that the partners are comparable.

Collecting information from the partners

Many benchmark studies rely entirely on survey questionnaires to gather information from the partners. As a starting point of a formal benchmark study, the surveys provide a baseline of data. However, surveys rarely give you the insights you need into innovative activities. The questions focus on what is already known, rather than what is new and unexpected. Telephone interviews and visits to each partner's location provide richer information than can be developed from surveys alone. The on-site visits are preceded by questionnaires so that everyone is prepared, but the true insights come during the discussions. A good interviewer leaves openings for the partner to describe innovations that have proved successful or even those that have failed.

During a benchmark study of exemplary customer service organizations, we learned about the customer partnering process that became the basis for the partnering innovation that Compaq adopted. Details of the partnering process are described in Chapter 6: Developing Relationships with Customers and Stakeholders. In a benchmark on organizational structure in departments, Comtech learned of a process in which teams were assigned to develop information on three releases of a product simultaneously to avoid the peaks and valleys of traditional release processes.

Analyzing the information and preparing a benchmark report

After all the questionnaires, interviews, and on-site visits are complete, someone must be responsible for compiling the results and developing a report of the findings. Sometimes the reports include general recommendations and specific recommendations for each partner.

Most benchmarking reports that Comtech develops are organized around best practices identified through the study. Rather than looking at recommendations for one organization, all the partners learn from the collection of best practices discovered. Remember that the goal of a benchmark study is to look at how other organizations function and to learn from their experiences.

Acting on the benchmark results

If by participating in a benchmark study, you hope to influence innovation in your organization, you need to prepare a plan of action in direct response to the benchmark findings and recommendations. The action plan means that you are directly benefiting from the study and returning value for the investment your organization has made.

An often overlooked benefit of participating in benchmark studies is in the sense of community it creates among benchmarking partners. Most professional organizations in the information-development industry are open to all—including information and training developers, editors, technical illustrators, managers, and so on. Opportunities for groups of managers to convene to discuss issues that concern them all, compare their different approaches, and, together, arrive at a core of possible solutions, are rare.

Many large corporations see the value of professional exchange among managers and bring their training, documentation, and customer service managers together for periodic meetings and seminars. However, while such meetings are effective means to ensure that similar styles, approaches, and processes are being used across different groups within a corporation, they fail to inform managers about how competitors and others in the industry are addressing the same issues. Moreover, for the managers at single-site documentation groups—or at smaller companies—there are few opportunities to meet formally with peers.

By participating in benchmarking activities and attending partner conferences, managers create new opportunities for sharing information and extend their immediate circle of professional peers. Inasmuch as they have agreed to participate in the same bench-marking study, managers at the partner companies share similar concerns and are eager to learn more about the benchmark topic—most important, they are eager to learn from each other. Benchmark participants often report that the expertise and insight they gain from knowing what peer and competitor organizations are doing is invaluable.

Creating an Innovation Council

The CIDM has been actively supporting the establishment of Innovation Councils to help information-development managers find new ways of working and new ways of deliver-ing valued information to customers. Palmer Pearson, a senior manager at Cadence Design Systems, developed the Innovation Council concept when he invited colleagues from a wide variety of non-competitive companies in the area to attend monthly meetings. At the meetings, the managers discuss innovations that they are implementing or they have learned about. The managers are convinced that pursuing innovative ideas is a key ele-ment in their success and the success of their organizations.

The Innovation Councils are purposely small, allowing for an intimate idea exchange among a small group of interested managers. The leaders recommend that when a council gets larger than 12 members, a new council be formed. The purpose of the council is to enhance the business acumen of managers by keeping them abreast of innovations in the field and in neighboring areas. One council, for example, includes a representative from a local college that provides a certificate in technical communication management. The managers involved bring new ideas being developed or considered in their own organiza-tions and ideas that come from their ongoing investigation of the business environment.

Innovation Councils also allow managers informally to benchmark their activities with others in the field. Information benchmarking helps everyone judge the position of an organization with respect to their colleagues. More formal benchmarking, which could grow out of an Innovation Council, sets up a regular and organized information exchange among partner companies. A best-in-class benchmark study specifically works to identify industry leaders and ask them to share information about their innovations. Competitive benchmarking does the same among competitors in a particular product or industry area.

Summary

Developing innovative approaches to information design and development is essential but never easy. You first have to generate lots of new ideas and weed out the less promis-ing ones. You then have to gain the support of senior management, peer management, and your own staff members, all of whom may be quite comfortable with the status quo. Before you can focus on the innovation project itself, you have to be prepared to overcome the barriers.

Consider these approaches as you get started:

✔ Identify the obstacles that are likely to make an investment in an innovation unsuccessful. Develop tactics for overcoming the cognitive, resource, motiva-tional, and political hurdles that will come between you and an innovation.

✔ Understand the threats that lie ahead for your organization in the form of disruptive competition. Outsourcing, offshoring, and reductions in force are all consequences of disruptive innovation. Face the threats calmly and plan innovations that will increase the perceived value of your staff and your organization.

✔ Remember that the most significant innovations are those that increase the value of your information products for the customers. Get out of the office and learn what your customers are doing with information in their environment. With strong customer understanding aligned with corporate strategies, you have a significant position from which to plan and implement innovations, even those that may initially encounter opposition.

✔ Consider the effect of operational innovations for tackling the internal costs and timelines. Identify opportunities for short-term gains that accumulate progressively to result in significant operational changes. If you can identify improvements that make the greatest differences in productivity and that minimize the learning curve and culture shock, you are more likely to build a team culture that embraces innovation instead of fearing it.

✔ No improvement is easy. Create a full-scale project around the innovation you think is most valuable and achievable by your organization. Be prepared to devote resources in the form of people and time to its success, even if new money is not immediately forthcoming. Initial successes that are well documented and communicated to the larger organization may lead to new funding later. Be certain that your project has measurable goals and clear expectations.

✔ Support your internal efforts by knowing what your competitors are doing and by learning from best-in-class colleagues. Successful innovations rarely occur in a vacuum. If you're the only one pursuing a new idea, be careful that the idea isn't out of touch with reality.

By encouraging your staff to become advocates of business-oriented, productivity-enhancing innovations, you will build recognition for your activities throughout your organization. The more you are recognized as an effective innovator, concerned with delivering value and reducing costs, the more highly regarded your organization will be.

Part 3

Project Management

"The project development life cycle has
five phases – from planning to development."

Chapter 14

An Introduction to Project Management

> Project management is the process of managing, allocating, and timing resources to achieve the desired goals of a project in an efficient and expedient manner. The objectives that constitute the desired goal are normally a combination of time, cost, and performance requirements.
>
> —Adedeji B. Badiru 1993

In most enterprises, information-development projects occur in the context of larger projects. In product-development organizations, information developers support the product's customers, especially those who install, configure, administer, modify, and use the product in their workplace. They also support internal customers who train and support customers, install and repair equipment in the field, or market and sell the product. In other organizations, information developers are responsible for information to support internal activities. They work with others engaged in defining policies and establishing procedures. Information developers are the writers, editors, and production coordinators for projects that are guided by subject-matter experts or even committees or teams of experts. Some information developers may themselves be the subject-matter experts. They are responsible for creating a process and preparing training and reference materials to support those who perform the process. Only rarely are information developers completely responsible for a project independent of some larger entity or activity. In those cases, the information developers may be developing policies and procedures for their own organizations, such as style guides, process flows, and information architecture.

Because most information developers are part of larger projects in the enterprise, those responsible for managing the information-development project will find themselves interacting with other people who also have project management roles. In a product-development project, a technical expert such as an engineer or programmer may have project management responsibilities, or a product manager may manage the specification and release of a product to external customers. In an operations organization, people responsible for a process may lead projects. For example, a human resources director may lead a project to develop an employee handbook. A finance director may be responsible for the corporation's annual report. The city manager may be in charge of producing a report to the city council. The development of a new business proposal may be in the hands of a proposal manager or a sales professional. And so on, and so on. The list of projects in any enterprise that may involve information developers is endless.

In some cases, individual information developers report directly to someone in the larger project who is responsible for the project management. In that case, the information developers are likely to understand how to manage themselves, because the manager of the larger project rarely knows anything about the activities required for the successful development of information products. Individual information developers managing their own work in the midst of a larger project will find that many of the activities outlined in Part 3 must be included in their workflow.

Within the context of these larger projects, you are also likely to work in an organization dedicated to information development. In a professionally run information-development department, individual information developers may still be responsible for their own project work. As projects require more than one person, the role of project manager becomes more well defined until some individuals have the sole responsibility for managing information-development projects. The range of activities included in that role is the primary subject of Part 3.

Depending upon the type of organization in which you find yourself, consider these questions:

- ✔ Where do you fit in as a project manager for an information development project?

- ✔ Are there other project managers? What is your relationship to them?

- ✔ Are you new to project management? Do you have lots of experience in managing projects?

- ✔ Are you a portfolio manager (discussed in Part 2), which means that you are responsible for the direction of all the projects in your organization?

- ✔ Are you responsible for all or a subset of the projects in your organization?

- ✔ Are you managing only yourself or several people involved in information development?

- ✔ Do you manage a team of people working on the project?

- ✔ What if you are managing only your own project in which you are the primary or only contributor?

✔ Are you and your team responsible for all the project activities related to information development?

✔ Do you have colleagues who manage localization and translation, graphics development, final production and delivery?

✔ Are you responsible for one or multiple deliverables?

✔ Are you the only information-development project manager, or do you have colleagues managing other parts of the information-development process?

✔ Are you responsible to a portfolio manager or department head who is managing all the projects in which your organization is engaged?

✔ Are you a department head who manages all the projects in your department?

No matter what your position in the larger enterprise or within your own organization, if you are responsible for seeing that an information-development starts, proceeds, and ends successfully, you have the role of project manager even if you don't have the title.

The Purpose of Information-Development Project Management

As an information-development project manager, you have two, often competing, goals: the efficient management of the project so that you reach your goals, and the successful development of information products that excel at meeting the needs of your customer and adding value to the product or service that your information supports. If you can achieve both these goals without one overtaking the other, you will have a successful career as a project manager.

In working toward your project goals, you may find yourself with the responsibility for leading a team of individuals and helping them collaborate, innovate, and adhere to standards, process, and deadlines. At the same time, good project management does not mean imposing troublesome processes on people trying to do good work. To facilitate good work, you must take responsibility for handling the process details for your team so that they have the time and opportunities to achieve the best results possible.

With every project you manage, you face these and often more challenges:

✔ Having enough resources to get the work done by the deadline

✔ Having sufficient budget to support your team's process effectively

✔ Developing a team that is efficient, supportive, and innovative

✔ Collaborating with people elsewhere in your organization who have their own constraints to manage

✔ Knowing what information will provide value to your customers

✔ Balancing the demands for quality with limitations on your resources and your process

✔ Ensuring that the information-development process is wisely implemented

✔ Knowing when an activity is essential and when it can be purposefully neglected

Your ability to meet the challenges with the assistance of your team members and other colleagues in the organization will test your skill as a project manager. My intent in Part 3 is to give you a process to follow that will help ensure success. As you implement the process, described as the phases of an information-development project life cycle, you will discover what works well in your organization and what doesn't. I encourage you to view the project life cycle as a series of best practices, developed over many years with the help of many skilled and talented information project managers and industry leaders. Like any best practices, however, you will need to mold them to your own style, the needs of your team, and the needs of your larger organization.

The Information–Development Life Cycle

The information-development life cycle encompasses a series of processes that must be completed to deliver effective information to customers while meeting project deadlines. Although the information-development life cycle can be divided into any number of phases, I find that a five-phase model is convenient. You will, of course, find models that have four, five, six, or even seven phases. The processes are usually similar; they are just split into different parts to support the way a particular organization defines its workflow.

The five-phase information-development life cycle parallels many popular approaches to product development. For example, the Jim Highsmith, leading proponent of Agile Project Management, outlines five phases from Envision and Speculate through Explore, Adapt, and Close, a design very similar to the model I recommend to you. Because the model I present is familiar to people managing other types of projects in your organization, you will find it easier to introduce this model and explain its effectiveness and versatility to your colleagues.

Figure 14-1 illustrates the five phases of the information-development life cycle model.

Figure 14-1: The five-phase information-development life cycle

In the following tables, you learn what activities occur in each phase, who is responsible for them, and what deliverables need to be produced. You also learn what project management deliverables you will produce and include in your Project Management Folder. Setting up your Project Management Folder is, in fact, one of your first activities. Either in your file system or your content management system, define a primary folder for your project, divide it into the five phases, and use it as a permanent repository of all the decisions and actions you take to manage your project.

Information Planning

During the first phase of your information-development project, you and members of your team as needed work together to define what you are going to deliver by the end of the project and why. The major deliverable of the first phase is the Information Project Plan, a critical document that outlines the goals and objectives of the larger project of which you are a part. Without a project plan in place, the rest of the project is likely to flounder for a lack of direction.

Never short change the planning phase, even if you find it tempting to argue that you have "no time for planning." Even short projects, completed in a few days or weeks, need to be planned. In fact, you will find from experience that short projects are more difficult to manage than longer ones because all the required activities must be compressed.

By planning your project, you help to ensure that you have sufficient resources to meet the deadlines, are able to produce final deliverables that meet your customers' needs, and maintain the morale of your team members. Without adequate planning, you put all of these at risk. Table 14-1 details the activities, deliverables, and responsibilities associated with the Planning Phase of the project.

Table 14-1: Phase 1: Planning

Phase	Schedule	Activity	Deliverable	Responsibility
Planning*	Approx. 10% of total project time	Understand the goals and objectives of the larger project Understand the information needs of the audience Define the scope of the high-level deliverables Develop the project schedule Define the requirements for localization and translation Define the requirements for final production Acquire people and other resources to support the project Evaluate project dependencies and risks and plan mitigations Communicate the project vision and goals to all stakeholders	Project Plan	Project Manager with team members as needed, including Information Architect, Production Coordinator Localization Coordinator, and Information Developers

Continued

Table 14-1: Phase 1: Planning (Continued)

Phase	Schedule	Activity	Deliverable	Responsibility
		Set up the Project Management folder	Project Management folder	Project Manager
		Manage a review of the project plan	Final approved Project Plan	Project Manager and Project Stakeholders

Find information on the Planning Phase in Part 3, Chapters 15 and 16.

Information Design

Even though Information Design is officially the second phase of the information-development life cycle, it often begins long before the current project begins. Information Design is the product of continuous improvement in your understanding of the audiences for your information and the business requirements of your organization. Design work may begin with the completion of the previous project, after you begin to get feedback from customers and colleagues in your larger organization. Nonetheless, the information design for your particular project occurs in the second phase of the project, at which time you and your team members decide exactly how the content will be structured and how the final deliverables will be organized and will look.

In a large organization, you may work with an information architect who is chiefly responsible for the design of information products. In a small organization with more limited resources, you may wear both hats: project manager and information architect. In either case, you will want to involve the entire team in the design process so that they assume ownership for the success of the design as they develop the content. Table 14-2 details the activities, deliverables, and responsibilities associated with the Design Phase of the project.

Table 14-2: Phase 2: Design

Phase	Schedule	Activity	Deliverable	Responsibility
Design*	Approx. 20% of total project time	Conduct a user and task analysis Understand the goals of the product or service through the use cases Understand the marketing goals and objectives Define the usability goals Define the information structure Define the writing style Define the deliverables	Design Plan Annotated Topic List Deliverables Structure Authoring Guidelines	Information Architect with other team members as needed, including Localization Coordinator, Production Coordinator and Information Developers

Phase	Schedule	Activity	Deliverable	Responsibility
		Manage a review of the project design	Final approved Design Plan	Information Architect and Project Manager
		Monitor the project Re-estimate resources, budget, and schedule Report progress Communicate the Design Plan to all stakeholders Add to the Project Management folder	Revised Project Plan Progress Reports	Project Manager

* Find information on the Design Phase in Chapter 17.

Information Development

The third project phase embodies the majority of project time, encompassing all the work to develop the content and ensure that it supports the product and the customers. During the development phase, the content creation work is in the hands of the information developers on your team, assisted by specialists in graphics or other media required for the project.

As information developers create and update content, they should also be engaged in quality assurance activities, with developmental, copy, and production editors or with peers. They should also interact with subject-matter reviewers and those responsible for testing the information against a product. Depending upon the size of your project and the resources available, the roles and responsibilities during information development may be shared among specialists or handled by the core information developers themselves. Nonetheless, your responsibility as project manager is to make certain that quality assurance activities occur during Phase 3 of the project.

During the Development Phase, the project manager monitors the project, alert to changes in scope or project dependencies that increase the risk of the project and require action to keep the project on track. You may find yourself revising the project plan, changing the project scope, looking for additional resources, or accommodating time that is added to the schedule by delays in the larger project. As you track the project schedule and workload, you make constant adjustments to ensure that the project is completely successful. You record actual time spent on tasks so that you can compare them with your initial estimates in the Project Plan. As you monitor and adjust the project, you record your observations and actions in periodic progress reports to your management and other stakeholders. Table 14-3 details the activities, deliverables, and responsibilities associated with the Development Phase of the project.

Table 14-3: Phase 3: Development

Phase	Schedule	Activity	Deliverable	Responsibility
Develop-ment*	Approx. 60% of total project time	Develop new content Revise existing content Modify the annotated topic list Revise draft content based on feedback from SMEs Report progress	Draft Content Final Content Topic List Weekly Progress Reports	Information Developers with assistance from the Information Architect and the Subject-Matter Experts
		Edit content for structure and style Edit content for compliance with templates	Edited drafts	Editors, including Development, Copy, and Production
		Revise the design in response to information changes	Revised Design Plan	Information Architect with assistance from Information Developers and Editors
		Index the content and develop keywords for electronic search	Index Keywords	Indexer Information Architect
		Monitor the project Re-estimate resources, budget, and schedule Report progress Identify and resolve problems Add to the Project Management folder	Revised Project Plan Meeting Minutes Progress Reports	Project Manager

* Find information about the ongoing management responsibilities during the Development Phase in Chapters 18, 19, 20, and 21.

Production

The activities that occur in your Production Phase, which includes both production of final deliverables and delivery to the appropriate organization that send information to its users, will depend heavily on the tools and methods you are using to produce the final deliverables from your project. Most organizations today produce PDFs for print output and delivery to web pages or CDs, rather than producing traditional camera-ready copy. Organizations using desktop publishing systems typically can produce PDF and HTML output directly from their source assemblies. Organizations producing output for various types of help systems often have to use a third-party tool to generate the help output.

Organizations using XML and a content-management system can usually automate the generation of PDF, HTML, and various help output. Using XSLT processing with XML assembles such as DITA maps, you can automate the production of many outputs, including PDF, HTML, and various help systems.

The more automated your production processes are, the less time you will need to devote to Phase 4 Production. In my experience with a traditional publishing system, production and delivery tasks were allocated about 20% of total project time. With XML and content-management systems, that time allocation could be reduced to 10% or less. Some organizations, using XML processing and carefully designed style sheet, plus lots of experience, are able to produce all their deliverables in multiple languages in a matter of minutes rather than days.

Table 14-4 lists typical production and delivery activities associated with Phase 4 of the information-development life cycle. Whether or not you need to include any or all of these activities will depend upon your tools and production process.

You may not be completely responsible for managing production. Large organizations often have a production coordinator who manages the production tasks. However, as project manager, you are responsible for coordinating with the production lead to ensure that your project moves successfully through Phase 4.

Table 14-4: Phase 4: Production

Phase	Schedule	Activity	Deliverable	Responsibility
Production*	Approx. 10–20% of total project time	Plan and manage production activities Plan and manage localization and translation activities Select and contract with production vendors	Production Plan Contracted vendors	Project Manager Production Coordinator Localization Coordinator Purchasing
		Run pre-production tests on draft content	Successfully completed tests Content ready for final production	Production Coordinator Information Developer
		Obtain final approvals	Approved content ready for production	Project Manager Stakeholders including legal representative
		Prepare index Prepare keywords for search	Index Keywords	Indexer Information Developer
		Develop production checklist Conduct production edit	Production checklist Final content	Production Editor Information Developers
		Deliver content to localization and translation vendors or work with your local localization coordinator Deliver content to your production coordinator	Final content in multiple languages Final content fully rendered in the appropriate output deliverables in all languages	Project Manager Production Coordinator Localization Coordinator

Continued

Table 14-4: Phase 4: Production (Continued)

Phase	Schedule	Activity	Deliverable	Responsibility
		Perform final quality assurance checks on all rendered output	Final information ready for delivery	Production Editor Production Coordinator
		Deliver content to parties responsible for delivery to customer Archive final deliverables	Final delivered output (print, CD, HTML, help)	Production Coordinator Project Manager

* Find information about the ongoing management responsibilities during the Production Phase in Chapters 22 and 23.

Evaluation

Phase 5: Evaluation takes place after all the deliverables are complete and handed off to those responsible for delivery to the customer either by shipping paper or CDs, embedding Help systems in products, or implementing content on websites. Evaluation is the last step you must complete to review the project's successes and challenges, plan for improvements, and ensure that the information your team has produced has value to the customer and meets their needs.

Phase 5 activities should take place as soon after the deliverables are complete as possible. If you delay too long, team members and other stakeholders will be engaged in new projects and may have forgotten issues that they wanted to address. Table 14-5 details the activities, deliverables, and responsibilities associated with the Production Phase of the project.

Table 14-5: Phase 5: Production

Phase	Schedule	Activity	Deliverable	Responsibility
Evaluation*	Following the completion of the project deliverables	Accumulate and analyze project data Compare estimates with actuals Hold "lessons learned" review meetings with team members and other stakeholders Prepare the project's final report Write project-specific personal reviews of team members Ask team members to review the project manager	Project Final Report Team Member Reviews	Project Manager, with assistance from Information Developers and other Stakeholders

Phase	Schedule	Activity	Deliverable	Responsibility
		Survey customers	Customer satisfaction survey	Project Manager
		Interview stakeholders who have close contact with customers, such as product support, sales, and training	Customer Feedback Forms	
		Collect customer feedback	Problem Resolution Report	
		Identify problems and correct		

Find information about the ongoing management responsibilities during the Production Phase in Chapter 24.

Your Role as an Information-Development Project Manager

In the activity tables in this chapter, you recognize that the information-development project manager has many responsibilities. In small organizations, the project manager may be responsible for managing all the activities in Phases 1 through 5. In large organizations, the project manager may hand off direct management responsibilities to individuals coordinating major parts of the life cycle. For example, as project manager in a large organization, you may work with independent organizations responsible for localization and translation, production and delivery, graphics, information architecture, and the user experience, including user and task analysis, usability testing, and customer feedback. In either case, you remain responsible for the final success of your project, along with your immediate team of information developers.

In some organizations, you will have the role of a "lead writer," an individual who manages the project and contributes to the information development. You may be a lone writer on a project, with project management responsibilities in addition to all your other duties.

While you are wearing your project manager hat, no matter how many different roles you play, your primary management responsibilities will likely include these:

✔ Develop the project plan, including the schedule and the deliverables for the project

✔ Estimate and acquire the resources needed to complete the project by the deadline both in people, tools, and technology

✔ Revise the project plan as you gain more knowledge of the project

✔ Monitor the progress of the project, watching for changes in scope, schedule, and dependencies

✔ Communicate the goals and vision of the project to all stakeholders, including members of your information-development team

- ✔ Coordinate with others involved in localization, translation, production, delivery, user experience, and user support, and others involved in the successful completion of the project

- ✔ Ensure that the product is completed on time and on budget and at the level of quality needed for the enterprise and the customer

- ✔ Evaluate the success of the project, the opportunities for process improvement, and the performance of team members

- ✔ Gather customer feedback, identify and resolve problems, and plan for the next project

These management responsibilities exist even if you are working alone and are responsible for developing the information yourself. No wonder you never have enough time to write. However, if you have a team of information developers working with you on the project, you have a leadership role to play as well.

Establishing your relationship with your team members

If you are new to project management or you find yourself managing a team of people for the first time, you need to think about your responsibility to the team. Effective project managers are not only number crunchers or paper pushers. They assume leadership responsibility to ensure not only that deadlines are met but that the quality of the work produced by the team delights customers and exceeds their expectations.

Certainly, information-development projects require good management. You do, after all, usually have real deadlines and limited resources and budget. But, at the same time, you want to encourage your team to innovate in the design and development of information products. So much information developed does not meet customer needs because it is inaccurate, unreadable, or inappropriate for the customers' needs. That information, a product of many years of pursuing technical details rather than user quality, needs to be updated. But if the only goal you have for your project is meeting the deadline, then badly done information continues to get updated.

As a leader, you need to develop a relationship with your team members that encourages them to change the status quo. Within the constraints of time and budget, you can ask them to find ways to do something better than it's been done before.

If you develop a collaborative team environment, in which people genuinely contribute to each other's work, you have the possibility to create better information products. Perhaps not as much better as you would wish, but better than what was produced in the past.

At the same time, you don't want complete anarchy on your project team. Individual team members who insist on pursuing their own personal agendas need to be reined in. You need to know what people are doing so that they don't spend too much time going in the wrong direction.

As a result, you have to be involved with the work of the team, communicating the project vision and removing barriers to change. By becoming a leader-manager, you will help contribute to the goals of the larger enterprise and establish the value of your team's work.

Establishing your relationship with other project managers

In many information-development projects, you will find yourself working closely with other parts of your enterprise that themselves have projects to complete. Many of these projects will have project managers. For example, you may be leading a small team of information developers working on a new policy and procedure for storing records. Your team is part of a large team of records management specialists from throughout the enterprise. The team leader is a records manager at headquarters, an expert in the subject, and the project manager responsible for developing a first-class storage process. Although this team leader is managing the larger project, he is not an expert in information development. Your team is responsible for writing the policy and the procedure so that they are understandable and usable by people at company locations. Many of those people are not records specialists. As a result, they need information that is written for generalists, not specialists.

In your relationship with the larger project manager, you should work together to determine each of your areas of responsibility. If you are responsible for ensuring that the policy and procedure follows corporate standards, meets usability and accessibility requirements, and is readable by the target audience, then your primary responsibility in these areas should be well-defined and recognized by the members of the larger team. If you are responsible for ensuring that the final deliverables are published according to corporate procedures, that too should be clearly defined. If you are responsible for handing off the deliverables to a production and delivery team with its own project manager, you will need to communicate with the production manager as well.

As you begin your project, you will need to understand all the relationships that you must manage between your team and other parts of the organization. Sometimes managers of the larger projects assume that they are responsible for all aspects of project management, including information products. If that assumption exists, you will need to discuss the extent of your responsibilities and how you will work together effectively to produce a good outcome of the project.

If you find yourself in a conflict with another project manager over responsibility for information deliverables, you may need to escalate the problem to your manager. If another project manager insists that you deviate from standards, produce information that is clearly inappropriate for the target audience, and does not fulfill the requirements you have for usability, readability, and accessibility, you should first try to solve the problem yourself and then find the appropriate stakeholders to help represent you in negotiations. I believe that information design and delivery should be the responsibility of information-development professionals if that is the job they have been given. However, disputes will arise. The best way to counter potential disputes is to establish good lines of communication from the outset. In most cases, you will find that your colleagues in other organizations are cooperative and don't want to assume responsibility for areas outside their expertise.

The more support you have from senior management, the easier it will be to define your areas of responsibility. However, you will frequently find that senior management does not know much about information development. It will be up to you to find opportunities to explain what you do and how good information-development processes and sound project management will reduce development costs and ensure quality for customers and users of information. Remember that the business value of your work is the most important to management, not a focus on the niceties of writing and editing.

You will also need the support of your team members. If some team members are reluctant to follow your lead and are prone to take direction from subject-matter experts and outside project managers, you need to communicate with them and with your own management. Team members should work toward the same goals in a collaborative work environment, not head off on their own. Loyalty and allegiance to the team and the success of the project as defined by the team will be essential. Building that loyalty and allegiance may take time and persuasion. Following good change management processes will help you build the team and unite them around a unified vision of the project.

Characteristics of a successful project manager

Given all of the diverse roles you are expected to play as an information-development project manager, you will find several important skills that will help you to be successful:

- Communication
- Project planning
- Attention to detail
- Vision
- Leadership

Communication skills help you form allegiances among your team and with other stakeholders with whom you must interact. You will need to communicate with your management and other senior managers, with members of peer organizations, with other project managers, with customers and customer representatives, and with your own team. Communication skills do not necessarily come easily to people who chose information development as a profession. The field is dominated by introverts with a decided preference for working alone at the computer.

If you are a strong introvert, you will have to work seriously on developing communication skills to be effective as a project manager. These skills include leading meetings, giving presentations, handling one-on-one discussions, and writing communications carefully.

Project-planning skills must begin with a thorough understanding of the information-development life cycle and all the activities required to bring about a successful project. If you haven't had experience in information development, you may want to review the process presented in this book and work with other people in your organization who have had experience. If you are managing a project for the first time and you have no colleagues internally, look into establishing a network with other information-development professionals in your area. You may want to join the Society for Technical Communication as well. Information about membership is available at http://www.stc.org/.

Not only do you need knowledge and experience with the life cycle, but you also need experience or help in project estimating, tracking, and scheduling. Following the processes in the remaining chapters will help you learn how to estimate the scope of the project and apply standard estimating to the project. However, the best way to achieve accurate estimates is to have information on existing projects. Consult with colleagues and find records on previous projects if they are available. Once you have an initial estimate and acquired resources to meet your schedule, be prepared to update the estimates as the project changes.

A critical characteristic of successful project managers is **attention to detail.** An information-development project is complex, with many details to keep track of. If you don't have an innate ability to pay attention to details, you should get help from team members who are detail-oriented and make lots of to-do lists. It's best not to burden team members who have their own work with sorting out all the details of project progress, scope changes, and re-estimates of resources and schedules. As much as possible, you want to relieve that burden by doing the estimating and tracking yourself.

Project management in information development must be more than recordkeeping. Without a vision of the project that articulates what you hope to accomplish in terms of business and customer goals, your team will simply engage in updating existing information that may never meet anyone's needs and remains terribly inefficient to produce. Senior management is looking to your projects to be efficient, productive, and accountable for the quality of the results. If you understand the customers for your information products well, you can plan how to meet their needs with increasingly effective information. If you understand the objectives of your organization, you can make decisions about what to include and what to leave out of the information you deliver. If you recognize the need for continuous productivity improvement, you will find ways to help your team adopt the best practices that you learn here and from other experts in the field.

To develop an innovative **vision** for your project, you will need to listen to the way in which your senior management articulates the goals for the enterprise as a whole. You also need to pursue information about your competitors and information about industry best practices and design and technology innovations. If you are not aggressively trying to learn from experts in the information-development profession, you will not know how to frame and articulate a vision for improvement.

Finally, you need **leadership** skills to communicate the project vision to team members and stakeholders and to motivate your team in making that vision a reality. The core value of leadership is respect for and encouragement of change, of being dissatisfied with the status quo and continuously searching for ways to deliver more value. Leaders work with people, not in an authoritative manner, but through collaboration because they recognize that the results will come only from innovative and responsible team members.

Summary

Embarking on your new role as an information-development project manager starts with the five phases of the project life cycle. In Chapters 15 through 24, you learn about the processes that a project manager needs to follow in planning, encouraging innovative design, ensuring the development proceeds smoothly, moving the project through to the final deliverables, and evaluating the success of the project after it is complete.

If this is your first project or if it is the first managed project in your organization, you will need to learn about all the project phases, but you will also have to make choices. You may find it impossible to do all of the activities I recommend. If so, don't neglect planning.

Planning is the most critical activity you will engage in. If you don't have a plan, you'll never know what you've achieved by the end—what you had hoped or something less. Please don't listen to people who tell you that there is no time to plan. Without an adequate plan, you're most likely to waste considerable time and energy going in the wrong direction.

Most new project managers complain that they don't have any data to use for estimating the resources they'll need to complete the project by the deadline. You can never have too much data. Even if no one in the organization has ever tracked how much time and effort a project takes, begin tracking with your first project and continue tracking religiously. The data you accumulate will make you a better estimator and manager on every subsequent project you do.

Finally, work on your people skills. If you are a strong introvert, consider playing the role of an extrovert. The more you practice communicating, the easier it will become.

"How big is this project anyway?
When do we need to get started?"

Chapter 15

Starting Your Project

> Plans are nothing; planning is everything.
> —Dwight D. Eisenhower, quoting 19th century Prussian
> General Helmuth von Moltke

The first phase of an information-development project addresses the thinking you need to do to ensure the success of your project. Once you have understood the nature and scope of the larger project of which you are part and envisioned the goals, scope, resources, and schedule of the project you are managing, you will be ready to develop your Information-Development Project Plan, described in detail in Chapter 16: Planning your Information-Development Project.

Planning is essential to the success of any project, large or small. In fact, the shorter the project and the more difficult its schedule, the more essential the need for careful planning. The planning document is the culmination of all the research and information gathering you have done at the beginning of your project. It records the information you have collected and the decisions you have made with respect to that information.

Information planning takes real time. You need time to ask many of the questions outlined in this chapter and get as many answers as possible. You need time to communicate what you've learned to your team members and enlist their help in deciding how to respond. You need time to compose a brief but comprehensive planning document so that it communicates your vision of the project and how you and your team intend to fulfill it.

I estimate that new information planning should take approximately 10% of the total time devoted to the project, as shown in Figure 15-1. That means a 10-month project requires one month of planning; a 10-week project requires one week of planning. No one is wasting time during the planning phase. Information developers gather project information from product developers or other subject-matter experts and, as project manager, you accumulate information about the larger project of which you are a part. You and the team spend time planning, and you write the results of your planning in the Project Plan.

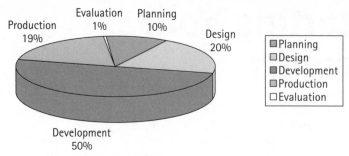

Figure 15-1: Percentage of total project time for Phase 1 planning activities[1]

Of course, straightforward revisions of existing projects may require a lower percentage for planning activities, but that percentage should never be zero. If you have a Project Plan for a previous release, you can update it with the new information and use that information to estimate your schedule and resources. Even so, don't discount the planning. Even with a revision, you need to know how much time you need. If you have a small project, something with a one or two week deadline, you should think about what you need to accomplish and how the project fits into your overall workload, considering you may be responsible for multiple projects. Even a one-week project will require at least a day for information gathering and planning. Although you won't need a lengthy Project Plan, you will need to indicate how much of your workweek will be consumed by this project.

Why you shouldn't confuse writing the plan with real planning

Because people trained as information developers often view their primary work as writing, they have a tendency to confuse planning with creating the Project Plan document. You create the document itself primarily to communicate in writing with the project stakeholders and your own team members. As you know, "writing it down" provides some guarantee that the stakeholders have a common understanding of the project and its goals from the beginning. If the plans are not written down and communicated, it's very likely that the various stakeholders will have different ideas about how to proceed.

Despite the importance of "writing it down," actually doing the planning must come first and have primary importance. In creating a Project Plan, you are trying to foretell the future without the benefit of a crystal ball. You will often find yourself planning without what you might view as adequate information about the product or service. No

---

[1] Note that the percentage for Phase 5: Evaluation is set at zero. The Evaluation activities take place after the project has been completed and are not included in the total time estimated for the project.

amount of planning guarantees that your project will be a success because unanticipated events do happen. Although even good planning doesn't provide guarantees, failure to plan almost always leads to serious problems in reaching consensus about results, having sufficient resources ready, and being able to deploy those resources in a timely manner. As an example, recall what happened in 2005 during the hurricane season along the Gulf Coast of the United States. Some plans existed but were incomplete, the plans that existed were not followed, resources were unavailable, and even those that were available could not be deployed in a timely manner. The result was chaos and death.

Most likely, no one will die if you don't plan effectively. But your own experience should tell you that chaos happens more frequently than we'd like to admit.

Planning makes you more knowledgeable about the project on which you are embarking. Planning helps you to convince yourself that you're not alone.

Why bother to plan?

Even though planning is never perfect, project managers learn quickly that they have many sound reasons to plan:

✔ Plans help you anticipate the activities and events most likely to occur during the project. These likely, expected activities and events enable you to produce a sound schedule of milestones and other scheduled deadlines and start out with the resources you need to handle both scope and schedule for the project.

✔ Plans help you anticipate unlikely events, which you learn more about in the discussion of the Dependency Calculator in Chapter 16: Planning your Information-Development Project. Planning requires you to think about the range of events, expected and unexpected, that have occurred on projects in the past. Planning allows you to be prepared for the most likely deviations to the project plan.

✔ Plans provide the basis for your resource calculations so that you are adequately prepared for the coming workload. They also provide the basis for resource contingency planning. For example, it's been my experience that software development projects often encounter technology barriers that slow development. If I am planning a project in which the developers are pursuing new technologies, I will prepare for that contingency in my plan by using checkpoints to reevaluate my schedule and assigned resources. I may even have some resources in mind that I could use if one of the more unlikely events happens to occur.

Of course, even the best laid plans can go astray when completely unexpected events occur on your project. You cannot hope to control everything, especially when you must interact with many other parts of your organization and diverse teams of individuals in other departments. Remember, you don't judge a plan by its predictability about the future but whether or not it provides value to you and your team and stakeholders. If you are engaged in extreme or agile projects, you learn quickly that predictability is not a hallmark of these methodologies. In fact, they acknowledge that some development problems are so difficult that they are quite unpredictable. Yet, extreme or agile methodologies require planning, and they require re-planning at each mini-stage of the project. Agile project managers don't dismiss planning (or shouldn't if they apply the methods correctly), but they assume that planning takes place in a highly changeable environment.

Why you can't shortchange planning

Are you already considering your short-turnaround project? Do you hear yourself thinking, "I could have this project finished in the time it will take to plan."

Don't fall into the "I don't need to plan" trap. You can waste much time writing the wrong content or producing the wrong deliverable. For example, you may be asked by a customer to create a software manual that describes all the screens and fields in the product but never tells the users how to do their work using the software. If you stop a moment to consider the request, you may be able to plan a much more useful and successful outcome. A manual that tells people how to use the software in the context of their real work will be much more helpful than page after page of screen descriptions.

Inadequate resources (not enough people to accomplish the project by the deadline) are one of the most serious consequences of a lack of planning. Managers simply assume that one individual is sufficient to develop one document, regardless of its length, complexity, amount of change, or the schedule. The first time many project managers plan and estimate the resources required for their project, they are amazed that they cannot accomplish the project with the resources already assigned.

Don't shortchange your users, your team, or yourself by neglecting the planning process. Remember that planning is more important than the plan, but if creating the plan gets you to stop and think, it's worth every hour you devote to it.

Best Practices in Project Planning

As an information developer, your role is to produce information that helps others to be successful. Their success may involve performing activities that support their use of a product or their performance of a business process. As a project manager, your role is to ensure that information development happens—on schedule, at the right cost, and with the quality expected by your organization and the people who use your information.

At the outset, you should ask a number of key questions about your project. You want to know when to start the planning process and who should be involved. You need answers to your questions about scope and business goals. You need information about the audience and other stakeholders in your results, both internal and external.

This chapter identifies eight best practices associated with planning your information-development project and knowing how it aligns with the larger project of which you are a part:

- ✔ Understanding the project starting point
- ✔ Characterizing the project environment
- ✔ Identifying project goals

✔ Analyzing project scope

✔ Identifying project stakeholders

✔ Developing a project communication plan

✔ Understanding the project schedule

✔ Identifying the project risks

If you work in a large organization, the senior management of your information-development organization may be able to provide you with much of the information you need to respond to the planning practices. Your portfolio manager may have already engaged in preliminary planning with members of product and marketing management. You may even have a hand in the early estimates of project scope. However, if you are in a small organization, you may find yourself responsible for understanding the early requirements of your project. You may even be responsible for ferreting out the very existence of the project in the organization.

The earlier in the production-development life cycle that you can begin to pose questions and gather information, the more likely you will be able to develop a responsible and effective information product. Being involved early enough to support a reasonable planning and design effort will further ensure project success. But even if you are unable to find out enough about a possible project until late in the product-development life cycle, consider that answers to the key questions about your project are essential to its success.

Best Practice–Understanding the project starting point

Depending on the relationship between information development and the rest of the product-development process, you are likely to manage projects that begin at several points in relationship to the larger projects that you support. Consider the following possible starting points:

✔ The larger project has just started; you're involved in the early stages.

✔ The overall project plan is already in place, and the early feasibility and planning phases have been completed.

✔ The project is well under way; you have just barely enough time to meet the deadlines.

✔ The project is nearly complete; you do not have enough time to produce all of the proposed deliverables by the deadline.

All projects have a starting point, although sometimes that starting point is difficult to fix in a chaotic organization. In your role as a project manager, you have to begin somewhere. In the best cases, you begin when other participants in the project begin. That best beginning point is when the feasibility of a proposed project is being investigated. The following sections identify the pros and cons of starting your involvement at various points in the life cycle of the larger project.

Early feasibility stage

Why is project feasibility the best point to begin? At the early stages in which a project is under consideration for management support and funding, you are able to research the importance of information to the project's success. You are able to understand the importance of the project to your organization. And, you are able to determine how challenging your team will find the subject matter, the deadlines, and the work environment. You will know if the risks are manageable or require a new way of working.

Early planning stage

In a reasonably well-organized product-development life cycle, you can begin to plan your information-development project once the proposed product-development activities have been judged feasible and the product is in the early design stages. At this point, the overall project has been approved by marketing and product development. You may even discover that basic use cases have been created, outlining how the proposed product or product updates will support customer goals.

As an information project manager, you will need to gather all the information available about the customers, goals, objectives, and proposed features and functions of the product. You may have time to conduct user site visits or some usability testing of the previous release of the information. Your own planning can take place in concert with the product planning and design activities. Although changes are likely to occur later in the life cycle, you can ensure that your own project is aligned with the delivery schedules for the product.

Of course, it's always preferable to be involved when you have an opportunity to better understand the customer and propose ways in which information can enhance the customer's performance. However, if you become involved in the planning stages of the product, you still have an opportunity to influence the usability and learnability of the product. You may be able to influence the design of the user interface so that processes are straightforward and easy to explain. You may be able to ensure that the terminology is consistent with industry and customer standards and is consistent throughout the product interface and the information. You may even be able to ensure that the features and functions being designed for the product correspond to valid customer use cases, rather than to the whims of developers.

Late planning stage

In many less well-organized product-development groups, information development is not notified of new product-development activities until the development activities are well along. The features and functions have been defined, the user interface has been developed, and terminology has been selected, all without a thought to the requirements of information development.

Unfortunately, in this case, your role as information project manager is more circumscribed because many of your options have been taken away. You often can no longer influence the usability and learnability of the product. You may have to work with terminology that is inconsistent and fails to conform to the standards of the industry or to the language preferred by the customers.

However, just because you are informed late in the product-development life cycle does not mean that you can afford to ignore your own planning activities. Even with the shortest of deadlines, your chances of creating usable and useful information and meeting the project release schedule will be enhanced by planning.

Too often, information developers who are called in late believe that they have no time for planning. They simply want to start writing . . . something. Unfortunately, in most cases, they discover that the "something" is not what the users really need to know. Great time and effort is wasted explaining the obvious because writers find it easiest to learn what is least difficult about using the product. The information to support use of the product in the customer's real world is neglected.

Impossible planning stage

Some projects are so late to involve information development that few options are available. Sometimes all that can be produced is the most rudimentary information, simply to get some customers past the initial implementation of the product. In fact, in such cases, you must insist that it is best to develop a small, interim set of information rather than create a large, useless, untested set.

You will find it worthwhile, however, to spend time planning what to do. You may ask yourself and your colleagues the following questions:

- ✔ Is the product likely to be offered only to early adopter customers first?

- ✔ Is it better to provide a modicum of information to support early adopter customers and focus on less skilled customers after the initial release?

- ✔ Are initial customers more likely to be supported by support services? Is your first audience actually the support professionals from your own organization? If so, can you work with them to discover exactly what information is critical to their success?

- ✔ If new customers are involved, can you provide "getting started" tutorials and procedures that will take them through the initial stages of using the product and deliver more advanced information later?

You may discover that you can provide some small amount of critical information initially and deliver more well-organized and in-depth information later. To understand more about your options, consult the discussion of the Technology Adoption Life Cycle in Chapter 5: Understanding the Technology Adoption Life Cycle.

Best Practice—Characterizing the project environment

The way you manage a project will depend upon the nature of your organization and the methods used to manage the product-development life cycle. Most information development is associated with projects rather than with ongoing business functions. If your project is part of an organization that develops commercial products, you are most likely to be involved in projects that have external customers to whom you may have difficulty gaining access. If your project is part of an internal technology development initiative, you are likely to be developing for internal customers to whom you may have considerable access.

In either case, you must evaluate the characteristics of the larger project with which you are associated. In some organizations, projects are managed formally, sometimes with several layers of project management and business rules governing the progress of the project. In other organizations, projects are loosely managed, depending on the management styles of individual development groups or even individuals.

In the past several years, some organizations have moved away from traditional product-life-cycle management to embrace more rapid methods, referred to as *agile*, *lean*, or *extreme* forms of product development. Table 15-1 catalogs the differences between traditional and agile projects.

At the beginning of a new project or a project in a new organization, you will need to consider the following questions about the type of project you are working with.

✔ Is the product being managed using a traditional (waterfall) project management process? Is the product being managed using extreme, lean, or agile techniques?

✔ Is there little or no formal project management in place?

✔ Is the project large or small? Large projects may have many people involved, across multiple locations with multiple layers of management. Small projects may involve only a few individuals at one location.

Table 15-1: Traditional versus Agile Project Environments

Traditional Project	*Agile Project*
Follows a plan	Responds to change
Structured, well-defined process	Continuous innovation
Products delivered at the end of the process	Working products updated continuously
Comprehensive process documentation	Minimal process documentation
Requirements defined exhaustively	Requirements continuously adapted to changing customer
Contract negotiations	requirements
Long development schedule	Customer collaboration
Reliable results	Reduced development schedules
	Unexpected results

Note: The environment described in this table relies upon the arguments in Jim Highsmith's Agile Project Management *(Addison-Wesley 2004).*

Process maturity levels

In this discussion of the preliminary product information you need for your Project Plan, I have assumed that you are working with at least a Level 3: Managing and Repeatable product-marketing and -development organization (see Chapter 2: The Information Process Maturity Model for a description of the Information Process Maturity Model). At Level 3 and above, the larger organization skillfully articulates its vision for the product, understands the customer's needs, and has developed at least an initial description of the product architecture.

However, if you find yourself working with a Level 1 or 2 organization, you may find that you will have to develop your understanding based on incomplete, vague, or even nonexistent information. The project vision, including goals and scope definition, may never have been articulated. It could be that nothing has been written down about the project, or what is written may be focused on technology rather than customers. I recall a product description for a hospital information system that focused primarily on the need for a larger mainframe computer rather than on the system to be developed or the customers to be served.

In this case, find out whatever you can during your information gathering and state your assumptions about the project. Your statement may be the only one that is ever written. Even if it turns out to be incomplete or wrong, at least it provides you with a starting point for the development of your own vision for your part of the project.

Common project characteristics

Despite differences in the way projects are managed in an organization, they all have certain characteristics in common:

- ✔ Projects have a beginning and an end.

- ✔ Projects have goals to achieve and deadlines to meet throughout their life cycle.

- ✔ Projects must be managed if they are to achieve their goals and meet their deadlines.

- ✔ Project management takes real time and cannot be regarded as an add-on to other work.

- ✔ Projects change frequently during their life cycles and project managers must accommodate the changes in their plans, schedules, and resourcing.

- ✔ Some projects will fail and some succeed. Often the failures are doomed from the start.

By understanding the core responsibilities that you undertake as a project manager, you will be able to adjust to any number of innovative approaches to product development without panicking.

Critical project differences

Most information development occurs in the context of traditional projects. A traditional project moves through a series of established phases from feasibility and planning through design, development, testing, and implementation. The project has a series of key milestones or decision points, often referred to as gates, that must be cleared before the project can move on to the next phase. At each gate, project managers and developers typically produce documents showing the progress made and the issues remaining on the way to the goal of a completed project. If the managers who must approve the phase gates find that the project has encountered problems, they will either ask that the phase be redone or they might cancel or postpone the project.

Traditional projects call for a traditional approach to information development. As a manager of the information project, you will often be asked to provide your own status reports at the end of each phase in addition to developing documents that outline your plans for the information-development project.

Throughout Part 3 of the book, I reference the relationship between the larger project and the information-development project in terms of project phases. Being part of the phases of a traditional project-development process often means that the needs of information development are more likely to be taken into account. The reporting required of traditional project management means that you will know when project changes are under consideration before they occur rather than long afterward.

Traditionally managed projects are often quite successful in reaching their goals and in meeting their deadlines. However, they often suffer from high cost, delayed schedules, and final products that are no longer useful to the customers. The new forms of project management that evolved in the 1990s attempted to address the problems that seemed inherent in the traditional approaches.

In each case, the traditional phases are abandoned in an attempt to reduce costs, shorten schedules, and produce outcomes more responsive to the changing needs of the customers. Typical of these new methods are innovative and short development cycles in which all of the planning, design, development, testing, and deployment occur within each project cycle iteration. Project teams, often virtual teams with members working across the globe, at home, and in the office, work quickly to develop an idea in response to customer requirements and immediately review their new but limited design with customers. In response to customer feedback, they start the entire process over again. During each cycle, they supposedly come closer to the customers' goals for the project.

Agile project-development methods present challenges for information developers and those managing the information-development projects. The product-development teams may create plans and make decisions about the features and functions of the product without writing anything down. Unless the information developers are part of the development teams, they find it very difficult to discover what is being designed and developed.

It is most important for information project managers to recognize that traditional project management methods must be adapted to agile projects. It may be impossible, for example, to produce traditional documentation at all. In fact, traditional documentation that explains to users how to perform a series of tasks using a product may not even be appropriate, as discussed in Part 2 on portfolio management.

If you find yourself managing in the midst of an extreme, agile, or lean product-development project, consider these possibilities:

- ✔ Spend a greater percentage of your project time on planning, especially if you believe that new approaches to information products may be necessary. Neglecting to plan is clearly the wrong approach. You cannot afford to waste time going in an unproductive direction.

- ✔ Severely limit the scope of the information products so that your team can also produce basic customer information in short time frames. Because the product is likely to be only partially developed, consider the least information that customers will require to provide feedback.

Through planning, be prepared when the time comes for a product release. You will probably have to produce "final" information products quickly. If you have developed a framework or model of your deliverables, you will be able to plug in the required content more successfully than if you try to begin writing too soon. Early writing of supposed final content will be frustrating since the content is likely to change many times.

As a project manager, you may encourage your team members to use this new product-development environment to introduce innovations. Minimalist documentation that focuses only on essential information may be most appropriate. Quick reference tools for primary tasks, conceptual insights into the product architecture for early adopters, graphical overviews, and a focus on complex rather than simple tasks may provide customers what they need most rather than lots of text.

The problem with managing traditionally

Traditional project management is a product of the construction industry. If you're building a skyscraper, a bridge, a battleship, or a missile, your primary objective as a project manager is to keep the project under control. You must begin with precise plans and develop a schedule that coordinates the various activities so that they occur in exactly the correct sequence. You must know the tasks to be performed in great detail so that you can track any changes in schedule due to unexpected delays or greater difficulties than expected in performing scheduled work.

If you have a fully specified project in which the plan is clearly defined, traditional project management techniques will be productive. However, most product-development projects are not fully specified, even under the best of circumstances. In most cases, the goals are not well understood, often because the customer does not know exactly what is needed to meet his or her needs.

Competitors, shifts in the market, financing, and other slippery forces require major changes in project objectives along the way. The need to beat the competition and enter the market at the right time places constraints on product development that don't often exist in traditional construction projects.

Because many information-development projects exist in a competitive and changeable environment, information-development project managers must learn to be spry. That means making flexible plans and having a talented staff that can respond to change and create innovative information products under the most difficult circumstances.

The deadline dilemma

Information developers will often boast about their ability to meet their deadlines at all costs. They are willing to put in extra hours to ensure that all the i's are dotted and the t's crossed. Thus, they feel threatened by projects with shortened schedules and projects in which change is celebrated rather than castigated.

As a result, information developers often succumb to the pressure of rapid development projects by cutting corners rather than innovating. They produce the same weighty tomes, only in less time. As a result, they eliminate planning, testing, and quality assurance, devoting their time exclusively to writing more words. The resulting information products frequently fail to meet the needs of users and contain information that is out of date and incorrect. Deadlines are met but the information is useless.

The approach you must take as an agile project manager is entirely different. Here are some ideas to consider:

- ✔ **You must take a leadership role in the project itself.** You must be thoroughly involved in the work of a collaborative team of information developers. Your job is not to make the decisions about the content but to provide your team with the time and tools.

- ✔ **Your team must focus on innovation**. The same old approaches to information development simply will not work. Your role is not to design the innovation but to encourage its design among the team.

- ✔ **Your team must be aggressively responsive to the customer**. The internal customer is likely to be the product-development leader who is pushing for innovation in his or her own team. The external customer, who should be participating in the rapid development effort, must also become involved in identifying the most effective information products.

- ✔ **You must remain focused on the overall business objectives.** In traditional projects, it has seemed easy enough to meet the deadlines by compromising quality. As a consequence of a reduction in quality, teams often have little if any traditional quality assurance built into their processes except by individual information developers reviewing their own work. Documentation is rarely tested against the product or the customer and even basic copyediting for consistency, clarity, and readability is neglected. As a result, the costs of supporting the product soar after it is released.

As project manager, you are obligated to ensure that value is provided to the customer rather than simply to ensure that something gets out the door.

As a project manager in an agile project, you don't need to know how to address all the technical issues that your team members must understand. Your role is to understand the business context of the project and balance that context against the deliverables.

Best Practice—Identifying project goals

At the beginning of the project, you must understand thoroughly the business and technical goals of the larger project in which you are taking part. In Part 2, you learned about the Technology Adoption Life Cycle. At different stages of a product's life, you will identify different customer and product goals. A development effort that seeks to develop a completely new product in response to market and competitive pressures will have different goals than one that is designed to meet the continually emerging requirements of a mature market.

In discussions with senior and peer managers, consider the following questions about the goals of the larger project:

- ✔ Is the project entirely new, something that your organization has never produced or questions it has never addressed?

✔ Is an existing product or service to be completely redesigned and brought up to current standards?

✔ Is the project an update of an existing product or process, focused on adding new features and functions or addressing existing process problems? Are the updates designed to keep pace with the competition or are they exploring new ideas for the product or service?

✔ Is the project focused on fixing problems and otherwise maintaining an existing product or service?

✔ Is your project designed to improve the quality of the information delivered to customers, with or without other changes to the product or services?

In addition to core questions about the nature of the project in relationship to other projects, you also want to ask about the business goals:

✔ Is the project a speculative attempt to enter a new market or serve a new audience and, therefore, likely to undergo many design changes?

✔ Is the project well defined by stakeholders with clear ideas about the required features and functions?

✔ Is the project intended to drive the company into a new market with new kinds of customers?

✔ Is the project serving the needs of existing customers who are themselves heavily involved in driving the changes?

These questions are not intended to exhaust the subject but to indicate how you might approach discussions with your internal stakeholders.

By understanding the business goals of the project, you will be prepared to help your team formulate the goals for the information-development project and present them to your decision makers.

Best Practice—Analyzing project scope

Be aware that your definition of project scope at this stage is likely to be high level and more vague than you might like. You may find that different parts of your company have not yet reached agreement about the product vision. Different stakeholders may have very different points of view about the project, often colored by their own internal interests in technology or political positioning. You may find, for example, that product developers view the project, not in terms of producing something that customers will value, but in terms of the technology they will have an opportunity to develop. They may be more interested in learning a sleek new skill set than in understanding the customers' requirements.

If you're in doubt about handling the political morass yourself, consider discussing the project with key decision makers. Use your own network of mavens and influencers to evaluate the information you gather. Malcolm Gladwell defines mavens and influences in *The Tipping Point* (Little, Brown 2002). You may also need an astrologer, a psychic, and a crystal ball. Recognize that projects for which the scope is difficult to identify are likely to be high risk projects that will require contingency planning on your part. If the managers of the larger project have engaged in their own contingency planning, find out as much as you are able about their defined risks and plans. If you believe that the larger project is quite risky and no plans have been made to account for the risks, you will need to apply an even higher risk assessment to your project.

Best Practice—Identifying project stakeholders

Every project has a complex set of stakeholders both internal to your company and external as customers. At the beginning of your project, you need information about the stakeholders and the degree to which they will weigh in on your decision-making process.

Here are the steps in a stakeholder analysis:

1   List all the potential stakeholders who you believe will have some influence on the success of your project. Stakeholders may be in your department, in other parts of your company, and outside customers and customer representatives.

2   Briefly describe each stakeholder. If you find you don't know a stakeholder well enough to describe him or her, enlist others in your organization to help.

3   For each group of stakeholders you identify, consider the following questions:

   • How important is this stakeholder's agreement to project decision making?

   • How knowledgeable is this stakeholder at this time about the information-development process?

   • Is this stakeholder already in agreement with the goals of your project? Or must the stakeholder be persuaded about the viability of the goals?

   • What affect will this stakeholder have on the success of the project?

4   Using the stakeholder analysis table in Figure 15-2, evaluate each stakeholder on your list. Using the scales indicated on the table, rank each stakeholder's position. A 1 indicates a low level of knowledge, support, or agreement. A 5 indicates a high level.

5   Add up the rankings so that you have a total for each stakeholder.

Stakeholders with high numbers are likely to be very important to the success of your project and require regular communication. Those with low numbers are less important but still need to know what is happening on the project.

Stakeholder	Level of Interest (1-5)	How interested is this person in your project?	Positive or Negative (1-5)	Is this person positive about your project?	Support (1-5)	How much support is this person likely to give to your project?	Total

Figure 15-2: Stakeholder analysis table

Internal stakeholders

Internal stakeholders may include

- ✔ your department manager

- ✔ peer project managers

- ✔ writers, editors, graphic artists, production and localization coordinators, and other members of your own team

- ✔ subject-matter experts in product development, training, customer support, and others who may provide source information, review your project plans, or review your team's draft information

- ✔ senior management responsible for financial, legal, marketing, sales, and other decisions that may impact the success of your project

External stakeholders

External stakeholders may include

- ✔ users of the information your team develops

- ✔ buyers of your company's products and services

- ✔ senior management among your company's clients

- ✔ dealers, distributors, and others who purchase and resell your products

These lists are not intended to be complete but to give you an idea of whom you might include in your stakeholder analysis.

If you are not personally familiar with each of the stakeholders, consult others in your own organization, especially your department manager and other project managers. If some stakeholders are new to the organization, consider discussing your project plans with them individually, especially if they have a significant effect on project success.

If you have important stakeholders who are uninformed about your project and you believe that they may be difficult to communicate with, look for a project champion who can help you analyze the communication problems and perhaps intervene on your behalf.

Analyzing the internal politics means charting the relationships you have with each internal stakeholder and the relationships they have with one another. By carefully planning your communication to internal stakeholders, you are likely to be more successful in promoting your project plans and winning agreement. Joel DeLuca, in *Political Savvy* (Evergreen Business Group 1999), provides an excellent method for analyzing the stakeholders involved in a decision-making process.

Analyzing the external stakeholders will almost always require communication with people outside your own organization. You may need to collaborate with sales or marketing to understand clients, both from a buyer and a user perspective.

Best Practice—Developing a communication plan

After you have analyzed the potential stakeholders, develop a plan to communicate with them about your project. A communication plan should include a list of the activities you will use to communicate. For example, some stakeholders will be part of your team and require regularly scheduled meetings to review project status. Other stakeholders may need regularly scheduled progress reports that point to progress, plans, and problems. Some stakeholders will need presentations about your project at key milestones such as the initial project plan, the information design phase, and at regular intervals during development and production.

Communicating with key customers will depend upon the relationships you build. If you are working closely with a group of customer partners, they will want to know about your project and design plans and the progress of the development activities. If your team is changing the design of the information or how it is delivered, you may need to ensure that customers are informed about the change. In some instances, customers may require that they approve new designs or plans to change how they receive product or services information.

Stakeholder analysis and a well-formed communication plan are particularly significant if you are managing a project that is new or represents a significant departure from previous practices. Your analysis and your communication plan will be simpler if your project is more typical of the normal "new release" of an existing information set.

In both cases, however, knowing who has a stake and planning how you will communicate with them will help to ensure your ability to meet your project goals. The better your analysis and planning, the less likely you will have last minute surprises.

Best Practice—Understanding the project schedule

Depending upon the manner in which product development is managed in your organization, your questions regarding schedules are apt to produce very different responses. In organizations that follow traditional project management processes, schedules are often developed in early phases of the project. Once a project is formally announced, deadlines for final and interim deliverables are set and communicated to all team members.

In organizations that follow agile project management methods or any other of the rapid development initiatives, final deliverable schedules may be established to account for customer and market pressures but interim schedules may be only vaguely defined or not defined at all.

Iterative development processes often require that some outcome be announced on a regular but short schedule, such as every 30 or 60 days.

Even with formal project schedules in place for the larger project, the interim schedule dates may be quite flexible, especially if the development project encounters problems. As a project manager responsible for information development, you are expected to react as schedules are changed. I discuss the process of changing schedules in Chapter 17: Implementing a Topic Architecture Track and provide a method for calculating the effect of change on the information-development project.

At the beginning of the project, your role is to find out as much as you can about the proposed schedule. Consider the following questions:

✔ Is a final release date associated with some external event like an annual meeting, a user group meeting, the end of the fiscal year, or the end of a quarter?

✔ Is the final due date of the project associated with external requirements, such as a change in legislation, regulation, or contractual obligation?

✔ Are the interim milestone dates calculated in relationship to a set of standard project phases? For example, is the end of the first phase calculated at 25% of the total calendar time? Or, are interim dates associated with other events in the corporate calendar?

✔ Does your product-development organization use a methodology to calculate the amount of time (in person hours) required to complete the project? For example, has a project manager used function points to be delivered to estimate a software-development project? Does a hardware-development effort use previous project data to develop a new estimate of work?

✔ Is the schedule heavily influenced by political pressure from people outside the development organization? Do you hear through the grapevine that the project schedule is ridiculously short because of outrageous demands from senior executives?

Depending upon the answers that you can extract from project stakeholders, you will be able to establish a schedule and milestones in your own project plan. Details of that plan are discussed in Chapter 16: Planning Your Information-Development Project.

Best Practice—Identifying the project risks

It should be apparent from the types of questions you need to ask about the project schedule that you will also need to evaluate the risks associated with the project you are about to manage. You can often tell very early, sometimes simply by the history of projects in your organization, that you will be managing a more or less risky project.

Just how should you define the riskiness of a particular project? Once again, you have questions to ask of the stakeholders:

- ✔ Are those managing the larger project known for maintaining schedules or breaking the schedule regularly?

- ✔ Does the project involve a new technology? Does it require a major change in the way your organization does its work?

- ✔ Is the project adequately staffed to meet the schedule? Has someone estimated the level of effort and the scope of the larger project in sufficient detail to have a solid sense of the schedule?

- ✔ Are there many new people involved in the project? Is the project being conducted by a new development group? Is the development being done in a new offshore location with no previous experience working for your organization?

- ✔ Are the project team members all in the same location or widely dispersed geographically? Are team members working in different time zones?

- ✔ Is information development considered a regular part of the project and included among the larger project's deliverables and schedule?

- ✔ Are you dealing with information coming from multiple outside sources, including other companies with or without their own information developers?

Chapter 16: Planning Your Information-Development Project explains how to use the Dependency Calculator to take project risks into account. However, at this early stage of the project, you need to accumulate information about possible risks. Once again, make use of your stakeholders, especially those who are experienced in the management of projects in your organization. They can provide insights into the potential pitfalls associated with a particular development project that may affect the success of your project.

Projects that are at risk generate a predictable set of problems:

- ✔ Deadlines are often exceeded because the direction of the project has changed or problems are encountered in developing the product or service that were not originally taken into account.

✔ People involved in developing the product or service are so pressed to meet unreasonable deadlines that they are unavailable to provide source material or be interviewed by information developers.

✔ Project management is so vaguely defined that developers write nothing down or update what they originally wrote about the project design.

✔ Project planning is so casual that decisions are made and never communicated to all the stakeholders, including you.

✔ Goals and objectives are so poorly defined that the project continues to change direction, requiring everyone involved to scrap work and start over.

If you fail to do your analysis of risk and understand what you're likely to encounter during the course of the project, you won't be ready to make quick adjustments or even pursue new directions.

You may decide that a project needs more time to develop before you can assign resources. A poorly defined project may need only one information developer assigned part-time to attend development meetings and keep track of progress. Once you get a report that the project is beginning to move ahead, you can assign more significant resources. You may want to begin with a significantly smaller team than you believe you will eventually need so that you don't leave people with too little to do. Then, you can prepare to move more people onto the project once the workload increases. Very often, information developers work on multiple projects at the same time with a clear understanding of priorities and deadlines. However, as a project manager, you need to keep tabs on those assigned to multiple projects, especially new team members. People inexperienced at juggling schedules may end up with one or more projects that are in serious trouble.

In some cases, you may have a "SWAT" team in reserve, including one or more senior developers who are adept at moving into a project quickly and efficiently. However, I would not rely upon a swat team to replace sound strategy and good estimates.

Summary

As you begin a new assignment as project manager, you must carefully assess the larger project of which you are a part. Consider the following:

✔ At what stage of the larger project have you been asked to become involved? Early, middle, or late?

✔ What methods are being used to conduct and manage the larger project?

✔ What are the overall goals of the larger project and how do these goals relate to those you will establish for the information-development project?

✔ Do you have a sufficient understanding of the larger project scope to estimate the resources needed for the information-development project?

✔ Who are the internal and external stakeholders who will have some relationship to how your project is conducted and how should you best communicate with them? Have you accounted for the political nature of the project in your stakeholder analysis?

✔ What is the schedule for deliverables in the larger project and how carefully has this schedule been calculated?

✔ Have those managing the larger project taken the scope and risks into account in calculating the schedule?

✔ How risky is the larger project likely to be?

With answers to these questions in hand or at least a basic set of assumptions, you are ready to develop the Information Development Project Plan.

"We have lots of ideas about this project.
Let's get them into our project plan."

Chapter

16

Planning Your Information Development Project

> Planning is bringing the future into the present so that you can do something about it now.
>
> —Alan Leiken

You may not have all of the information about your project that you would like to as you begin your Information Development Project Plan, but if you have answers to most of the questions outlined in Chapter 15: Starting Your Project, you should be ready to put your plan in writing. Writing the Project Plan is essential to good planning. It is not enough to have ideas about your project in your head. No one can see them there. If the plans are not expressed clearly and communicated effectively to team members and other stakeholders, you can predict with great accuracy that each individual will have a different idea about the project.

An effective Project Plan is primarily a communication tool, although you may find that committing your thoughts to paper clarifies your thinking. You use the plan to communicate the vision, goals, scope, and cost of your project to your team members, members of the team developing the product or service, those involved in marketing and product design, the users of the information, and others whose agreement you need. The Project Plan is an important step toward reaching consensus, making decisions, and obtaining the funding you need.

When you produce a plan for your project, you are relying on your predictions about the future. Niels Bohr, Nobel Prize-winning physicist, summed up the difficulty when he remarked: "Prediction is difficult, especially if it's about the future." Your plan is only as good as your analysis has been. Consequently, the more difficult and risky your project, the more likely that your plan will be incomplete and inaccurate. In fact, the plan you develop at the beginning of the project is least likely to be accurate because it is based on the least information. As the project progresses and you gain information about it, you will be able to revise your plan and increase its predictability.

Managers, especially inexperienced ones, frequently voice their frustration in developing their project plans. They just want to begin "doing," whether that's writing or designing, rather than spend precious time planning.

Despite arguments to the contrary, experienced managers who develop detailed project plans report that the effort pays for itself. Project planning, when done in the context of user requirements, decreases information development time and affects the distribution of activities during the information-development life cycle.

Begin your Information Project Plan by reviewing the template shown below. The template outlines the table of contents for the plan and provides brief reminders noting the required information for each section. Next, use the information in the best practices sections of this chapter to complete each section of the template. This chapter shows brief examples of section content from the semi-fictional MarketTarget project.

Remember that the outline is a guide for your planning. You should customize the outline to meet the needs of your own organization and the type of project you are managing.

- ✔ **Updated releases of legacy products** call for short and simple plans that allude to earlier decisions about information-development vision, goals, audience, and usability.

- ✔ **Redesigns of legacy products** require plans that describe the new project goals and explain how the information will change as a result.

- ✔ **Redesigns of legacy information** need detailed plans focusing on audience and usability that justify the costs of redesigning the information to make it more accessible, easier to understand, or more relevant to the audience's goals.

- ✔ **New projects** require detailed plans explaining the vision and goals, the audience and usability requirements, and how they can best be met during the project.

Note that the outline for the Project Plan focuses on business, customer, and stakeholder issues rather than on scheduling the work. You will get to scheduling in Chapter 17: Keeping Your Project On Track, but only after the Project Plan is developed and approved. The Project Plan presented in this chapter reflects the values of innovative, customer-driven projects that drive agile development. If your projects are focused on the continual updating of legacy information, and you find yourself adding more content but not increasing value, you can construct a short form of the plan.

If you want to pursue innovation and direct projects that are quick moving and exciting, consider putting most of your planning effort into the design of the project. Then follow by calculating dependencies and providing time and cost estimates that demonstrate how you can increase efficiency and decrease project costs at the same time that you deliver better content.

The following section provides a template for an Information-Development Project Plan. The bracketed text provides instructions for completing each section of the template. Examples from the MarketTarget project are used to illustrate each portion of the plan. You can also find this template on the book's website.

Information-Development Project Plan Template

[Enclosed in brackets you will find instructions for completing the plan. Replace or delete all of the instructions and add your own content to the template.]

[Name of your project] Project Plan

[Project code name or other identifier]

Revision history

[Provide an account of the revisions you have made to this project plan.]

Date	Version	Author	Explanation of Change
	[Version number]	[Author names]	[Explanation of the change]

[Use version numbers until you have final approval. Afterward use point changes for minor updates. If you prepare a major update, as you might if you have a complete change in the project scope, create a version 2.0. Include a brief explanation of the change.]

Contact information

[Provide all the contact information for the information project manager.]

Prepared by
Organization
Location
Email address   Phone number

Approved by

Name	Position	Date
[Names of the project approvers]	[Their positions]	[The date signed]

[Include sufficient lines for all the required approvals. Under each line, add the approver's name, position, and the date signed.]

Project Overview

[In this section of your project plan, provide all the background information needed to identify the goals of the larger project, the audience for the product and for your information, and the project usability requirements.]

Description of the larger project

[Briefly describe the larger project of which information development is a part.]

Project vision and goals

[Briefly describe your understanding of the larger project's vision and the goals that your organization has set for this project.]

Project scope

[You have a number of options in describing the scope of the larger project. You may want to list the features and functions that will be included in the product-development effort. You may include descriptions of project hardware to be designed and built and software to accompany the hardware. You may describe a project in which information is the most important outcome, such as a project to develop policies and procedures. Provide sufficient information to identify the project scope characteristics that you will need to guide your estimate of the scope of the information-development project.]

Project schedule

[Provide an overview of the larger project's schedule, emphasizing key schedule dates that you must take into account. Include information about milestones that you must meet, deadline dates associated with these milestones, and information about the relationship of a particular milestone to the general availability of the product or service.]

Milestone	*Actual or Relative Date*
[List the external project milestones that drive your information development schedule, such as feasibility approval, approval to develop, functional requirements freeze, functional testing, systems testing, beta release, final release, and so on.]	[List an actual date or a date relative to the general availability date for the product release. Provide information about how the date was determined and who is responsible for setting the date.]

Project budget

[Provide information about the overall project budget if it is available, especially if your information-development budget is established as a percentage of the whole project budget.]

Description of the customer for the product and the information

[Provide as detailed a description as you can for the customer for the product and the audiences for the information. If you have several audiences, include each of them in this section, with accompanying descriptions. Be certain to include in your description your team's assessment of the potential information needs of each audience group. If you have outside sources for your information, such as customer site visits, usability studies, market analyses, or feedback from training and customer support, include that information here.]

Usability requirements

[State the usability requirements of the larger project and the information. This section is especially important as an extension of the customer and audience descriptions in the previous section. You may want to emphasize ease of access to information by an experienced user or the step-by-step coverage for a more inexperienced user. Use the information you have gleaned from any customer studies to inform your assessment of the audience's usability needs.]

Information Development

[In this section, you move from the larger project to the information-development project that you are managing. The purpose of this section is to facilitate the stakeholders' understanding of the information requirements and the scope of your development efforts.]

Information-development project vision

[Briefly describe the information set you intend to develop for this project. If you are updating existing information, note that here. If you are developing information products for an entirely new project, describe what you hope to achieve in the design. If you intend to make significant changes to an existing suite of information products, describe why the changes are being made.]

Project schedule

[Provide an overview of your proposed schedule for the information development activities. Once again, describe each milestone and explain the reason for the actual or relative date. Relative dates describe the dependencies of your schedule on external events that guide the larger project. In this section, note any schedule dependencies for your project.]

Milestone	*Actual or Relative Date*
[List each milestone that you propose for the information development project. Define what will be needed to complete the milestone. Don't include in this list your internal milestones such as developmental editing, quality assurance, or copyediting. Those will be included in the detailed project schedule.]	[Either provide an actual date for the milestone to be completed or state the time relative to an appropriate external milestone from your previous list. You may need to explain the date and any special requirements.]

Project budget

[Provide information about your project budget. You may have a total budget amount that you must stay within, or your budget may be calculated as a percentage of the larger project budget. Your budget may be based entirely upon headcount assigned to the project, or it may include additional funding for translation and localization and the production of final deliverables and their distribution.]

Information-development project scope

[Briefly describe your project's scope, outlining the major deliverables. Then, construct your detailed table of deliverables, similar to the table given here as an example.]

Information-Development Project Scope

Deliverable title	Part#	Delivery method	Languages	%Changed	Page/Topic Count
[Title of the document or other final deliverable]	[Part number]	[Method of delivery, including print, PDF, HTML, Help, or other]	[Languages into which the document will be translated]	[The percent of change you anticipate in the deliverable from the last release. For new deliverables, the percent change is zero.]	[The estimate page or topic count for the deliverable.]

Roles and responsibilities of the project team

[Include in this section a list of all the members of your information-development team and a list of all others associated with the project throughout your larger organization. Describe the roles and responsibilities of each team member on the project.]

Information-Development Project Team

Role	Name	Location	Responsibility
[The role of a member of your project team]	[Name]	[Location Email Phone]	[Describe the responsibilities of this team member on your project team.]

Larger Project Team

Role	Name	Location	signoff/Responsibility
[The role of a member of the larger project team]	[Name]	[Location Email Phone]	[Describe the responsibilities of this team member to your project team. If the team member has signoff responsibility, note that here.]

Project dependencies and risks

[In this section, analyze your project dependencies and the risks you associate with the project. Note the critical items that you need to ensure that you can successfully deliver the information required by customers on schedule and with the quality you have determined.]

Project assumptions

[State the assumptions you have used to estimate the resources required to meet the project scope and schedule. If you must have access to the product by a certain date for the information to be ready on time, state that assumption here. If you are expecting certain freeze dates or timely reviews, note those assumptions as well.]

Risk analysis

[State the risks that you and other members of the larger project have associated with this project. Complete the following risk analysis table for the project, explaining both your prevention and contingency plans.]

Risk Description	Probability	Potential Impact	Prevention Plan	Contingency Plan
[Describe the risk.]	[Indicate if it is high, medium, or low.]	[Indicate if it is high, medium, or low.]	[Explain what you will do to prevent the problem from occurring.]	[Explain the actions you will take if the problem occurs.]

Dependencies calculation

[Show your dependencies calculation in this section. See Figure 16-1.]

Average Hours/Page	5.0

Dependency	Ranking	Factor
Product Stability	○1 ○2 ◉3 ○4 ○5	x 1.00
Information Availability	○1 ○2 ◉3 ○4 ○5	x 1.00
Prototype Availability	○1 ○2 ◉3 ○4 ○5	x 1.00
Subject Matter Experts	○1 ○2 ◉3 ○4 ○5	x 1.00
Review	○1 ○2 ◉3 ○4 ○5	x 1.00
Writing Experience	○1 ○2 ◉3 ○4 ○5	x 1.00
Technical Experience	○1 ○2 ◉3 ○4 ○5	x 1.00
Audience Awareness	○1 ○2 ◉3 ○4 ○5	x 1.00
Team Experience	○1 ○2 ◉3 ○4 ○5	x 1.00
Tools Experience	○1 ○2 ◉3 ○4 ○5	x 1.00

Hours/Page Projection: 5.00

Figure 16-1: Dependencies calculator

Project total hours and costs

[Provide your final calculations of the size, scope, and cost of each project deliverable and of your project as a whole. Break these totals down by the items in your deliverables table.]

Deliverable Title	Page/Topic Count	% Change	Total Hours	Total cost
[Name the deliverable.]	[Indicate your count of pages, topics, or other metric.]	[Indicate the percent change expected in the deliverable.]	[State the total hours you have calculated to produce this deliverable.]	[State the total cost you have calculated for this deliverable.]

Project resource requirements

[You have already listed the participants in your project in the roles and responsibilities section of the Project Plan. Include here any other resources you require for the project. If you need access to a lab or access to the product or other materials, note that information here.]

Best Practices in Developing Your Project Plan

Now that you have reviewed the template, you will find five best practices associated with the development of a sound Information-Development Project Plan:

- ✔ Envisioning the information-development project
- ✔ Defining the information-development project scope
- ✔ Defining the roles and responsibilities of the team
- ✔ Calculating project dependencies and risks
- ✔ Estimating the project resource requirements

Best Practice—Envisioning the information-development project

As you begin your Project Plan, you must ensure that you have a clear understanding of the larger project of which you may be a part and that you have a well-articulated vision of the role of your own project in meeting the needs of your organization and your customers. Your vision for your project must be focused on providing value for your customers, both inside your organization and among those who use your product or service to gain their own value.

Your responsibility in articulating a vision for information development begins with an understanding of the vision of the larger project. That project may be intended as a simple upgrade of an existing product, or it may encompass a new and innovative approach to providing customers with valuable tools and services. You are responsible for gathering the information you need to understand fully the project vision and then use that information to inform your vision for information design and development.

Describe the larger project

Your first responsibility in formulating your information project plan is to understand and articulate to your team members the goals and objectives of the larger project. Without a clear picture of how the larger project is being defined and developed, your team will be unable to ensure that their own work on the information-development project is aligned with larger goals.

In the opening section of your Project Plan, briefly but clearly describe the larger project of which your work is a part. The larger project may include the development of hardware and software, policies and procedures, an organizational website for internal or external use, the description of a service, or any of the myriad projects with which technical communicators might become involved.

State the larger project's vision

Be certain that your project description is brief but contains sufficient detail so that you can use the description to confirm your understanding of the project. You may need to describe the vision and focus of an entirely new development effort. You may be working with an existing product or service that will be enhanced with new features. You may be involved with a project to articulate a new policy for your organization or to develop a procedure to enforce that policy. The more clearly you understand the goals and objectives of the larger project, the better you will be at developing a sound plan for the information-development project you manage.

Ask the following questions to better understand the vision for the project:

- ✔ Is the project the development of an entirely new product or service for your organization? How is the market for this product being discussed or described?

- ✔ Is the project a regular update of an existing product or service that maintains the initial design and adds additional features and functions?

- ✔ Is the project a redesign of an existing product or service that includes a significant change in the architecture, the interface, or other aspect of the product?

- ✔ Are their significant external constraints governing the project? For example, is the project being designed to meet the requirements of a particular customer? Is the project required because of legislative, legal, scientific, or other external influences?

- ✔ Does the project primarily involve a redesign of the information and is not associated with the release of a new product or service?

- ✔ What are the key benefits that will be provided to the customers?

- ✔ Will the project provide your organization with a competitive advantage?

An Example of a Project Vision Statement

Designed to help small retailers manage their marketing more effectively and target their products to well-defined segments of their customer community, MarketTarget is a software program that provides a database of customer information and gives a small retail business the ability to analyze the database and select appropriate customer segments for sales promotions. Unlike similar products that have been developed to aid the marketing efforts of large and sophisticated financial institutions, MarketTarget is designed to meet the database marketing needs of a small retail business. MarketTarget represents an entirely new flagship product for our company.

State the larger project's goals

State as accurately as you are able the goals of the larger project if you have not already included the goals in your project description. In the project description, you should have learned whether the product, service, or even information being developed is entirely new, a significant redesign of an existing product, or a new version of an existing product. In the project goals, describe your understanding of the manner in which the project fits into the overall goals of your organization.

For example, you may want to consider the following questions to better understand the project goals:

- ✔ Is this project being undertaken in response to customer requirements or pressure from competitive products?

- ✔ Is the product out of date, providing an opportunity for a redesign?

- ✔ Is the company entering a new market with this product?

- ✔ If the product is for internal customers, what business problem is it being designed to solve?

- ✔ Must the product be redesigned because of customer dissatisfaction or usability problems?

- ✔ Is the redesign product, service, or information influenced by technology changes?

You may find that it takes some research before you can clearly define the project goals. In fact, individuals from different parts of the organization may have different perspectives about why the project is being undertaken. You may find it advantageous to talk to people who may have different perspectives, including marketing, sales, product development, operations, services, and training. Each of these groups may have a perspective on the product and its customers that will add to your understanding.

An Example of Project Goals

The goal of the MarketTarget project is to introduce a new automated process to a customer who either has never targeted its marketing efforts at all or has conducted its campaigns solely using manually generated data. This customer is interested in optimizing its sales campaigns and is intrigued with the possibility of using a computer database to assist in this effort. However, the customer may not have a clear idea of how a database of its own customers might work and how it might be used effectively.

Define the larger project's scope

Describe as completely as possible the size and complexity of the project. In an analysis of project scope, you may want to include a detailed table of new features and functions to be included in the project, along with the relative size and complexity of each and their impact on the information-development project you are managing. You may note that the project involves either a small effort conducted by a small, close-knit group or a large effort involving many people representing many parts of a global organization.

Chapter 15: Starting Your Project lists some of the issues involved in an early analysis of the larger project scope. Here are additional questions that will help you to address project scope:

- ✔ Is this a major new project involving many groups and individuals throughout the organization?

- ✔ Is this a minor new project involving only a few people in a single, close-knit group?

- ✔ Does this project involve a major or a minor revision or redesign of an existing project?

- ✔ Does this project involve a significant new technology or changes to technology that require re-architecting the product?

Record the answers you have researched to these and other questions involving project scope and summarize the results in this section of your plan. Remember that although the project vision and goals may remain relatively stable throughout the project life cycle, the scope will often undergo major redefinition along the way.

The estimate of the resources you need for your information-development project will be closely aligned with your analysis of the scope of the larger project. If the scope changes, you will have to re-estimate your resources. Without a clear definition of scope and a statement in your Project Plan of your working assumptions about scope, you will not have a basis on which to revise your Project Plan and your resource and schedule estimates.

An Example of Project Scope

The MarketTarget project is a completely new project being created by a small development team in close conjunction with the company's leaders in finance and marketing.

The project involves building a user interface and all the accompanying functionality to allow users to

✔ create a database of customers with appropriate metadata about each customer

✔ update the database

✔ perform administrative functions to control the database (privileges, backups, etc.)

✔ analyze the database for trends in purchasing behaviors

✔ analyze the database for other similarities

✔ extract customer information from the database based on selected criteria

✔ produce mailing lists and labels for selected customers

✔ record information about the affect of targeted marketing campaigns

Because all these functions are being built for the first time, they will require careful testing to ensure that they perform as intended. However, the basic technology is well understood by the product developers which may allay our risk concerns about the new customer requirements and functionality being addressed.

Describe the customers for the product and your information

Describe the customers for whom you will develop information. At best, a customer or audience analysis should be based upon your team's direct experience with potential customers for your information. At the very least, you should summarize information gleaned from other sources, including marketing, sales, support, training, and product development.

In Chapter 7: Developing User Scenarios, I describe the process that an information architect might use to gather information about customers. This process is a distillation of the detailed approach to gathering audience and task information that I developed with Ginny Redish in *User and Task Analysis for Interface Design*.

Your user and task analysis should provide you with detailed information about the audience, including answers to these and related questions:

✔ Does your audience include everyone from administrators, planners, technicians, and other technical experts to ordinary end users of consumer products?

✔ What are the real goals of each audience group with regard to your product or service? Do people use your product as an integral part of their full-time tasks on the job or is the product something that they may use occasionally?

✔ What is the level of experience of each audience group with your product or service or a similar one? Are they already experts on your product or a related one or are they complete novices?

✔ What is the willingness of the audience members to use information to support their goals in using the product or service? Are they likely to use product information to support their task performance, or are they more likely to be trained on the job or call someone for assistance?

✔ How much time might audience members have to devote to learning new information? Is your product used as part of a busy work schedule?

✔ What is the expected level of training on the product or service to be provided by your company? Are audience groups provided with training to use the product or service, or do they learn on their own?

If you don't have direct contact with your audience, you are at a distinct disadvantage in planning your information-development project. However, you or your information architect can use surrogates to gather some audience information. Talk to others who have contact with users, including trainers, support service personnel, sales personnel, and consultants. If your organization engages in usability studies, consult with the usability professionals. Find out if marketing, product management, or product-development staff have visited users of the product, not just buyers. Work with your information architect to gather as much information as you can and then state your assumptions in this section of your Information Plan.

An Example of a Customer Analysis

The MarketTarget users are most likely to be small business owners or their staff members. Many of the small business users may be unfamiliar with using computers in the workplace. They are unlikely ever to have used a database program or one that helps them analyze their customers. They may be concerned that using a computer may take time away from their primary task of running their businesses and may be unwilling to use written information to learn how to use their new marketing software.

Staff members may be somewhat more familiar with computers, although most likely someone in the business uses a computer only to handle accounting data. They may be unfamiliar with database programs and software designed to analyze customers. They may also have older computers that do not have a more recent Windows operating system.

Evaluate the usability requirements

Evaluate the usability issues for your project that may require you to make changes in your information design. For example, many organizations are now required to provide web-based information that is accessible by people who have limited vision or are blind. The disabilities requirements mean that alternative text must be provided to describe graphics and illustrations so that a person who cannot see the images can understand what they contain.

Based on your audience analysis, you may want to change the presentation of your information in print and online to accommodate older readers. You may also have feedback on previous projects from users that information is not easily accessible or is difficult to understand once they find it. You may learn that the topics have not provided the information that users need to be successful.

Summarize the result of the usability analysis performed by your information architect and explain how it will affect the information you plan to deliver.

An Example of a Usability Analysis

MarketTarget's users are likely to have several usability issues associated with the product and the information. As business owners, they may be reluctant to use computers in the workplace to solve business problems. They may be skeptical of the value of analysis based on a database of customer information. They will need to be convinced that they can use the product to create successful marketing and sales campaigns.

Staff members may prefer to use their own manual methods of accounting for customers because they are familiar and simple to them. They may resent a layer of automation imposed upon them that appears to require more, rather than less, work.

Develop your vision for the information-development project

At this point in your plan, you move from a description of the larger project to the information-development project that you are planning. The information you and your information architect and other team members have gathered about the larger project and the customers will help you articulate your own vision for the information-development project. In the previous sections of your plan, you have clarified your understanding of the larger scope of the project and its customers. In this section of the plan, you develop your own vision for delivering effective, high-quality information to your customers.

A project vision will help you and your team members maintain their focus throughout the project. It will establish what you intend to build and why. It will keep you from going down unproductive paths and creating information that no one needs. It will help you make key decisions as you respond to the inevitable changes that will occur during the project.

Too often, information-development projects are stuck in the past.

"We've always included an 'about this manual' section at the beginning of our documents, although we don't think anyone ever reads it."

"We don't have time to make the information correct; we just have to meet the deadline."

"That information has already been translated. We know it's wrong and unusable, but we can't change it now. It would be too expensive to translate again."

"We have to include all these long descriptions of the technology, even if the users will never read or understand them. The product developers insist."

A project vision that is developed and owned by you and your team members will help to quiet the naysayers and provide ammunition in your negotiations with members of the larger team.

Consider the information vision Comtech developed for the MarketTarget project.

An Example of an Information-Development Project Vision

The primary customer for MarketTarget information is the small business owner who is responsible for developing marketing and sales campaigns for her company. She is typically very busy and reluctant to spend time reading manuals and learning about software. Yet, we believe she will need a basic manual that explains how to reach her goal of more effective sales campaigns and increased revenues and profits for her company. The basic business owner's manual will be minimal, targeted at less than 60 pages with many illustrations. It will focus on real-life business examples that show how other business owners have used the customer database analysis tools to understand their own customers more effectively.

The content will be based upon typical customer scenarios (use cases) that illustrate the utility of each product feature in terms of received value. The text will contain examples of using the product features to reach a business goal with a minimal amount of step-by-step task instruction. The task-oriented, step-by-step instructions will be included in the product's help system rather than in the basic manual.

By stating up front that we would base the manual for the business owner on customer scenarios, we were able to resist demands that we document every field on every screen of the software. We based our decisions on a thorough analysis of the intended customer for the product, enabling us to reinforce our vision with solid customer knowledge. Because of our close connection to the customers, we were able to gain the support of the product manager from marketing for this vision of the information. That support proved critical when the lead developer demanded a feature-based approach.

In a project that involved a hand-held device for handling contact lists and phone numbers, calendars, spreadsheets, and other consumer-oriented features, our customer study led us to recommend an interactive manual on the company website and delivered to the customer on CD-ROM, instead of the long, boring, and dysfunctional PDF version of a manual that accompanied the product's initial release.

Clearly, your information architect and your team members will play an important role in articulating your information-product vision. However, as project manager, you are responsible for communicating that vision to project stakeholders and protecting the vision throughout the life of the project. Although the details of your implementation of the vision may change over time and must respond to changes in the larger project's scope, schedule, and budget, the vision itself should not be subject to project vicissitudes.

All of the information you have considered so far in the development of your Project Plan reflects your overall understanding of the goals of your project and the alignment of your goals with those of the larger project.

For complex projects involving information innovations or redesign, you may need more complete documentation of your reasoning in support of your vision for the information-development project. For simple upgrades or revisions of existing products, these initial sections of your Project Plan may be quite brief.

The information you develop and summarize up to this point in your Project Plan is designed to

- ✔ confirm your understanding of the product's vision, goals, and customers

- ✔ articulate your understanding of the information needs of those customers as they relate to the product

- ✔ communicate the vision of you and your information-development team to the other project stakeholders, including the customer

In the next section of the Project Plan, you will turn your vision into a set of specific milestones and deliverables.

Best Practice—Defining the project details

After you have established with your team the vision you have for the information-development project, it is time to define the project details. These include your project milestones, deadlines, schedules, and budget in addition to a list of the project information deliverables.

Project milestones

If your organization uses a formal project-development methodology, the project milestones may already have been well defined. In some cases, these may be referred to as "project gates." If so, list those milestones or gates in your own plan. If you have well-defined deliverables associated with each pre-defined project milestone, include those on your detail summary. For example, you are likely to be required to deliver this Information-Development Project Plan as one of your first deliverables in a traditional project management framework.

You do not need to include details about the delivery of draft topics or sections for editing or review at this point in your planning. You will be able to address the delivery of individual topics or sections as part of your detailed planning in Phase 2 of the project. At this point, you should communicate with your team about the milestones to expect and acknowledge to your stakeholders that you understand your commitment to them.

If your organization does not require certain milestone deliverables from information development, list your own milestones using the five-phase information-development life cycle model presented in this book. Include brief definitions of each planned milestone and what you expect to deliver. You may also want to include certain key information life cycle milestones that are not part of the five-phase model, such as the localization and translation, because they are so critical to your schedule although they may not be part of a larger project-development methodology.

Project deadlines

List the primary project deadlines of the larger project. Include dates for beta releases and the date for general availability of the product. In a service-oriented project, include the date on which the new service will be made available to the customer. In an information-oriented project, include the date that the new information or training must be delivered to the customer. Each of these critical dates depends upon the type of project you are working with, but you will have at least one of them to plan against.

In your deadline statement, indicate who in the organization has set the deadline. Is this a deadline imposed by marketing, product development, information technology, or the customer itself? If there is an external event such as a product show, a regulation change, or a commitment made externally, include a statement about that event. If the project starts to slip in ways that affect the deadline, you can remind yourself and your team about the urgency of the project and decide how to respond if the original deadline must be met.

Project schedule

Based on your discussions with product and development managers, you should have information on the proposed schedule for the larger project. Your information-development schedule must be aligned with the larger project schedule so that you are able to ensure that information deliverables are completed appropriately.

In your project details, list dates associated with the milestones you identified earlier, as illustrated in Table 16-1. If you do not have firm dates for the external milestones, provide information about how the dates were estimated and who is responsible for setting the dates. Because these dates may change, you may want to indicate them in terms of weeks prior to a final availability date.

Table 16-1: External Milestones

Milestone	*Actual or Relative Date*
General availability—the scheduled release date for the final product deliverables including the information. Note whether the information must be available in all languages by this date.	General availability is scheduled for June 30, 2007 in all languages.

Next list the dates for your high-level internal milestones. These are the specific publication dates that you have added to the standard set of larger project milestones. (See Table 16-2.) Again, you may want to specify the dates in terms of days or weeks needed prior to an external milestone or associated with an external milestone.

Table 16-2: Major Deliverable Milestones and Schedule Details (example)

Milestone	*Actual or relative date*
First Draft Review—must contain all information needed for each topic or section of a document. The topic or section has been reviewed by the developmental editor and has undergone a basic copyedit before being submitted to the external reviewers.	First Draft Review occurs for each topic after the initial draft is complete and has been reviewed by the developmental editor. The first draft review must be complete four weeks prior to the first product testing date. We expect the first draft review of a single topic to be completed in one business day. Durations for multiple topics in a collection will be determined prior to submission for review.
Second Draft Review	
Final Draft Review*	
Final Draft Approval	
To Translation	
To Production	
To Distribution	

Note: It is essential that documentation be thoroughly reviewed during the scheduled review periods. The final draft review is intended to discuss any open issues and resolve any items that remain open before the final draft approval. Late review comments increase the risk to schedule and quality.

It requires research to identify the dates you want to associate with your own milestones. If you want to schedule your localization and translation tasks, you may need to consult with a localization coordinator in your organization or an outside vendor to estimate how much time will be needed for their tasks. You need to know, in addition, if all translated versions of your documentation must be delivered simultaneously with the product or project delivery or if some translations can be scheduled for later delivery. Making that decision requires that you know the proposed global release plan for the project. In many cases, different country- or language-specific versions are delivered on later schedules.

Schedule dependencies

As part of your assessment, note any external dependencies on which schedule milestones are dependent. Consider external dependencies only at this point, because your internal dependencies will be stated in your detailed Project Plan. For example, you cannot begin the review process until your substantive editing is complete on a topic or chapter. Your internal dependencies are reflected in the workflow you set up.

Your external schedule dependencies should be those that are influenced by people or events outside of your control. For example, you cannot complete a topic or section of a book and move it to translation or final production until the reviews are completed and the final drafts approved. Note in your schedule dependencies that there is an interrelationship between reviews and approvals and final draft stage.

You know that you cannot complete task-related topics for a software product until the user interface is complete. State that relationship in your schedule dependencies notes.

Project budget

If you know the overall budget for the project, include the information here. Knowing the project budget is extremely useful for your planning. The budget for most development projects is a direct reflection of the number of people working on the project. In many organizations, the budget for information development, exclusive of translation and localization, is supposed to be not more than a certain percentage, perhaps 10%, of the total project budget.

Seven to ten percent of the total project budget appears to be average, based on data I have collected through surveys of publications managers. The percentage has gone down a bit in the past five years in some industries, but 7% to 10% remains a useful point of comparison.

In some industries, however, you will find vastly different percentages. In telecommunications, for example, information development averages 3% or less of development costs, even when the development is primarily software. In organizations developing hardware, the information development budget may be lower because of the high cost of hardware design.

I believe that one of the reasons for offshoring information development to the same low-cost economies in which products are being developed is related to keeping this percentage intact. If product development or engineering costs are lowered by one-third or one-half by moving them to a low-cost country, then the cost of information development remaining in the high-cost country will increase the percentage allowed for information.

An Example of a Project Budget

The MarketTarget project has a total budget of $2.5 million for one year. Product management hopes that the information development budget can stay under 10% of the total budget, excluding the cost of translation into French.

Budget dependencies

Just as you added notes for schedule dependencies, do the same for budget dependencies. Consider what might cause changes in the overall budget that would have a direct bearing on the budget for your project. If the project budget is increased by addition of developers because of a schedule crisis, if the developers need additional expertise or time to master a new technology, if additional requirements are added with development budget increases, note how that will affect your proposed budget. If the project you are supporting grows larger and your budget remains the same, you will have to make decisions later in the project to reduce your scope. At this point, note any budget dependencies that you think likely to influence the success of your project.

Proposed information deliverables

In projects that are updates of existing information, the information deliverables are most likely to remain the same as they have been in the past. However, if your project vision calls for innovations in the information design or you are developing a completely new project, you need to describe the final deliverables that your team intends to develop, as illustrated in Table 16-3.

Your list of information deliverables should include all the media types that you intend to develop and all deliverables that will be prepared for each media type.

An Example of Information Deliverables

The MarketTarget project requires four basic deliverables: a printed manual for the business owner, a basic training program for the business owner, a help system for the individual using the software to develop a marketing or sales campaign, and a training program for the same end user. These deliverables were developed in one language for a US audience.

We have used two techniques to estimate the scope of these deliverables. Based on our analysis of the business owner, we conclude that the print manual had to be no larger than 50 to 60 pages of text and illustrations. We base this estimate on our understanding of the business owner's basic reluctance to read about a software system. We also know that the company installation team would provide training. The simple training program should take no more than one hour for the business owner.

The help system scope is related to the feature list developed earlier to define the scope of the product and from our analysis of the audience. By starting with the feature list, factoring in an analysis of the projected number of software screens proposed for the product, and considering the need for a minimalist approach, we anticipate producing 75 to 80 help topics, including task topics, a very few concepts, and a set of glossary definitions.

Table 16-3: Deliverable List for the MarketTarget Project Example

Document Title	Part#	Delivery Method	Languages	%Changed	Page/Topic Count
Business Owner's User Guide	1234	Print	EN FR	0	50–60
Business Owner's Training Manual	1234tr2	Print	EN FR	0	20
End-User Help System	1234hp	HTMLHelp	EN FR	0	75–80 topics
End-User Training	1234tr2	Print	EN FR	0	20

For an existing project, include a list of the changes to be made to the existing chapters, sections, or topics. Since you won't be changing the overall approach to information delivery, your detailed list should focus on the details, to the extent that you know them at this point. In some cases, you'll only have enough information to estimate a percent change in an existing book or set of topics.

"The user's guide will require changes to approximately 50% of the pages or 50% of the topics. It will also require an increase of 15% new pages or topics."

For a new project or a significant redesign effort, include a list of the intended deliverables and your initial estimate of scope for each deliverable as illustrated in Table 16-4.

Table 16-4: Large-Project Deliverables List Example

Deliverable Title	Part#	Delivery Format	Languages	%Changed	%Reuse	Topic/Page Count	Scope of Effort
List the final items to be delivered	Add the part number if applicable	List the formats used to deliver this item	List of languages into which the item will be translated	Indicate how much of the original version of the item is likely to be changed. Zero indicates	Indicate the percentage of the deliverable that will reuse information created for another deliverable	Indicate your preliminary count of the topics or pages in the final deliverable	State the anticipated level of effort, such as new; minor, moderate, or major changes to the original; use previously developed doc; and so on

Note: If you are estimating an existing project use their actual topic or page counts of the existing items, either documents, help topics, or web pages. For new information products, review the recommendations in the Scope estimates discussion below. In estimating your scope of effort, take into account the discussion of quality level to indicate the required level of work for the deliverable.

Scope estimates

You may be asking yourself where the page or topic counts in the tables come from. Estimating the scope of any project is always difficult, especially if the project is new and has no earlier examples for comparison. At this early point in the information-development life cycle, your estimates are even more difficult because you often know very little about the emerging project. In the best case, you can base your early estimates of scope on previous similar projects. Some techniques you might use early in the project are

✔ an estimate based on previous projects of similar scope and audience

✔ comparison with competitor's documentation for similar products

✔ a metric based on other project indicators, such as the size and staffing of the development project

✔ project feature or function metrics or project use cases

✔ an estimate of the acceptable scope of the information for the audience (as we determined for the MarketTarget business owner above)

✔ a pure guess-timate

Previous projects are always the best indicators of scope, as long as you account for changes in audience or additions to the list of features and functions. The more experience you have with previous projects, the better you will be able to evaluate the scope of the new project. If you haven't kept all previous projects in your head, it may be advantageous to have a database of previous projects, possibly in the form of updated deliverables lists like the one provided here.

Competitor's documentation may give you a useful starting point when your organization is developing a competing new product. If you are in an industry in which products are publicly reviewed and compared, you may find it important to compare favorably with a competitor in scope of coverage in the document, that is, unless you believe that the competitor's manuals are unusable and filled with marketing fluff and descriptive text rather than tasks.

Other project indicators may give you a basis for scoping the information-development project. Just as you may have used the size of the larger project's budget as an indicator of your own project, you can use the same information to judge the number of resources you can apply to this project. If you have roughly $100,000 to staff your project and the project will take six months, you know that you have budget for only one or two people. Don't estimate 2,000 new topics or an 800-page manual. You don't have the resources to produce this much new information in such a short amount of time. If industry averages for writing new topics comes in at roughly 4–5 hours per page or 8–10 hours per topic, consider just how much scope your small team can absorb.

Feature/Function or use case metrics allow you to establish a strong link between your own scope estimate and the initial estimates on which the product development project manager is basing scope. Remember that at early project stages, the product development manager is having the same difficulty estimating scope that you are. If you have a fairly well-structured product-development methodology, you may be able to review the use cases that have been developed. Use cases describe the way the users will interact with the product, giving you an idea of the tasks they will need to perform and the conceptual and reference information they may need. If your organization doesn't create use cases, look at the feature/function lists that might be in the product planning documents. As long as these lists reflect user-desired features and functions rather than technology features and functions, they may be an early indicator of scope. The MarketTarget project described in this chapter included eight significant user features that had to be addressed by the product. Given 5–6 pages per feature, we felt confident in our estimate of a minimalist business owner's manual at 50–60 pages.

Acceptable length is an interesting scope estimating method, one that we applied to the MarketTarget project. We considered just how much a member of the intended audience might be willing to look at without being intimated. Even if you consider that most people don't really want to read the documentation, you also know that consumers or executives or people busy with other work will want to read even less. For example, my oven comes with a small user guide with about 16 pages, not too annoying and small enough to keep in the drawer next to the oven. Unfortunately, I needed it for a year or two after purchasing the oven because the controls were designed for the space shuttle. Thirteen years later, I don't consult it unless I get some weird error message on the LCD display. However, if I were a specialist in repairing ovens, I might be willing to carry around something more bulky.

A **pure guess** is certainly not your first choice in a responsibly developed Project Plan. However, you are sometimes forced by circumstances into committing to something on the spot. The best defense is a strong offense, as you know. Pushed into such a corner, ask that the inquisitor provide you with some very specific data, such as the number of function points to be included in the software, as a basis of your estimate. Your inquisitor is likely to disappear for days or months. If you have to estimate something, just make sure you have stated your assumptions. A responsible pure guess should be accompanied by a rationale for the estimate. If you've guessed that your information should be comparable to a competitor's information, state that assumption. If you've guessed that your information should be limited by the number of resources you can afford to assign, state that assumption as well.

As you consider previous projects, remember to consider design changes you wish to implement. One manager, pursuing a minimalist agenda, produced a new information set that was 25% of the previous information set. The new design emphasized user goals and tasks instead of product descriptions and conceptual information.

Level of effort

Estimating the level of effort required to produce each document or other deliverable for your project will help to establish not only the work to be done to develop the content but the activities that will need to take place to ensure quality. No one wants to define a quality level that is less than perfect for a project, but, in reality, we make many decisions about the design of our processes, our staffing levels, the types of quality assurance we undertake, and even the types of information we decide to include. All of the decisions affect the quality of an individual project and the quality of a department's overall work.

Although you may want to refer to quality levels in your notes on the scope of effort required in your milestone list, the quality levels should be developed and implemented throughout your information-development organization rather than for an individual project. However, an individual project may call for one of several quality levels that you have defined. For example, consider a Level 1 Quality definition established by an organization for its least important products.

> **Quality Level 1**—meets minimal documentation standards. Includes tasks only, with no conceptual information. Tasks are checked for accuracy and completeness but are not reviewed externally. Basic copyediting is conducted. Level 1 documents contain no indexes, examples, or screen shots. They do contain any applicable warnings or cautions.

Although writers may be unhappy with this level of quality, you may have a product that is nearing end-of-life, has few customers and little revenue, or has existing customers who need only minimal information about changes to the product. You will find more information about decisions regarding relative quality in Chapter 21: Managing Quality Assurance, which discusses the decision making required to manage an entire portfolio of information products in a corporation.

A high level of effort may be best associated with a premier product or services, something designed to attract new customers and revenues or a product that has crossed the chasm and is now being offered to mainstream customers who demand outstanding levels of support. You might define a Level 3 Quality for these customers.

> **Quality Level 3**—meets top-level documentation standards. Includes all information required by the customers, including tasks, concepts, and reference information as defined by user profiles. All information is checked for accuracy and completeness, including complete reviews by subject-matter experts, usability testing, and thorough developmental editing. All deliverables include a complete index or other search mechanisms. All illustrations, warnings, and cautions are included as needed.

In reviewing the wide range of total time devoted to information-development projects, I find it is quite obvious that different quality standards are being imposed on projects, even given differences in the difficulty in developing the content. Organizations that include careful developmental and copyediting and other quality assurance activities differ from those in which information deliverables are rarely reviewed by anyone other than the writer and the technical experts. These organizations clearly produce different levels of quality. In short, you get what you pay for.

The point is that you need to be up front about your quality decisions. You are most likely constrained by the funding you receive and the headcount you have available for the work. Level of effort simply indicates what you are including in your process and what you are leaving out.

Best Practice—Defining the roles and responsibilities of the team

Just producing a list of people working on your project may be sufficient for your Project Plan, but thinking about your own team members, external team members who will provide resources to your team, and other project stakeholders will ensure that you are planning for the right audience. Getting the right people to be associated with your project is essential for your success. In one project, we had all the product developers providing information to the writers except for the one key person who alone understood how the customers were going to use the product in their business environment. Critical information-focused decision making at early stages of their product implementation was missing from the information set until we identified the missing stakeholder and obtained a commitment from him to become involved.

Getting the right people on your team and getting commitments from key stakeholders will directly influence your project. You need people who understand the needs of the information audience, understand the technology being developed, and know how to communicate information effectively and efficiently to the right audiences. We know that we generally find that about 20% of the people engaged in our projects produce about 80% of the work. The more difficult your project is to accomplish, the more you need talented people who are above average. You need people who are not only capable of doing the work required but are also committed to doing the work well.

To find the right people, you need to know the nature of the project you are proposing and the level of effort required to complete it. You don't need your top technical star or your most talented information designer to work on a simple maintenance project requiring Level 1 Quality. You may do better with a well-supervised intern instead. You want to reserve your best players for the most demanding projects. You can place your less experienced staff members alongside your best players to help increase their skills.

Getting the right people extends to your technical product developers and other reviewers who provide input to your information development. The wrong "technical expert" can doom the development of complex information. Without the assistance of people with genuine understanding of how the product works, you won't get the depth of information you need. Without the assistance of people who really understand the users, you're likely to write information that they don't need either.

As you think through your resource requirements with respect to team members and stakeholders, consider which people are critical to the success of your project, which are essential, and which are interested but not essential. If you don't have the participation of the most critical individuals, the project is doomed. If the essential people don't participate adequately, they can delay the project but you can probably find ways around them. If the nonessential people don't participate, you won't see any direct impact. However, if you don't communicate with them effectively, they can quickly become critical.

Team members

List the members of the project team, including yourself as project manager and the other information developers. All these team members are either critical or essential to your project. For each team member, briefly describe the responsibilities for which the individual will be responsible during the project. See Table 16-5 for an example.

Table 16-5: Information Development Team Example

Title	Name	Location	Responsibility
Info developer	Harry M	Denver	Create topics and submit to production
Info developer	Sarah H	San Jose	Create topics and submit to production
Information architect	Jackie O	San Jose	Design the content for all deliverables based on audience and task analysis. Develop the annotated topic list and deliverable maps of the content for print and help.
Editor	John R	Phoenix	Review all drafts for adherence to templates, style, content, and design
Illustrator	Aru R	Bangalore	Prepare all illustrations and perform QA
Manager	Cami R	Denver	Provide all resources, track project, produce all project documentation

Information resources

List the information resources on which your team will depend to complete the project. Resources might include documents, but, more importantly, they should include people. List the key people in the larger external project team. Be certain to include the project manager, head of the development team (product, service, or information), program manager or marketing representative, and any others who will be directly involved in supporting information development. Indicate who has signoff responsibility.

One organization creates what they call a "dance card" for new information developers that list the people they need to go to for various types of information. (See Table 16-6.) Consider the key product developers, the user-experience or usability team members, people in the training and support organizations who work closely with the customers, and even the customers themselves.

Table 16-6: Product Development Team Example

Title	Name	Location	Signoff/Responsibility
Program manager	Tim Q	Boston	Signoff—yes
Marketing	Georgina P	Boston	Signoff—yes
Engineering project lead	Harry S	Bangalore	Signoff—yes
QA manager	Gaurav R	Bangalore	Signoff—yes

Best Practice—Calculating project risks and dependencies

Analyze the risks that are most likely to be associated with your project. A risk analysis provides you with critical information that you will need to estimate the resources required to complete your information-development project by the scheduled deadlines. You will use the information you develop in the risk analysis to apply the dependency calculator to your project.

Project assumptions

State the assumptions you have made in developing your Project Plan. You may assume, for example, that the product developers have sufficient time allocated in their schedules to review the information thoroughly and provide feedback to the information developers. You may assume that your team will have access to the product or the user interface in sufficient time to complete drafts. You may assume that product development has been frozen in enough time to complete the information and hand it off to translation and localization or that you have reviews completed in enough time to meet final deadlines. Whatever assumptions you have started with should be stated here.

An Example of Project Assumptions

This list specifies the assumptions made in planning this project:

- ✔ Product developers must perform timely and thorough reviews of the information topics as they are developed.

- ✔ New and modified features of the product are thoroughly defined in the product specification.

- ✔ Names of final approvers of the information are verified and any changes in this role are communicated to the project manager promptly.

- ✔ Information developers will have access to the completed hardware and software approximately four weeks prior to the first review.

Risk analysis

You will find a number of important issues that affect the risks associated with your project. A risk-free project means that you can calculate the resources you will need to complete the project you initially define, and you are pretty much guaranteed that these resources will be adequate. Nothing unusual will happen during the course of the project to change the deliverables, restructure the content, rewrite topics, or change the scheduled deadlines. Everything in a risk-free project will proceed as originally expected.

Given this description of a risk-free project, you already recognize that most projects are not risk-free. Most information-development projects are accompanied by a significant amount of risks, primarily due to the risks involved in the product or service development project that you are supporting.

High-risk projects are those with many unknowns: the exact nature of the technology may be relatively unfamiliar to the product-development team, technology problems may be difficult to understand or to resolve, the customer requirements may be poorly defined, customers and their internal representatives may be unclear about the goals of the project, schedules and deadlines may be redefined, and competitors may introduce new challenges. You may have risks within your own organization: information developers may be unfamiliar with the technology used in the product, they may be using new tools to develop information, you may need to redesign the information deliverables, the audiences may be poorly defined, team members may leave the project unexpectedly, and you may encounter challenges in communicating with globally dispersed team members. In each case, your project will be more difficult to accomplish than you may originally have hoped.

The point of risk assessment (see Table 16-7) is to attempt to predict the degree of risk and plan for changes in the project that are most likely to occur.

Table 16-7: Risk Analysis Example

Risk Description	Probability	Potential Impact	Prevention Plan	Contingency Plan
Inaccurate or late documentation because an accurate prototype is not available when needed.	Medium	High	Engineering must be prepared to ensure that the prototype is delivered according to schedule.	The document will have to be revised and reshipped to the customer after the release date.

Applying the Dependency Calculator

I designed the Dependency Calculator to assist information developers and project managers in applying their assumptions and risk assessment directly to their estimate of the resources required to complete the project by the deadline. Too often, project estimates are based on the same metric, such as hours per page, for every project. A project manager decides that the average project in his or her organization takes four hours per page to complete even though it's clear that every project is somewhat different. The project manager is then surprised that the project being estimated turns out to be more difficult than expected, which means that an insufficient number of people have been assigned to work on the project. As a result, the beleaguered team begins making quality cuts in the information to make the deadline.

The Dependency Calculator is a straightforward way to acknowledge that a particular project is unlikely to be average. Some projects are simpler and take less time to complete, and others are more difficult. Even within a single project, elements of the project may have different levels of difficulty or different risks and assumptions associated with them. In that case, you may want to make more than one dependency calculation. In most cases, however, at this early phase in the information-development life cycle, a single assessment for the entire project is sufficient to provide you with an indicator of the adequacy of your resource estimates.

The 10 dependencies that make up the calculator are based upon the experience of many information developers and managers in assessing their projects. However, you may have dependencies in your organization that are not included. For example, if you are working with team members at a new location in a developing country, you may have communication factors that are not included among the experience dependencies in the calculator. In that case, I recommend adding one or two dependencies or replacing existing dependencies that you don't need to factor into your calculations.

Figure 16-2 again illustrates the current Dependency Calculator. An interactive version of the calculator is available at http://www.comtech-serv.com/dependency_calculator.htm.

Average Hours/Page	5.0		
Dependency	**Ranking**		**Factor**
Product Stability	○1 ○2 ⊙3 ○4 ○5	x	1.00
Information Availability	○1 ○2 ⊙3 ○4 ○5	x	1.00
Prototype Availability	○1 ○2 ⊙3 ○4 ○5	x	1.00
Subject Matter Experts	○1 ○2 ⊙3 ○4 ○5	x	1.00
Review	○1 ○2 ⊙3 ○4 ○5	x	1.00
Writing Experience	○1 ○2 ⊙3 ○4 ○5	x	1.00
Technical Experience	○1 ○2 ⊙3 ○4 ○5	x	1.00
Audience Awareness	○1 ○2 ⊙3 ○4 ○5	x	1.00
Team Experience	○1 ○2 ⊙3 ○4 ○5	x	1.00
Tools Experience	○1 ○2 ⊙3 ○4 ○5	x	1.00
	Hours/Page Projection: 5.00		

Figure 16-2: The Dependencies Calculator

In analyzing your project's dependencies, consider that you need to consider five external and five internal dependencies. The external dependencies describe risks outside of the publications organization and the project manager's sphere of control. The external dependencies are

✔ product stability/completeness

✔ information availability

✔ prototype availability of the product

✔ availability of subject-matter experts

✔ effectiveness of reviews (information inspections)

The four internal dependencies are those that may be within the project manager's ability to influence. The internal dependencies represent the publications staff's

✔ technical experience

✔ writing and document design experience

✔ audience understanding

✔ tools experience

✔ team experience

Table 16-8 lists each dependency and provides an explanation of its meaning.

Table 16-8: Project Dependencies

Dependency	*Interpretation*
Product stability/ completeness	The amount of change the product is likely to undergo during the course of the project. A product that changes drastically during the development cycle, especially when major changes occur late in the development cycle, its likely to require frequent changes to the draft information.
Information availability	The existence and quality of written information about the product, including marketing studies, requirements definitions, specifications, user profiles, task analyses, and other information that will help the publication team understand the audience and product as quickly as possible. A lack of planning information may mean that information development is being included early in the product development cycle. It also may mean that little planning has been done for the project—a good predictor of numerous and late development changes.

Information availability is one of the factors that may contribute to a reduction in hours per page for revision projects. If the information-development team is working with a high-quality earlier version of the information, then the job of learning the new information and fitting it into the existing structure is easier. |
| Prototype availability | The existence and quality of a prototype of the product under development. The existence of an early prototype often indicates the sophistication of the development team in their planning. It also reflects on the stability of the product design. The lack of a prototype may make it difficult for the information-development team to develop an accurate use model (a picture of how the user will perform a task using the product). |

Dependency	*Interpretation*
Subject-matter expert availability	The availability of the subject-matter experts (SMEs) to aid the information developers. SMEs, or technical experts, are important sources of information for the publications team. If they are too busy, unavailable, or lack knowledge about the development effort, they may impede the progress of the information-development effort. Cooperative SMEs who look upon writers as equal development partners can greatly reduce the difficulty of an information-development project.
Review experience	The likelihood that reviews will be thorough, complete, and timely. Thorough and timely reviews or technical inspections of draft information are essential to the success of an information-development project. Poor reviews can impede progress or require costly reworking at the end of the project life cycle.
Technical experience	The degree of technical experience your team has with the new product. Every information-development team has a learning curve on a new technical project. The writers and others involved in the project must learn the technical matter to be addressed. While not every team member needs to be technically skilled in the subject matter, team productivity increases if some team members are reasonably familiar with the technical subject matter.
Writing and design experience	The amount of writing and design experience your team has with the type of information. Especially when they are involved in designing information that represents a new approach to the information and the audience, information developers may experience a design learning curve. When the information types are familiar and the team experienced in their design, this learning curve may be reduced. This dependency may also be used to account for changes in the tools and technology available to the information-development team. For example, if you are moving to topic-based authoring from a book-based model, you will experience a migration cost for your project as people learn new ways of designing and authoring information.
Audience understanding	The degree to which your team understands the audience requirements. Although we expect the information-development team to conduct a user and task analysis at the beginning of the planning phase of a project, the more they already know about the intended audiences, the more effective they may be in designing and conveying information.
Team experience	The amount of experience your individual staff members have working on teams. If you are a project manager of an information-development team, the experience you have in working with the individual team members and their experience working together effectively may enhance your ability to complete the project on time and on budget. The less experience you have working with your team members, or they with one another, the higher the risk.
Tools experience	The knowledge that the information developers have of their own tools. If they have been using the same tools, such as a desktop publishing system or a product to develop online help, for many years, their tools experience will support the current methods of developing information. However, if you are moving to new tools for your project, your team members will experience a learning curve. For example, you may be using an XML editor or a content-management system or both for the first time. Although these tools are introduced to save costs in the long run, they may cause an initial project to be slower than expected.

Each dependency is evaluated on a scale of 1 to 5 in which 3 represents the average, most normal case in your organization. By selecting a 1 or 2 for a dependency, you indicate that you expect that dependency to be working to the advantage of your project. If you select a 4 or 5 for a dependency, you indicate that at least one aspect of your project may be more difficult than normal for your organization. For example, you may select a 5 for Tools Experience when you are introducing an XML editor and a content-management system to your project. You are planning for extra learning time for team members to become familiar with the new tools. If you select a 1 for Product Stability, you are indicating that your team is working with a well-understood product development effort, perhaps just maintenance fixes of the existing product.

I recommend using the interactive, online version of the Dependencies Calculator. However, it is possible to calculate the dependencies by hand. Use the same multipliers for all the dependencies except the first, Product Stability. Product Stability is a doubled factor because I have found it is a major risk factor in most projects. Table 16-9 shows the composite scores associated the product-stability factor, and Table 16-10 shows the composite scores associated with all other factors.

Table 16-9: Composite Score for the Product-Stability Dependency

Factor you select	Composite score
1	0.80
2	0.90
3	1.00
4	1.10
5	1.20

Table 16-10: Composite Scores for All Other Dependencies

Factor you select	Composite score
1	0.90
2	0.95
3	1.00
4	1.05
5	1.10

Once you have selected all the factors for each dependency, multiply all the composite scores together in one long sequence. For example, you may have a set of multipliers as follows:

$$1.10 \times 1.10 \times 1.00 \times 0.95 \times 0.90 \times 0.95 \times 1.05 \times 1.00 \times 1.00 \times 0.90 = 0.92$$

This multiplier indicates that your project is slightly less difficult than average. Any number below 1.00 indicates an easier than average project. Any number above 1.00 indicates a more difficult than average project. The value 1.00 indicates an average project.

Contingency planning

Discovering that you have a project that is significantly more difficult than average, with a dependencies calculation at 1.2 or 1.3 or higher, you will need to plan to mitigate the risks. If you have high dependency factors for your internal dependencies, you may need to consider ways to ensure that your team gets the training it needs in the product, the information design, the tools, or the customers for their information. You may want to add a person to your team who has more experience in the area in which you see a weakness.

If your external dependencies are high, consider what you can do to mitigate them. For example, if you have factored in a high dependency for SME availability, you may be able to negotiate for better communication and cooperation with the product development manager. If you have a high factor for information availability, you may insist that your most skilled team member attend product development meetings to gather information about the product. If you believe that you cannot include early actions to mitigate the risks, especially for highly unstable product development efforts, you will need to include a contingency plan in your schedule. Remember that the manager for the larger project is probably planning for contingencies as well.

Let's look at two possible project scenarios, one for an average project and one for a high-risk project.

> **Average project**—With an average project, you expect to have information products completed by the end of the sixth month and in time for the announced project deadline and the schedule for general availability of both product and information for the customers.

> **High-risk project**—With a high-risk project, you add a contingency factor to your schedule, indicating, as you did with your milestone schedule, that the deadline to complete the information development is dependent upon the final successful testing of the last features and functions of the product. You may state that your information will be completed four weeks or some amount of time following an external project event.

To accommodate the potential problems associated with a high-risk project, you should consider creating two plans: a commitment plan and a work plan, as illustrated in Figure 16-3.

The commitment plan reflects the actual delivery date you have promised to the stakeholders. The work plan is the real plan you have developed for the project. It includes an amount of time set aside to manage the risk. It acknowledges that you expect to encounter problems that will require additional project work or course corrections. In information-development projects, the additional work usually includes writing new topics that were not part of the original plan or revising topics because the product has changed and the original content is no longer correct.

The plan you commit to, the commitment plan, includes the reserve time in the schedule that you have set aside to account for the contingencies of new topics and revisions.

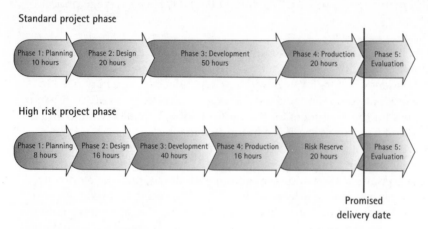

Figure 16-3: You should plan time in your work schedule to allow for the unexpected.

If you don't include a reserve for a high-risk project, you have few alternatives when the risks play out. If you haven't planned for the changes likely to occur in the project, you can either reduce the quality of the work, reduce the scope in a way that may impact customer quality, or take the risk out on the staff in the form of unpaid overtime. Both alternatives do, in fact, occur on many information-development projects. In fact, both of them occur on the same projects. Only when the schedule is extended to accommodate unforeseen or unlikely risks in the product-development process do information developers gain the time they need to accommodate the project changes and delays.

A high-risk project is not only subject to changes in scope. Most likely you encounter barriers to completing the information caused by missed prototype delivery schedules, incomplete or late information reviews, inadequate information from developers, and incorrect information based on inadequate understanding of the product architecture. On projects that are not related to product development, delays can occur for many of these reasons as well as the politics of getting a policy or a procedure approved. Often the key stakeholders have not been included in the communications about the project, and they introduce new issues near the end of the development process.

You are also likely to encounter projects in which senior management has already established a risk reserve. They just haven't told you about it. They set an ambitious and short project schedule, knowing full well that the risks make achieving the deadlines highly unlikely. In many cases, project schedules are set to satisfy the demands of stock analysts rather than to account for the real time it will take to complete the project successfully.

Best Practice–Estimating the project resource requirements

By this point in your project analysis and planning, you have two critical pieces of information at hand to use in estimating the resources you will need to complete the project successfully and by the deadline: your estimate of scope (number of topics, number of pages) and your dependencies calculation. By multiplying the two together, you know the total number of hours your team requires to complete the project.

For example, the project outlined in Table 16-11 indicates that your team will produce four deliverables: a 50 to 60–page user guide, two 20-page training manuals, and 75 to 80 topics for a help system.

Table 16-11: Sample Deliverable List for the MarketTarget Project

Document Title	Part#	Delivery Method	Languages	%Changed	Page Count
Business Owner's User Guide	1234	Print		0	50–60
Business Owner's Training Manual	1234tr2	Print		0	20
End-User Help System	1234hp	HTMLHelp		0	75–80 topics
End-User Training	1234tr2	Print		0	20

You know that a user guide for a new software product, in your experience, averages 4.0 hours/page to complete. The four hours accounts for time spent

- ✔ learning the product
- ✔ creating the content with several drafts
- ✔ editing the content
- ✔ incorporating reviews
- ✔ testing the information against the product
- ✔ managing the project
- ✔ getting everything ready for production
- ✔ coordinating translation activities
- ✔ managing production activities

Note that only about half the allotted time is for actual content creation.

However, you have applied the Dependencies Calculator to this project, finding that the project is much more difficult than average.

✔ Product stability = 4 (1.10)

✔ Information availability = 4 (1.05)

✔ Prototype availability = 4 (1.05)

✔ SME availability = 4 (1.05)

✔ Review experience = 4 (1.05)

✔ Writing experience = 2 (0.95)

✔ Technical experience = 2 (0.95)

✔ Audience awareness = 3 (1.00)

✔ Team experience = 4 (1.05)

✔ Tools experience = 5 (1.10)

You have calculated the total dependencies factor as 1.39, definitely a high-risk project. Next, apply the dependency factor to your average hours per page:

4.0 x 1.39 = 5.6

Because of the risks involved, your new hours per page for this project is 5.6. Given a 60-page user's guide, the total time you need is

5.6 x 60 = 335 hours

You want to use the same hours per page for the 40 pages of training material you are going to create:

5.6 x 40 = 224 hours

Now look at the help system. In your experience, building a comprehensive help system with entirely different information than you will put into the user's guide requires approximately 3 hours per help topic. You believe that the dependencies factor for the help is the same as for the documents and training material. Your calculation produces the following result:

3.0 x 1.39 x 80 help topics = 334 hours

If you add together all of the pieces of the project, you have the total hours for the information-development project:

335 + 224 + 334 = 893 total project hours

In reviewing the initial project estimate, you have completed the following calculations:

1 Calculated the total number of pages and help topics you will develop for the project.

2 Consulted your project histories and previous experience to determine the average time your team needs to develop user guides, training materials, and a help system.

3 Calculated the dependencies or risk factor associated with this project.

4 Multiplied the total number of pages times the average hours per page times the dependencies factor for each deliverable type.

5 Added together all the total hours for all deliverables to calculate the total number of project hours required for this project.

Note that your calculations do not include project time that is the responsibility of people outside of your immediate team. You have accounted for neither the time spent in reviews by the product developers nor the time they spend communicating with the information developers. You have not calculated the time spent by graphic designers or illustrators if they are needed to prepare materials for the information products. You have not included the time spent on translation or localization either. Each of these activities may be included in your estimate either as distinct line items in your calculations or as additions to the average hours per page or topic that you use to determine the total hours for the project.

However, it is always risky to add in hours for people whose schedules you cannot control. You probably have no history of their work on information-development projects. They may, as in the case of translation and localization, be outside vendors who will estimate their own part of the project and give you a schedule and cost to meet your deadlines.

Estimating staff requirements

Based upon your total hours calculation for the project, estimate the number of full-time equivalent staff members you will need to complete the project by the deadline. In the sample project calculation, you found you needed 893 total project hours to produce the deliverables you have identified in your project scope analysis.

A staff member is typically able to devote 30 to 35 hours per week to project work, allowing for minor personal or sick leave or other non-project work that must also be accomplished. Based on a 35-hour week, an 893-hour project required 25½ weeks to complete for one person doing all the tasks, including the project management, research, design, writing, editing, revising, translation coordination, and final production. A 25-week project represents about half a year, without time off for holidays or vacations.

If you must complete the project in less time or cannot devote one individual to the project full-time, you will have to divide the project among several staff members, each doing a specialized part of the job. Your project may include you as the project manager, an editor, an information architect, a translation coordinator, and a production specialist. Your entire team will take responsibility for the total of 893 hours among them. In the next chapter, you learn how to create a project spreadsheet that shows you how to divide hours and schedule among all the information-development team members.

You have already included all your team members in the roles and responsibilities section of your Project Plan. You have yet to assign them to specific tasks and schedule milestones for the project.

It is highly likely that you will discover that you do not have sufficient staff to accomplish all the tasks represented by the total project hours you have calculated before the deadline. In this case, you will need to take one of three courses of action:

✔ Add staff members to the project until you have sufficient hours scheduled to cover your project estimate.

✔ Decrease the scope of the project (i.e., number of deliverables, topics included, and so on) until you have deleted sufficient hours to accomplish the project with the staff you have assigned.

✔ Decrease the quality of the project by omitting steps in the process, such as editing, reviews, testing, quality revisions, and other activities that will diminish the effectiveness of your information products for the customers.

Obviously, the last course of action is generally not considered desirable by most information developers. It is far better to add people to the project or decrease the scope than to take steps that affect the quality of the work. However, in real projects, quality is often compromised to meet schedules and maintain minimal staff. Many projects omit critical time for quality assurance activities, such as developmental editing, testing the topics against the product, coordinating content with other information developers, or responding to project scope and content changes. Whenever you omit critical activities from your organization or from particular project schedules, you are affecting the quality of the products.

You know that sometimes lower levels of quality are acceptable in elements of an information-product portfolio because certain products and projects are deemed less important to company or organizational goals and commitments to customers. Use the information you have developed through portfolio analysis to help you make the appropriate quality, scope, and resource decisions as project manager of a particular project.

Calculating project costs

Total project costs always begin with the cost of the people assigned to the project. Once you have the total number of hours needed to complete the project, you can easily calculate the people costs by multiplying the total hours by the average cost per person. To do so, you must use the fully burdened rate for staff members in your organization. The fully burdened rate includes salary, all taxes and benefits, and all overhead. In the US, a typical fully burdened cost of a full-time information developer is $125,000 per year, based on an average salary of $62,500 per year. Given holidays, vacation, and other time off, the average worker in the US works about 1,900 hours per year. That comes out to approximately $66 per hour. If you multiple your total hours by $66 per hour, you will have a reasonable estimate of the people costs of your project.

893 total project hours x $66 dollars per hour = $58,938

If you have a finance organization, they should be able to tell you the hourly rate for the people in your department. If it's significantly less or more than $66 per hour, ask how the calculation is done.

If you have other costs associated with your project, such as printing, distribution, and translation costs, you will need to add those based upon your own calculations or the estimates you receive from outside vendors.

Deciding if you can make the deadline

Consider how to interpret the total hours you have calculated. Do you have enough time and resources to complete the project? How can you tell?

Let's say that your project is scheduled to be completed, start to finish, in six months, and you have one information developer assigned to create all the materials you need. In six months, you have about 875 available hours, if no one is ever sick or takes time off and can devote about 80% of total time to the project each week.

25 weeks x 35 hours per week = 875

That means you can easily complete your project in six months, especially if you assign the project management time needed to yourself.

But what happens if the project must be completed in four months? Instead of 25 weeks, you have only 15 weeks to work or 525 hours. What do you do?

- ✔ Ignore the calculations and pray that you can still complete the project with one information developer (and nothing happens to that individual).

- ✔ Increase the time available by doing some of the writing yourself.

- ✔ Find another writer to bring into the project after the initial planning phase.

- ✔ Assume that the project deadline will be extended (which may actually happen given the high-risk associated with the shortened development effort).

- ✔ Decrease the scope of the project by eliminating one or more of the deliverables to bring the project down to 525 hours.

- ✔ Decrease the quality of the information development by eliminating quality assurance and other parts of the process and reducing the hours per page or help topic below your average.

Rather than engage in wishful thinking or compromising quality, your best alternative is to increase the staffing on the project. If that isn't possible, you should next consider decreasing the project scope. If the project is extended, as you suspect it will be, you may be able to add the deliverable back into the project. However, if the project is extended, it is highly likely that the product will have changed, requiring the addition of topics or the revision of topics with changed content.

Preparing to adjust your estimates

At this point in your planning process, you have a complete plan and an initial estimate of the hours required and the cost of those hours for your project. You've followed a sound estimating process based on careful and thorough consideration of the goals, scope, and complexity of your project. You've also accounted for your experience with previous projects and analysis of project histories to arrive at your average number of hours per unit of deliverables.

Nonetheless, your project estimate is not without risk. You have a considerable number of unknowns to account for, including your estimates of scope and risk. You must prepare to adjust your estimates as the real project emerges.

Your initial estimate for the project is probably the most unreliable estimate you can make because you are dealing with so many unknowns. The unreliability of estimating causes many people to decide it's not worth doing at all. Or they develop an initial Project Plan and estimate the project, only to put all of their work on a shelf and forget it as the real project emerges. They move from a doubtful estimate to complete reactionary chaos with no idea if they can actually make their deadlines without severely compromising quality.

The alternative is to understand the nature of project planning. Projects change. The best laid plans go astray. Your first estimate is the worst one you'll do because you know the least. You need to go back to your estimate whenever you find that a change is occurring in the project and redo the estimate. Here are some early warning signs:

- ✔ The developers are too busy and frantic to talk to your information developers. You hear rumors of big problems with the technology.

- ✔ Marketing and others are continually adding new features and removing others, increasing the instability of the project.

- ✔ The prototype you expected last week is nowhere in sight.

- ✔ Your information developers tell you that they can't nail down the feature list, and no one has an interface design for them to study.

- ✔ The VP of development has unexpectedly announced his immediate retirement.

- ✔ Half of the developers have quit.

- ✔ The overall churn on the project has increased, meaning a lot more meetings and grumbling in the hallways.

In any event, be prepared to quickly reestimate your project and make adjustments to scope, quality, schedule, and resources as you respond to change. Chapter 17: Implementing a Topic Architecture and Chapter 19: Managing as the Project Changes cover the process as you move through the design and development phases of the project.

Summary

In this chapter, you learned to develop a Project Plan to document your vision for the project and its audience and scope. You included details in the Project Plan to provide a clear and comprehensive roadmap of the project deliverables, schedule, budget, and required resources. You stated your basic assumptions about the project, assumptions that relate to the risks that you face and the dependencies that will affect the success of your project. You also calculated the risks and dependencies associated with your project and used the results of your calculations to estimate the total hours and budget for your project.

Finally, you analyzed the estimate in relationship with milestones and deadlines and decided whether the project can, indeed, be accomplished with the resources that you have assigned. If you found that the resources are not adequate, you consulted the portfolio analysis that you or a more senior information-development manager in your organization developed. Through portfolio analysis, you can decide if you need to cut back on your vision for the information, your definition of project scope, or your resource estimate so that your project is aligned more satisfactorily with organizational goals and commitments.

Now you are ready to start the design work and add detail to your management plan.

"We can organize the content topics into a map."

Chapter 17

Implementing a Topic Architecture

> An information architect must learn about business goals and context, content and services, and user needs and behavior; and then work with colleagues to transform this balanced understanding of the information ecology into the design of organization, labeling, and navigation systems that provide a solid but flexible foundation for the user experience.
>
> —Peter Morville

Applying your information architecture to individual projects and areas of content requires that every team member understand the organization-wide Information Model. Your Information Model provides the framework for creating, managing, and publishing the content needed by information users in your organization and among your customers. The primary components of an Information Model are

- ✔ the standard patterns and the specific content plans for the output your organization will deliver to users

- ✔ the information types that encompass the information to be authored

- ✔ the internal structure of those information types, embodied in content units

- ✔ the metadata schema used to describe each information object or content topic you create

- ✔ the strategy you will use to optimize the management of content in your organization's repository, avoiding duplication of topics and proliferation of topics that provide little value to the customer

Mature information-development organizations have always developed consistent structures for the content they create and deliver, although they have rarely referred to those structures as an Information Model. Nonetheless, these organizations have agreed to follow standards to develop their content, often by defining consistent sets of tables of contents for particular documents such as installation, administration, user, and other guides. In many cases, information developers in these mature organizations have followed standards to structure their content, especially step-by-step procedures and reference data.

Immature organizations, typically at Levels 1 and 2 of the Information Process Maturity Model have left the design of documents to individual authors. They have often limited their structure to the formats of the output by defining layout and typography. Although heading levels and bulleted list styles may be consistent across a library of documents and some aspects of the writing style specified, the content is rarely standardized. Rather, the approach to the content is left in the hands of the authors to use their own best judgment in approaching a subject area for a specific audience.

If your goal includes moving your organization to more mature processes and information models, then you cannot afford to leave the information planning to individuals working in isolation. Each project manager or lead information developer in your organization must take responsibility for working with the entire team of information developers to plan the particular content for the next product release in collaboration with other project managers. Project managers or lead information developers must work in close collaboration to ensure that the most efficient and effective way of developing content takes place, even though individual teams may be working on different parts of the product. Each manager must not only apply the standards of the Information Model to the deliverables of his or her project but must also coordinate with other project managers to take advantage of opportunities to share content across product lines.

Whether you are managing an individual project or managing the portfolio of projects your organization supports, you must apply the standards of your organization's Information Model to your project's information plan. In the previous chapters, you learned about developing the components of an information plan, and you learned to estimate the scope and degree of difficulty of your projects. Estimating the project scope includes estimating the number of topics that must be written new or revised. Deciding which topics your authors need to write requires that you apply the overall Information Model to the particular demands of each project and use the outcome of your initial planning to coordinate with the leaders of other teams.

Best Practices in Implementing a Topic Architecture

As you begin to plan your project, you must initially research the overall plans for the larger project in which your team takes part. The project may involve the development of new versions of existing hardware or software, entirely new products or services, or updates of existing policies and procedures.

Your team is typically responsible for defining each of the project deliverables in the project information plan and extending the information plans into detailed analyses of the topics to be written for each deliverable. Those topics might be combined to form chapters and sections of print or PDF deliverables or sections of a website or help system. If you are implementing DITA (Darwin Information Typing Architecture), the topics are written independently and then assembled into DITA maps to create the deliverables.

Once you understand the user scenarios that your information deliverables must support, your team can develop tutorials to support the scenarios and begin to list the tasks that will enable the users to reach their goals. From an initial list of tasks, your team decides which conceptual, descriptive, and reference topics are needed to support task performance. These are added to form a complete annotated topic list for the project. If you are working with existing content, your annotated topic list includes notations that separate the topics into action categories such as new, major modification required, minor modification required, no change, and remove.

With your topics defined and the work to develop them estimated, you should map your topics into the hierarchical and networked structures you require for the deliverables. A DITA map enables you to create a table of contents for a print, PDF, web, or help system output and establish parent/child relationships among the topics. These parent/child relationships become simple links in HTML or help systems. You can also establish more complex relationships in your DITA maps by using DITA relationship tables, thus creating links between tasks and reference items or other defined relationships.

With your project detailed deliverables planned and recorded in an annotated topic list and your DITA maps, you can provide your content plans to the reviewers for feedback. As your team begins to develop the topics, you are likely to make frequent changes to your topic list and deliverables plan, adding, deleting, and modifying topics and changing hierarchical arrangements.

When you have topic under construction, team members must add index terms and metadata to the topics.

The following four best practices provide details to support the implementation of your information architecture to a specific project.

- ✔ Developing content plans for each project deliverable
- ✔ Mapping hierarchies and creating related-topic links
- ✔ Developing indexes and assigning metadata

Best Practice–Developing content plans for each project deliverable

An Information Plan for a new project or set of new projects describes the strategy you plan to use to meet the needs of your customers. As a high-level analysis of the project requirements, the Information Plan identifies those who will need the information produced and defines the goals of the project in business terms. It also contains your initial estimate of the scope of the projects and the project costs in people, tools, and external contracts for activities such as localization, translation, printing, and delivery. As such, the Information Plan defines the business of the project.

The Content Plan defines the information you intend to plan, produce, and deliver to customers. At one level, your Content Plan describes the final deliverables of the project and identifies the particular audiences or job roles for which the deliverables are planned. For example, the content plan for a typical information-development project may include such final deliverables as printed installation instructions, an online help system embedded in the user interface of the product, a user guide provided as a PDF on a website or a CD, and an administrative guide available on the website in HTML format.

The Content Plan defines each deliverable by accounting for the content to be included. In a book-based Information Model, the content might be defined using multiple tables of contents, one for each deliverable. In a topic-based Information Model, the content might be defined as sets of stand-alone topics hierarchically arranged in tables of content or presented as a network of hypertext links on a website or a help system. In each case, the content itself is defined in the context of how it will be organized for use.

At the core of the Content Plan is the subject matter that the information developers will produce for the product or information release to the users. In many cases, a scheduled product release defines the information to be released since the information supports the use of the new or redesigned product features and functions. In some cases, especially when you release information to users independent of a product, you define the specific information that will be made available in the next delivery, whether that delivery is one new or revised topic or many.

If you are planning the content for an individual project or set of projects, you begin with your organization's Information Model. The Information Model provides you with the available selections for deliverables and provides standards for their design and packaging. For example, a well-defined Information Model describes the standard content organization for information that supports specific job roles. Your model should define the general sequence to be followed for all installation, administrative, troubleshooting, or other manuals. By following the standard organization of the content, you ensure that your information will be consistent with the organizational standards and be familiar to the user community.

The standard organization for an installation guide might define components like those in the illustration in Figure 17-1.

1   Overview of the installation process

2   Required equipment and other prerequisites

3   Installation procedures

4   Validation and verification of the installation

5   Troubleshooting the installation

Although the design standard dictates the larger structures for the parts of the installation manual, it does not account for the specific topics that your project teams need to develop. The process of developing the topics they intend to write requires that the project team review all product specifications and other information available to them and through meetings with product managers and development project leads. The process differs, of course, depending on whether your team is working on a new release of an existing product or process or the release of a new product or process.

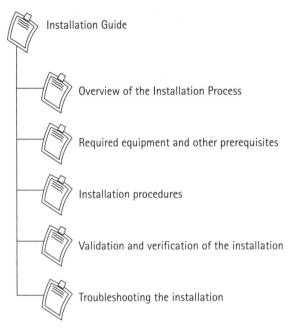

Installation Guide

Overview of the Installation Process

Required equipment and other prerequisites

Installation procedures

Validation and verification of the installation

Troubleshooting the installation

Figure 17-1: Standard organizational design for an installation guide

Begin with user scenarios and use cases

To develop the Content Plan for your project, the best place to begin is by understanding the user scenarios and use cases that the product or service supports. In Chapter 7: Developing User Scenarios, you find a complete discussion of how to identify user scenarios, based on your customer research and the information available to you from product marketing or other internal resources. If your marketing organization and the product developers create use cases to guide product development, you have a good resource. Use cases should describe how a new product feature or function is intended to be used. A user scenario describes the users' situation of use in terms that extend an otherwise development-oriented use case.

Developing Use Cases

In a project involving a small business phone system, Comtech worked closely with product marketing to understand the user scenarios involved with administering the system. Marketing explained that the system was designed to be administered by a person in the organization who probably had little or minimal experience doing administration tasks. Most small businesses identified as potential buyers of this new product had no one dedicated to handling the phone system. In fact, any previous system they might have had required a technician from the manufacturer to change settings. The new system was being developed so that the administrative tasks could be done by an inhouse employee of the business.

continued

Developing Use Cases *continued*

The use cases described how the inhouse administrator would use the phone interface to set up the functions for all the phones in the company and apply individual settings to particular phones. By typing codes into the phone using the keypad, the administrator could, for example, keep outsiders from using the phone in the lobby to make long-distance calls.

A typical user scenario would read as follows:

Janice wants to ensure that no one visiting the company uses the phone in the lobby to place long-distance calls. The company has had problems with long-distance charges on that phone. However, she does want to permit the visitors to use the phone for local calls and internal calls to people they are visiting. Janice consults the administrator's guide to find the instructions for restricting long-distance calling on a particular phone. She notices that she has to log in first and type a sequence of numbers and letters. It doesn't look easy. She'd like to practice before she tries a real change.

This scenario guided the development of the administrator's guide, which included a brief practice exercise or tutorial so that a new administrator would feel comfortable using the keypad to make a change. The tutorial also explained how to log in and where to find the administrator's login and password. The guide itself was organized into two major sections: the first provided step-by-step procedures to make changes to individual phones since these tasks were most likely to be performed by inhouse administrators. The second provided step-by-step instructions for making changes to all the phones at once, less likely after the installers developed the initial configuration and much more risky for an inhouse person.

As you can see, the scenarios guided the development of the information. The analysis resulted in a topic-oriented approach with very simple step-by-step instructions and minimal conceptual information. To further develop the content plan for the project required understanding the various functions that the administrator and the end users could control for themselves.

Develop an annotated task list

In most projects of this kind, you can next focus your analysis on the tasks that users will want to perform. The initial task list is derived from the user scenarios and use cases that are guiding the product development. As you and your team members research the project specifications and other information from marketing and development, you begin to assemble a list of the tasks that you will include in the product information. Tasks support the majority of users of most products. They simply want to get work done as quickly as possible and with minimal learning about the product. They are strictly "how do I ..." users. The topic-oriented list of tasks for the small phone administrators appears in Figure 17-2.

Topic Title	Info Type
Turning alarms ON and OFF	Task
Answering a ringing line with the hold button	Task
Changing the 'automatic connection to a ringing line'	Task
Turning call forwarding ON and OFF	Task
Changing 'call forwarding to an outside line'	Task
Changing 'ring telephone once when call forwarding'	Task
Choosing call waiting reminder tones	Task
Clearing personal station programming	Task
Changing the conference call capability	Task
Turning dial '9' access ON and OFF	Task
Allowing do not disturb	Task
Barging in on calls	Task
Changing the group listen	Task
Allowing the use of the DISCONNECT button with the headset	Task
Allowing the use of a DSS button as a hold button	Task
Allowing transfer hold using DSS button	Task
Putting a line on hold when selecting another line	Task
Allowing a program key	Task
Allowing extended key programming	Task
Changing the 'hands-free voice announce' capability	Task
Changing the hands-free dialing capability	Task
Changing the 'on-hold monitoring' capability	Task
Changing the full speakerphone capability	Task
Transferring calls without screening	Task
Designating night answer	Task
Designating a system manager master station	Task
Removing call waiting tones and preventing executive override	Task
Turning 'common audible ringing' ON and OFF	Task
Change the line-to-line connection	Task
Allowing system-wide music on hold	Task
Changing the night restrictions setting	Task
Setting the hour and minute for night restrictions to begin and end	Task
Setting an 'around-the-clock restriction'	Task
Changing the attendant night service key	Task
Setting the system speed dial codes	Task
System time and date	Task
Changing the duration for line-to-line forwarded calls	Task
Changing the call pickup timing	Task
Changing the first hold recall time	Task
Changing the second hold recall time	Task
Changing the line-to-line connection timer	Task
Changing the transfer return call time	Task
Changing the outpulse tone duration	Task
Modifying the day and night restrictions	Task
Changing the day restrictions	Task
Changing the night restrictions	Task
Programming system-wide account codes	Task
Using the activation button	Task
Programming frequently used account codes	Task
Programming account codes with speed dialing	Task
Reading BCD Values	Task

Figure 17-2: List of administrator tasks for the small business phone system

In a traditional book-based Information Model, individual information developers typically develop their own tables of contents for the books they are assigned to write. Each table of contents or topic list is independent of the other tables of content or topic lists, resulting in duplication of effort unless care is taken to coordinate the topics. In a topic-based Information Model, the topic lists are created by the entire project team, often in collaboration with other project teams at the management level. A single topic list for a project helps to ensure that topics are written once and used in their appropriate contexts. However, developing a single list requires a collaborative team in which the outcome of preliminary research into the project is shared among team members and the team decides what information is needed to support the user scenarios.

In most development organizations, the list of tasks evolves as the project develops and then changes. However, the initial task list provides you with an opportunity to estimate the scope of the project and assign resources.

One of the most important issues in developing your task topic list is developing a set of user-focused titles for the tasks. Too often, your first exposure to the tasks comes from the product specifications, rather than from well-defined user scenarios. Often, the names of the product functions are too technical or are specific to the product developers rather than the names that are easily identifiable to the users. You have to work carefully to develop titles that are meaningful so that users can match their goals to the titles in the documents and find the procedures that help them achieve their goals quickly. If you are working with legacy content, you may have to review and revise task titles so that they are more user-oriented.

Add conceptual and reference topics

After you have outlined the tasks and given them user-oriented titles, you need to consider what conceptual or descriptive information your users need to perform the tasks successfully. In many organizations, conceptual information, particularly descriptions of product functions, are copies from marketing reports and product specifications, not because they help users reach their goals but because the information is available. Much of the information included in user content is "nice to know" but not needed. If you cannot clearly link a description of a function or a basic concept to task performance, consider if it is needed in the information collection at all. If you are following a minimalist agenda in your content planning, you will minimize the amount of information that you expect users to read and understand before they get to the tasks they want to do. Unless your users are very experienced in their task domain and willing to spend time becoming expert in your product, they are likely not interested in concepts and descriptions unless they absolutely need them to succeed.

In traditional book-based projects, concepts are often duplicated among the books being developed. In many cases, the concepts are written differently by authors who are unaware that colleagues are writing the same information. Sometimes the conceptual information is treated differently, depending on the needs of the audience. However, the differences can often be incorporated into a single master concept rather than maintaining the information in multiple sources. In many book-based project environments, not only are concepts duplicated, but the product may never be completely explained because the conceptual information is developed independently by many authors rather than being treated as a whole.

In a topic-based environment, the collaborative project team should define the conceptual information that provides a good introduction to the product's functionality and features and support the expert users who want to understand more thoroughly how the product is supposed to work or users who need to understand basic functionality to perform tasks correctly. If the conceptual information is designed by the collaborative team, then individual concept topics can be used in collections as needed to support task performance.

Descriptions of product features and functions may be useful to users but only if they are directly related to tasks. In many cases, the functional descriptions can be provided as brief context statements in the task itself rather than in pages of descriptive text isolated in introductions or introductory sections of chapters. Many tasks need no context at all. They are adequately described by the title of the task and a short description.

Some of the best conceptual information describes the way in which a product is architected, which may help expert users better understand how parts of their systems work together. Process descriptions that explain the interrelationships of system parts and the flow of information through a system are often genuinely useful to provide an overview of a set of tasks.

An ideal concept contains a standard set of content, beginning with a definition of the concept itself. For example, a beginning concept statement might read as follows:

Call log

A call log stores information about up to the last 10 missed, 10 received, or 10 dialed calls. It also adds the total duration of all calls. When the number of calls exceeds the maximum, the most recent call replaces the oldest.

This statement answers the question: "what is . . . ?" and, in the final sentence, explains how the call log functions.

Following the key definition, a concept may contain an extended definition, usually covering the parts of the concept, and a set of examples that explain to the user how the concept applies to the product he or she is using. Concepts may also contain non-examples, accounts of what the concept is not.

The following paragraph explains what a timed profile is and provides examples of its use:

Timed profile

Timed profiles can be used to present missed calls. For example, suppose you attend an event that requires your phone be set to Silent before the events starts, but you forget to return it to Normal until long after the event. During this time, you have missed several calls because the ringing tone was silent. A time profile can prevent this by automatically returning your phone to the default profile at the time you specify.

Conceptual information like this is best presented immediately in the context of the text and before the procedure.

Reference information, unlike much descriptive and conceptual, is often essential to task performance. For example, a technician needs a table that lists and describes the environmental requirements of a site before installation can begin. A planner needs a list of parameters about a product to use when deciding on a configuration. An end user may want to know what the buttons on the remote control are used for. A programmer may want a list of commands and examples of their use.

When reference information is embedded in a concept or a task, it is often more difficult for the user to find and the author to update. By separating reference information and maintaining it independent of other content, you increase its accessibility and make it easier to change.

Your information architect may want to define additional information types with templates that facilitate the authors. I created a policy information type for the American Red Cross so that policies would be consistently framed and contain information expected by the reader. A telecommunications company created an information type that enforced for authors and readers a standard presentation of a system alarm and its resolution. The semiconductor industry developed a standard information type for its technical datasheets, describing the functionality of a chip. A learning organization developed a standard for the tests given at the end of a technical course. Each of these information types encompassed key reference information needed by the user community.

Once you have identified the required conceptual and reference information, add it to your topic list, as illustrated in Figure 17-3. You can create one long list or different lists for different information types. If you use a single list, indicate which topics are tasks, concepts, or reference.

Topic Title	Info Type
Introduction	Concept
What features can I change?	Concept
Understanding this guide's conventions	Reference
Equipment	Reference
Using the telephone to modify features	Concept
Before you begin modifying features	Reference
Understanding the features section	Concept
Using the ON-OFF lamps	Concept
Changing the feature values	Concept
Using the keys, buttons, and feedback tones	Concept
Using the pound and star keys	Reference
Using the dialpad keys and buttons next to the lamps	Reference
Understanding the feedback tones	Reference
Modifying telephone and system features	Concept
Tutorial for modifying a telephone feature	Tutorial
Logging on	Tutorial
Targeting a station	Tutorial
Identifying the feature	Tutorial
Changing the lamp settings	Tutorial
Exiting the management code	Tutorial
Logging off	Tutorial
Tutorial for modifying a system feature	Tutorial
Logging on	Tutorial
Identifying the feature	Tutorial
Setting a multiplier	Tutorial
Changing the feature value	Tutorial
Exiting the management code	Tutorial
Logging off	Tutorial
Overview of telephone features	Concept
Telephone features	Reference
Turning alarms ON and OFF	Task
Answering a ringing line with the hold button	Task
Changing the 'automatic connection to a ringing line'	Task
Turning call forwarding ON and OFF	Task
Changing 'call forwarding to an outside line'	Task
Changing 'ring telephone once when call forwarding'	Task
Choosing call waiting reminder tones	Task
Clearing personal station programming	Task
Changing the conference call capability	Task
Turning dial '9' access ON and OFF	Task
Allowing do not disturb	Task
Barging in on calls	Task
Changing the group listen	Task
Allowing the use of the DISCONNECT button with the headset	Task
Allowing the use of a DSS button as a hold button	Task
Allowing transfer hold using DSS button	Task
Putting a line on hold when selecting another line	Task
Allowing a program key	Task
Allowing extended key programming	Task
Changing the 'hands-free voice announce' capability	Task
Changing the hands-free dialing capability	Task

Topic Title	Info Type
Changing the 'on-hold monitoring' capability	Task
Changing the full speakerphone capability	Task
Transferring calls without screening	Task
Designating night answer	Task
Designating a system manager master station	Task
Removing call waiting tones and preventing executive override	Task
Overview of system features	Concept
System features	Reference
Turning 'common audible ringing' ON and OFF	Task
Dropping out of a conference call	Concept
Change the line-to-line connection	Task
Allowing system-wide music on hold	Task
Night restrictions	Concept
Night restrictions table	Reference
Changing the night restrictions setting	Task
Setting the hour and minute for night restrictions to begin and end	Task
Around the Clock restrictions	Concept
Around the Clock restrictions table	Reference
Setting an 'around-the-clock' restriction	Task
Changing the attendant night service key	Task
Setting the system speed dial codes	Task
System time and date	Task
System timing	Concept
Changing the duration for line-to-line forwarded calls	Task
Changing the call pickup timing	Task
Changing the first hold recall time	Task
Changing the second hold recall time	Task
Changing the line-to-line connection timer	Task
Changing the transfer return call time	Task
Changing the outpulse tone duration	Task
Overview of telephone and line restrictions	Concept
The restriction table	Reference
Restriction levels	Reference
Modifying day and night restrictions for telephones and lines	Concept
About modifying telephones and lines	Concept
Modifying the day and night restrictions	Task
Day and night restrictions section	Reference
Restrictions features	Concept
Changing the day restrictions	Task
Changing the night restrictions	Task
Overview of SMDR	Concept
Call record	Concept
Field descriptions on call record	Reference
Date record	Concept
Using account codes for SMDR	Concept
Programming system-wide account codes	Task
Using the activation button	Task
Programming frequently used account codes	Task
Programming account codes with speed dialing	Task
Reading BCD Values	Task
Station Characteristics Form	Reference

Topic Title	Info Type
System Characteristics Form	Reference
Automatic Night Restrictions	Reference
System Timing	Reference
Day and night restrictions	Reference
SMDR Account Codes	Reference
System Speed Dial Codes	Reference
System Speed Dial Buttons	Reference
Glossary	Reference

Figure 17-3: Concepts and reference added to the topic list

Add annotations to your topic list

For a new product or service, all the topics in your topic list will be new. For updates to an existing product or service, you will have a combination of new and existing topics. Annotate your topic list, indicating which topics are new and which already exist. For the existing topics, indicate which topics require modifications for the next release of the product or service. If necessary, you may want to indicate which topics require major and which require minor modifications.

If your organization is creating standalone topics stored in a file or a content management system, indicate in your annotations the topic file name and the person responsible for creating and modifying the topic. The information you provide in the annotated topic list will be important to identify the project scope and estimate its resource needs.

Figure 17-4 shows an annotated topic list that contains topics for the job role of administrator.

The complete list of topics for your project marks the essential content to be developed and constitutes the core of your project Content Plan. It also associates the content with particular job roles and ensures that content is not duplicated across job roles or product or service types. For example, you may have a task, such as logging on the system, that is performed by everyone from end users to administrators and maintenance technicians. That task is listed only once in your annotated topic list. You may have different logon procedures for different versions of your system. You may want to create one topic that encompasses the differences, or you may want to create multiple topics if the differences in the procedures are sufficient to justify the near duplication. Deciding when to create a "master" topic containing all variations and when to create independent "sibling" topics is a decision for your information architect.

Your annotated topic list provides you with a good start on your Content Plan but the list alone is not sufficient. You need to determine how the topics will be used in the final deliverables of your project.

Master Annotated Topic List

Group	Topic Title	File Name	8/21/06	Topic Owner	Modify?				
					New	Major	Minor	None	Unused
Getting Started	Introduction	IntroductionConcept.dita	Concept	Chris		X			
Getting Started	What features can I change?	FeatureChangeConcept.dita	Concept	Chris		X			
Getting Started	Understanding this guide's conventions.	ConventionsReference.dita	Reference	Chris			X		
Getting Started	Equipment	EquipmentReference.dita	Reference	Chris			X		
Getting Started	Using the telephone to modify features	ModifyFeaturesConcept.dita	Concept	Chris			X		
Getting Started	Before you begin modifying features	BeforeModifyingReference.dita	Reference	Chris					
Getting Started	Understanding the features section	FeaturesUnderstandingConcept.dita	Concept	Sue		X			
Getting Started	Using the ON-OFF lamps	OnOffLampsConcept.dita	Concept	Sue		X			
Getting Started	Changing the feature values	FeatureValuesConcept.dita	Concept	Sue		X			
Getting Started	Using the keys, buttons, and feedback tones	KeysButtonsTonesConcept.dita	Concept	Sue					
Getting Started	Using the pound and star keys	PoundStarKeysReference.dita	Reference	Ann			X		
Getting Started	Using the dialpad keys and buttons next to the lamps	DialpadKeysReference.dita	Reference	Ann			X		
Getting Started	Understanding the feedback tones	FeedbackTonesReference.dita	Reference	Ann			X		
Getting Started	Modifying telephone and system features	ModifyPhoneSystemFeaturesConcept.dita	Concept	Ann					
Getting Started	Tutorial for modifying a telephone feature	ModifyPhoneFeatureTutorial.dita	Tutorial	Ann		X			
Getting Started	Logging on	LoggingOnTutorial.dita	Tutorial	Bill		X			
Getting Started	Targeting a station	TargetingStationTutorial.dita	Tutorial	Bill			X		
Getting Started	Identifying the feature	IdentifyPhoneFeatureTutorial.dita	Tutorial	Bill		X			
Getting Started	Changing the lamp settings	ChangeLampSettingsTutorial.dita	Tutorial	Bill		X			
Getting Started	Exiting the management code	ExitManagementCodeTutorial	Tutorial	Bill		X			
Getting Started	Logging off	LoggingOffTutorial.dita	Tutorial	Bill					
Getting Started	Tutorial for modifying a system feature	ModifySystemFeatureTutorial.dita	Tutorial	Joe	X				
Getting Started	Logging on	same as above	Tutorial	Joe					
Getting Started	Identifying the feature	IdentifyingSystemFeatureTutorial.dita	Tutorial	Joe			X		
Getting Started	Setting a multiplier	SettingMultiplierTutorial.dita	Tutorial	Joe			X		
Getting Started	Changing the feature value	ChangeFeatureValue.dita	Tutorial	Joe			X		
Getting Started	Exiting the management code	same as above	Tutorial	Joe		X			
Getting Started	Logging off	same as above	Tutorial	Joe					
Telephone Features	Overview of telephone features	PhoneFeaturesOverviewConcept.dita	Concept	Jane	X				
Telephone Features	Telephone features	PhoneFeaturesReference.dita	Reference	Bob	X				
Telephone Features	Turning alarms ON and OFF	AlarmsOnOffTask.dita	Task	Jane				X	
Telephone Features	Answering a ringing line with the hold button	AnswerRingingLineTask.dita	Task	Jane				X	
Telephone Features	Changing the 'automatic' connection to a ringing line'	ChangeAutoConnectionTask.dita	Task	Jane	X				
Telephone Features	Turning call forwarding ON and OFF	ChangeCallForwardOnOffTask.dita	Task	Jane	X				
Telephone Features	Changing 'call forwarding to an outside line'	ChangeCallForwardOutsideLineTask.dita	Task	Bob	X				
Telephone Features	Changing 'ring telephone once when call forwarding'	ChangeCallForwardRingTask.dita	Task	Jane				X	
Telephone Features	Choosing call waiting reminder tones	ChooseCallWaitReminderTonesTask.dita	Task	Jane				X	
Telephone Features	Clearing personal station programming	ClearPersonalStationTask.dita	Task	Bob				X	
Telephone Features	Changing the conference call capability	ChangeConCallTask.dita	Task	Bob				X	
Telephone Features	Turning dial '9' access ON and OFF	Dial9OnOffTask.dita	Task	Bob				X	
Telephone Features	Allowing do not disturb	AllowDoNotDisturbTask.dita	Task	Harry				X	
Telephone Features	Barging in on calls	BargingInTask.dita	Task	John				X	
Telephone Features	Changing the group listen	ChangeGroupListenTask.dita	Task	John				X	
Telephone Features	Allowing the use of the DISCONNECT button with the headset	AllowDisconnectHeadsetButtonTask.dita	Task	Harry				X	
Telephone Features	Allowing the use of a DSS button as a hold button	AllowDSSButtonHoldTask.dita	Task	Harry				X	
Telephone Features	Allowing transfer hold using DSS button	AllowTransferHoldTask.dita	Task	Harry				X	
Telephone Features	Putting a line on hold when selecting another line	AllowLineonHoldTask.dita	Task	Harry				X	
Telephone Features	Allowing a program key	AllowProgramKeyTask.dita	Task	Harry				X	
Telephone Features	Allowing extended key programming	AllowExtendedProgramKeyTask.dita	Task	Jane				X	
Telephone Features	Changing the 'hands-free voice announce' capability	ChangeHandsFreeAnnounceTask.dita	Task	Jane			X		
Telephone Features	Changing the hands-free dialing capability	ChangeHandsFreeDialTask.dita	Task	Bob				X	
Telephone Features	Changing the 'on-hold monitoring' capability	ChangeOnHoldMonitorTask.dita	Task	Jane				X	
Telephone Features	Changing the full speakerphone capability	ChangeFullSpeakerPhoneTask.dita	Task	Harry				X	
Telephone Features	Transferring calls without screening	TransferCallsNoScreeningTask.dita	Task	Harry				X	

Group	Topic Title	File Name	8/21/06	Topic Owner	Modify? New	Major	Minor	None	Unused
Telephone Features	Designating night answer	SetNightAnswerTask.dita	Task	John					X
Telephone Features	Designating a system manager master station	SetMasterStationTask.dita	Task	John			X		
Telephone Features	Removing call waiting tones and preventing executive override	RemoveCallWaitingTonesTask.dita	Task	John					X
System Features	Overview of system features	SystemFeatureOverviewConcept.dita	Concept	Eric	X				
System Features	System features	SystemFeaturesReference.dita	Reference	Eric			X		
System Features	Turning 'common audible ringing' ON and OFF	CommonRingingOnOffTask.dita	Task	Eric			X		
System Features	Dropping out of a conference call	DropOutConCallConcept.dita	Concept	Bob	X				
System Features	Change the line-to-line connection	ChangeLineConnectionTask.dita	Task	Bob	X				
System Features	Allowing system-wide music on hold	AllowMusicHoldTask.dita	Task	Bob	X				
System Features	Night restrictions	NightRestrictionsConcept.dita	Concept	Bob	X				
System Features	Night restrictions table	NightRestrictionsReference.dita	Reference	Bob		X			
System Features	Changing the night restrictions setting	NightRestrictionSettingTask.dita	Task	Bob	X				
System Features	Setting the hour and minute for night restrictions to begin and end	SetNightRestrictionBeginEndTask.dita	Task	Bob		X			
System Features	Around the Clock restrictions	RoundClockConcept	Concept	Bob					
System Features	Around the Clock restrictions table	RoundClockReference.dita	Reference	Ann					X
System Features	Setting an 'around-the-clock restriction	SetRoundtheClockTask.dita	Task	Ann		X			
System Features	Changing the attendant night service key	ChangeNightServiceKeyTask.dita	Task	Chris		X			
System Features	Setting the system speed dial codes	SetSysSpeedDialCodesTask.dita	Task	Chris	X				
System Features	System time and date	SetSystemTimeDateTask.dita	Task	Chris					
System Features	System timing	SysTimingConcept.dita	Concept	Chris					X
System Features	Changing the duration for line-to-line forwarded calls	ChangeDurationForwardCallsTask.dita	Task	Chris	X				X
System Features	Changing the call pickup timing	ChangeCallPickupTimingTask.dita	Task	Sue					
System Features	Changing the first hold recall time	ChangeFirstHoldRecallTimeTask.dita	Task	Sue	X				
System Features	Changing the second hold recall time	ChangeSecondHoldRecallTimeTask.dita	Task	Sue		X	X		
System Features	Changing the line-to-line connection timer	ChangeLineConnectionTimerTask.dita	Task	Sue			X		
System Features	Changing the transfer return call time	ChangeTransferReturnCallTimeTask.dita	Task	Sue			X		
System Features	Changing the outpulse tone duration	ChangeOutpulseDurationTask.dita	Task	Sue			X		
Restrictions	Overview of telephone and line restrictions	RestrictionsOverviewConcept.dita	Concept	David	X				
Restrictions	The restriction table	RestrictionTableReference.dita	Reference	James	X	X			
Restrictions	Restriction levels	RestrictionLevelsReference.dita	Reference	James	X				
Restrictions	Modifying day and night restrictions for telephones and lines	DayNightRestrictionsConcept.dita	Concept	James	X				
Restrictions	About modifying telephones and lines	AboutModifyPhonesLinesConcept.dita	Concept	James	X	X			
Restrictions	Modifying the day and night restrictions	ModifyingDayNightRestrictionsTask.dita	Task	James	X				
Restrictions	Day and night restrictions section	RestrictionsFeatureSectionReference.dita	Reference	James	X				
Restrictions	Restrictions features	RestrictionsConcept.dita	Concept	David	X				
Restrictions	Changing the day restrictions	ChangeDayRestrictionTask.dita	Task	David	X	X			
Restrictions	Changing the night restrictions	ChangeNightRestrictionTask.dita	Task	Sally	X				
Station Message Detail Recording	Overview of SMDR	SMDROverviewConcept.dita	Concept	Ryan					
Station Message Detail Recording	Call record	CallRecordConcept.dita	Concept	Ryan				X	
Station Message Detail Recording	Field descriptions on call record	CallRecordReference.dita	Reference	Ryan					
Station Message Detail Recording	Date record	DateRecordReference.dita	Reference	Ryan					
Station Message Detail Recording	Using account codes for SMDR	UseAccountCodesSMDRConcept.dita	Concept	Ryan					
Station Message Detail Recording	Programming system-wide account codes	ProgramSysAccountCodesTask.dita	Task	Ryan		X			
Station Message Detail Recording	Using the activation button	UseActivationButtonTask.dita	Task	Ryan				X	
Station Message Detail Recording	Programming frequently used account codes	ProgramFreqAccountCodesTask.dita	Task	Ryan				X	
Station Message Detail Recording	Programming account codes with speed dialing	ProgramSpeedDialAccountCodesTask.dita	Task	Ryan				X	
Appendix	Reading BCD Values	ReadBCDValuesTask.dita	Task	Ryan			X		
Forms	Station Characteristics Form	StationCharFormReference.dita	Reference	David			X		
Forms	System Characteristics Form	SysCharFormReference.dita	Reference	David			X		
Forms	Automatic Night Restrictions	AutNightRestrictionsReference.dita	Reference	David			X		
Forms	System Timing	SysTimingReference.dita	Reference	David			X		
Forms	Day and night restrictions	DayNightRestrictionsReference.dita	Reference	David			X		
Forms	SMDR Account Codes	SMDRAcctCodesReference.dita	Reference	David			X		
Forms	System Speed Dial Codes	SysSpeedDialCodesReference.dita	Reference	David			X		
Forms	System Speed Dial Buttons	SysSpeedDialButtonsReference.dita	Reference	David			X		
Glossary	Glossary	GlossaryReference.dita	Reference	David			X		

Figure 17-4: Annotated topic list for the system administrator guide

Best Practice—Mapping hierarchies and creating related-topic links

In a traditional book-based environment, authors often develop the tables of contents for their assigned book independently, although they may follow standards for the organization of certain commonly developed books. In a topic-based environment, the team determines which topics belong in which collections, as appropriate for the job roles and product variations they support. Topics are assigned to one or more deliverables, including print, PDF, help, web, or other.

If you are producing traditional books, you may still use a topic architecture and assign topics to chapters and sections in your tables of contents. Only when you write the topics, however, do you decide how the topics are related to one another. You may have some topics that appear at the first order of headings in a chapter and other topics that appear as second or third order headings. As a result, you cannot reuse a topic in a chapter in another book without changing its heading level.

If you produce topics by mapping them into collections, you can establish the hierarchy of the topics through the map functions. DITA maps, supported by the Darwin Information Typing Architecture, enable you to arrange your topics at first, second, third, or other levels in the hierarchy without resetting the heading level in the topic itself. The heading levels are determined when a style sheet is applied to your topics to create final deliverables. A same topic may be at the first level of heading in a print or PDF output but at a second or third level of heading in a website or a help system.

Figure 17-5 shows the hierarchical arrangement of topics for the telephone system administrator's guide as a print or PDF output. Note that, in most cases, the concept topics come before the tasks.

The hierarchical arrangement of topics provides a linear sequence if you are producing a complete book, as you would for printing or developing PDFs. The same hierarchical arrangement, however, can be used to develop links between topics for a website or a help system. In a simple hierarchy, you establish parent and child topics by virtue of the hierarchy itself. If you have a presentation that allows hypertext links, the parent topic contains a link to the child topics, and the child topics link back to the parent topic. You can expand upon this simple hierarchy by using the DITA function called a relationship table.

Develop a relationship table

If you plan to deliver a print publication or a PDF of that print publication, you will not need a relationship table. A relationship table allows you to establish links for online delivery in a help system or an HTML-based website of technical content. A relationship table allows you to link topics in ways that go beyond the basic parent/child relationships that you establish in your hierarchical presentation in your map.

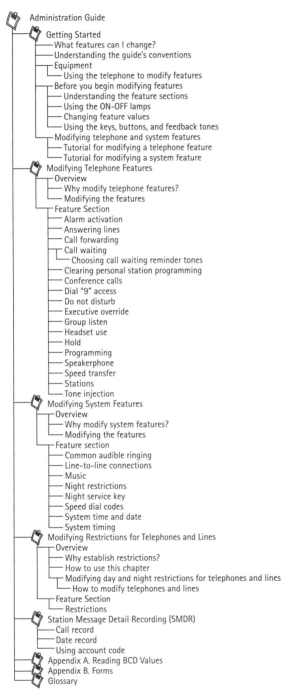

Administration Guide
    Getting Started
        What features can I change?
        Understanding the guide's conventions
        Equipment
            Using the telephone to modify features
        Before you begin modifying features
            Understanding the feature sections
            Using the ON-OFF lamps
            Changing feature values
            Using the keys, buttons, and feedback tones
        Modifying telephone and system features
            Tutorial for modifying a telephone feature
            Tutorial for modifying a system feature
    Modifying Telephone Features
        Overview
            Why modify telephone features?
            Modifying the features
        Feature Section
            Alarm activation
            Answering lines
            Call forwarding
            Call waiting
                Choosing call waiting reminder tones
            Clearing personal station programming
            Conference calls
            Dial "9" access
            Do not disturb
            Executive override
            Group listen
            Headset use
            Hold
            Programming
            Speakerphone
            Speed transfer
            Stations
            Tone injection
    Modifying System Features
        Overview
            Why modify system features?
            Modifying the features
        Feature section
            Common audible ringing
            Line-to-line connections
            Music
            Night restrictions
            Night service key
            Speed dial codes
            System time and date
            System timing
    Modifying Restrictions for Telephones and Lines
        Overview
            Why establish restrictions?
            How to use this chapter
            Modifying day and night restrictions for telephones and lines
                How to modify telephones and lines
        Feature Section
            Restrictions
    Station Message Detail Recording (SMDR)
        Call record
        Date record
        Using account code
    Appendix A. Reading BCD Values
    Appendix B. Forms
    Glossary

Figure 17-5: Administrator's guide hierarchical mapping

Take, for example, the instructions for logging onto the telephone system. At the beginning of every task, the system administrator must log on. However, the steps to log on are identical every time they are used. In the printed System Manager's Guide, we chose to print the logon instructions on the inner back cover of the book for easy access. The logon instructions are also presented in the "Tutorial for modifying a telephone feature," presented in the "Getting Started" chapter. This tutorial provides a basic instruction and practice for the system manager to get started with a simple "modifying" task. The tutorial references the instructions in the back inside cover for future reference. Each time a task is presented, it begins with "Log on" but provides no additional information.

In an electronic version of the System Manager's Guide, we would choose to link from each task to the logon instructions. In a help system, this basic instruction might appear as an expansion link presented in the context of each task for someone who is new or does not remember how to logon. Figure 17-6 shows how the logon instructions might appear as an expansion link.

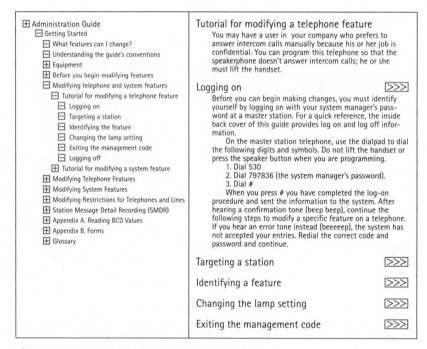

Figure 17-6: Logon instructions as an expansion link

To create such a link, you can develop a relationship table that links all tasks to the single logon instruction. You list all the tasks in the task column of the three-column relationship table and list the logon instruction as a reference item in the reference column. Nothing appears in the concept column. Table 17-1 shows the resulting row of a relationship table.

Table 17-1: A Relationship Table Showing a One-Way Link from Tasks to the Logon Reference

Concept	Task	Reference
	Turning alarms ON and OFF Answering a ringing line with the hold button Changing the "automatic connection to a ringing line" Turning call forwarding ON and OFF Changing "call forwarding to an outside line" Changing "ring telephone once when call forwarding" Choosing call waiting reminder tones Clearing personal station programming Changing the conference call capability (and so on)	Logging onto the phone

In the next relationship table, two tasks refer to the three reference tables. To establish the appropriate links, you might set up a relationship table as follows in Table 17-2.

Table 17-2: A Relationship Table Showing Two-Way Links between the Restriction Tasks and the Relevant Tables

Concept	Task	Reference
	Changing the day restriction Changing the night restriction	DKTS Plus equipment only–restriction table PBX/Centrex restriction table Lamp settings to establish subitems and restrict levels

Use relationship tables when the relationships among your concept, task, and reference topics are complex. By setting up relationship tables, you establish a mechanism to easily change relationships without having to modify links in individual topics. Maintaining links is much simpler.

Relationship tables are a DITA feature and are created in the DITA map.

Best Practice–Developing indexes and assigning metadata

Indexing and assigning metadata are integral steps in the development of your final deliverables. They should be planned as part of your project and implemented as your information developers create individual topics. Indexes support print or PDF publications, as well as the electronic indexes included in help systems. They are designed to enable users to find information at a more detailed level than is typically available in a table of contents.

Metadata allows you to categorize your topics to match the way your users might be searching for information on a website or in a help system. You may categorize your topics by their information type (task, concept, reference, or other), by the audience for whom they are intended, by the products that they support, or by many other categories.

Plan and develop an index

To plan and develop an index, begin with an analysis of the user scenarios. By profiling your potential users for the information, you can better understand how they are likely to look for information in an index. An index should contain terms that users are likely to look up, which means that most primary (first-level) index items will be nouns. Actions that support the tasks that are included among your topics become secondary (second-level) index items. For example, in the System Manager's Guide, you might index "call forwarding" as a primary index item but introduce the task of "turning on and off call forwarding" as a secondary index item.

It is best to index a representative sample of the topics for a particular user group and let that sample guide the rest of the indexing. By establishing a model for the index items, you can ask the information developers to index each of the topics they write according to the model as the topics are completed. Although you may have to return to the index during a final review of the assembled deliverables, you will have a good start. You may have to create additional index transformation when you review the draft index. For example, you may have an index item that refers to all the forms in the deliverable, such as the SMDR form as a secondary index item under the primary "forms" index item. You may also want to index the SMDR form as a secondary index item under account codes, since this form is used to establish account codes. The two index entries may look as follows:

> account code
>
>> SMDR form
>
> forms
>
>> SMDR account codes

For a library of deliverables intended for multiple users, you may find an advantage in establishing a master index. A master index establishes the basic terms with which individual topics will be indexed. You may decide, for example, that all forms should be indexed according to their names as primary index items and listed as secondary index items under a "forms" primary item. You may decide that index items should include technical terms for the administrators and popular terms for the end users, even though the terms refer to the same features. For example, you may use SMDR as an index item for administrators because you believe that the abbreviation is commonly known among experienced users, especially those users who will be involved with setting up Station Message Detail Recording (SMDR).

A master index helps to ensure that information in all the deliverables involving a particular technology are indexed similarly. Users who need to access more than one deliverable can thus depend upon the indexes to be consistent.

Plan and add metadata

By adding metadata to your topics, you enhance the ability of the users to find information through searches online. In the small business phone system, for example, Comtech wanted to assign job roles to each topic, defining the topics as appropriate for end users of the telephone system or telephone system administrators or both. In the audience category of metadata, we assigned values for telephone user and system administrator and provided for either one or both to be selected by the information developer.

You might also decide that you need metadata that accounts for different telephone system products or multiple releases of the same product lines. You might want metadata that distinguishes basic tasks from advanced or more difficult tasks using a metadata attribute called "importance." Likewise, you may want to allow users to search only for tasks or only for reference information among the topics, reducing the number of topics returned.

User scenarios are once again the source for your decision making about metadata. Consider how your various user communities may want to sort the information in your topics. Typical metadata categories include audience, product, release, version, and platform. You may need additional categories, depending on your user community and the nature of your information.

Your information developers must assign metadata as they create their topics. However, much of the metadata you will want to include may be automatically inserted by your content management system or through default settings in your editor. It is important, however, to review the metadata being established to ensure that it is accurate. Incorrectly entered metadata will return topics on searches that are not relevant to the users' needs.

For more information on determining metadata requirements, see *Content Management for Dynamic Web Delivery* and *The Content Management Bible*, listed in the bibliography.

Summary

Developing a comprehensive content plan and annotated topic list requires that you use the standards established in your corporate or departmental Information Model. The Information Model defines the information types and their allowed content units and provides standard outlines for classes of deliverables, including user and administrative guides, programming manuals, command reference manuals, and any other collection of topics that your users require. Your Information Model also defines the metadata you will use to categorize the topics your team develops and provides a master list of index terms used across the information libraries.

With an understanding of the standards established in the Information Model, you are ready to embody those standards in your project plan. Your Information Plan sets the overall deliverables for your project, including all the print pieces and PDFs you will create, as well as the content you will deliver to websites, help systems, and other deliverables. You must apply the standard set to your individual project. The Information Model prescribes the information types and their allowed content for your organization. You must decide which information types apply to your project and how you will apply them to the content your team creates.

As your project takes shape, your information developers research the use cases that describe how users of all types will take advantage of the product or service functionality that your larger organization is developing. Based on the product use cases, they define user scenarios that lead to a comprehensive list of tasks that users need to perform to reach their goals. Based on the task list, your team defines the concepts that users need to understand to perform successfully and the reference information they will need to look up. With this information in hand, your team develops an annotated topic list and a set of maps that show how the topics will fit into the final deliverables.

As topics are developed and the product or service evolves, the topic list is likely to change, sometimes daily or weekly. It becomes a working statement of your intended outcomes for the project.

As your information developers create new topics or modify existing topics, they add or modify the metadata that describes the topics and add index items that are used to produce indexes for print and online deliverables.

The lists and notes you develop as you plan the content for the project becomes the key source of data for the information project manager. The project manager uses the content plans and topic lists to estimate the resources required to complete the project by the deadline and produce the level of quality required by customers. The project manager also uses the evolving list of topics as an indicator of changes in the scope and length of the project. As the lists change, so must the estimates of resources and determinations of the ability of your team to meet the deadlines.

"Where is the project heading?
The tracks are getting ahead of us."

Chapter 18

Keeping Your Project on Track

> Time is the scarcest resource and unless it is managed nothing
> else can be managed.
>
> —Peter F. Drucker

In preparing your Project Plan, you completed the Planning
Phase of your project, at least from the perspective of project
management. Depending upon the complexity and uncertainty
of the Planning Phase, you have used about 10% of the total time
of the project for planning activities. Throughout the design and
Development Phases, you are likely to spend between 5% and
20% of total project time managing the information-development
activities. The percentage of project management time depends
upon how difficult the project will be to manage.

At the beginning of the Design and Development Phases, you
and your team members already have the baseline information
you need to proceed with design and development. The major
activities associated with design and development are handled
by the information architect and the information developers
assigned to the project.

The information architect is responsible for identifying the
topics that must be created to produce each of the major deliverables that are on your schedule. For example, the architect
knows what information is typical of your organization's user
and installation manuals. If this is a new project, the architect
determines what topics need to be created to complete the user
and installation manuals, usually based on an analysis of the
information requirements of end users and technicians. The

architect will also analyze the features and functions to be designed for the new or updated product. During the Design Phase, the architect develops a list of the topics needed to support the product customers, including the tasks to be performed, the concepts needed to help understand the tasks, and the reference information that enables the customer to accomplish the tasks correctly. The architect then associates the topics with each deliverable, producing tentative tables of contents. If you do not have a dedicated information architect, you might assign these duties to a senior information developer.

Although the information architect is primarily responsible for the topic list, in most projects, the architect works closely with the individual information developers. The information developers are usually closest to the product developers and adept at gathering information about the features and functions being developed. The information developers consult the project requirements and specifications and discuss the product with the engineers, developers, or other subject matter experts.

As the topic list is being planned by the information architect and the information developers, the architect brings to bear upon the decision making the information gathered about the customers' requirements:

✔ Does the customer need minimalist information to get started, possibly through quick reference material?

✔ Does the customer need to understand why he or she would want to use this product?

✔ Does the customer need comprehensive technical information about the product's architecture to be successful?

Knowledge of the customers and their specific needs guides the development of the topic list during the Design Phase of the project. For more discussion of the architect's role in the Design Process, consult Chapter 17: Implementing the Topic Architecture.

As the project moves from the Design to the Development Phase, the primary work is shifted to the information developers and specialists in graphics and illustration, multimedia production, and other information types that will be included in the final deliverables. The information architect typically monitors the development activities, making changes in the project architecture in response to changing project requirements. During the Development Phase, the information developers research and create content. The content is reviewed for completeness and accuracy by subject-matter experts and effectiveness and usability by peers and developmental editors. Finally, by the end of the Development Phase, the content has been edited for compliance to corporate styles and approved for release to production. Figure 18-1 illustrates the importance of the Development Phase.

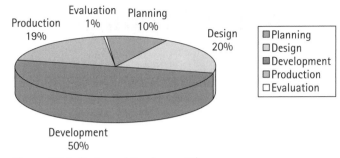

Figure 18-1: Design and Development Phases

The Role of the Project Manager during the Design and Development Phases of the Project

The role of the project manager during the Design and Development Phases is to monitor the progress of the project and ensure that resources continue to be sufficient to complete the work according to the key milestones and deadlines. Monitoring progress and ensuring the schedule can be met are by no means simple and straightforward for most information-development projects. The project manager is also responsible for negotiating, facilitating, and resolving issues and conflicts.

In all but the most stable information-development projects, the project begins to change almost as soon as the ink is dry on the Project Plan. The information architect gathers additional information about customer information requirements, the product development shifts direction in response to customer and competitive factors, and the information developers themselves learn more about the details of the project from the developers. That information helps deepen their understanding of the scope and complexity of the project. All these changes to the project must funnel through to the information project manager, who makes adjustments to the schedule, the deliverables, and the risk analysis for the project.

Not only are you, as project manager, monitoring the changes occurring, but you are responsible for communicating the impact of the changes on resources, budget, and schedule to your management. As project manager, you have regular responsibilities for reporting the status of the project to the various stakeholders. As you gather information from all of your sources, you prepare regular status reports and alert stakeholders to the critical decisions they must make if the project is to stay on track and deliver value to the customers.

In this chapter, you learn how to set up a system for tracking your project and reporting progress. In the next chapter, you learn how to monitor the project and respond to changes.

Best Practices in Project Tracking and Reporting

Based on the assumption that information architecture and information development are being handled by your team members, as project manager, you will find six best practices associated with tracking and reporting the status of your project to the stakeholders:

✔ Developing a resource-tracking spreadsheet

✔ Moving from tracking deliverables to tracking topics

✔ Ensuring adequate resources are assigned to topic development

✔ Developing topic milestones

✔ Reporting progress

✔ Building your Project Management folder

Best Practice–Developing a resource-tracking spreadsheet

The project resource-tracking spreadsheet provides you with a useful initial view of the project. The spreadsheet helps ensure that you have correctly assigned staff members to the project so that it can be completed by the deadline. In Chapter 16: Planning Your Information Development Project you calculated the total hours necessary to complete your project, which had four deliverables: a business owner's guide and training materials and a help system and training materials for an end user. Based on your previous experience with similar projects and your analysis of risk using the dependency calculator, you estimated that the project would take 893 hours to complete. You have about three months, or 13 weeks, to complete the project, because you have learned that the end of March is the date for the final deliverables to be sent to translation.

The resource-tracking spreadsheet provides you with a graphical view of the project and your staff resource allocations.

To prepare your resource tracking spreadsheet, you need to know

✔ each deliverable that your team will produce

✔ the total number of hours you have estimated are needed to complete the project

✔ the final project deadline and any key interim milestones and their deadlines

✔ the number of staff you have estimated that you will need for the project, including information developers, editors, illustrators, production specialists, information architect, and yourself as project manager

With this information in hand, you are ready to prepare your spreadsheet. Either use a spreadsheet program such as Microsoft Excel or construct a spreadsheet by hand using old-fashioned ledger paper, whichever you are more comfortable doing. In either case, you have a visual representation of your project that is quick and easy to construct. As the project changes, you can easily update your spreadsheet with additional information.

Developing a complete plan with resources assigned to the entire schedule

In developing your project resource spreadsheet, follow these steps:

1   Create a basic spreadsheet shell for your project with the weeks or months of the project along the top row (the X axis).

2   Label the first column, on the far left (the Y axis), with each of your deliverables.

3   Under each deliverable in the left column, add the names and the roles of each person assigned to the project. If some staff members will work on more than one deliverable, add their names to each deliverable.

The spreadsheet shell you construct will look like the one in Figure 18-2.

In the spreadsheet illustrated in Figure 18-2, your project is three months long and has four deliverables. You have assigned five major roles to each deliverable: project manager, information architect, information developer, and editor. Because the deliverables will be translated into French, you have also assigned a localization coordinator to the project. The actual localization and translation work will be handled by an outside service provider. Although you need to account for this activity in the schedule, you won't add it to your project resource spreadsheet.

Next, you need to review how much time you have in the months you are able to assign for this project.

1   Look at each week or month of the project and eliminate any holidays.

2   Count the number of days remaining in each week or month.

3   Multiply by six or seven hours per day, which represents the average hours per day that a full-time staff member can devote to project work.

4   Note the total hours per week or month at the bottom of each month column on the spreadsheet.

The hours you have noted on the spreadsheet, as illustrated in Figure 18-2 under the item called Hours/Period, are the hours available for your project for each person assigned full time. They provide you with a reference point for each week or month of the project. If someone is assigned half-time to your project, you can take that into account by allocating only half the hours available for a full-time person. If people on your team are taking vacations during the project, you can easily deduct that time from their schedule for project work.

Projected Hours Spreadsheet

Factor	Skill Level	Projected Jan-07	Projected Feb-07	Projected Mar-07	Projected Apr-07	Projected May-07	Projected Jun-07	Projected Jul-07	Projected Total Hours	Hours/Unit
Business Owners User Guide		Page Count= 60		Metric= 5.60						
	Project Manager	0.00	0.00	0.00	0.00	0.00	0.00	0.00	0.00	0.00
	Information Architect	0.00	0.00	0.00	0.00	0.00	0.00	0.00	0.00	0.00
	Information Developer 1	0.00	0.00	0.00	0.00	0.00	0.00	0.00	0.00	0.00
	Information Developer 2	0.00	0.00	0.00	0.00	0.00	0.00	0.00	0.00	0.00
	Editor	0.00	0.00	0.00	0.00	0.00	0.00	0.00	0.00	0.00
	Localization Coordinator	0.00	0.00	0.00	0.00	0.00	0.00	0.00	0.00	0.00
	Subtotal	0.00	0.00	0.00	0.00	0.00	0.00	0.00	336.00	5.60
Business Owners Training Guide		Page Count= 20		Metric= 5.60						
	Project Manager	0.00	0.00	0.00	0.00	0.00	0.00	0.00	0.00	0.00
	Information Architect	0.00	0.00	0.00	0.00	0.00	0.00	0.00	0.00	0.00
	Information Developer 1	0.00	0.00	0.00	0.00	0.00	0.00	0.00	0.00	0.00
	Information Developer 2	0.00	0.00	0.00	0.00	0.00	0.00	0.00	0.00	0.00
	Editor	0.00	0.00	0.00	0.00	0.00	0.00	0.00	0.00	0.00
	Localization Coordinator	0.00	0.00	0.00	0.00	0.00	0.00	0.00	0.00	0.00
	Subtotal	0.00	0.00	0.00	0.00	0.00	0.00	0.00	112.00	5.60
End User Training Guide		Page Count= 20		Metric= 5.60						
	Project Manager	0.00	0.00	0.00	0.00	0.00	0.00	0.00	0.00	0.00
	Information Architect	0.00	0.00	0.00	0.00	0.00	0.00	0.00	0.00	0.00
	Information Developer 1	0.00	0.00	0.00	0.00	0.00	0.00	0.00	0.00	0.00
	Information Developer 2	0.00	0.00	0.00	0.00	0.00	0.00	0.00	0.00	0.00
	Editor	0.00	0.00	0.00	0.00	0.00	0.00	0.00	0.00	0.00
	Localization Coordinator	0.00	0.00	0.00	0.00	0.00	0.00	0.00	0.00	0.00
	Subtotal	0.00	0.00	0.00	0.00	0.00	0.00	0.00	112.00	5.60
End user Help System		Topic Count= 80		Metric= 4.17						
	Project Manager	0.00	0.00	0.00	0.00	0.00	0.00	0.00	0.00	0.00
	Information Architect	0.00	0.00	0.00	0.00	0.00	0.00	0.00	0.00	0.00
	Information Developer 1	0.00	0.00	0.00	0.00	0.00	0.00	0.00	0.00	0.00
	Information Developer 2	0.00	0.00	0.00	0.00	0.00	0.00	0.00	0.00	0.00
	Editor	0.00	0.00	0.00	0.00	0.00	0.00	0.00	0.00	0.00
	Localization Coordinator	0.00	0.00	0.00	0.00	0.00	0.00	0.00	0.00	0.00
	Subtotal	0.00	0.00	0.00	0.00	0.00	0.00	0.00	333.60	4.17
	Total Hours	0.00	0.00	0.00	0.00	0.00	0.00	0.00	0.00	
	Hours/Period	154.00	140.00	154.00	147.00	154.00	147.00	147.00		
	Full-Time Equivalent/Period	0.00	0.00	0.00	0.00	0.00	0.00	0.00		

Figure 18-2: Initial Shell for the Project Resource Spreadsheet

Now that you have a complete template for your resource spreadsheet, you can begin filling in the details. In Figure 18-3, you see one possibility for staffing the project. Since the content of the Business Owner's Guide and the End-User Help System will be significantly different, two writers (called Information Developers 1 and 2) are assigned to accommodate the work required. The two writers will work on both the guides and the training, reducing the total time required for writing and review.

Note that the project manager has a role to play in the entire project, represented as approximately 10% of the total time for each project. If you know how to set up an equation in your electronic spreadsheet, you can automatically calculate the 10% for project management as you add or delete time for the other team members.

In addition, note the time provided for editing. The percentage of time for editing will depend upon the skill and experience of the information architect and the information developers. Editing time, both for developmental editing and copyediting, may represent between 5% and 25% of information development time. Set the percentage you believe is needed for your project, depending upon your previous experience with editing.

If your process calls for a production specialist or an illustrator, add additional rows in your spreadsheet for these responsibilities. Include a production specialist at the last 5% or 10% of the project, depending on the type of production work required. Include a localization coordinator and add a row in the spreadsheet for this activity. Include an illustrator if your project requires graphics to be developed. In some projects, graphics are simply screen images often handled by the information developer. In other projects, graphics are technical illustrations developed by a professional illustrator or graphic artist. The roles and responsibilities of individuals who you may want assigned to your project are described in Table 18-1.

Table 18-1: Potential Roles and Responsibilities to Assign to Your Project

Role	Description
Project Manager	Responsible for developing the Information Project Plan, estimating the total time required to complete the project by the milestones and deadlines of the larger project, ensuring that sufficient resources are assigned to the project, coordinating the project activities with other team members, tracking the project's progress, and making adjustments to schedule, assignments, and deliverables based on changes to the project's schedule and scope.
Information Architect	Responsible for planning the content to be created for the project, in cooperation with the project manager and the information developers. Ensures that standards are followed in the development of final deliverables. Creates the annotated topic list and revises as necessary.
Information Developer	Works with the information architect to develop the plan for the content and the deliverables. Researches the product, service, or other subject matter and creates final deliverables. Collaborates with other information developers throughout the project to ensure that topics are created only once and reused appropriately among the deliverables.

Continued

Table 18-1: Potential Roles and Responsibilities to Assign to Your Project (Continued)

Role	Description
Editor - Quality Assurance (developmental, copy, and production edits)	Ensures that the content follows standards and meets the usability goals of the project and the audiences. Reviews topics in various draft stages for compliance with style, language, format, and other guidelines. Performs final edits to ensure that the deliverables are ready for production.
Illustrator or Graphic Designer	Prepares illustrations or other graphics for the final deliverables. May be responsible for screen shots in conjunction with the information developers. May also be responsible for the design of the final deliverables (information design), including print, PDF, HTML, Help, and other deliverable formats.
Translation Coordinator	Responsible for coordinating the work of internal or external localization and translation experts. May work with a localization vendor or an internal localization group. Ensures that content is ready for localization, terminology is well established, schedules are communicated, and all materials are ready for final production in all languages.
Production Specialist	Prepares the final deliverable for distribution to end users. Coordinates with print production vendors and other distributors. Ensures that help systems and HTML pages are ready for deployment to product or websites. Reviews final deliverables to ensure compliance with design standards and guidelines. Handles content and document management activities related to the repository.

Each role and responsibility required for your project is added as a row in the spreadsheet associated with each deliverable. Although you may use the same information in both the guides and the training materials, you will need to allocate a role and time to develop slides for the instructor to use in a classroom setting, to create laboratory exercises, and to design tests. Your experience in estimating the amount of reusable information and allocating time for the remainder of the activities ensures that you have not over- or underestimated the project.

All the activities of people involved in your project will take up the total time you have calculated for each project. Remember that in calculating total time, you used an hour per page or hours per topic that was based upon the experience of your organization. You may find it easier to add all of this experience to the total hours per page or topic for your project, rather than calculating separate totals for each activity. However, in some projects, you will need independent calculations of the amount of time for editing, graphic development, production, or translation coordination. In that case, work with the people responsible for each activity and add the time needed to your spreadsheet.

In Table 18-2, I provide guidelines for the allocation of percentages of total project time to the roles summarized in Table 18-1.

Table 18-2: Allocation of Total Project Time

Role	Percentage of Total Project Time
Project Manager	10% to 15% of total project time, depending on the complexity of the management tasks. Consider the management of a geographically dispersed team, the addition of new offshore team members, the requirements of new tools and technology, and other factors that will add to the time required for project management tasks. Consider the expertise of the information developers and the subject-matter experts.
Information Architect	10% to 20% of total project time during the Planning and Design Phases of the project. Consider the time needed to design the architecture for a new project or redesign legacy information to meet a new structured authoring environment.
Information Developer	50% to 60% of total project time. Essentially, you assign to information development all the project time not already allocated to project management, architecture, quality assurance, or production.
Quality Assurance	5% to 25% of information-developer time. Unlike the allocations for project management and information architecture, the time you allocate for developmental quality assurance is directly calculated based on the information developer's time. If you have a team of new information developers with little experience in the type of information you are creating, you may need a higher percentage for review of their content development. If you have an experienced team, the time for developmental quality assurance may be quite low. You may want to allocate time for copyediting separately from developmental editing, based on a total number of pages expected in final deliverables. If the information developers use a controlled language tool, you may be able to decrease the percentage of time or the total hours for copyediting to a minimum.
Illustrator or graphics specialist	You may want to include time to create graphics in your total project estimate, especially if your project data represents all time, including graphics, and your graphics needs are minimal. In many software projects, the majority of time for graphics involves the development of screen images. These images are often created directly by the information developer. However, in projects that require professionally developed illustrations, including hardware drawings, you may want to estimate graphics separately and include graphics time as an independent activity in your spreadsheet. If you are not adept at estimating graphics projects, include the graphics specialists in your project estimating.
Translation Coordination	If translation coordination is an integral part of the activities in your organization, you may have already estimated time for coordination in your hours per unit estimate. However, if translation coordination is handled by another part of the organization or calculated separately, work closely with the specialists in this area to estimate the total time required for that work and when it needs to begin so that deadlines can be met.

Continued

Table 18-2: Allocation of Total Project Time (Continued)

Role	Percentage of Total Project Time
Production Specialist	Like graphics and translation, you will find it best to gather data from your production staff to estimate the percentage of time needed to complete the production activities for your project. If you include this time in your unit estimates, you may find that your project requires about 10% of total time for production.
	The allocation for production will depend upon the number and complexity of production tasks. In many organizations, the information developers do their own production work. If the production is largely automated using standard style sheets, the production allocation may decrease from more manual production.
	If you have a dedicated production team, once again work with them to develop an estimate of the time and resources they will need to deliver the project on schedule. If production activities must occur for interim deliverables, be certain to add time and resources for these activities to your schedule.

Although you may want to plan for the time required for subject-matter experts to review the content, because these individuals do not report to you or allocate their time to your organization, you should not allocate time for this activity in your project spreadsheet. Allocate only the time interval in which the reviews are likely to take place.

As you can see, allocating resources to your project spreadsheet will take careful planning. The more historical data you have about the allocation of time to activities in the past, the better you will be at estimating the future. Work with your team members to arrive at the best, most comprehensive schedule of work and milestones as you can at this early stage of the project. Be aware that some of the time for project management, quality assurance, and architecture will have been estimated when you performed your dependency calculations. If you calculated that your project would be more difficult than the average to complete, you have in fact planned for the extra time needed to manage the project or to ensure that quality goals are met.

By reviewing the sample data in Figure 18-3, you can start your own project estimate by using this example of how one project was estimated. Use the example as a guideline, but remember that your own historical data is your best guide.

Note that in the project resource spreadsheet, you have made an attempt to schedule some of the activities that surround information development. You need to associate these activities with various project milestones that you identified in your Information Project Plan. If you know that you will need two weeks in production to prepare the deliverables for final release, you need to ensure that information development, graphics, translation, and editing are complete prior to the beginning of production in all languages. If you have scheduled translation to begin after final production in the source language but before the final release date, show that activity in your spreadsheet by assigning your translation coordinator to the project during that period in which translation activities are most likely to occur. If you will bring translated content back inhouse, making your production team responsible for assembly of final deliverables in all languages, show that activity on the production row of your spreadsheet.

Projected Hours Spreadsheet

Factor	Skill Level	Projected Jan-07	Projected Feb-07	Projected Mar-07	Projected Apr-07	Projected May-07	Projected Jun-07	Projected Jul-07	Projected Total Hours	Hours/Unit
Business Owners User Guide		Page Count= 60		Metric= 5.60						
	Project Manager	10.00	10.00	10.00	0.00	0.00	0.00	0.00	30.00	0.50
	Information Architect	30.00	5.00	5.00	0.00	0.00	0.00	0.00	40.00	0.67
	Information Developer 1	80.00	80.00	80.00	0.00	0.00	0.00	0.00	240.00	4.00
	Information Developer 2	0.00	0.00	0.00	0.00	0.00	0.00	0.00	0.00	0.00
	Editor	0.00	5.00	10.00	0.00	0.00	0.00	0.00	15.00	0.25
	Localization Coordinator	0.00	0.00	10.00	0.00	0.00	0.00	0.00	10.00	0.17
	Subtotal	120.00	100.00	115.00	0.00	0.00	0.00	0.00	335.00 **336.00**	5.58 **5.60**
Business Owners Training Guide		Page Count= 20		Metric= 5.60						
	Project Manager	2.00	2.00	2.00	0.00	0.00	0.00	0.00	6.00	0.30
	Information Architect	8.00	4.00	4.00	0.00	0.00	0.00	0.00	12.00	0.60
	Information Developer 1	0.00	35.00	40.00	0.00	0.00	0.00	0.00	75.00	3.75
	Information Developer 2	0.00	0.00	0.00	0.00	0.00	0.00	0.00	0.00	0.00
	Editor	0.00	5.00	5.00	0.00	0.00	0.00	0.00	10.00	0.50
	Localization Coordinator	0.00	0.00	10.00	0.00	0.00	0.00	0.00	10.00	0.50
	Subtotal	10.00	46.00	57.00	0.00	0.00	0.00	0.00	113.00 **112.00**	5.65 **5.60**
End User Training Guide		Page Count= 20		Metric= 5.60						
	Project Manager	2.00	2.00	2.00	0.00	0.00	0.00	0.00	6.00	0.30
	Information Architect	8.00	4.00	0.00	0.00	0.00	0.00	0.00	12.00	0.60
	Information Developer 1	0.00	0.00	0.00	0.00	0.00	0.00	0.00	0.00	0.00
	Information Developer 2	0.00	35.00	40.00	0.00	0.00	0.00	0.00	75.00	3.75
	Editor	0.00	5.00	5.00	0.00	0.00	0.00	0.00	10.00	0.50
	Localization Coordinator	0.00	0.00	10.00	0.00	0.00	0.00	0.00	10.00	0.50
	Subtotal	10.00	46.00	57.00	0.00	0.00	0.00	0.00	113.00 **112.00**	5.15 **5.60**
End user Help System		Topic Count= 80		Metric= 4.17						
	Project Manager	10.00	10.00	10.00	0.00	0.00	0.00	0.00	30.00	0.38
	Information Architect	12.00	6.00	0.00	0.00	0.00	0.00	0.00	18.00	0.23
	Information Developer 1	0.00	0.00	0.00	0.00	0.00	0.00	0.00	0.00	0.00
	Information Developer 2	82.00	82.00	82.00	0.00	0.00	0.00	0.00	246.00	3.08
	Editor	0.00	15.00	15.00	0.00	0.00	0.00	0.00	30.00	0.38
	Localization Coordinator	0.00	0.00	10.00	0.00	0.00	0.00	0.00	10.00	0.13
	Subtotal	104.00	113.00	117.00	0.00	0.00	0.00	0.00	334.00 **333.60**	4.18 **4.17**
	Total Hours	244.00	305.00	346.00	147.00	154.00	147.00	147.00	895.00	
	Hours/Period	154.00	140.00	154.00	147.00	154.00	147.00	147.00		
	Full-Time Equivalent/Period	1.58	2.18	2.25	0.00	0.00	0.00	0.00		
Total Hours										
	Project Manager	24.00	24.00	24.00	0.00	0.00	0.00	0.00	72.00	
	Information Architect	58.00	19.00	5.00	0.00	0.00	0.00	0.00	82.00	
	Information Developer 1	80.00	115.00	120.00	0.00	0.00	0.00	0.00	315.00	
	Information Developer 2	82.00	117.00	122.00	0.00	0.00	0.00	0.00	321.00	
	Editor	0.00	30.00	35.00	0.00	0.00	0.00	0.00	65.00	
	Localization Coordinator	0.00	0.00	40.00	0.00	0.00	0.00	0.00	40.00	

Figure 18-3: Complete project resource spreadsheet example

Developing an iterative plan

A single project spreadsheet for an entire complex project may not be sufficient to indicate the details of the schedule, especially those that must take place to meet interim milestones. It may be best to divide your project spreadsheet into a number of subsections, each representing a project iteration or phase. You may find it convenient to place each iteration or phase of the project on a separate page so that you can provide a more detailed estimate of activities.

For example, if you set up your overall project spreadsheet in terms of months, you may want to convert your interim spreadsheet pages into weeks so that you can anticipate the activities at a more granular level. You may want to track, for example, the completion of the first draft iteration of content development so that you know when drafts of chapters or topics must be completed before the first milestone. You may want to indicate how information developers will interact with graphics or developmental editing during the construction of the first drafts of content, ensuring that you have adequate resources for these tasks at the time they are needed in the project.

In handling the details of your project schedule, you may decide to work with project-scheduling software rather than a spreadsheet. Project-scheduling software allows you to create a detailed schedule for every activity in your plan. If you have learned to use schedule software efficiently, by all means use it for your detailed plans. However, I recommend using a simple spreadsheet to show in general terms how you will staff your project so that your allocated time matches your estimated time to complete.

Developing a holistic plan

In most cases, starting with a top-down estimation of total hours needed will yield a better estimate than trying to add up bits of time for a variety of activities. Although it may seem contradictory, top-down estimates of information-development projects are more accurate because it is easier to envision the entire project rather than try to judge accurately the time for small individual tasks. However, you can improve your estimating if you look at your project from several perspectives. By discussing the project with estimating specialists (if you have those in your organization), you will learn to appreciate how they estimate their own tasks.

The more actual data you have accumulated from previous projects, the better your estimates of new work are likely to be. Keeping track of time for information development, project management, editing, graphics, production, translation coordination, and other activities that occur in your organization will help you recognize just where all the time goes and help you develop better overall estimates in the future.

🏆 Best Practice—Moving from tracking deliverables to tracking topics

In the discussion of the estimating spreadsheet, I recommended dividing more complex projects into several separate spreadsheets to represent project milestones. In that way, you can more carefully allocate resources and track progress for the phases of the project and the deliverables. However, the traditional estimating spreadsheet is still built upon final deliverables, the deliverables you used to do your overall estimate of the project. Because much historical data we have available is based upon final deliverables, these data are still extremely useful in helping you perform a top-down, holistic estimate of the project.

However, technical information development is moving quickly away from developing "books" to a topic-based information architecture in which the final deliverables are assembled from independently developed modular content. Each topic might represent a small segment of content, based upon the development of concepts, tasks, and reference information or other topic types that your information architecture team has specialized for your content. Project managers are beginning to ask information developers to create a set of topics that will later be assembled into various outputs, including books (print or electronic), help systems, web pages, on-device information such as PDAs, and others. In many cases, the topics will be used in more than one output, depending upon the overall configuration of the information architecture. You may even be sharing topics between information development and education or training.

With the move toward topic-based information development and reuse of topics across multiple deliverables, you will need to allocate adequate resources for topic development in your project schedule. Each topic or group of topics needs resources allocated and a schedule for several interim milestones, including design, writing, editing, review, assembly, and final production.

Breaking deliverables down by topics

In Chapter 17: Implementing a Topic Architecture, you learn how to develop an information architecture for your project that specifies the topics to be created or revised from previous work in the subject domain. As project manager, you need to coordinate closely with the information architect so that you know exactly what topics have been designed for each final deliverables. In fact, as the project moves through the Design Phase, you will use the information architect's analysis of topics to ensure that the time estimates for the final deliverables are adequate. In most cases, the list of topics changes during the Design Phase, often requiring that the total estimates for final deliverables be refigured up or down.

By allocating time in your project schedule for topics or groups of topics covering a subject domain of the content, you are better able to produce a detailed schedule and ensure that the team is making progress toward interim milestones and final deadlines.

Estimating resources based on topics is similar to creating a detailed work breakdown structure for your project. However, the work breakdown in this case is based upon the content to be developed rather than on the subtasks to be performed by the team members. Within each topic resource allocation will be the work required to research the subject and the audience; draft the appropriate content; edit it for clarity, completeness, and consistency; assemble it into a hierarchy or into a networked relationship with other topics; and produce the final deliverables. Certainly, you have broken down some of that work already based on allocations of time in your project to quality assurance, information development, and production. But the allocation of time and resources to topics focuses specifically on ensuring that time is allocated to the information that adds the most value to customers.

In developing content for books or help systems, information developers often find it easy to include details of content that derive from product specifications, project requirements, or other technical sources. Detailed step-by-step procedures for filling out on-screen forms absorb many hours because they are obvious to the writers and fill up the available time. Too often, content is included simply because it is easily available or easily written. By developing content topics that are specifically designed to meet customer needs for information, you can avoid developing content for your project that is unneeded by customers. Topics , if they are well designed and carefully monitored during development, ensure that writers concentrate on precisely what customers need rather than on what they know about or can easily obtain.

Topic-based information development also aids in the management of the project by ensuring that time is spent adding content that has value for customers rather than going off on tangents and producing content that customers find irrelevant to their environment and superfluous in guiding their successful use of a product or service. One of the most discouraging aspects of project management is discovering that an information developer has spent precious project time developing content that is neither useful nor valuable to the information customer. Topic-based authoring means that you plan the content much more carefully at the outset and write to your plan.

Once you have a detailed and thoroughly annotated list of the topics to be developed for your project, you have a greater measure of control over what gets developed. I recall many years ago discussing project management with a manager of software development. She mentioned that the most difficult part of her job was to keep the programmers from going off on tangents and creating features and functions in the code that weren't needed. She wanted to ensure that they first developed what was required before adding new ideas to the mix. The same is true with topic-based authoring. Although you certainly plan for changes to occur in the content that eventually is included in the project, you want to ensure that your information developers are focused on adding customer value rather than adding content that no one needs to know.

Developing topic-based estimates using top-down metrics

At this stage of your project, you have a complete project spreadsheet that outlines the deliverables for the project and allocates time and resources for their development. You may have already created a set of spreadsheets that address independently the work required in each stage of your information-development life cycle. Your next step is to create a spreadsheet that accounts for the topics outlined by the information architect.

The spreadsheet in Figure 18-4 illustrates how to break the project into three information types: tasks, concepts, and reference topics. Note that the project includes 75 tasks, 25 concepts, and 15 references topics. The project manager has estimated the hours needed to produce the topic types: five hours per topic for tasks, seven hours per topic for concepts, and six hours per topic for reference material. These estimates are the product of historical data from previous projects and the same dependency calculations discussed in Figure 18-3. The total time required to complete the project is estimated at 640 hours, beginning the first of October and ending in mid-January.

Once again, the total time represents not only the work required for the information developer to research, write, and revise the topics but also the work required to manage the project, edit the topics, and assemble them into final deliverables.

In reviewing the spreadsheet in Figure 18-4, note that the project manager and the information architect complete the Planning Phase of the project (approximately 64 hours or 10 percent of the total project time) before involving the information developer in the Design Phase (approximately 128 hours or 20 percent of the total project time). The project manager time represents 10% of the total time on the project, and the editor represents 15% of the information developer time, except in the Design Phase, extending through the end of October into early November.

Projected Hours Spreadsheet

Factor	Skill Level	Projected Oct 1-15	Projected Oct 16-31	Projected Nov 1-15	Projected Nov 16-30	Projected Dec 1-15	Projected Dec 16-31	Projected Jan 1-15	Projected Total Hours	Hours/Unit
Tasks		Topic Count= 75		Metric= 5						
10%	Project Manager	3.50	5.30	8.79	6.21	7.59	1.23	1.50	34.12	0.45
	Information Architect	35.00	8.00	12.00	0.00	0.00	10.00	0.00	65.00	0.87
	Information Developer	0.00	40.00	66.00	54.00	66.00	2.00	0.00	228.00	3.04
15%	Editor	0.00	0.00	9.90	8.10	9.90	0.30	0.00	28.20	0.38
	Production Coordinator	0.00	5.00	0.00	0.00	0.00	0.00	15.00	20.00	0.27
	Subtotal	38.50	58.30	96.69	68.31	83.49	13.53	16.50	375.32 / **375.00**	5.00 / **5**
Concepts		Topic Count= 25		Metric= 7						
10%	Project Manager	1.50	2.80	0.00	1.00	0.00	7.21	3.30	15.81	0.63
	Information Architect	15.00	8.00	0.00	10.00	0.00	10.00	0.00	43.00	1.72
	Information Developer	0.00	16.00	0.00	0.00	0.00	54.00	20.00	90.00	3.60
15%	Editor	0.00	0.00	0.00	0.00	0.00	8.10	3.00	11.10	0.44
	Production Coordinator	0.00	4.00	0.00	0.00	0.00	0.00	10.00	14.00	0.56
	Subtotal	16.50	30.80	0.00	11.00	0.00	79.31	36.30	173.91 / **175.00**	6.96 / **7**
Reference		Topic Count= 15		Metric= 6						
10%	Project Manager	1.00	2.00	0.00	0.00	1.00	1.00	3.26	8.26	0.55
	Information Architect	10.00	8.00	0.00	0.00	10.00	10.00	0.00	38.00	2.53
	Information Developer	0.00	10.00	0.00	0.00	0.00	0.00	24.00	34.00	2.27
15%	Editor	0.00	0.00	0.00	0.00	0.00	0.00	3.60	3.60	0.24
	Production Coordinator	0.00	2.00	0.00	0.00	0.00	0.00	5.00	7.00	0.47
	Subtotal	11.00	22.00	0.00	0.00	11.00	11.00	35.86	90.86 / **90.00**	6.06 / **6**
	Total Hours	66.00	111.10	96.69	79.31	94.49	103.84	88.66	640.09	
	Hours/Period	60.00	66.00	66.00	54.00	66.00	54.00	54.00		
	Full-Time Equivalent/Period	1.10	1.68	1.47	1.47	1.43	1.92	1.64		

Figure 18-4: Topic-based estimating

Once the Design Phase is complete, the information developer focuses on developing the tasks. Concepts and reference topics are developed after the tasks are completed. The production coordinator handles the assembly of the topics into the project deliverables during the second week of January. The time allocated to all members of the project team equals the 640-hour total for the project, and the project is completed by the January 15 deadline.

Evaluating topic complexity and risk

The time to produce the topics differs because the work required, at least for this project, also differs. An information developer may find that developing the task-based information and creating step-by-step procedures is relatively straightforward for the particular products but the time required to develop conceptual information is much higher. Developing conceptual information may require a much better understanding of the customers' needs. It may require research and conversations with many more technical experts. It may require that information be structured to support task performance rather than simply copied from product requirements or specifications. Producing useful conceptual information is often quite difficult.

The same difficulty may surround the development of the factual information that customers need to reference. Reference content may require considerable research and frequent revision before it is ready for release to customers.

As project manager, your responsibility is to know what topics need to be created for your project, how much time it has taken to create similar topics in the past, the dependencies involved with the topics, and the milestones associated with their development.

The amount of time calculated for the entire project must be the same as the amount of time calculated for the development of the topics. Topic-based estimates allow you to begin to break the work down into measurable milestones.

Using multiple estimating techniques

The estimating techniques I have presented in this chapter are based upon the concept of top-down estimating. Top-down estimating encourages you to look at the project as a whole, rather than to focus on all of the individual parts. Once you have estimated the amount of time needed for the whole project, based on your historical data and your dependency calculations, you can divide the project into phases in the development life cycle or topics to be developed.

Some managers ask why they should not try to estimate all the details of the work to be completed and add them up to derive the total project time. In some discussions of project estimating, authors have suggested adding the time to create each element of the final deliverable, including time for the title page, introduction, table of contents, each chapter and appendix, and so on and so on. The problem with this kind of bottom-up estimating is the very nature of information-development projects.

Let's compare an information-development project with a construction project. Construction projects often use bottom-up estimating to arrive at the cost and duration of the project. If, for example, you ask a contractor to estimate the cost of building you a new home and the time it will take until the project is complete, the contractor begins with the blueprint of the home and the bill of materials developed by the architect. Based on detailed historical data for each small task to be accomplished, the contractor is able to add up all the estimates and give you a final cost and schedule. Unless you change the design, you know precisely how much you will pay. Unfortunately, the schedule isn't so simple. Unknowns such as weather or undependable workers often wreck havoc with the schedule.

Why can a construction project be estimated from the bottom up quite successfully? The key is the blueprint and the bill of materials. A construction project is fully specified to the smallest detail right from the start. The contractor has extensive historical data from hundreds of previous projects across the industry and recent data based on the changes to the cost of materials and labor in the local area. Although there are risks and dependencies to be accounted for, the wealth of historical data makes the estimating process quite systematic. Construction estimators also apply top-down estimating when they compare their bottom-up estimates with the average price per square foot for homes in your area.

Information-development projects are different. They are rarely completely specified. In product-development organizations, new projects are designed based on vague or poorly articulated customer requirements. The projects are incompletely specified from the perspective of technical functions. The product developers have new technologies, unforeseen complications in legacy code, system integrations that are unpredictable, and so on. Product features are added and subtracted based on continuing interactions with the customers and the competition. In service- and information-centered organizations, projects often begin with committees and consensus building, making it difficult to specify the dimensions of the final outcomes. As a result, very little of the historical data about individual actions can be usefully applied to the new project. There are simply too many unknowns.

On the other hand, experienced information developers often have a reasonably good sense of how long a project is likely to take. The more historical data available for entire similar projects, the more likely the new estimates will be accurate enough for the initial top-down estimates. Data on past projects and good dependencies calculations make the initial estimates reasonable, with a measurable amount of doubt thrown in.

If you try to use detailed data to estimate bottom-up, you're likely to encounter two possibilities:

✔ Your initial bottom-up estimate greatly over- or underestimates the time needed to complete the project. Because it is so difficult to estimate how long undefined individual tasks will take, bottom-up estimating is remarkably unreliable. The larger the team involved, the less accurate the estimating. Individual contributors tend to assume they will require less time for collaboration with colleagues than actually occurs.

✔ Your projects are so completely defined that your bottom-up estimates will work. Some organizations occasionally have completely defined information-development projects, a rarity that project managers can be thankful for. More often, organizations assume that a project is well defined, only to discover half way through the project that all the unanticipated complications have actually occurred.

If you are inexperienced in estimating projects or if you have little historical data to back you up, consider estimating your project in more than one way.

1 Talk with information developers in your organization and gather informal information about previous projects. Ask them for their own top-down estimates of the work required for your new project.

2 Gather all the details you can about the milestones and deliverables for the project.

3 Develop a bottom-up estimate by asking people how much each type of work may take to complete, including such standard activities as content authoring, editing, and production.

4 Compare your top-down and bottom-up estimates and make adjustments. You might begin somewhere in the middle between a too high and a too low estimate for your project.

5 Compare your project with similar projects in the past, taking into account all of the factors measured by the dependency calculation.

6 Recognize that your estimate at the beginning of your project is the least accurate estimate you will do through the course of the project.

7 Be prepared to re-estimate the project as soon as you have data from the first activities.

Recognize that no early estimate is perfect. Be prepared to adjust course as needed during the life of the project.

Estimating a new project based on revisions of existing content

The examples I have reviewed in this chapter have assumed that all new content is being created for your project. Of course, most projects in information development begin with existing content. A new release of a product is planned, a new version of an existing product is being created based partially on the existing design, an existing service is being expanded, or an existing policy and procedure has to be updated because circumstances have changed. Most information-development projects include writing new content, deleting old content, and updating and revising existing content.

The process of estimating a new project based on existing information is similar to the top-down estimating you have learned, but it includes breaking the tasks into more categories.

Table 18-3 shows a spreadsheet for estimating a new product release that includes updating existing information. To produce an estimate for your project, consider the following steps, based on a topic- rather than a book-based architecture:

1 Begin with a list of existing topics in your spreadsheet.

2 Add the new topics to be developed based on the analysis done by your information architect and information developers.

3 Mark any topics that will be deleted or replaced by new topics.

4 Indicate which of the existing topics will need to be revised. Note that some may require major revision and some minor.

5 Ensure that each topic is assigned a type (concept, task, or reference).

6 Assign the hours per topic for new, major, and minor revisions. You might estimate, for example, that a new task topic will require five hours of project time, a major revision four hours, and a minor revision one hour. Use historical data from previous projects whenever possible, noting that you are using average hours per topic.

7 Add all the estimated hours to product a total estimate for your project.

8 Compare this bottom-up estimate with your top-down project estimate for new release projects with similar scope. Be certain that you have applied a new dependencies calculation.

9 Adjust your bottom-up, topic-based estimate to bring it into line with your top-down estimate and your dependencies calculation.

If you are not using a topic-based architecture for your projects, you can develop your new estimate based on total pages of the deliverables. Estimate new and revised pages and divide the revised pages between major and minor revisions (see Table 18-3).

Table 18-3: Estimating a Project with Existing Content

Topic	New	Major Revision	Minor Revision	No Change	Delete
Topic 1—Concept		X			
Topic 2—Task	X				
Topic 3—Task			X		
Topic 4—Task				X	
Topic 5—Concept			X		
Topic 6—Task					X
Topic 7—Reference		X			
Topic 8—Task	X				
Topic 9—Task		X			
Topic 10—Reference		X			
Hours/topic	5.6	5.0	2.1		
Total hours	11.2	20.0	4.2		
Total of all topics	35.4				

You may want to add some time to account for topics with no change or even for deleted topics. You may need time to delete topics from your deliverables, especially if you are mapping topics to more than one deliverable.

Adding external estimates for editing, graphics, production, and other activities done outside your organization

As you create a total project estimate and develop your project spreadsheets, you may need to include estimates of time and relationship to milestones and deadlines provided by others in your organization or by independent departments. If your historical data include only your own staff members, you will need estimates from outside organizations, both internal and external. Some of the time estimates need to be included in your spreadsheets. Others may provide data only about overall costs.

Some or all of the following activities may have to be estimated by internal or external resources:

✔ development of illustrations or other graphics by an internal graphics department or an external graphics vendor

✔ editing, indexing, or production by an external resource

✔ localization and translation by an outsource vendor

In each case, you must provide a statement of work or specification to the outside organization to obtain a reliable estimate of time and schedule. For graphics development, you must provide an estimate of the number and complexity of the illustrations or other graphics that are to be produced for your project. For editing, indexing, or production activities, you must provide estimates of topics or page counts, size and scope of final deliverables, and number and type of media. I discuss quality assurance and production activities later in Chapter 21: Managing Quality Assurance and Chapter 23: Managing Production and Delivery, but you need the estimates of work during the Planning and Design Phases. Just as you must work with incomplete information during the Planning Phase of the project, you will probably have to provide similarly incomplete information to your vendors. The more careful your specifications are, the more accurate their estimates will be.

Localization and translation is particularly difficult to estimate at the Planning Phase of your project because these estimates require word counts and descriptions of final media. By all means, use previous projects as a guideline and be prepared to revise your estimates as the project develops. If you have moved to topic-based information development, your estimates of new and revised topics are likely to provide more detailed and accurate information to your localization vendor.

If your organization employs staff members who coordinate outsourced graphics, localization, or other activities, work closely with them during the Planning Phase. If not, work directly with the vendors and be as open as possible about the unknowns and risks of your early estimates. As your project changes and you begin to build a more accurate picture of the requirements for your vendors, be certain to update them accordingly. The result will be vendors who will work as partners rather than adversaries for your project.

One of the most important reasons for asking your vendors, coordinators, or collaborators from other parts of your organization to provide you with their estimates of time and cost is the impact on your own project management time. I have recommended allocating between 10% and 15% of total project time to project management. If you simply calculate the time required by your immediate staff and fail to take into account the time for outside activities, you will find that you have shortchanged yourself on project management time.

One manager told me that he found it difficult to manage his projects after his company outsourced graphics. Previously, the manager of the internal graphics team had handled the project management for this part of the project. Although the outsource vendor provided project management of its resources, the manager found himself spending far more time than he had previously to coordinate the graphics production.

Estimating a new project that includes using topics in more than one deliverable

In the sample spreadsheet in Figure 18-3, you started with a project that had two deliverables: two user guides based upon the needs of two different audiences. You then calculated the number of topics needed for the project, including the topics for both user guides. The assumption thus far has been that the two user guides are completely independent of one another, containing different sets of topics for each audience. But what if the two user guides are not independent? What if you have the opportunity to use some content in both guides? How does that affect your estimate?

Your information architect, working in a topic-based environment, should already have informed you that some topics will be used in both deliverables. That information will be noted in the annotated topic list. The outlines of your final deliverables should also note which topics are used in more than one place.

When you develop your topic spreadsheet, you can quite easily include time estimates only for unique topics to be developed or revised. However, that may mean that your topic-based project estimates will not be synchronized with your estimates for the deliverables as a whole. The conflict will be most apparent as you move from a final-deliverables estimating method to a topic-based estimating method. All your historical data are based on final deliverables rather than on topics.

Returning to your deliverables project spreadsheet, you have two options:

✔ Remove all duplicate project hours from one of the deliverables

✔ Remove a portion of the duplicate project hours from each deliverable

Your decision of which option to use will depend upon how you intend to assign resources to the project. If you want to assign all the reusable topics to one information developer who is responsible for the set of topics for one deliverable, then remove the hours from the second deliverable. If you want to spread the assignment to create the reusable topics among all the information developers, allocate the time required between the deliverables. You learn about this and other alternatives for estimating reuse and allocating resources in the next section.

Figure 18-5 illustrates the initial deliverables-based project spreadsheet (see Figure 18-3) redone to account for pages that will be used in both deliverables. In this example, you decide that the help system provides the best opportunity to reuse content in the business owner's guide and the training materials. You believe that 50% of the 80 help topics will be reusable among the owner's guide and the two training pieces. You maintain the hours assigned to the writer of the help system, but you deduct hours from the time required to produce the owner's guide and the training materials. Half of the help topics can be reused in the end-user training materials, but only about 20% can be reused in the owner's training materials. About 10% of the help topic content can also be used in the business owner's guide. Instead of 336 hours to complete the business owner's guide, you now need only 302 hours. The time for the training guides is reduced from 112 hours each to 90 hours for the business owner's and 56 hours for the end-user's training guides. The total number of hours for the project has decreased from 894 to 782, an overall reduction of 13%.

One issue to remember, however, is that you must complete the help system earlier than planned so that it is completely edited before you can complete the owner's guide and the training materials. Your plan is to use the material from the help system in the owner's guide and training materials. For example, if you are using FrameMaker to create the guide and training, you could include the help-based content by saving it as separate files and using the text inset feature. If you are using an XML editor, you can include the help topics as part of the content maps for the other deliverables.

Of course, the two information developers who were originally working separately to develop the content for the business owner and the end user now have to plan the content together. Note that in the spreadsheet, neither information developer is working full-time on the project but each is working more than half-time. You still cannot do this project with one writer.

Also notice that you have decreased the time for the editor because there is less content to edit.

Best Practice–Ensuring adequate resources are assigned to your project

In the sample project spreadsheet in Figure 18-5, the number of people assigned to the project is adequate to complete the project by the deadline. Two information developers are assigned to two deliverables and the work is divided quite evenly between them. If you review the topic-based spreadsheet in Figure 18-4, you also find that two writers working full-time on the project are able to produce all the topics. But what happens when you have a project that cannot be accomplished by the deadline with the resources that you have allocated.

If you do not have adequate resources to complete the project, you have four alternatives during the Planning Phase of the project:

✔ You can add additional resources to the project to meet the deadline.

✔ You can ask staff members to work extra hours to meet the deadline.

✔ You can ask for more time to complete the project, effectively adding available hours to the project.

✔ You can decrease the scope of the project by removing some of the activities you have already scheduled, such as a deliverable or a set of topics.

Projected Hours Spreadsheet

Factor / Skill Level	Projected Jan-07	Projected Feb-07	Projected Mar-07	Projected Apr-07	Projected May-07	Projected Jun-07	Projected Jul-07	Projected Total Hours	Hours/Unit
Business Owners User Guide	Page Count= 60		Metric= 5.60		Reuse= 10%				
Project Manager	10.00	10.00	10.00	0.00	0.00	0.00	0.00	30.00	0.50
Information Architect	30.00	5.00	5.00	0.00	0.00	0.00	0.00	40.00	0.67
Information Developer 1	80.00	70.00	62.00	0.00	0.00	0.00	0.00	212.00	3.53
Information Developer 2	0.00	0.00	0.00	0.00	0.00	0.00	0.00	0.00	0.00
Editor	0.00	5.00	5.00	0.00	0.00	0.00	0.00	10.00	0.17
Localization Coordinator	0.00	0.00	10.00	0.00	0.00	0.00	0.00	10.00	0.17
Subtotal	120.00	90.00	92.00	0.00	0.00	0.00	0.00	302.00 / **302.40**	5.03 / **5.60**
Business Owners Training Guide	Page Count= 20		Metric= 5.60		Reuse= 20%				
Project Manager	2.00	2.00	2.00	0.00	0.00	0.00	0.00	6.00	0.30
Information Architect	8.00	4.00	0.00	0.00	0.00	0.00	0.00	12.00	0.60
Information Developer 1	0.00	15.00	40.00	0.00	0.00	0.00	0.00	55.00	2.75
Information Developer 2	0.00	0.00	0.00	0.00	0.00	0.00	0.00	0.00	0.00
Editor	0.00	2.00	4.00	0.00	0.00	0.00	0.00	6.00	0.30
Localization Coordinator	0.00	0.00	10.00	0.00	0.00	0.00	0.00	10.00	0.50
Subtotal	10.00	23.00	56.00	0.00	0.00	0.00	0.00	89.00 / **89.60**	4.45 / **5.60**
End User Training Guide	Page Count= 20		Metric= 5.60		Reuse= 50%				
Project Manager	2.00	2.00	2.00	0.00	0.00	0.00	0.00	6.00	0.30
Information Architect	8.00	4.00	0.00	0.00	0.00	0.00	0.00	12.00	0.60
Information Developer 1	0.00	0.00	0.00	0.00	0.00	0.00	0.00	0.00	0.00
Information Developer 2	0.00	0.00	25.00	0.00	0.00	0.00	0.00	25.00	1.25
Editor	0.00	0.00	3.00	0.00	0.00	0.00	0.00	3.00	0.15
Localization Coordinator	0.00	0.00	10.00	0.00	0.00	0.00	0.00	10.00	0.50
Subtotal	10.00	6.00	40.00	0.00	0.00	0.00	0.00	56.00 / **56.00**	2.30 / **5.60**
End user Help System	Topic Count= 80		Metric= 4.17		Reuse= 0%				
Project Manager	10.00	10.00	10.00	0.00	0.00	0.00	0.00	30.00	0.38
Information Architect	12.00	6.00	0.00	0.00	0.00	0.00	0.00	18.00	0.23
Information Developer 1	0.00	0.00	0.00	0.00	0.00	0.00	0.00	0.00	0.00
Information Developer 2	82.00	82.00	82.00	0.00	0.00	0.00	0.00	246.00	3.08
Editor	0.00	15.00	15.00	0.00	0.00	0.00	0.00	30.00	0.38
Localization Coordinator	0.00	0.00	10.00	0.00	0.00	0.00	0.00	10.00	0.13
Subtotal	104.00	113.00	117.00	0.00	0.00	0.00	0.00	334.00 / **333.60**	4.18 / **4.17**
Total Hours	244.00	232.00	305.00	0.00	0.00	0.00	0.00	781.00	
Hours/Period	154.00	140.00	154.00	147.00	154.00	147.00	147.00		
Full-Time Equivalent/Period	1.58	1.66	1.98	0.00	0.00	0.00	0.00		
Total Hours									
Project Manager								72.00	
Information Architect								82.00	
Information Developer 1								267.00	
Information Developer 2								271.00	
Editor								49.00	
Localization Coordinator								40.00	

Figure 18-5: Revised spreadsheet reflecting the reuse of content among deliverables

Each of the decisions has consequences for the success of your project. Review the spreadsheet in Figure 18-6. It shows a project with inadequate resources assigned. The project that was originally scheduled to be completed mid-January now has to be completed by mid-December. You've lost almost 200 hours from the project. You need to try reallocating the hours to find if you need additional help.

First, you need to move the 30 hours assigned in the last week of the project to the production coordinator into December 1–15 to complete the production work on the project. You also need to adjust the information architect hours in support of production. At this point, the architect is helping to prepare the final deliverables.

The information developer is now short by 98 hours originally scheduled for the end of December and early January. In addition, you had assigned 66 hours for information development in the two weeks through December 15. Because production has to occur in the last week, you need to cut back on the information developer's time during that week. If you add all the additional hours to your single information developer, he or she ends up working 99, 108, 77 hours in each of the final two-week periods, including the 108 hours during the November two-day holiday, because that's the last period that the writer has to complete the writing before production starts. Figure 17-6 shows the spreadsheet with the unfortunate writer asked to work 77 hours in one week and extra weekends to accommodate the holiday. You have a problem that can't be handled by dumping all the time on one person.

Unfortunately, I frequently find projects assigned just this way because someone assumed that one information developer could handle the job no matter what changes occur on the project. Either the individual works huge amounts of overtime to complete projects or begins cutting out critical activities from the schedule, often without letting anyone know.

Assigning overtime to the team

Many projects expand the hours available by assuming that team members will work the extra time required to keep the project on track. In most companies, full-time, direct employees are not paid for additional hours worked, although they may be compensated with extra time off after the project is completed. However, asking people to work regularly scheduled overtime to keep a project on track has significant consequences.

Occasional overtime is an accepted fact of life for information development, especially as deadlines near. But scheduled overtime takes a toll on the health and morale of dedicated employees. If you're willing to sacrifice your staff to your project, make sure that you're also willing to put in the extra time yourself. If everyone is working late, you need to be there to put in an extra effort. Also consider the possibility that you will lose more time than you've gained if staff members are sick more often or decide to look elsewhere for employment. Scheduled overtime is not a responsible tactic.

Neither, of course, is unscheduled overtime. When projects are incorrectly estimated or not estimated at all, project managers simply assume that a particular team member will complete an assignment, even if the assignment should have been given to two or three people rather than one. I'm always amazed at the assumption that one information developer can complete one book by the deadline, no matter how much work is involved. When the deadlines are missed or quality is compromised, the information developer is blamed, not the person who made the assignments in the first place.

If you are carefully estimating and planning your projects, you should be certain to assign adequate resources.

Projected Hours Spreadsheet

Factor	Skill Level	Projected Oct 1-15	Projected Oct 16-31	Projected Nov 1-15	Projected Nov 16-30	Projected Dec 1-15	Projected Dec 16-31	Projected Jan 1-15	Projected Total Hours	Hours/Unit
Tasks										
10%		Topic Count= 75		Metric= 5						
	Project Manager	3.50	5.30	12.59	6.21	6.30	0.00	0.00	33.89	0.45
	Information Architect	35.00	8.00	12.00	0.00	10.00	0.00	0.00	65.00	0.87
	Information Developer	0.00	40.00	99.00	54.00	33.00	0.00	0.00	226.00	3.01
15%	Editor	0.00	0.00	14.85	8.10	4.95	0.00	0.00	27.90	0.37
	Production Coordinator	0.00	5.00	0.00	0.00	15.00	0.00	0.00	20.00	0.27
	Subtotal	38.50	58.30	138.44	68.31	69.25	0.00	0.00	372.79 **375.00**	4.97 **5**
Concepts										
10%		Topic Count= 25		Metric= 7						
	Project Manager	1.50	2.80	0.00	7.21	4.30	0.00	0.00	15.81	0.63
	Information Architect	15.00	8.00	0.00	10.00	10.00	0.00	0.00	43.00	1.72
	Information Developer	0.00	16.00	0.00	54.00	20.00	0.00	0.00	90.00	3.60
15%	Editor	0.00	0.00	0.00	8.10	3.00	0.00	0.00	11.10	0.44
	Production Coordinator	0.00	4.00	0.00	0.00	10.00	0.00	0.00	14.00	0.56
	Subtotal	16.50	30.80	0.00	79.31	47.30	0.00	0.00	173.91 **175.00**	6.96 **7**
Reference										
10%		Topic Count= 15		Metric= 6						
	Project Manager	1.00	2.00	0.00	0.00	5.26	0.00	0.00	8.26	0.55
	Information Architect	10.00	8.00	0.00	0.00	20.00	0.00	0.00	38.00	2.53
	Information Developer	0.00	10.00	0.00	0.00	24.00	0.00	0.00	34.00	2.27
15%	Editor	0.00	0.00	0.00	0.00	3.60	0.00	0.00	3.60	0.24
	Production Coordinator	0.00	2.00	0.00	0.00	5.00	0.00	0.00	7.00	0.47
	Subtotal	11.00	22.00	0.00	0.00	57.86	0.00	0.00	90.86 **90.00**	6.06 **6**
	Total Hours	66.00	111.10	138.44	147.62	174.41	0.00	0.00	637.56	
	Hours/Period	60.00	66.00	66.00	54.00	66.00	54.00	54.00		
	Full-Time Equivalent/Period	1.10	1.68	2.10	2.73	2.64	0.00	0.00		

Figure 18-6: Project without adequate resources

Adding additional resources

Adding additional resources to the project from the beginning is a perfectly acceptable and even desirable solution as long as those resources are available to your organization. Adding resources may require that you move a person from another project, hire an additional employee, or hire a contractor to work on your team for the duration of the project. In each case, you have to plan how you will handle the additional resource load.

First, remember that adding more people to a project will not only increase the amount of project management and quality assurance time, but you may have to increase the percentage of time you have allocated for project management and quality assurance. Managing a larger group of people often adds coordination responsibilities. The more people you have to manage, the more possible interactions among the staff, potentially leading to conflict or inadequate collaboration, especially if the staff is geographically dispersed. If you have estimated 10% of total time for project management with a small staff, you may want to increase your project management time to 12% or 15%.

If you move an existing staff member onto your project from another project or add a new employee or contractor, you should recalculate your dependencies. If you have a completely new employee or contractor, you add risk and time to the project since the new team member needs training on subject matter, tools, and other project essentials. If you add an information developer from another project, you need to review that individual's experience with your project content. Of course, the earlier you add additional resources, the better. Each team member has an opportunity to learn about the project content at the same time. However, your more experienced team members may need extra time in their schedules to accommodate their responsibilities for orienting and training new staff.

Figure 18-7 shows the new spreadsheet with an additional information developer assigned. The new person starts slightly later than the first information developer and takes responsibility for the conceptual and reference information. Although the schedule works out well, you still have a problem to consider. To write the conceptual and reference information, an information developer needs to understand the tasks. You may need to rethink how you allocate work assignments when you consider leveling your resources as discussed later in this chapter.

Adjusting the schedule

The least damaging way to adequately staff your project with existing, experienced resources is to adjust the schedule. By extending the schedule, you staff the project with the people you had originally planned to use; you meet the deadlines with regular, full-time work; and you maintain the appropriate level of quality and scope for the project. In fact, many product-development projects are estimated with a contingency for additional time and extensions to deadlines, using the original staff. Contingency planning that calls for schedule extensions is the least damaging to the project's success.

Projected Hours Spreadsheet

Factor	Skill Level	Projected Oct 1-15	Projected Oct 16-31	Projected Nov 1-15	Projected Nov 16-30	Projected Dec 1-15	Projected Dec 16-31	Projected Jan 1-15	Projected Total Hours	Hours/Unit
Tasks		Topic Count= 75		Metric= 5						
10%	Project Manager	3.50	5.90	12.24	6.21	6.30	0.00	0.00	34.15	0.46
	Information Architect	35.00	8.00	12.00	0.00	10.00	0.00	0.00	65.00	0.87
	Information Developer 1	0.00	40.00	66.00	54.00	33.00	0.00	0.00	193.00	2.57
15%	Information Developer 2	0.00	0.00	30.00	0.00	0.00	0.00	0.00	30.00	5.08
	Editor	0.00	6.00	14.40	8.10	4.95	0.00	0.00	33.45	0.45
	Production Coordinator	0.00	5.00	0.00	0.00	15.00	0.00	0.00	20.00	0.27
	Subtotal	38.50	64.90	134.64	68.31	69.25	0.00	0.00	375.60	9.69
									375.00	**5**
Concepts		Topic Count= 25		Metric= 7						
10%	Project Manager	1.50	3.04	3.80	3.30	4.30	0.00	0.00	15.94	0.64
	Information Architect	15.00	8.00	10.00	10.00	10.00	0.00	0.00	43.00	1.72
	Information Developer 1	0.00	16.00	0.00	0.00	0.00	0.00	0.00	16.00	0.64
15%	Information Developer 2	0.00	0.00	33.00	20.00	20.00	0.00	0.00	73.00	24.01
	Editor	0.00	2.40	4.95	3.00	3.00	0.00	0.00	13.35	0.53
	Production Coordinator	0.00	4.00	0.00	0.00	10.00	0.00	0.00	14.00	0.56
	Subtotal	16.50	33.44	41.75	36.30	47.30	0.00	0.00	175.29	28.10
									175.00	**7**
Reference		Topic Count= 15		Metric= 6						
10%	Project Manager	1.00	1.00	0.00	3.68	2.50	0.00	0.00	8.18	0.55
	Information Architect	10.00	8.00	0.00	0.00	20.00	0.00	0.00	38.00	2.53
	Information Developer 1	0.00	0.00	0.00	0.00	0.00	0.00	0.00	0.00	0.00
15%	Information Developer 2	0.00	0.00	0.00	32.00	0.00	0.00	0.00	32.00	32.00
	Editor	0.00	0.00	0.00	4.80	0.00	0.00	0.00	4.80	0.32
	Production Coordinator	0.00	2.00	0.00	0.00	5.00	0.00	0.00	7.00	0.47
	Subtotal	11.00	11.00	0.00	40.48	27.50	0.00	0.00	89.98	35.87
									90.00	**6**
	Total Hours	66.00	109.34	176.39	145.09	144.05	0.00	0.00	640.86	
	Hours/Period	60.00	66.00	66.00	54.00	66.00	54.00	54.00		
	Full-Time Equivalent/Period	1.10	1.66	2.67	2.69	2.18	0.00	0.00		
Total	Project Manager	6.00	9.94	16.04	13.19	13.10	0.00	0.00	58.26	
	Information Architect	60.00	24.00	12.00	10.00	40.00	0.00	0.00	146.00	
	Information Developer 1	0.00	56.00	66.00	54.00	33.00	0.00	0.00	209.00	
	Information Developer 2	0.00	0.00	63.00	52.00	20.00	0.00	0.00	135.00	
	Editor	0.00	8.40	19.35	15.90	7.95	0.00	0.00	51.60	
	Production Coordinator	0.00	11.00	0.00	0.00	30.00	0.00	0.00	41.00	
									640.00	

Figure 18-7: Adding an additional resource to your project

If you are part of a larger project that is wildly adding staff after the beginning of the project, rather than extending the schedule, you are part of a project that is in big trouble. Adding staff during the project contributes to breakdowns of training, experience, and coordination. Projects are much better off when they extend the schedule, even though this action also has consequences:

✔ An extended schedule may result in a missed window of opportunity for selling the product.

✔ Competitors may get to the market first.

✔ Staff members assigned to the project become unavailable for subsequent projects, affecting many schedules.

✔ Staff members assigned temporarily to the project cannot get back to their regular jobs, affecting the overall quality and timeliness of the organization's work.

Despite the consequences, schedule adjustments are routine, even expected, in many product- and service-development organizations. Unfortunately, information development rarely has the clout in the organization to ask for a schedule extension. As a project manager, you have to try to anticipate delayed schedules by communicating with the product developers. By interacting regularly with your colleagues in the larger project, you can often learn about problems that are likely to delay the project before they are officially announced. If you work in an organization that is known for underestimating its projects, you often find yourself playing the same game. You know that the resources assigned to the project are inadequate to meet the announced deadline, but you know that in 100% of the projects in your organization, the deadline is always extended. Not a very responsible way to manage projects but the reality in many organizations, nevertheless.

Note that the activities required of a project manager might be viewed as micromanaging by some staff. If you are concerned about the possibility of this response, discuss the responsibilities and their affect on the information developers and explain the business reasons behind them. You are trying to protect the project and the information developers and ensure that they have adequate resources to support their work.

Changing the scope

A more responsible way to manage your project resources is to decrease the scope of the project. When your estimates demonstrate that the project cannot be completed by the deadline with the assigned resources, you should look carefully at the project deliverables and decide if something can be eliminated or postponed. If you have planned two user guides for the project, one for the business owner and the other for the individual using the product, can you delay one of these deliverables, thereby extending the schedule? Can you remove topics from the schedule and exclude less critical content from the initial deliverables? Can you minimize the content further, perhaps by decreasing the number of illustrations or eliminating some of the conceptual information?

Each scope change does, of course, have consequences for the quality or the timeliness of the information you are delivering to the customer. Fewer illustrations or screen shots may make learning more difficult for the customer. Eliminating conceptual information may make the tasks more difficult to perform correctly. You may increase the costs of product support after the release because the information is not adequate for some customers.

Deliberately decreasing the scope of the deliverables is, however, a much preferred solution to inadvertently or deliberately decreasing the quality of the deliverables. When there is more work to do than there are resources to do it, quality suffers, even when people work lots of extra time to meet the deadlines. Quality assurance is eliminated or shortchanged. Reviews are put under pressure or omitted. Technical information is regurgitated from inadequate sources rather than restructured or rewritten to better meet the needs of the customers. The quality of the work suffers quietly.

As a project manager, you have two responsibilities: to deliver the project on time and to deliver the quality required by your customers. Inadequately estimating and staffing a project jeopardizes both responsibilities.

Leveling your resources

An additional consideration to keep in mind as you review the staffing for your project is resource leveling. Resource leveling refers to the process of ensuring that everyone assigned to the project has approximately the same workload. As a project manager, you want to ensure that everyone is working about the same amount of time, not that some people are overloaded and others have time to waste. Remember that people will fill the time available to them, often by adding unnecessary details and complexity if they don't have enough to occupy them productively. Every team member should be tasked to add value for the customer but not to bloat the content because they have more time than needed.

If you are developing your workload assignments using topic-based estimating, you are more likely to balance the workload for each team member than if you assign work by deliverables. Without adequate estimates, using the old standby of one information developer, one book, it is extremely likely to overload one person while under loading another.

Review your topic-based work assignments. Consider with your information architect and developers the complexity of the topic assignments. Ensure that one person does not have all the difficult topics and another all the easier ones, although your more senior people will be expected to handle more challenging assignments. Recognize that you might not know enough at the beginning of the project to evaluate the complexity of the proposed topics. Be alert to the need to adjust the estimates and ensure level workloads. Be alert both to signs of overloading and under loading by setting up and monitoring interim milestones. Talk to your team members regularly and help them to recognize when they have too much or too little to do.

Figure 18-8 shows a small project with 13 task, 3 concept, and 2 reference topics. You have estimated during the Planning Phase that you need 98 hours to complete these topics based on the metrics of 5, 7, and 6 hours per topic. During the Design Phase, you have asked your team members to estimate the scope of topics, including their size and complexity. You've asked them to assign values from .5 to 4.5 hours per topic. As a result, your new estimate for the project increases from 98 hours to 131 hours. That indicates that you either have to eliminate topics, negotiate more time for the project schedule, negotiate additional people for the project, or ask people to work faster than they've estimated. That might mean reducing the quality of the final deliverables or finding a more efficient way to work.

You decide to increase efficiency in the project and normalize the hours assigned to each topic and return to your initial estimate of 98 hours. However, you need to be certain that each staff member if assigned a similar amount of work. Since the topics differ in difficulty, it would be easy to assign all the difficult topics to one writer and all the easy topics to another. That's not equitable and is likely to jeopardize project quality. By leveling the hours among the three writers, you ensure that the work is shared fairly among them.

In the spreadsheet in Figure 18-8, Alice, Dave, and Tom are assigned different numbers of topics based on the scope of each topic. Alice is assigned nine topics, Dave is assigned six topics, and Tom is assigned four topics. Alice and Dave share the development of one topic. They each work approximately 32 hours on the project. Tom is obviously assigned the more difficult topics than Alice or Dave, but their workload is approximately the same.

Resource Allocation Spreadsheet

Priority	Topic	Avg Hours/ Topic	Topic Scope	Topic Hours	Normal- ized Hours	Alice Tasks	Alice Hours	Dave Tasks	Dave Hours	Tom Tasks	Tom Hours
1	Task 1	5	0.5	2.5	2.0	1	2.0		0.0		0.0
2	Task 6	5	0.5	2.5	2.0	1	2.0		0.0		0.0
14	Concept 3	7	1	7.0	4.0		0.0	1	4.0		0.0
3	Task 4	5	1	5.0	4.0	0.5	2.0	0.5	2.0		0.0
4	Task 9	5	1.5	7.5	6.0	1	6.0		0.0		0.0
18	Reference 1	6	3	18.0	12.0		0.0	1	12.0		0.0
5	Task 11	5	1	5.0	4.0		0.0	1	4.0		0.0
15	Concept 2	7	0.5	3.5	2.0	1	2.0		0.0		0.0
6	Task 12	5	4.5	22.5	18.0		0.0		0.0	1	18.0
7	Task 7	5	0.5	2.5	2.0		0.0		0.0	1	2.0
8	Task 2	5	0.5	2.5	2.0	1	2.0		0.0		0.0
13	Reference 2	6	0.5	3.0	2.0		0.0		0.0	1	2.0
9	Task 3	5	1.5	7.5	6.0	1	6.0		0.0		0.0
10	Task 8	5	2	10.0	8.0		0.0	1	8.0		0.0
12	Concept 1	7	1	7.0	4.0		0.0	1	4.0		0.0
11	Task 10	5	0.5	2.5	2.0	1	2.0		0.0		0.0
17	Task 13	5	2	10.0	8.0	1	8.0		0.0		0.0
16	Task 5	5	2.5	12.5	10.0		0.0		0.0	1	10.0
		98	24.5	131	98		32.0		34.0		32.0

Figure 18-8: Resource leveling

Best Practice—Developing topic milestones

After you have created your spreadsheets, allocated your resources, and made specific topic-based assignments, you are ready to monitor the project through the Development Phase of the project. Assume that all of the high-level and detailed planning of deliverables, topics, and schedules has taken you through the end of the Design Phase of the project. Now the third phase, development, has begun. During the Development Phase of the project, the role of the project manager is focused on monitoring the progress and making adjustments.

During the Development Phase, the details of topic-based schedule may conveniently be managing by the information developers themselves. Because they are closest to the product- or service-development activities, they should be able to tell you what topics should be scheduled at what point during development.

Assigning topics to milestones related to the product development feature schedule

If you are managing an information-development project associated with the development of a hardware or software product, you may find it advantageous to correlate the schedule and milestones of your topic development with the schedule for feature development in the product-development project. Both traditionally managed product development and agile product development methods often focus on the development of features and functions within software. Similarly, hardware is often developed as an assembly of components that have to be explained for use or for maintenance. By organizing the schedule of topic development to align with the development schedule for the product, you are more likely to succeed in meeting interim and final milestones and deadlines.

Without this alignment, you will find that writers waste time working on topics for which there is little source information or prototypes rather than working productively on those parts of the product that are ready to be written about for the customers. Unfortunately, one of the most serious disconnects with poorly managed software development occurs when the user interface is the last item to be defined, drafted, tested, and completed. Because much end-user information is directed toward interactions with the product interface, delaying interface design often results in rushed and problematic information.

To assign topics in alignment with the development of product features requires that you gather detailed information about the product-development schedule. By discussing the schedule with the project managers working on the product, with individual developers, and with people in quality assurance or testing, you can relate your schedule to theirs more effectively. Often a liaison with the testing organization will give you insight into the schedules. Attending project meetings for the larger project can help you track what is being worked on, how the development is going, what is being delayed for later iterations, and what is on schedule. Equipped with this information about the larger project, you can help the information developers schedule their own work more effectively. With the right group of information developers, you can give them the responsibility to gather the feature schedule data and report back to you.

An added advantage of the alignment with the development schedule is the ability to send small groups of topics to the appropriate developers for review at the same time they are engaged in the development of the features or functions with which the topics are associated.

Developing tracking milestones (focusing on small increments of work)

With traditional book-based information development, tracking is usually associated with major milestones, such as the following interim events:

- ✔ Information plan for the project

- ✔ Annotated tables of contents for the final deliverables

- ✔ First draft of a complete book or completed chapters or sections for edit and review

- ✔ Second draft of a complete book for copyedit, final review, and approval

- ✔ Drafts made available for field testing

- ✔ Final draft ready for production edit

- ✔ Final production deliverables

Despite the long history associated with milestones like these, they present problems. They are difficult to track because it is difficult to tell how much work is complete and how much is left to be done (percent complete calculations are described in more detail below). They often require long editing and review cycles because of the large amount of content. They are often sent out as a whole to reviewers who are expert only in small portions of the content. Reviewers find it difficult to review content that they don't understand themselves, often leading to late or absent review comments or conflicts among the reviewers themselves.

You will find it beneficial to provide guidelines to reviewers that explain your expectations for the review. Reiterate that you expect them to focus on the technical content not the writing style. Their expertise is required to ensure that information is correct, but writing style, layout, and even the degree of completeness is the responsibility of information development.

Table 18-4 provides an example of a traditional set of milestones.

Table 18-4: Project Milestone Tracking

Deliverable	Percent of Total Time	Milestone Date
List of project deliverables with a brief description if necessary	Note the percentage of the total project time that is represented by this deliverable.	State the date on which the deliverable is due.
Example: complete first draft of User Guide	60% of total project time	June 21

Agile project management suggests that we might handle interim milestones differently. A topic-based architecture provides an excellent vehicle to track work completed and work left to be done at a more granular level, giving project managers greater insight into the status of the project and its progress toward successful completion.

Table 18-5 illustrates the milestones that you might associate with individual topics.

Table 18-5: Topic-Based Milestone Tracking

Deliverable	Time Allocated	Milestone Date
Annotated topic list	30% of total project time	Date calculated according to allocations of time to the phases of the project
A group of topics associated with a product feature or function	Calculated according to hours per topic Could include a dependencies calculation	Individual dates per topic group

Deliverable	Time Allocated	Milestone Date
Within each topic group, schedule First draft complete Editing and review complete Second draft complete Editing, reviews, and approvals complete final draft complete	Interim hours calculation applied 60% of total content-development hours per topic 5–10% for editing; review time not included *20% of total hours per topic 5–10% for final editing *5% or less for final corrections	Individual dates based on topic schedules
Development of final deliverables maps	Part of the initial 30% Planning Phase of the project	Date calculated according to allocations of time to the phases of the project
Final production of deliverables in multiple media	5–20% of total project time	Date calculated according to allocations of time to the phases of the project Calculated back from final release date

The example percentages in Table 18-5 help to ensure that you have sufficient time in the schedule to produce interim deliverables. Note that the first draft deliverables should take 60% of the total content-development time allocated to the schedule. If interim drafts must be delivered before 60% of the time has been scheduled, you must either reschedule the interim milestone or modify the resource allocation for the first draft. You may actually need more staff resources prior to the first draft deliverable of a group of topics than you need for the second or final draft.

Defining measurable milestones

It is not enough simply to assign a due date to an interim deliverable. Anyone can hand in a version of content on a particular date without ensuring that the requirements of that interim deliverable have been met. Too often information developers simply hand in whatever they have completed at a particular date, rather than ensuring that they have reached an appropriate goal with the content produced thus far.

As a project manager, you must avoid treating milestones as dates on the calendar. Milestones are defined events in the information-development life cycle. Milestones must be measurable.

Working with your project team, define how you will judge whether or not a milestone has been reached. For example, you might define the first draft of a group of topics this way:

- ✔ Ninety percent of the information in each topic is complete.

- ✔ Graphics are sketched or described, and screen shots are listed in place.

- ✔ The topics are submitted for review within the appropriate map or maps, showing where they will be placed in a larger context (a table of contents).

- ✔ A spelling and grammar check has been run on the topics.

A first draft at this level of completion is ready for a development edit by team members or a developmental editor or ready for review by product developers.

You may argue that it may be difficult to decide if a draft topic is 90% complete. What exactly would that mean? You and your team members need to decide what a complete draft might contain. For example, a complete first draft might include a small number of queries to the product developers but should contain sufficient content to indicate the direction of the task, the concept, or the reference information.

Each of the milestones you set for the content development should be defined as carefully as you define the first draft milestone. The resulting measurable milestones, accompanied by a percentage of total hours worked, will help you in managing the milestone scheduled. For example, difficulties in the product- development process may cause certain draft topics to be incomplete at the time they are due. If your information developers have already used all their first draft time to produce an incomplete set of topics, you know you will have to increase the amount of time allocated before the next milestone to complete the missing topics. Measurable milestones provide invaluable information about the progress of your project.

Remember that just because someone claims that a task is finished does not mean that it is correct. Be sure that you do not record a task as complete until the work has been reviewed by at least one other person and that person understands the requirements for measuring completion.

Assessing progress

Clearly you have an opportunity to assess the progress made on your project when you reach an interim milestone. However, you are likely to discover that the due date for a group of topics is too late to ensure that your team is making adequate progress on the work. You don't want to discover at the last minute before a milestone that the required work is not complete. At the last minute, there is nothing you can do to recover.

Meeting with team members

Project management requires that you hold regular meetings with your team members to discuss progress and problems with the project. Informal meetings one on one or brief team meetings will help everyone become aware of the need to alert the manager or other team members to events that may affect the project. One information developer who is unable to research a topic effectively because developers or other experts are unavailable may cause the entire project to head off track. Information developers should be encouraged to report problems as well as successes. If problems are addressed early, it might be possible to rearrange the schedule, revise the priorities for topic development, or add additional resources in time to make a difference.

One-on-one meetings with new team members will also help you assess the quality of their work before it becomes a problem. You might suggest that you review completed topics together to ensure that the new team members understand the style guide, are correctly following the design pattern for the content, or are using the tools correctly.

It might also be helpful to foster a collaborative environment by inviting more senior team members to review their approach to the content during a team meeting. In this way, all members of the team gain insight into the best way to handle the information. Such reviews are similar to the code review, instituted by product developers. Topic reviews serve the same purpose—to ensure that the development is going in the right direction before it is too late to change. Be aware that you will save considerable rework time on the project by setting the direction early.

Developing a tracking system

Although there is really no substitute for regular meetings among team members and with the project manager, you will also need a tracking system that allows you to account for the time expended on your project and the progress made.

A simple weekly report that is well structured and takes minimal time to complete will help you and your team members track the project. I have long used the weekly project reporting form in Figure 18-9 to record activity on one or multiple projects.

The best method for keeping track of activities and progress is to complete the form each day or as you complete work during the day. If you neglect to complete the form daily, you will not remember what you've done nor how much time you spent doing it by the end of the week.

Note the project you are working on, the type of activity, the time spent, and a brief description of the activity. At the end of the form, note the following critical information about work status:

✓ Which tasks were you not able to start that should have started?

✓ Which tasks were you supposed to have completed but were unable to complete?

✓ Was the amount of time you spent on the tasks in line with what was estimated and expected?

✓ Which tasks took longer or less time than you expected them to take?

If each team members conscientiously considers these factors in each weekly report, you will have a reasonable start at judging the progress of your project. However, the progress reports alone will not be sufficient. Regular meetings with individuals and the staff as a whole are absolutely required to provide you with the adequate information.

Remember that no one has to account for all of the time in a day or a week on the project reporting form. Everyone is expected to spend time on activities that are not part of the project.

Remember that if you ask your team members to complete weekly project reports, you must complete one yourself, and you are obligated to review their reports and enter the information into your tracking spreadsheet. You will find that some team members are confused about time reporting and tracking their progress. They may include information in their reports that is too vague or incomplete or too detailed. They may not know what to do with time spent on departmental activities outside their project work. By reviewing progress reports early in the project, you have the opportunity to offer suggestions about the information you would like to see.

Your tracking spreadsheet begins with your topic estimating spreadsheet and information about the interim milestones and the amount of time allocated to each topic group set. As you receive progress reports, note how much time has been spent on a particular topic or group of topics and whether an interim milestone has been met. Your tracking spreadsheet might look like the one in Figure 18-10.

Weekly Time Report   **Week of: Jan 15**_____

Name: Pubs Writer--George

Project	Subject Area	Project	Draft	Activity	Time	Description
Sun						
Mon						
X100	Install	Starting the install	1	Writing (W)	2 hr	Began developing the installation topics. Starting provides the pre-requisites for the process
	Install	Working on the install	1	Writing (W)	4 hrs	Worked through all the details of the lengthy install process with Jim. Discussed the total number of images that we might want to includes
Total Project time					6 hrs	
Tues						
X100	Install	Working on the install	1	Writing (W)	6 hrs	Continued the development of this section. Will divide into six topics to handle the detail of each component of the install separately.
Total Project time					6 hrs	
Wed						
X100	Install	Acquiring APS	1	Writing (W)	1 hr	Split the Working topic into six topics. Forwarded the info on the additional topics to Info Architecture for approval.
		Acquiring SVP	1	Writing (W)	Same	
		Acquiring XSL	1	Writing (W)	Same	
		Installing the first package	1	Writing (W)	Same	
		Installing the second package	1	Writing (W)	Same	
		Completing the install	1	Writing (W)	Same	
	Install	Reconfigured the Map	3	Writing (W)	2 hrs	Restructured the install map with the Info Architect to reflect the new topic division
	Install	Acquiring APS	1	Writing (W)	4 hrs	Completed the development of the first draft of this topic. Send to the reviewer to check several of the details.
Total Project time					7 hrs	
Thur Holiday						
Fri Holiday						
Sat						

Progress

Project	Subject Area	Topics	Draft	Activity	Hours Allocated	Hours Used This Week	Hours Used to Date	Date Due	% Complete
X100	Install	25	1	Writing	65	19	19	Feb 15	25 %

Plans

Will continuing working on the Install topics all next week. Should have the first draft of all topics completed and ready for first draft review by info architecture and engineering.

Problems

- None at this time.

Figure 18-9: Weekly project reporting form example

Projected Hours Spreadsheet

Factor	Skill Level	Projected Oct 1-15	Actual Oct 1-15	Projected Oct 16-31	Actual Oct 16-31	Projected Nov 1-15	Actual Nov 1-15	Projected Nov 16-30	Actual Nov 16-30	Projected Dec 1-15	Actual Dec 1-15	Projected Dec 16-31	Actual Dec 16-31	Projected Jan 1-15	Actual Jan 1-15	Projected Total Hours	Actual Total Hours	Actual Minus Projected	Projected Hours/Unit	Actual Hours/Unit
Tasks			Topic Count= 75					Metric= 5												
10%	Project Manager	3.50	5.00	5.90	6.00	12.24	0.00	6.21	0.00	6.30	0.00	0.00	0.00	0.00	0.00	34.15	11.00	-23.15	0.46	0.15
	Information Architect	35.00	30.00	8.00	10.00	12.00	0.00	0.00	0.00	10.00	0.00	0.00	0.00	0.00	0.00	65.00	40.00	-25.00	0.87	0.53
	Information Developer 1	0.00	0.00	40.00	35.00	66.00	0.00	54.00	0.00	33.00	0.00	0.00	0.00	0.00	0.00	193.00	35.00	-158.00	2.57	0.47
	Information Developer 2	0.00	0.00	0.00	0.00	30.00	0.00	0.00	0.00	0.00	0.00	0.00	0.00	0.00	0.00	30.00	0.00	-30.00	5.08	0.00
15%	Editor	0.00	0.00	6.00	0.00	14.40	0.00	8.10	0.00	4.95	0.00	0.00	0.00	0.00	0.00	33.45	0.00	-33.45	0.45	0.00
	Production Coordinator	0.00	0.00	5.00	0.00	0.00	0.00	0.00	0.00	15.00	0.00	0.00	0.00	0.00	0.00	20.00	0.00	-20.00	0.27	0.00
	Subtotal	38.50	35.00	64.90	51.00	134.64	0.00	68.31	0.00	69.25	0.00	0.00	0.00	0.00	0.00	375.60 **375.00**	86.00	-289.60	9.69 **5**	1.15
Concepts			Topic Count= 25					Metric= 7												
10%	Project Manager	1.50	1.00	3.04	3.00	3.80	0.00	3.30	0.00	4.30	0.00	0.00	0.00	0.00	0.00	15.94	4.00	-11.94	0.64	0.16
	Information Architect	15.00	22.00	8.00	6.00	0.00	0.00	0.00	0.00	10.00	0.00	0.00	0.00	0.00	0.00	43.00	28.00	-15.00	1.72	1.12
	Information Developer 1	0.00	0.00	16.00	12.00	0.00	0.00	0.00	0.00	0.00	0.00	0.00	0.00	0.00	0.00	16.00	12.00	-4.00	0.64	0.48
	Information Developer 2	0.00	0.00	0.00	0.00	33.00	0.00	20.00	0.00	20.00	0.00	0.00	0.00	0.00	0.00	73.00	0.00	-73.00	24.01	0.00
15%	Editor	0.00	0.00	2.40	2.00	4.95	0.00	3.00	0.00	3.00	0.00	0.00	0.00	0.00	0.00	13.35	2.00	-11.35	0.53	0.08
	Production Coordinator	0.00	0.00	4.00	5.00	0.00	0.00	0.00	0.00	10.00	0.00	0.00	0.00	0.00	0.00	14.00	5.00	-9.00	0.56	0.20
	Subtotal	16.50	23.00	33.44	28.00	41.75	0.00	36.30	0.00	47.30	0.00	0.00	0.00	0.00	0.00	175.29 **175.00**	51.00	-124.29	28.10 **7**	2.04
Reference			Topic Count= 15					Metric= 6												
10%	Project Manager	1.00	1.00	1.00	1.00	0.00	0.00	3.68	0.00	2.50	0.00	0.00	0.00	0.00	0.00	8.18	2.00	-6.18	0.55	0.13
	Information Architect	10.00	12.00	8.00	8.00	0.00	0.00	0.00	0.00	20.00	0.00	0.00	0.00	0.00	0.00	38.00	20.00	-18.00	2.53	1.33
	Information Developer 1	0.00	0.00	0.00	0.00	0.00	0.00	0.00	0.00	0.00	0.00	0.00	0.00	0.00	0.00	0.00	0.00	0.00	0.00	0.00
	Information Developer 2	0.00	0.00	0.00	0.00	0.00	0.00	32.00	0.00	0.00	0.00	0.00	0.00	0.00	0.00	32.00	0.00	-32.00	32.00	0.00
15%	Editor	0.00	0.00	0.00	0.00	0.00	0.00	4.80	0.00	0.00	0.00	0.00	0.00	0.00	0.00	4.80	0.00	-4.80	0.32	0.00
	Production Coordinator	0.00	0.00	2.00	2.00	0.00	0.00	0.00	0.00	5.00	0.00	0.00	0.00	0.00	0.00	7.00	2.00	-5.00	0.47	0.13
	Subtotal	11.00	13.00	11.00	11.00	0.00	0.00	40.48	0.00	27.50	0.00	0.00	0.00	0.00	0.00	89.98 **90.00**	24.00	-65.98	35.87 **6**	1.60
Total	Total Hours	66.00	71.00	109.34	90.00	176.39	0.00	145.09	0.00	144.05	0.00	0.00	0.00	0.00	0.00	640.86 **640.00**	161.00	-479.86		
	Hours/Period	60.00		66.00		66.00		54.00		66.00		54.00		54.00						
	Full-Time Equivalent/Period	1.10	1.18	1.66	1.36	2.67		2.69		2.18										
	Project Manager	6.00	7.00	9.94	10.00	16.04	0.00	13.19	0.00	13.10	0.00	0.00	0.00	0.00	0.00	58.26	17.00			
	Information Architect	60.00	64.00	24.00	24.00	12.00	0.00	10.00	0.00	40.00	0.00	0.00	0.00	0.00	0.00	146.00	88.00			
	Information Developer 1	0.00	0.00	56.00	47.00	66.00	0.00	54.00	0.00	33.00	0.00	0.00	0.00	0.00	0.00	209.00	47.00			
	Information Developer 2	0.00	0.00	0.00	0.00	63.00	0.00	52.00	0.00	20.00	0.00	0.00	0.00	0.00	0.00	135.00	0.00			
	Editor	0.00	0.00	8.40	2.00	19.35	0.00	15.90	0.00	7.95	0.00	0.00	0.00	0.00	0.00	51.60	2.00			
	Production Coordinator	0.00	0.00	11.00	7.00	0.00	0.00	0.00	0.00	30.00	0.00	0.00	0.00	0.00	0.00	41.00	7.00			

Figure 18-10: Project-tracking spreadsheet example

In the example, we can see that one month of the project is past, and the project is slightly behind schedule. The project manager estimated 66 hours for the first two-week period, but the information architect used more hours than predicted (64 instead of 60). During the second two-week period, the estimate called for 109.34 hours, but only 90 hours were used. It looks like the information developer got a slower start, and the editor didn't have much to do as yet. It should not be difficult to get the project back on the estimated hours budget in November.

Not only will you have to track time spent and milestones completed for your own project staff, you will also need to meet with others involved with the project to ensure that their work is progressing adequately. If you are responsible for measuring the progress of graphics development, editing, localization and translation, production, or reviews and approvals, you will need to know how these activities are progressing even when they are under the management of people outside your team. You may want to ask people who are coordinating these activities to provide you with progress reports as well.

Using automated workflow systems to track progress

If you are working with a content management system in which the status of a topic or section of a document is recorded in the topic metadata by the information developer, you can use direct reports from the workflow system to help track progress. The reporting system should tell you if

- ✔ a topic is in progress

- ✔ the first draft of a topic has been completed

- ✔ a topic is in developmental editing or copyediting

- ✔ a topic is in review

- ✔ a topic has been revised and the second draft completed

- ✔ the topic is being tested and validated

- ✔ a topic is in final draft and has been delivered for translation or final production

Any number of useful data can be recorded by setting up status reporting in an automated workflow system. You should have sufficient information to help you judge at any point in time if your project is on track and progressing adequately toward interim milestones and final deliverables. The automated workflow system reporting won't give you all the information you need. You will still need to meet with team members and ask them to complete their weekly progress reports on time.

Best Practice—Reporting progress

If you have been meeting regularly with team members, receiving their weekly progress reports, adding the data to your project tracking spreadsheet, and reviewing data from the workflow system status reports, you should have all the information you need to produce monthly or more frequent reports to your management about the progress of your project.

You should work hard to ensure that, in addition to your own accountability, your management and the managers of the larger development project take responsibility for the success of your project. They need to provide assistance if you ask for it and become actively involved in helping to resolve conflicts and problems that may have emerged.

It is not enough simply to send your project report by email to your management. You must meet with the project sponsors and stakeholders to deliver your report and ask for their assistance if needed. If you don't ask for help in a timely manner, you will be held accountable if the project fails to deliver. If your project is high risk, you need to meet frequently with management to ensure that you are on the right track and are pursuing appropriate solutions to problems.

Scheduling your reports

Short projects, those that take three months or less to complete, require that you report progress weekly or every other week at the most. Larger projects generally require monthly reporting, although if your project is especially high risk or if you are experiencing problems with schedule, resources, or quality, you may need to report more frequently.

Developing a project-reporting template

The typical information you need to provide in your periodic report as project managers is as follows:

✔ Summarize the current progress of your project, including what has been accomplished to date and whether or not the project is following the plan and is presently on track.

✔ If your project is not on track and is experiencing problems, state the reasons for the problems.

✔ Note the actions that you have taken to solve the problems.

✔ Explain what actions, if any, you need your management to take to support you in resolving the problems.

✔ Summarize the project tracking information, including the progress on topics or groups of topics, interim milestones, and deadlines.

Remember that the best project managers are not afraid to ask for help as soon as they need it. Waiting too long, ignoring project problems, or demanding that everyone put in exorbitant overtime all happen on projects. Just don't let them happen on yours.

Figure 18-11 provides an example of a typical progress report and provides an example the information that might be included in each section.

Progress Report

BB Information Development
April – May 15

Progress (April 1 – May 31)

Most of the work on BB this month was for usability. Activities included:

- April 13, Pubs delivered the first draft of the Usability Assessment Plan.

- April 29, Austin delivered review comments on the Usability Assessment Plan.

- May 3, Pubs and Austin discussed the Usability Assessment Plan review comments. Because BB's beta release is postponed, we proposed that an Interface Questionnaire that Austin could use to gather customer feedback on the interface in time to make any needed interface changes before beta is released.

- May 3, Pubs faxed Austin a draft Interface Questionnaire.

- May 7, Pubs received Austin's review comments on the Interface Questionnaire.

- May 14, Pubs and Austin discussed the Interface Questionnaire review comments and faxed Austin information on usability test labs in the Boston area..

- May 16, Pubs delivered hard and soft copy of the Interface Questionnaire.

Plans (June – July)

In the next two-month period, we plan to concentrate our efforts on the following ways:

- Return the revised Usability Assessment Plan to Austin for final approval.

- Meet with Austin to determine what content to write up for BB's beta release and to revise the deliverables schedule per the new beta release date.

- Deliver beta documentation.

- Discuss reviewer comments and finalize beta documentation as needed.

Concerns

BB has been changing enough that we felt it would not be prudent to spend project hours trying to determine which features do and do not work in each version. For this reason, we have not attempted to draft any documentation for the features that work, preferring instead to wait until beta release, when the software modules are working, to begin drafting.

We will need to meet with you to determine a new schedule for deliverables, given the new beta release date.

Financial Report

As shown in the following table, we have used just under 60% of the budget but estimate that we are only a little over 50% complete. The discrepancy in hours results from the delay in the software schedule and resulting lack of information. When we recognized this fact last month, we stopped work on the project with the exception of the interface and usability assessments. We hope to make up the 6% discrepancy when work begins again. We will keep Austin informed about the feasibility of this plan once the software is ready to be documented and we can finalize the scope of the documentation.

Deliverable	Projected Hours	Actual Hours	Percent Hours Used	Percent Complete
New Ideas	**2468.75**	**1442.25**	**58.42%**	**52.41%**
Information Plan	321.00	321.00	100.00%	100.00%
User Guide CP	124.25	124.25	100.00%	100.00%
User Guide D1	116.25	67.25	57.85%	50.00%
User Guide D2	72.00	0.00	0.00%	0.00%
User Guide D3	46.50	0.00	0.00%	0.00%
Install CP	40.50	40.50	100.00%	100.00%
Install D1	82.00	6.00	7.32%	0.00%
Install D2	54.50	0.00	0.00%	0.00%
Install D3	40.75	0.00	0.00%	0.00%
Help CP	397.25	397.25	100.00%	100.00%
Help D1	540.25	367.50	68.02%	40.00%
Help D2	225.00	0.00	0.00%	0.00%
Help D3	135.00	0.00	0.00%	0.00%
Usability/Interface	273.50	118.50	43.33%	50.00%

Figure 18-11: Periodic progress report example

Best Practice—Building your Project Management folder

At this stage of the project with Phase 1: Planning and Phase 2: Design completed and work being accomplished in Phase 3: Development, you have produced a variety of work products to help you manage the project. All the information you have developed to plan and manage your project should be maintained in a Project Management folder. That file might be part of your content management system or might be maintained separately in your file system. If you have a content management system, use the version control capabilities of the system to record the changes that you make in your planning, design, and development documents. If you are using a file system, develop a naming convention to show changes to initial plans and use comments to record your notes.

At this point in your project, the Project Management folder should include the following:

✔ Your initial project plan with appropriate approvals and changes as necessary

✔ Your initial project spreadsheet with changes

✔ Your topic spreadsheet with all topics assigned and scheduled with changes

✔ Your resource leveling spreadsheet

✔ The weekly progress reports of your team members

✔ Minutes of team and other meetings

✔ Your project reporting spreadsheet

✔ Your periodic project reports

✔ Other reports provided to you by stakeholders outside your team

When your project is complete, this information will prove essential in producing a final report and evaluating the success of the project. The information will also be invaluable for future projects.

Summary

In this chapter, you dove into the heart of project management. Through the development of a series of project spreadsheets, you have learned to

✔ develop a framework for creating a project spreadsheet

✔ add people time to your spreadsheet based on traditional book-based deliverables

✔ add people time to a spreadsheet developed for a topic-based project

✔ handle a project in which you had initially assigned inadequate resources by adding an additional information developer to the project

✔ consider the effect of adding resources to a project as well as rethinking the scope of the project or assuming that more time will be available because the deadline is usually extended

✔ recognize that you have not allocated time fairly among information developers and learn to use a leveling technique for your project

In addition, you have a standard form that you and your team members might use to track project time, a standard spreadsheet that you can use to record the time everyone is using on a project, and a standard format for a periodic project report that you are now prepared to write for your management and other stakeholders on the project.

Many people who suddenly have project management thrust upon them believe they don't have a "number" sense. I have tried to make the spreadsheets obvious and easy to use. You can use a spreadsheet program that will calculate sums and products for you if you learn to add a few formulas, or you can calculate your spreadsheets on paper by hand. You can also find the project spreadsheets presented in this chapter on the book's website for downloading.

The next chapter continues your project management education by learning to handle the inevitable changes that occur on most projects.

"We better rethink this project."

Chapter 19

Managing as the Project Changes

> Good design begins with honesty, asks tough questions, comes
> from collaboration and from trusting your intuition.
>
> —Freeman Thomas

During Phase 3 of the information-development life cycle, your
project is well underway. Information developers are creating
content, editors and reviewers are suggesting changes, topics
are being validated for accuracy and completeness, and localization and production coordinators are anticipating the move into
Phase 4. If you are using a topic-based architecture, topics may
already be scheduled for translation as soon as they are in final
draft.

As project manager in your role as coach and team builder,
you are involved in keeping the team moving ahead. Not only
do you want to ensure that every individual is working to his
or her potential, you want to anticipate changes that may occur
in the project and ensure that you make the decisions needed for
the project to progress.

At this point, team members know their assignments and are
chiefly responsible for managing their own workload. In scheduling the topic interim milestones, you have worked together to
predict which areas of the project are likely to be completed
first, which are most likely to change, and which will be completed at the very end. However, the information developers
must be alert to subtle changes in the product-development
schedule. As they interact with developers, they must pay close
attention to indicators that the project is changing.

As project manager, you are responsible for being alert to the
not-so-subtle changes. In your meetings with product managers

or senior managers, you must watch for changes in customer requirements or redefinition of existing requirements. If you or other team members (for example, the information architect or the usability experts) are able to interact with the customers during the development life cycle, you are likely to refine your understanding of their information needs or discover new needs that require a change in the direction of your project.

Best Practices in Managing Change

As you work together with your team members and other stakeholders on the activities of the Development Phase of the project, six best practices will help you assist the team in meetings its goals and responding effectively to the inevitable changes that occur on information-development projects:

- ✔ Managing the team
- ✔ Tracking the change
- ✔ Responding to change
- ✔ Initiating change
- ✔ Analyzing ongoing project risks
- ✔ Communicating about change

Best Practice–Managing the team

An important role and responsibility for the project manager is keeping the information-development team motivated, productive, and on track with the project. If you have done your best to assemble the right team for the project, you will have a disciplined and enthusiastic team to work with. Keeping them that way is not always easy, especially when projects grow increasingly chaotic. It's easy to assume that everything is going well, especially when inexperienced team members are reluctant to report challenges.

You may have been communicating frequently and effectively during the planning and Design Phases of the project. You must be particularly alert to maintain communications during the Development Phase at times when information developers often retreat into their cubicles or spend most of their time with the product developers. Better communication will promote continued time unity and a positive spirit on the project.

In Chapter 16: Keeping Your Project on Track, you developed a process for progress reporting, and you considered having frequent meetings with your team members. If you find that meetings are falling off because everyone is frantically working, consider 15-minute stand-up meetings daily rather than longer meetings. Restrict the meetings to critical questions:

✔ Have you accomplished what you planned?

✔ If not, what are the barriers you've encountered?

✔ How do you plan to get back on track?

Don't try to solve the problems in the brief meetings. Just learn what they are. Arrange to talk individually with anyone who is encountering a problem. Help everyone to know that reporting problems early, before they become catastrophes, is the only acceptable course of action. The longer you wait to address a problem, the greater the problem becomes. It's much better to fix problems when they are small.

Encourage team members to rely upon one another for help. They also need to talk about their work, especially to ensure that they are not duplicating efforts. Team members frequently learn about project issues that affect all the work, not only their own. They need to make opportunities to share understanding and design solutions.

If you have worked with individual team members before, you will know who needs more regular consultation and who can manage independently most of the time. With new team members, you will have to judge the amount of assistance each may need. If they are less able to manage their own time and schedules, you may need to provide more help initially, until they learn how to handle the demands of the project. Just don't let them get too far into trouble before taking action. Consider the following activities for new team members who are challenged to manage their own time:

✔ Work with the new team member to develop a detailed personal schedule working back from the due dates for each project milestone.

✔ Schedule daily progress meetings of 10–15 minutes each until you are assured that information development tasks are on track. Cut back to once a week meetings when you are confident the new team member is able to meet deadlines.

✔ Watch carefully if a new team member has multiple projects to balance. It is easy for someone to devote extra time to a simple or interesting project, thus getting significantly behind schedule on more challenging projects.

✔ Encourage a new team member to hand off content to editing and team review earlier rather than later. New team members are often concerned about work that seems unfinished. As a result, they hold onto content too long, only to be informed that their content does not meet expectations. By then, they are left with too little time to change course or make substantial corrections.

Ensure that everyone on the team understands the project plan, especially the audience and vision for the project. Work with less experienced team members to review the design and style standards for the project so that their work is on track from the first. Make time to do quick reviews of topics. It's better to find misunderstandings with the templates or styles before tens of topics need to be changed.

Be especially alert to information developers who are willing to let someone else do their jobs. In most cases, these individuals are afraid of making mistakes. They are most secure simply copying content provided in specifications or dictated by product developers. Watch for signs that some of your team members are content to use someone else's words or include information that is inappropriate for the audience. You can usually tell when people have written something that they don't themselves understand. If you do early, quick reviews of the topics written by team members you don't know or those you suspect may take the easy path of "cut and paste," you're more likely to catch the problems early enough to solve them. After the project is 100% or 50% complete, you won't have time to get the changes made.

Resolving problems

Because it is so important to keep your team functioning and productive, as project manager, you are responsible for removing barriers. Some team members will be quite successful on their own in removing barriers, but many will not be successful nor will they try. At that point, you must intervene. Consider the following barriers:

- ✔ Requested information about the product is unavailable or has been delayed.

- ✔ Product developers are too busy to talk with your team members.

- ✔ Changes to the user interface or parts of the product that will affect the information are not communicated to your team in a timely manner.

- ✔ Review schedules are not respected.

These are only a few of the barriers that organizations construct around projects. Some of them occur because the right processes are not in place to communicate among all the team members. Others occur because individuals do not understand their responsibilities with respect to information development.

As soon as some or all of these barriers appear, consider your best course of action. Do you meet with individuals who are obstructing your team's progress and discuss how they can keep to schedules or provide needed information more easily? Do you meet with the managers of your primary sources of information to explain what your team needs to be successful? When do you escalate your concerns to higher levels of management?

Calm negotiation early in the project can often lead to more responsive processes and behaviors. However, if a problem seems intractable, escalate it in the organization and flag it in your periodic progress report.

Be certain that your team members know that you have worked to resolve the problem to the best of your ability. If you can't reach a resolution, explain to them what happened and why. Have a plan for handling the problem even if you don't have the support you need.

Handling difficult team members

Although you'll find yourself working on problems between your team and others in your organization, primarily product development, you are also likely to have difficult members of your own team. When we put team members together for weeks or months, even years, you are going to have conflict at some time.

You may find team members with personal agendas in conflict with the group's purpose and vision of the project. You may have team members who do not get along well. You may have team members who gossip or report the misdeeds of others.

None of these problems can be solved without active intervention. You need to find the root cause of the problem as quickly as possible. In most cases, informal conversations with those involved are enough. In some cases, however, you need to find more formal solutions, including replacing an individual who is disrupting the team.

Developing a self-managed team

Every manager appreciates team members who are good at managing their own work, but teams don't become self-managing on their own. They need to know what it means to be self-managing because some people interpret that directive as no management. Self-management means that individuals take responsibility for the direction of the team at different points in the project, depending upon expertise. For example, if you are working out the graphic design of a web page with your team members, you would expect the graphic artist to lead the team through the decisions.

To ensure that your team members work effectively together and take responsibility for all the work of the team, review the following with them:

- ✔ Everyone knows the information-development process steps and how to achieve them

- ✔ Each team member develops a personal work schedule to meet the deadlines

- ✔ Team members share information about project status and technical content

- ✔ Everyone feels responsible for doing the best job possible, not waiting for others to correct one's work

- ✔ Team members help one another achieve quality in the work rather than accusing some of lacking competence

- ✔ Team members work together to solve problems that they can solve and bring others to the attention of the project manager

Establish guidelines for conducting meetings, resolving disagreements, and tabling issues that can be considered later. Get answers to questions quickly so that people don't waste time going in the wrong direction on the project.

Promoting quality goals

Everyone on the team is responsible for the quality of the product you develop. Promoting quality among team members means ensuring that everyone knows the rules:

- ✔ Review your own work before anyone else sees it.

- ✔ Apply spelling and grammar checking tools before sending work for review.

- ✔ Present your work for review by the team, especially when you have had difficulty understanding a technical point.

✔ Review the work of colleagues in a nonjudgmental way.

✔ If possible, verify that your technical content is correct by testing it against the product.

✔ Don't depend on product developers or subject-matter experts for all the content; investigate the subject matter yourself and gain expertise.

✔ Know exactly for whom you are writing even if it means describing a hypothetical audience. Audience knowledge resolves most conflicts over content and style.

✔ Work closely with editors to meet your organization's standards for topic development and minimalist writing style.

✔ If you don't understand how to use your organization's information architecture, find someone who does and learn from them.

✔ Recognize that your work will appear in context with work from other team members.

The more dispersed the team, the less direct communication you have with team members or other stakeholders, the more difficult it will be to maintain the quality of the product. When your team is developing topics and reusing topics among different deliverables, you need to work together more closely than if everyone is assigned to produce unique deliverables. As project manager, you need to foster team work and collaboration by promoting the vision of the project and ensuring that people learn how to work together effectively.

Best Practice–Tracking change

If you developed your project tracking spreadsheet, as outlined in Chapter 16: Keeping Your Project on Track, you are already charting the progress of each team member on a weekly basis and resolving problems with quick daily meetings. Despite all the best efforts, projects rarely proceed from start to finish without change.

As project manager, you are responsible for anticipating changes that will affect the work of your team and your ability to meet deadlines. If you cannot anticipate all changes that might occur, you need to work with your team members and stakeholders to uncover changes to the project as quickly as possible so that you can take action to ensure that you are still able to meet the schedule and maintain the quality of the work.

Anticipating change

You may think it illogical to assume that you can anticipate when changes are likely to occur in your project, but you'll find that in most product-development projects, changes are frustratingly predictable. Many projects are planned without sufficient understanding of customer requirements, with overly optimistic schedules, and unreasonable deadlines designed to satisfy demanding, poorly informed executives. Managers of the larger projects that your team supports know that schedules will slip as requirements are redefined or difficult technology problems encountered. They may have planned for these contingencies in their project schedules.

Your experience working with particular development organizations will give you considerable insight into which product-development teams are most likely to change direction and slip schedules. Even if you are working with a new development group, you can often predict the level of change by the uncertainty and risk surrounding the project plan. If there is no project plan at all, you can usually count on major disruptions to the project.

I have also found that schedule slips occur just before major milestone dates. No one seems to want to admit in advance that the milestone will not be made. They wait until the last minute to announce the change, hoping that the bad news will be viewed as inevitable.

Schedule slips also occur in some projects because of external pressures. I rarely find development and product managers who will announce a schedule problem right before a quarterly report to stockholders is due. The announcements come about two weeks after the report is out.

By communicating with product and project managers in marketing and development, you can hope to learn about problems occurring in your project. You may learn that part of the development team is having difficulty getting a new technology to work or having difficulties with integration. By following the interactions of marketing with customers, you will learn early about changes in direction, requirements redefined, and problems with early prototypes that demand change.

If you have already aligned your topic-based deliverables with the product features or function development, you may have anticipated the areas that are most risky for your part of the project and scheduled them for later completion. As soon as you become aware of potential product schedule changes or changes in requirements or design, be ready to adjust your own schedule and communicate the changes with your team members.

You may, of course, be part of a larger project that has instituted a change control process. When changes occur in the project, they are recorded in a change management system. As part of most change-management systems, the project manager must estimate the effect of the change on the budget, resources, and schedule of the project. In the best systems, the effect of the change on all the project stakeholders, including information development, are taken into account. As the manager of the information-development project, you may be asked to estimate how the change will affect the work of your team. Sometimes changes are not accepted for a system because they adversely affect the ability of all the stakeholders to meet the deadlines or maintain the quality of the product.

Evaluating the impact of change on the project

When your team faces changes in the larger project, they typically take three forms:

- ✔ Topics that have already been developed have to be rewritten.

- ✔ Topics already developed are no longer part of the product and will be deleted from the final deliverables, usually along with some topics yet to be developed.

- ✔ New topics have to be developed that were not part of the original schedule.

In each case, the schedule you have developed for your project no longer applies.

The greatest danger you face as project manager is having your team members respond to changes in the product without consulting you or the rest of the team. Information developers often want to be responsive to requests for change that come from the developers they work with every day. It's simple enough to say "yes" to a request to add a few topics to accommodate a new feature or function. Unfortunately, those requests often add up and have a disturbing effect on the project as a whole.

What happens, for example, when one information developer decides to add 20 new topics to the 100 topics planned for a deliverable? The 20 new topics represent a potential 20% increase in the workload of everyone involved in producing the deliverable. The editor will have 20 additional topics to edit; reviewers will have 20 additional topics to review; translators will have additional work added to their schedules. The information developer might argue that it only took her two or three days to add the topics. As a result, six other members of the information-development project have just had additional work added to their already overstuffed schedule. Plus, the larger information deliverable will now cost more to produce, print, and distribute (if it is being distributed in paper).

It's not up to individual team members to make change of scope decisions. You need to work out a process with your team to ensure that they don't accept work without consultation. Insist that they inform their development colleagues that they have to check with the project manager before accepting the additional work. By shifting the responsibility to you, they take the pressure to accept a change off themselves.

Use the form in Figure 19-1 to institute your own change control procedure if one doesn't already exist in your organization. Communicate the effect of change to all your team members so that they understand the consequences of hasty decisions.

Best Practice—Responding to change

If you have anticipated a high-risk project as part of your original project estimate, you may have sufficient contingency time in your project plan to accommodate some degree of change. For example, you may have calculated project dependencies based on an unstable product, leading to a higher than average hours per page. A typical unstable project may result in a 20% increase in time allocated for the project. With this extra 20% available, your time may be able to absorb the changes and keep the project on track because the additional time for revisions is already built into the schedule.

If, however, you have scheduled your project tightly, usually because of inadequate resources assigned, you will be unable to absorb the changes into the schedule. At that point, you have to estimate the affect of the changes on the project estimates and the schedule.

Consider the following situation, illustrated in Figure 19-2. You had planned your project based on an estimate of 75 task topics. On December 1, you find that new features have been added to the project, increasing the number of tasks by 25. In addition, five of the tasks already written are no longer needed for the project. However, most of the time to create them has already been expended. Your hours per topic rate by the end of the project will be affected by the work already done and discarded.

Publication Change Request

Project name _____

Initiated by _____ Date _____

Description of the change_____
_____
_____
_____

Publication affected_____
_____
_____

Number of affected pages_____
_____
_____

Estimate of page-count increases/decreases_____
_____
_____

Estimate of illustration increases_____

Start date _____ Completion date _____

Effect on resources
   people sources_____
   schedule _____
   localization and translation_____
   manufacturing_____

Approval signatures
   publications and manager_____ Date_____
   development manager_____ Date_____
   marketing manager_____ Date_____
   other_____ Date_____

Figure 19-1: Publications change-request form

Re-estimating the project

In your original estimate, you had figured 5.0 hours per task topic. When you review the tasks to be added, you note that they can be accommodated in the original architecture for the topics and deliverables but will require additional time for writing, editing, production, and translation coordination. Your planning and Design Phases represent about 30% of total project time, leaving you with 3.5 hours per task topic.

Multiplying 25 topics by 3.5 hours per topic requires an addition of 73.25 hours to your project schedule.

At the same time, you won't have to complete production and translation coordination for 5 topics, which represents an additional 10% of time that you have not yet expended. You can deduct .5 hours for each of the deleted topics, or 2.5 hours. That means you subtract 2.5 from 73.25 hours, leaving 70.75 additional hours for your schedule. Your new total time for the project is 725 rather than 640 hours. Given the addition of 25 topics and the removal of the production work for 5 topics, your new metric for 100 task topics becomes 4.6 rather than 5.0.

Since you have several people involved in the work, including the writers, the editor, the production coordinator, and the translation coordinator, you calculate that the 70.75 hours will take an extra two weeks. Unfortunately, the additional time required to write the 25 new topics will increase production time but decrease the number of days you have to do the production work. You may have to negotiate an additional week to allow for production quality control activities.

You have also learned that the additional work to write the new features amounts to a two-week slip in the development schedule. That means you're on track for the project although you know that other projects might be affected. You have five people who won't be available to work on their new assignments for an extra two weeks. Now you'll have to negotiate with the other project managers to gain some time.

The example here provides only one possible result of a change in scope to a project. Adding time to the schedule is often not an option. Use the following steps to evaluate the change and then consider what direction is best for your project:

1 Evaluate the proposed changes to the content in terms of topic changes, added, and deleted. Include graphics in your enumeration of the changes.

2 Review your original estimates and compare these with the actual time spent on your project to date.

3 If the hours per topic have increased or decreased since your original estimate, investigate why the change has happened. Consider that the dependencies may have changed.

4 Reevaluate the project dependencies and run the dependencies calculator again. Use the result to recalculate your hours per topic.

5 Use the new hours per topic to calculate the number of hours you will need to add to or take away from your project.

6 Take into account where you are in the information-development life cycle. It's easier to make changes during the Planning Phase than when you're ready for production to begin.

7 Evaluate the effect of the additional hours on your schedule and resources, taking into account the advantages and disadvantages discussed in the next sections.

Projected Hours Spreadsheet

Factor	Skill Level	Projected Oct 1-15	Projected Oct 16-31	Projected Nov 1-15	Projected Nov 16-30	Projected Dec 1-15	Projected Dec 16-31	Projected Jan 1-15	Projected Jan 16-31	Projected Total Hours	Hours/Unit
Tasks		Topic Count= 100		Metric= 4.60							
10%	Project Manager	3.50	5.30	8.79	6.21	7.59	1.23	1.15	8.10	41.87	0.42
	Information Architect	35.00	8.00	12.00	0.00	0.00	10.00	0.00	0.00	65.00	0.65
	Information Developer	0.00	40.00	66.00	54.00	66.00	2.00	10.00	53.00	291.00	2.91
15%	Editor	0.00	0.00	9.90	8.10	9.90	0.30	1.50	7.95	37.65	0.38
	Production Coordinator	0.00	5.00	0.00	0.00	0.00	0.00	0.00	20.00	25.00	0.25
	Subtotal	38.50	58.30	96.69	68.31	83.49	13.53	12.65	89.05	460.52	4.61
										460.00	**5**
Concepts		Topic Count= 25			Metric= 7.00						
10%	Project Manager	1.50	2.80	0.00	1.00	0.00	7.21	2.30	1.00	15.81	0.63
	Information Architect	15.00	8.00	0.00	10.00	0.00	10.00	0.00	0.00	43.00	1.72
	Information Developer	0.00	16.00	0.00	0.00	0.00	54.00	20.00	0.00	90.00	3.60
15%	Editor	0.00	0.00	0.00	0.00	0.00	8.10	3.00	0.00	11.10	0.44
	Production Coordinator	0.00	4.00	0.00	0.00	0.00	0.00	0.00	10.00	14.00	0.56
	Subtotal	16.50	30.80	0.00	11.00	0.00	79.31	25.30	11.00	173.91	6.96
										175.00	**7**
Reference		Topic Count= 15				Metric= 6.00					
10%	Project Manager	1.00	2.00	0.00	0.00	1.00	1.00	2.76	0.50	8.26	0.55
	Information Architect	10.00	8.00	0.00	0.00	10.00	10.00	0.00	0.00	38.00	2.53
	Information Developer	0.00	10.00	0.00	0.00	0.00	0.00	24.00	0.00	34.00	2.27
15%	Editor	0.00	0.00	0.00	0.00	0.00	0.00	3.60	0.00	3.60	0.24
	Production Coordinator	0.00	2.00	0.00	0.00	0.00	0.00	0.00	5.00	7.00	0.47
	Subtotal	11.00	22.00	0.00	0.00	11.00	11.00	30.36	5.50	90.86	6.06
										90.00	**6**
	Total Hours	66.00	111.10	96.69	79.31	94.49	103.84	68.31	105.55	725.29	
	Hours/Period	60.00	66.00	66.00	54.00	66.00	54.00	54.00	72.00		
	Full-Time Equivalent/Period	1.10	1.68	1.47	1.47	1.43	1.92	1.27	1.47		

Figure 19-2: Recalculation of project due to changes in scope

Preserving the project vision

It's quite easy to assume that your team can simply absorb any changes to the scope of the project. The addition of 25 new topics is a change of scope. It's best to carefully analyze the consequences of a change of scope before you make a commitment. What if the change of scope added months, rather than a few days to the schedule? Could that be easily accommodated with existing staff? What if the new work has to be done within the original deadlines? Is that why so much overtime happens on your projects?

The most damaging consequence of scope changes is quality. Will your writers have little time to validate their information against the product? Will you have to pressure editors to work faster and ignore anything but major problems? Will you have to compress production and translation schedules and hope that the work can be done in the remaining time?

Scope changes or changes in the complexity of your project do have consequences for the project vision. You have designed the project to provide value to the users of the information. Now the project is jeopardized because of late additions to the deliverables. You need to ensure that your vision for the project is maintained or negotiate a change to that vision. Changes that are likely to affect quality should be made only when all the stakeholders are aware of the consequences.

You need to discuss with yourself, your team, and sometimes your stakeholders just what the best course of action should be:

✔ Extending the schedule for the project

✔ Adjusting the interim milestone dates

✔ Adding resources to the project

✔ Adjusting the project scope

✔ Renegotiating project quality

Each of these choices has its own advantages and disadvantages to consider.

Extending the project schedule

By extending the project schedule, you preserve the original vision for the project and accommodate an increased project scope. However, you tie up staff who might be needed for other projects and you increase the cost of the project by continuing to expend hours on project activities. In fact, extensions to the project schedule can quickly become enormously expensive without delivering increased quality and value to the customer.

The disadvantages of an extended project schedule are many:

✔ By adding additional time to the project, the costs automatically increase to accommodate changes to the project deliverables or to correct deficiencies in the product design or execution.

✔ Staff who may have been scheduled to move to other projects are no longer available, possibly affecting the schedules of multiple projects.

✔ Some staff may no longer be available for the extended projects because of other required commitments, requiring staff changes late in the project life cycle and further adding to the cost of the project.

In addition to the effect on costs, extended projects are likely to have an adverse effect on the potential success of the product in the market. The competition may announce a similar product early, or the market requirements may change once again. The potential disadvantages of delaying the completion of the project must be carefully weighed against the advantages of gaining more time and ensuring that the project is successful.

Using sound project management methods to ensure that a project remains on schedule are clearly preferred to extending the schedule because project risks were never adequately evaluated in the first place.

Of course, you will also find advantages to extending the project schedule:

✔ The team that began the project is able to complete it with no need to disrupt the project schedule by adding and having to train new team members.

✔ The original goals may remain intact, allowing the team to produce the original deliverables at the quality required by the customer.

✔ Although the project extension may require rewriting completed topics or adding new topics, the extended schedule may accommodate much of the additional work.

Most project managers have experienced schedule extensions. In fact, extending the schedule is the most frequent change to a project in response to increased risk in the project and changes to the customer requirements. Perhaps the most unfortunate reason for extended schedules is flawed scoping and estimating of the project in the first place. Many projects are assigned deadlines by managers based more on wishful thinking than a competent analysis of the scope, risks, and deliverables. Project managers are themselves aware that the schedules originally agreed to can never be achieved, but they have capitulated to the demands of more senior managers who are apparently uninterested in listening to reason.

As a project manager in an information-development project, you need to be alert to the politics of projects in your organization. If you know that projects are regularly underestimated and the schedules are always extended, use this knowledge to plan your own schedule and make assignments. If you are confident that schedules will be extended because they are always extended, be careful about over-assigning people to your project early. You may be able to accomplish the project with a smaller team because you can assume that the team will have more time to complete the project than the initial schedule and deadlines indicate.

Adjusting the interim milestone dates

Many projects get off to a slow start, often because the requirements are not fully understood or effectively articulated. Often, such projects are likely to have extended deadlines because the amount of time allocated to the initial requirements definition or to estimating the technical challenges of the project are insufficient. For example, you may have a six-month project with one month allowed for requirements definition. However, after two months, the requirements are still not fully defined, and there are gaps in the product specification. Clearly this project is unlikely to be completed within the original six-month schedule. In fact, unless the scope is changed, the project will likely be twice as long as the executives had hoped.

How should you respond as you try to predict the interim milestones for the information-development project? You may have wanted to schedule writing tasks to begin at week 10 in the 26-week schedule or have the planning and design time complete at 30% of the total project time. It may be possible to begin some writing tasks without a complete specification, especially if you are updating existing information that is likely to have only minor changes. However, it would be best to delay the writing tasks until the requirements and specifications near completion.

If you try to maintain your original schedule of interim deliverables, you are likely to find that your staff is unable to get the information they need because too many decisions are waiting to be made about the product itself. You may make better use of your team's time by moving them to temporary activities on other projects than wasting time creating materials that will soon have to be rewritten. If your planning and design are thorough, to the extent that you have been able to complete those project phases with incomplete information, your team will be able to quickly write topics that have been fully defined in advance, even though the details are not yet fully known.

If the larger project is off to a slow and rocky start, it's best to slow the pace of the information-development project and shift the interim milestone schedule to avoid creating content too soon. Keep in mind that you want to minimize the time devoted to writing each topic. Writing a topic more than once simply wastes resources that are better used elsewhere. If the larger project is not adequately planned and specified, you are likely to find your team continually rewriting.

Adding resources to the project

Adding additional staff members to a project often appears to be a good tactic. At the beginning of the project, your estimates should lead you to providing the required resources to support the scope, quality, and schedule required. If your project needs two information developers rather than one, assigning two people from the beginning is the best course. Both individuals learn about the project together and support one another's activities throughout.

Adding additional staff members to a project later in the project schedule in response to change is considerably more problematic. A new person needs to come up to speed on the project, taking time away from the existing staff and adding a level of complexity to the project.

To calculate the affect of an additional staff member, you should return to the dependencies calculator and reevaluate your previous choices. A new information developer may have less experience with the technology of the project, may not know the tools, or may be skilled in areas of development that are different from your project's requirements. Or, the new information developer may be more experienced and skilled than the people currently assigned to the project. Dependencies may change either positively or negatively. In most cases, however, your team dependency will be affected. Adding people to an existing team disrupts the relationships already established, requiring everyone to make adjustments and potentially having a negative effect on the project schedule.

The positive affects of adding additional resources late in a project include

✔ adequately staffing the project in response to changes in the schedule, deliverables, scope, or complexity of the project

✔ adding skills that may enhance the quality of the project and ensure that the schedules are met

However, the problems often outweigh the positives:

✔ The complexity of the project increases with the presence of new people.

✔ Adding new staff late in the project schedule slows the project's progress as the new people come up to speed.

✔ A larger project, with more resources assigned, is more difficult to manage than a smaller project because of the need for more communication channels.

Frederick Brooks, in *The Mythical Man Month* (Addison-Wesley, 1995), notes that a late addition to a project team results in a loss of efficiency. For example, if a new person is assigned to 100 hours, for example, the actual productivity of that person may decline to 80 hours because of the learning curve and the need for the team to assimilate the addition.

It is also clear that new staff are not always available for a project. Everyone in the group may already be fully occupied with other projects. It may be difficult to hire someone from outside, even on a temporary assignment. No budget may be available for additional staff on the project. Even if additional budget becomes available, no one may be able to step into the project without considerable start-up costs and time.

Adjusting the project scope

One of the most effective tactics in responding to changes in the larger project is to adjust the scope of the information-development project. The goal of a scope adjustment is to maintain the original budget, respond to the new schedule if necessary, and avoid compromising the quality of the final deliverables.

In the original project described in Chapter 16: Planning Your Information-Development Project, the team had planned to develop a user's guide, a help system, and two sets of training material to introduce the new product to the customers. Depending upon the nature of the changes to the larger project, as project manager, you may want to look at the number of topics promised for the deliverables. It might be possible to cut the number of help topics developed and concentrate on those most likely to be needed early in a customer's use of the product. You might be able to reduce the set of training materials, producing only one set rather than two.

If you are pursuing a topic-based development plan, you should immediately review any topics that have not yet been scheduled, focusing instead on key topics that will get the customer started. You may want to deliver only essential task topics rather than conceptual information, relying on the training to cover the concepts.

In making scope adjustments, you must develop your plan for the changes and review these with your team members and with the key project stakeholders. You will need to know how much time you have already expended, the time required to change content that has already been developed, and the time needed to redefine the product architecture if necessary.

Negotiating project quality

No one wants to admit that project quality can be compromised when projects are not well managed. Even if you have determined that your project requires a level of quality that will support customer requirements for accessibility, accuracy, and usability, you may have had to short staff the project because of limited resources. When the project changes because of changed requirements in the larger project, a project already short on resources will be severely taxed.

Project managers and information developers often argue that they maintain project quality despite changes that affect the content. But whenever I press the case for quality, team members admit that they are forced to omit quality assurance activities that they would otherwise include. Under the pressure of changing requirements and unchanging deadlines, team members are unable to review and edit content internally, validate content against product interfaces, test content for accuracy, or ensure that content is relevant to user needs. Deadline pressure often results in removing from the project the very activities that are intended to guarantee quality. Worst of all, the quality assurance activities are eliminated without stakeholders being made aware of the consequences.

Negotiating project quality requires that team members report quality assurance problems to the project manager. If team members believe that critical activities must be removed from the schedule, the final decision should be made not by individual contributors but by the project manager in consultation with project stakeholders and team members. By reviewing quality issues as a team, you actively decide what quality assurance actions are omitted, rather than leaving the decisions to individual team members. You also have a record of your quality decisions, allowing you to allot extra time to correct the neglected areas at the next release.

The project manager's responsibility is to make clear the consequences of eliminating critical quality assurance steps. Failing to validate content against a product is likely to result in content errors, errors that will lead to increased costs of customer support and decreased credibility for the information as well as the product.

Rather than compromise quality, the project manager and team members should consider changing the scope of the project. By eliminating some deliverables, the quality of the remaining deliverables can be maintained.

Revising the project spreadsheets

A critical task that you must undertake as project manager is to revise the project spreadsheets to account for the changes to the project. First, however, you must estimate the impact of the changes on the work your team has already completed and the work that remains to be done. In many projects, changes to the customer requirements or product specifications result in changes to work that has already been completed. Traditionally managed projects are subject to rework. Agile projects assume in their very design that rework will occur as customers and stakeholders provide feedback on iterations of the design.

In agile projects, you must schedule the topics to be developed for each iteration. At the end of the iteration, you will re-evaluate the topics in light of customer feedback. Some topics that were complete are then scheduled to be revised or completely rewritten during the next iterations. Topics originally scheduled for the next iterations may have to be postponed to provide sufficient time to revise previous topics.

In traditional projects, project changes during the product-development life cycle result in changes to existing topics, the removal of some topics from the project, and the addition of new topics to account for new features introduced.

In either case, you have to estimate the impact of rework on your project in addition to accounting for new work that was not originally in your project plan.

The original project plan for an agile project has already accounted for more than one draft of the content being developed during the Development Phase of the project. That is, you have already accommodated a certain amount of time to rework existing topics in your plan. If changes to the project occur soon after many topics are in a first draft stage, then you should be able to accommodate minor and even some major revisions within your original estimate. You need to account only for those topics that will need complete rewriting or major revisions. With your information developers, review the impact of the project changes on all completed topics, accounting if possible for complete rewrites as well as major and minor changes. Refer back to Table 15-3 for an example of how to estimate the hours required for the rework, just as you might have originally estimated the hours for a project that included existing topics. At this point, you need to account only for the topics already written and requiring changes.

The farther along your information developers are in the Development Phase of the project, the greater the impact of project changes on rework. If you have topics that have been through editing, review, and revisions, representing a second or third draft, the cost of reworking those topics will be greater than for topics that are in first draft. If major changes occur to the topics, editing, review, and revision will have to occur all over again.

Making new assignments

Once you have identified changes that affect your project estimate, you need to adjust your project spreadsheet to account for the changes and amend your original project plan so that the changes are accounted for in future project estimates.

Best Practice—Initiating change

Many of the changes that require reestimating an information-development project result from changes in the larger project of which you are a part. In some cases, however, you and your team members will choose to initiate changes to the project. Possible sources of information-development initiated change are

✔ the discovery of customer requirements that were not anticipated earlier in the project life cycle

✔ recognizing a quality debt acquired during previous iterations of the information development that require correction

✔ iterating the design process to address usability problems encountered during the project

Each of these changes to the information design are best addressed as early in the project as possible and are ordinarily the responsibility of the information architect. If, however, you decide to make changes during the course of the project, you will have to refigure your estimates and find ways to add time, schedule, or resources to the project.

Note that some changes that you or your team members might initiate may be accommodated if the larger project becomes stalled. Rather than sitting idly by or spending time on other activities or other projects, you may choose to address problems with content that you could not afford to address before.

Identifying new customer information requirements

During the course of a project, your information architect may become aware of new customer requirements for information. You may learn that an audience assumed to be expert is less skilled and requires more basic information to use the product. You may find, as we did on a project, that users did not require task-oriented topics that explained how to use the interface. They were more interested in detailed reference information.

Although the information architect may identify the requirements, as project manager, you are responsible for deciding if the changes to the scheduled work can be accommodated without taxing the project schedule or the team members.

Review your project estimates and the progress reports from the team. Decide if you are on track with all assignments. If so, evaluate the impact of changing the information just as you would for any other change. Consider the customer value added by the changes in your approach to the content. If you continue on your original course with the project plan, you may introduce problems that will prove costly later.

Recognizing the quality debt trap

A quality debt on a project occurs when quality was compromised in past projects to meet deadlines or because the team members lacked knowledge of the audience or a complete understanding of the subject matter. As a result, your project may have accumulated a quality debt that needs to be addressed before you can adequately serve the needs of customers.

In reviewing existing content in a project, you will find information that is incorrect, poorly written, or structured in a way that makes the information less usable than it could be. In many cases, poor quality content is simply ignored by information developers. They are too busy adding new information or changing content that is affected by changes to the product or service they are documenting. They never seem to find time to evaluate the legacy content and decide if it needs to be revised, rethought, or eliminated.

Work closely with your information architect to evaluate the legacy content and decide on the best course of action. The earlier in the project you decide to attack legacy content, the more easily you can include the work in your project estimates. Even better, schedule the quality improvement activities during slower times between projects so that the content improvements are ready before the next round of updates.

If you find time during a project because of schedule delays, look for opportunities to reevaluate legacy content. An information architect, working closely with the information developer, was able to restructure legacy content in an administrative guide and reduce by 50% the total page count in the final deliverable. Not only was the information volume less, the value of the information to the customer was greatly increased because the content was more easily accessible. The information developer had transformed task-based information into tabular reference information that made search and retrieval of details more efficient and effective. The team also found that the newly restructured information was easier to maintain.

Choosing to improve the information design

Some organizations include usability testing of information deliverables in the project life cycle, often following the second draft of portions of the content. Usability testing readily reveals problems not only with the specific content but also with the overall design of the content. A test may demonstrate, for example, that the topics are not titled correctly, making retrieval difficult. Or, you may learn that information needs to be rearranged among deliverables because customers organize their enterprises differently than you had imagined.

As project manager, be prepared for changes in the project schedule as a result of usability tests and other types of customer feedback. If you have planned for a testing phase in your project, you should have time remaining in the schedule for an iteration of the content or even the overall design.

If you are unsure about outcome of the initial Design Phase because you lack customer knowledge in general, find ways to schedule testing earlier so that your team can incorporate the design changes needed.

Best Practice—Analyzing ongoing project risk

Your most important reason for tracking the time spent on project tasks is to uncover problems with the project in enough time to correct them. You may be happily tracking the time on a project, only to discover at a critical milestone that your team members are behind. You may have assumed that 80% or more of the topics would be at first draft when your team had expended 60% of the total project hours. At the 60% milestone, you discover that only 50% of the topics are at first draft. At this point, you need to address several key questions:

✔ Has the productivity of your team members been adversely affected by changes in the larger project? Changes may occur not only because of scope changes or redesigns but because the larger staff is facing their own challenges and have less time available than you had hoped.

✔ Are your team members having more difficulty than anticipated in completing their work? Are they not prepared for the technical content of the project or do they lack the skills needed to gather information or develop topics on schedule?

✔ Are your team members having difficulty with the tools they are using to develop information? Have you introduced new tools that are more difficult to use than anticipated? Are some team members simply more unfamiliar with your existing tools than you had judged?

✔ Have you had unexpected changes among your team members? Were people occupied on earlier projects that did not finish on schedule? Have they been absent from work more than you expected? Has someone on the team left the project at a critical time?

✔ Are your team members facing opposition from key stakeholders? Are you pursuing a new analysis of the users' information needs that some stakeholders find uncomfortable? Are you restructuring the information, pursuing minimalism, or otherwise changing how you develop content? Are stakeholders insisting upon a return to the old ways of doing things?

Each of these issues can increase the risk of your project and affect your ability to maintain your schedule and stay within your original estimates. In some of the situations I have described, you should re-evaluate the project dependencies. You may have a less experienced and skilled team than you had anticipated. Your team members may have encountered more difficult content than originally thought or find that the members of the larger project are not as available or communicative as you had hoped.

Analyze the situation as thoroughly as you can. Then determine how far behind the project has become, and decide how to recover. Be careful not to assume that the difficulties in the project will simply disappear on their own. Most likely, you will need to take positive action to correct the problems and get the project back on track.

Productivity challenges

Project managers all hope to have team members who are stars; capable of intelligently designing information to meet customer needs; interacting on an equal level with other stakeholders; writing effective, minimalist text that adheres to style and structure guidelines; and generally functioning as consummate professionals. We assume they can overcome challenges to their productivity and ask for help in overcoming barriers as soon as their own solutions seem not to work. Unfortunately, none of us has a team of all-stars all of the time. Team members face their own challenges in working productively. Even when they are skilled and motivated, they may encounter barriers that frustrate them and reduce their ability to get their work completed in a timely manner.

Your role as a project manager includes ensuring that your team members are performing to the best of their abilities and finding ways to help weaker members succeed. Doing so requires a more hands-on management style than many managers seem willing to assume, at the same time that they balance concerned participation with dangerous micromanaging. Since it usually remains your responsibility to see that the work is accomplished on time, on budget, and with an acceptable level of customer quality and value, you must be prepared to confront productivity challenges among your team.

Your first point of departure is communication. Frequent communication with your team members, in small group settings and individually, will help you to know about any barriers they are facing. In working with your information architect, you should be confident that the overall design of the content is on track with your understanding of user requirements. In working with the technical reviewers and the product-development manager, you can keep track of technical competency in the writing. You may hear that a team member is having difficulty understanding the technical content or is becoming overly dependent on the subject experts for source information. Other team members may alert you, sometimes indirectly, to a colleague that is having difficulty with the content. Your team editor should also alert you when someone's work is not up to par. You can watch the editing time itself. If you have budgeted a percentage of writer time for editing, note reported time that exceeds the schedule. If the editor is spending an inordinate amount of time with one writer's work, that may signal a problem that needs to be addressed.

You may also want to take part in at least spot reviews of the content being developed. If you have experience with the technical subject matter, you may be able to notice gaps in the information that might escape the editor. I find that reading through task topics carefully often reveals problems in understanding even if I don't know the content yourself. I find it quite possible to find missed steps, step sequences that don't make sense, information requirements that should be in prerequisites rather than buried in the steps themselves, and what should be preliminary tasks included in the context of later tasks. If you see writers struggling with the information or writers who appear to be overly dependent on poorly written technical specifications, you can address the problem before it becomes too large to handle.

You should be also working with team members to remove project barriers that are causing otherwise competent writers to lose time. If subject experts are not forthcoming with information in a timely manner or are difficult to find or understand, you may need to escalate the problem to more senior managers. You may also find in such delays early warning signs that the larger project is not going well. Small, early indicators of problems with the project should lead you to take action in discussing the schedule with your counterparts in the development team. Regular conversations with your team members about their progress and ability to work effectively can provide valuable information that may lead to major changes in scope or schedule for your project. The sooner you know about them, the easier it will be for you to accommodate the changes without affecting the quality of your team's work.

Technology challenges

Several technology-related challenges may interfere with the productivity of team members on your project:

- ✔ New team members may be unfamiliar with the tools your team is using for content creation and production.

- ✔ Everyone may be learning new tools, particularly if you are moving from a desktop publishing to an XML environment.

- ✔ Changes may make existing tools unfamiliar to everyone, including the introduction of new style sheets, document type definitions, or metadata configuration.

✔ Team members may be using the tools incorrectly, making their information diffi-cult to revise or adding production errors to the project.

✔ Team members may be modifying the tools themselves, making their information incompatible with their colleagues' information, and increasing production time and possibly affecting the cost of localization.

You may want to work with your editor and your tools specialists, if they are part of your team, to consider how to best work with team members who are unfamiliar with the tools or do not know how to use them correctly. Editors should review the proper use of styles or XML tags in draft documents until they are confident that team members use them properly. Tools specialists should provide training and support when new tools are introduced and check sample files from time to time.

When problems are not uncovered until the production phase of a project, they become much more expensive to correct than if they are discovered early. Team members can fairly easily correct their first drafts and correct problems before they permeate the entire topic set.

Tools that are used incorrectly also create problems for localization and translation. An incorrect XML tag might lead a translator to misinterpret a word, phrase, or section of text. For example, a translator coming upon text that is tagged as a syntax diagram but is actually the name of a user interface button may not know how to handle the translation.

Good writer guidelines, training, and reference material for the tools help team mem-bers avoid technology challenges that will have an adverse effect on the productivity of the team and the success of your project. Quality assurance activities simply ensure that the guidelines are followed. The result is a body of content that is consistently developed and more easily and efficiently modified for future product releases. In fact, quality assur-ance at all levels of content, including tool use, helps reduce the quality debt of your pro-ject for the future.

Staff changes

No one likes to experience staff changes in the middle of a project, but sometimes they are unavoidable. People get sick, have babies, take new jobs, and are needed for other pro-jects. You have already considered the possible effect on productivity of adding new team members after the project is underway. Losing existing team members has the same nega-tive effect on the success of your project.

If you can anticipate some changes in staff, like marriages, pregnancies, and looming additional projects, you may be able to plan for them from the beginning. If changes occur unexpectedly to your team, consider

✔ asking existing team members to absorb the workload, especially if the project is coming to an end

✔ asking that an experienced colleague be transferred from another project either permanently or until a replacement can be found

✔ using a temporary replacement by hiring a very experienced contractor, perhaps someone who you would not otherwise be able to afford on your project

✔ taking on a development assignment yourself or asking the same of your information architect, editor, or other individual familiar with the project

✔ reducing the scope of the project to accommodate the missing hours

✔ delaying the deadline of the project so that you can complete the work with the remaining team

As usual with changes of course on your project, the sooner you take action to mitigate the damage, the more quickly your project can get back on track or head in a new but otherwise acceptable direction.

Stakeholder requirements in conflict

Despite the communication you may have done as you were preparing the project plan, you are likely to find that some stakeholders have not understood nor accepted the goals of the information-development project. Even those who have approved a project plan and listened to your presentations about design changes or new ways of approaching the project may have failed to understand the implications of their agreement until they see topic lists or draft topics. In one case, a team had prepared a topic list following a carefully crafted minimalist approach, based on a thorough understanding of the users' goals and tasks. They had outlined their minimalist agenda in the project plan. However, once the developers saw the topic list based on the user and task analysis, they backed away from their agreement, creating their own list of topics based on the features, functions, and the menu design of the software product. The ensuing conflict over the content of the proposed topics stalled the completion of the project's Design Phase and delayed the beginning of the Development Phase until the conflict could be resolved.

Once again good communications are the key to avoiding conflict but may not be sufficient. As often as practical, schedule time for project reviews with the stakeholders, especially when architecture design changes are happening. You may need to repeat many times what your goals are and how you are accounting for user needs. Once people understand what you intend, some of their worries may dissipate.

If you anticipate many disagreements, especially if you are moving information in a new direction, be certain that you have a highly placed, politically savvy champion for your project. Take time to keep the champion informed about goals and progress at a high level. Ask the champion how to handle conflicting goals and requirements among other highly placed stakeholders. At key times, the champion may have to intervene directly with others at his or her level in the organization.

If you have sound user data backing up your design direction, publicize it among the stakeholders. Be certain that they know why you are rearchitecting the information-development project and how the new direction will benefit customers and the company. If you have an information architect, involve him or her in the discussions with the stakeholders.

Work with your team members on their political skills. Discuss the benefit of acting professionally and in the best interests of the customers. Convey your own understanding of the political situation and work with team members to mold their responses to challenges from stakeholders. Learning something about company politics and how they might best represent the team and your department will help them avoid conflicts and keep the project on track.

Best Practice—Communicating about project change

Communication is critical to successful project management, not only when a project already underway experiences changes but also when the project you are managing is different from earlier information-development projects. Not only must you communicate with your team members and colleagues within your department, but, most importantly, you need to establish a regular means of communication with stakeholders and your senior management.

I recommend writing project plans in part to initiate communication about the project. Although the project plan is primarily about planning the project, rather than writing the plan, the plan itself is an important communication tool at the beginning of the project and when you recommend a change in the project scope.

In addition, you may find that communicating through the plan alone is inadequate to reaching agreement and achieving consensus among team members. It's much too easy for a stakeholder to inform you that he or she "agrees with the plan," only to claim later that no communication ever occurred. New project managers, you may find it distressing to realize how little of what you try to communicate is either read or understood. My recommendation is to accept the fact that most communication is not successful, may be entirely misunderstood, and will have to be repeated frequently before a general understanding begins to emerge.

Communicating with team members

You might think that your own team members are listening when you discuss the plans for the project. After all, they are most affected by inadequate planning or by a lack of consensus around project vision and goals. It's wise to hold several short meetings in which you restate the project vision or ask others to state what they believe the vision to be. Writing the vision down, creating a short slide set, or developing a graphic that illustrates the project vision may help.

At regular intervals during the project, depending upon the project schedule, plan opportunities to reiterate the vision and discuss how you all intend to achieve your goals. If you are re-architecting the content, provide lots of examples of what a newly restructured installation or user guide might look like. If necessary, create a small mockup or storyboard of the planned redesign. If one of your goals is to reduce the volume of content in an information suite, discuss with your team how you will establish specific targets for reducing page or word count in each deliverable. Ask them to explain why they believe the re-architecting to be necessary. You may be surprised at the team members who don't see the point of the entire effort.

Communicating with stakeholders

Stakeholders often have strong personal goals associated with your project that may make communication difficult. Some stakeholders believe they are more capable of developing accurate information about the project than your team members. They may argue about content, style, format, wording, or even grammar and punctuation. They may have formed expectations about the scope, schedule, and activities of your project that are completely at odds with the reality.

At the same time, finding ways to communicate with stakeholders when they are busy and fully engaged with their own work is challenging. You can schedule meetings and have few if any of the participants that you believe should attend. You can distribute documents and have them ignored.

In Chapter 14: Introduction to Project Management, you learned about the importance of conducting a stakeholder analysis early in the Planning Phase of the project, as you planned your project. A stakeholder analysis will only bear fruit if you accompany it with a communication plan. A communication plan may include several activities or events that you will use to provide information to each group of stakeholders you identified.

Some stakeholder communication can take place in meetings; some must take place individually or in very small groups of two or three individuals. You may find it effective to ask the stakeholders at such meetings what they understand about your project. You may be quite surprised to find that they have little idea what your team is working on. Or, you may be equally surprised to find them well informed. However, the busier they are, the more they will know little about what you're doing. It's ultimately your responsibility to ensure that they are informed.

Individual meetings with stakeholders may help to further understanding in specific instances. Ask what they need to know about the work of your team. Be certain they understand your interest in usability and your understanding of the user community. In most cases, once the communication succeeds and stakeholders know what you want to achieve, they are cooperative and supportive.

Communicating with senior management

Communicating with senior management about your projects will depend upon the degree to which you are pursuing change. If your project is ordinary, representing the same basic processes and designs that you have always used, your communication task may be restricted to general reports about the progress of the department's work, rather than reports about individual projects. However, if your project is out of the ordinary, you may need to plan a communication strategy that demonstrates your alignment with corporate goals and objectives.

If your objective is to reduce project costs by fostering content reuse among deliverables, report at the end of the project your success by comparing the original information with the information from your project:

- ✔ number of total topics developed
- ✔ number of hours used to develop the reduced number of topics

✔ number of hours reduced for editing, reviewing, and testing topics because you have fewer topics to manage

✔ total amount of content (measured in words or pages) in the final deliverables of your project (each PDF, printed piece, web page, help topic, or other deliverable)

✔ total amount of content (again measured in words or pages) in the repository

✔ number of words translated

All these metrics, presented in one slide, demonstrate the gains you have made in productivity. If you can show reductions in actual costs, do so.

If the corporate goal is to increase the value to the customer of information, you may want to report successes in reducing customer content that no one seems to need, decreasing the verbiage, making content more easily accessible, or increasing consistency. Better yet, report on the feedback you are getting from customers on improvements in the information. Talk with trainers or support personnel who may have noted differences in the number of questions or types of calls they receive.

Whatever you report, keep it short. A few slides will go further in getting your point across than a long report.

Adding to your Project Management folder

As you re-estimate the project, make changes to your scope definition, reassign or add resources, add the revised planning documents to your Project Management folder. Include correspondence that details the changes, including minutes of meetings you have had with stakeholders, memos that indicate that changes are anticipated or have occurred, and notes that show how you have recalculated the resources required to complete the project by the deadline.

Summary

You already know that if you have a project that does not change during its development, you have a rarity indeed. Most projects undergo significant change, primarily from new information about customer requirements and changes of direction among the product developers or subject experts. In response to such changes, you have learned about

✔ working with your team members so that they understand the goals of the project and can contribute fully as professionals to the team

✔ tracking the progress of your projects, both in terms of changes that may occur in the larger project of which you are a part or changes within your own project

✔ reacting to changes in your project as quickly and thoroughly as possible, even anticipating changes by knowing the circumstances surrounding your project

✔ planning for your own changes to the project so that you can better serve the information needs of your customers

✔ continuing to evaluate project risks, including the loss of key team members

✔ communicating the changes to your team members, stakeholders, and even your senior management when necessary

The better you learn to anticipate and act quickly in response to change, the more confidence you will have in your own management abilities and the more confident your team members will be in your ability to manage effectively.

"We are all working together now."

Chapter 20

Managing in a Collaborative Environment

Time is the scarcest resource and unless it is managed nothing else can be managed.

—Peter F. Drucker

In the discussions of project planning (Chapter 16) and project management (Chapter 18) activities, I alluded to the importance of information developers and others involved in the development of information working together as a team. Unfortunately, teamwork is not always the norm among information developers. Many writers are hired by product development, marketing and marketing communications, customer support, manufacturing operations, and others responsible for providing instructional and policy-oriented content to external and internal customers. These writers often work independently, responsible for the entire task from planning and design through development and final production. I refer to these developers as operating at Level 1: Ad hoc of process maturity in the IPMM. Even the personality types of many writers, who are more often introverts, support a predilection to work alone even when they become members of information-development departments. Typically, each writer works independently on a book, help system, or other final deliverable, only occasionally coordinating work with others responsible for related publications.

As long as an organization is not worried about consistency in its publications, an environment that promotes individual contributors working in any way they choose is acceptable. As soon as an organization believes that its publications reflect upon its image to its outside customers or help to ensure consistent implementation of policies and procedures internally, the "branding" of corporate information (and thus the consistency of the publications) assumes a new level of importance. Such branding cannot take place without collaboration among those who develop organizational content.

As a project manager, your ability to work with a team of people toward a common goal is essential to the success of your project. If you have team members intent on going in their own directions, despite the needs of the team, your project will experience problems. Not only must you foster collaboration among your team members for the sake of the project, but you also need the support of the department or larger organization. If those who lead the larger projects of which your team is one part do not understand the benefits of collaboration nor cooperate in working with all the team members, your project may be seriously undermined.

I recall a situation in which we had a small collaborative team working on a software development project. The team was led, as usual, by our project manager and included an information developer and an editor. These three team members were all introduced to the software-development team and met with members of that team, including their project manager, often independently. At the end of the project, we learned that the software manager was disappointed that we had continually changed personnel on the project, forcing them to work with three different writers. They didn't understand that we were a team working together on their project. They were so accustomed to working with a single writer who did everything alone that they thought we kept changing personnel. Situations like this one are not uncommon because so many organizations hire information developers who never have the opportunity to work with a collaborative team.

Best Practices in Managing Collaboration

Managing your team in a collaborative environment involves two best practices I discuss here. You may find yourself asked to make a business case for using collaboration in the first place. Then, you need to work carefully with team members to create the best environment to foster collaboration, especially if you have a widely dispersed team. At a time of increasing globalization of the workforce, you are likely to find yourself managing teams with members in multiple locations, in multiple time zones, and representing multiple cultures and levels of experience with information development.

Consider the following best practices for collaboration:

- ✔ Making a business case for collaboration
- ✔ Creating a collaborative environment
- ✔ Recognizing that collaborations may fail

Before making a business case for a collaborative project team, consider the possible challenges of operating in a collaborative environment. Begin by considering several key questions regarding the project environment you are facing:

✔ Do your team members have a history of working collaboratively, or have they always worked independently, even if they are located in the same area?

✔ Are all the people whose work you are charged with managing part of the same department within the larger organization? Do they all report to the same departmental manager?

✔ If not, are the different managers willing to support a collaborative environment and support you as you try to get everyone to work together?

✔ Are you experienced managing a collaborative team yourself? Have you ever worked in a collaborative environment as an information developer or in another similar role?

✔ Are your team members all housed in the same building or on the same campus? Is it possible for everyone to meet together on a regular basis?

✔ Are your team members located in widely dispersed geographical locations so that meeting in one location as a team is rarely possible?

✔ Are your team members located in different time zones? Given different time zones, is it possible for the team to meet by conference call during everyone's regular business hours? Or, will some team members need to work early or late to take part in a meeting?

✔ Do your team members have at least one language, such as English, in common? Are they all native speakers? Are some difficult to understand when speaking to the rest of the team?

✔ Do your team members come from cultures that have different expectations about their involvement in the workplace? Do they regard deadlines or schedules differently? Are they willing to devote time to a project outside the regular work hours or work week if necessary?

✔ Do your team members have roughly the same level of experience in information development or in the subject area that they will be writing about? Are some entirely new to the field? Are they senior team members with considerable knowledge and experience that you can leverage?

✔ Do your team members know the tools and technology that you will be using for the project, including desktop publishing, XML, content management, help system development, and others?

✔ Are all the team members familiar with the work practices in your organization, especially the project phases that you will help them manage (planning, design, development, production)?

✔ Do your team members know and trust one another? Are they reasonably enthusiastic about working together in a new way?

If you answered "no" to any of these questions, you will find managing your project to be challenging, even if you do not try to foster a collaborative environment. However, you will find many advantages to collaboration that will help you frame your case to your own management. If you are embarking on a project in which you hope to reduce costs by reusing content among deliverables, collaboration will be a necessity. If you are managing a traditional project in which every individual can afford to work independently, you may find it possible to avoid collaboration, at least for some time.

Making a business case for collaboration

In the early 1980s, I was part of a group of engineering educators interested in fostering teamwork among our students and analyzing the effects of collaboration on innovation. We followed the progress of hundreds of student teams, not only studying the actions and success of each team but also evaluating the teams in terms of their personality types. We had administered the Myers-Briggs personality type test to every student prior to their enrollment in the experimental, team-centered engineering practices curriculum.

We learned that the best performing teams were the most diverse in personalities. Teams in which all the members had the same personality type struggled to find leadership and direction. Diverse teams explored more options in solving problems and tended toward more innovative solutions. Teams with variable leadership, in that the leaders changed during the course of the work assignment, depending upon personal expertise, were also more successful than teams that developed one dominant leader who told everyone else what to do.

The lessons learned in this experiment can be successfully applied to managing in a collaborative project environment.

Advantages

Although certainly challenging to manage, a collaborative project environment provides significant advantages to produce innovative results and ensure the success of your project. If one of your goals is to develop modules of content, using a topic-based, information-types architecture, collaboration is the only way to succeed.

Given a topic-based architecture, all members of the information-development team must work together to develop content that is correctly typed (as concept, task, reference, or other) and consistently developed following sound authoring guidelines. All the parts built by the team members must work together seamlessly to deliver practical and well-structured information to customers.

You could think about an analogy to building blocks. If the blocks aren't constructed according to standards, they won't fit together to make useful products. Everyone creating the blocks, from standard rectangular solids to specialized gears and widgets, has to agree to work together on design plans and design execution for the project to be successful.

Developing content in a topic-based environment promises increases in efficiency and productivity with related reductions in cost:

- ✔ By creating content once and using it in multiple deliverables, you reduce the amount of time required in your project to produce the content in the first place.

✔ By creating content once, you increase the overall consistency in your deliverables and promote the reliability of the content among the customers.

✔ By creating content once, you reduce the time needed for quality assurance, from editorial review through technical review and testing.

✔ Stand-alone modules of content used in multiple deliverables reduces the cost of localization and translation.

✔ Modular content produced using standardized markup language such as XML or SGML can be formatted quickly and easily by applying style sheets, reducing the time and cost of production tasks.

Topic-based content presents a more coherent message to the customer. Consistent branding of the information your organization develops and delivers will help customers recognize a reliable, well-executed information brand.

When your team members work together to plan and produce a set of well-defined, stand-alone topics, they avoid duplicating one another's work or creating content that is inconsistent. Because less time is spent doing the wrong thing, the team members should have more time to do things well, including finding innovative ways to better meet customer information needs. Team members will discover opportunities to pursue specializations that they find challenging, such as information architecture, information design and graphics, production coordination, quality assurance, and localization coordination.

A collaborative environment, once honed, should lead you to improved quality at reduced costs.

Disadvantages

In information development, a successful collaborative environment requires a centralized organization under a single senior management, or at least a reporting structure in which the managers are anxious to support collaboration. If all your team members report to different managers, you will find it difficult to maintain the cooperative esprit de corps necessary for successful team projects.

As the earlier discussion in Chapter 2: The Information Process Maturity Model argues, a centrally managed information-development organization is essential to move from Level 1: Ad hoc to Level 3: Organized and Repeatable. The same is true for most collaborations. If people can easily ignore the team and follow their own agendas, your project is unlikely to be successful. A centralized department structure, while providing the critical structure required for a collaborative project environment, presents a political dilemma in many organizations. As a central organization, its budget and staffing are more visible to senior management intent upon reducing costs and eliminating layers of management. The central organization is often outside the control and cost responsibilities of operational divisions. Division managers dislike paying a "tax" to keep the central organization operating, even though centralized functions reduce the costs for everyone.

Collaborative environments are sometimes opposed by team members who previously functioned independently. They often rebel against the controls needed to ensure that information is written in a consistent, structured manner. As a project manager, you may have most difficulty with more senior members of your team than with junior members. Making your business case among your own team members will be as important to your success as making your business case to management and other stakeholders.

Many information developers argue that a collaborative environment stifles creativity because the decisions made by the team must be implemented by everyone. On the other hand, opportunities for creativity expand among team members who have the time to consider innovations instead of duplicating the work of one another.

Some outspoken team members may argue that a collaborative environment makes it impossible for them to do their best work. They dislike the extra burden of having to build consensus for their ideas. However, my research suggests that that claim cannot be supported by experience with collaborative teams. All the work of the team tends to be stronger and more innovative than the sum of work of individuals working alone. In short, the whole is greater than the sum of its parts.

Best Practice—Creating a collaborative environment

If you have established a business case for developing a collaborative team for your projects, you need to consider the actions that will aid in your success.

Creating a collaborative environment is easiest when all your team members report directly to you and are co-located. You can focus on maintaining close communication among the team, with informal and frequent team conversations and one-on-one meetings between you and each team member.

However, if your team members are dispersed in different locations, represent different working environments acquired through mergers and acquisitions, include people in multiple time zones, and introduce people newly hired from communities and cultures with no tradition of technical information development, you face many challenges. In the face of such challenges, many project managers capitulate and divide responsibilities into more manageable pieces at the price of innovation, efficiency, and quality.

Developing a collaborative team

By following a simple plan, you can make rapid progress at developing a collaborative team to plan, design, and develop modular, topic-based information:

1   Select team members for your first excursion into collaboration from among those most enthusiastic about topic-based authoring and content management.

2   At the initial team meeting, describe your vision for the project and the goals that you hope to achieve. Explain the process you will use to encourage collaboration.

3   Begin the process at the planning stage. In many cases, once you describe your vision for the project in terms of topic-based authoring and information reuse, team members will go out to collect information about the project specifications and requirements for information development. Ask them to return in a specified amount of time with their results.

4   Meet with team members to review the information requirements for the project. Focus initially on what users will have to learn to do to use the product successfully. Create a comprehensive list of tasks that users will need to perform, including

tasks already in the existing information, tasks that must be revised for the new release, and tasks that are new or substantially changed.

5   Based on the list of tasks, identify supporting information. You may need conceptual, reference, and other background information to enable users to perform successfully. In identifying all your information types, you will produce a picture of the content that must be created for the project. By involving all team members in this collecting and sorting process, they all gradually become familiar with the universe of project content, rather than knowing only about the content they will develop.

6   Your collaborative team will produce a comprehensive content plan for the project. The content plan lists all the modules that will be needed, existing or not, and the list of final deliverables to be produced from the modules. Some groups use large tables or spreadsheets to organize the modules and indicate in which final deliverable they will appear. Assign ownership of the final deliverables to team members.

7   With a list in hand, work with team members to make assignments. Add the assignments to the table. Note which modules will be used in more than one deliverable, and ensure that the deliverable owners know when and where they must coordinate efforts. Modules to be reused must usually be negotiated between more than one author to decide upon the appropriate level of content.

8   If individual modules will require some variation in content to accommodate different deliverables, team members must decide how to handle the variations. If the variations are extensive, you may want to create more than one version of the module and maintain links between them. If the variations are discrete, you may be able to handle the variations inside the modules.

9   With assignments and responsibilities in hand, team members begin authoring, reviewing, and editing. It will be critical that they can review each other's work, especially for the shared information. Even if you have writing guidelines and a style guide in place, you will still need to ensure that consistency is maintained. Your role as leader is to help resolve differences and make final decisions.

10 Once the modular information is in development, work with team members to track the content. Be vigilant about changes in the project that necessitate new modules or different divisions among modules than initially planned. Monitoring project changes will often require that content be reassigned or assigned anew.

11 As the project develops, ensure that those responsible for the deliverables begin testing the modular content against the final list of topics. This list may be a Table of Contents for books or content presented in tabular form in help systems or linked form in websites. All of the outputs must be planned in advance and tested against the plan.

12 Finally, focus the team members on their responsibility of getting the integrated content reviewed for continuity. Consistency of your initial design and writing will go a long way toward ensuring the integrity of the content in context, but you still need a review process to spot gaps and overlaps.

Fostering collaboration among your team members

Effective collaboration begins with trust. If your team members trust one another to produce good work and understand their subject areas thoroughly, they are more likely to work together effectively. If no trust exists, either because people don't know one another or because they have had negative experiences in the past, your first task as project manager is to build or rebuild trust.

Collaboration in Theory and in Practice

Ginny Redish and I collaborated successfully on the publication of *User and Task Analysis for Interface Design* (Wiley 1997). Although we had known one another for many years, we had never before collaborated on such a complex project. We started by agreeing on our vision for the book and working through a detailed information plan. That information plan went through much iteration before we were both satisfied with the manner in which we were going to cover the subject.

When we began writing, we divided the first few chapters between us. Each of us would write one chapter and then send it to the other for revision. We weren't really editing each other's work so much as rewriting it. Although we had different native writing styles, we worked hard to develop a compatible joint style. We created a basic structure for each chapter after some trial and error that served us throughout the development of the book. We continued to work this way, trading chapters back and forth for more than a year. In the end, I think it would be impossible for anyone to know who had written which chapters originally. Because we trusted one another and were confident in our experience and expertise in the field, the collaboration was a success.

Developing a joint information plan for the content of the project is an excellent starting point. If you have a team member who has assumed the role of information architect, that individual should lead the information planning in consultation with team members who may be closest to the information resources provided by the product developers or other subject experts. Through a joint development of the information, begin by creating a team in which each individual is well informed about the purpose of the project and how you intend to meet the needs of the customers.

As you transform the information plan into the project plan and develop spreadsheets to account for the topic-development schedule, again ask your team members to contribute to the process. Ensure that everyone understands how you estimated the resources needed to meet the deadlines for the project and how you have planned the interim deadlines in the schedule. Ensure that everyone recognizes the limitations on time and resources so that they take responsibility for maintaining their own schedules and project time budgets.

As you develop the topic schedule, enlist the help of each team member in evaluating the possible complexity of each topic or set of topics. Consider that some new topics will be more complex than others and require more time to develop. Ask them to make the first estimate of the extent of work required to revise existing topics. Ask if the effort will be major (more than half the topic will change) or minor (less than half the topic will change).

As you go through the process of leveling work assignments, consider the expertise of individuals and discuss the possibility of assigning topics to others when certain team members are overburdened. If you have less experienced team members, consider tasking senior information developers to work with them as mentors. You may even want the more junior staff members to work as assistants to the senior subject experts. They learn both the subject and good work practices in addition to gaining the trust of the rest of the team.

Once the information-development work begins in earnest, develop small review teams. The experience of software developers in creating teams to review code has proven to be successful in removing errors early enough in the project that they are easier and less expensive to correct. A similar process can be used among information developers who can review each other's work as it develops. Not only does this process decrease errors, it builds trust and further removes the possibility of duplication of effort.

During the Development Phase of the project, you may have the advantage of a developmental editor who is responsible for looking across the topics and ensuring a consistent approach that fosters the reuse of topics across deliverables. If you don't have the advantage of a developmental editor, use your team members themselves to maintain quality at the developmental level. If your team grows accustomed to team work, the quality level of all the content will improve.

At this point, I can hear the objections. All of this collaboration will take too much time. Each information developer needs to stay locked in his or her cubicle, head down, developing content that you hope is unique. How can we collaborate when everyone is already so overtaxed with work assignments?

In my experience, a collaborative team is significantly more efficient than a set of independent contributors. The team, following a minimalist agenda, can decide to eliminate unneeded topics and streamline everyone's writing style. When everyone works alone, overwriting and duplication are much more likely.

Still, this description of an environment in which a collaborative team succeeds may seem "Pollyannaish," a confident leap of faith in the creativity and dedication of the team. Certainly, it is easiest to achieve a collaborative project environment when everyone works together in the same location. However, many project managers will argue that hardly anyone has that advantage today. More frequently, teams are dispersed and have little experience in working together and may even hold each other in contempt. Especially through mergers and acquisitions, outsourcing, and moving project teams offshore to emerging economies, distrust and disagreements can become impediments to your project's success.

Collaborating across geographical locations

Managing your project across different geographical locations increases the difficulty, not so much because team members are housed in different offices but because they may have no experience working together. In most cases, the geographic locations are the result of mergers and acquisitions, which means that you are not only working with people who don't know one another but also with people who come from different corporate cultures.

The best starting point for creating a collaborative environment is face-to-face meetings that are carefully designed to help people communicate on a personal level. A face-to-face team meeting is probably the most critical element in fostering teamwork. Video conferences or teleconferences, which work well for ordinary communication during a project, are incapable of equaling the benefit of an actual meeting.

Unfortunately, restrictions on travel may make a face-to-face meeting of your team impossible. If you have the facility available, arrange for a video conference so that people can learn what team members look like. Leave time for personal introductions. If you can't arrange a video conference, use a teleconference with the same time for personal introductions. In addition, create a team website where you post pictures and ask people to contribute personal introductions. Use the site as a communication center for the team where everyone can post questions, comments, interesting articles, slide presentations, and other community-building resources, in addition to links to the team's work products.

Most of the time, project managers who manage projects that cross geographical locations assign portions of the project to separate teams who may be able to work together collaboratively. However, if you want to build a collaborative environment that crosses geographies, consider using work assignments to that end. Develop work teams with subject area assignments that include members from different locations. Assist them in building communication tools and engaging in planning, design, and development activities together. The same techniques I suggest for co-located teams can be used by building cross-geography teams. Joint planning, development, and review activities work even when people are thousands of miles apart.

Technology Aids Collaboration

When Ginny Redish and I worked on the user and task analysis book, we had only three or four face-to-face meetings. Most of the time, we communicated by email or telephone. During part of the year, I was in Australia while Ginny was in South Africa, entirely different time zones. We marveled at the possibility of collaboration given our world travels. In fact, having the opportunity for one of us to review work while the other was sleeping was a small benefit.

As project manager, your task is to ensure that the communication activities take place regularly. Be watchful for team members that become isolated or team members who fail to communicate regularly and begin working independently. If you are convinced that a collaborative team will help your project succeed, you need to be thoroughly involved in the process.

Collaborating across corporate cultures

If you are beginning a project with team members from what were originally different companies, you may need to include several preliminary activities to foster cooperation. The original companies may have conflicting style guides or authoring guidelines. You may have entirely different processes for conducting projects. Team members in your location may have experience working collaboratively, while team members in other companies may have always worked independently. The organizations can be at vastly different levels of process maturity. A well-organized Level 3 department may be resented by those working in a loosely connected Level 2 department or an unconnected Level 1.

You may need to work closely with your department management before you begin a project that crosses company cultures. Building a common culture is difficult and time-consuming and requires work that must occur outside of a project's activities and deadlines. If the culture-building activities have not occurred, you may be unable to work as a collaborative team across organizations in which people are suspicious of the motivations, skills, and expertise of one another.

If your project is the first opportunity to build a common culture, consider finding an entirely new way of working. Collaborative teams help to break down walls because no one has been working in quite the same way previously. Move ahead with all the collaborative techniques I have been advocating, especially with team building activities, mentoring, and cross-geography assignments. Your goal remains to build quality and meet your deadlines. If these goals are not being met, you may have to consider other options, but you should find it most appropriate to assume that people are willing to cooperate if they recognize that they will achieve a better result by doing so.

Collaborating across time zones

Time zone differences are a symptom of diverse geographical locations. In the US alone, we work in four different time zones. If we include the central European time zone, US participants schedule meetings early in the morning while the participants in central Europe schedule the same meetings for late in the day. Add to that confusion staff in Asia or Australia and New Zealand, and your team is working around the clock.

Many teams accommodate the time zone differences by creative scheduling of meetings, with some team members agreeing to alter their work hours to fit into an international meeting schedule. Other teams manage communication by email and instant messaging or shared servers from which project work can be accessed by everyone. Adding to the mix, teams use collaborative tools, such as Adobe Acrobat, that allow people to add comments to each other's work and see everyone's comments at the same time.

Teams benefit from the 24-hour clock. One part of the team can develop information to meet a deadline, and the other part of the team can pick up the information and review and revise during the next set of office hours.

Teams also find the time zone differences to be challenging. Answers to critical, work-stopping questions may take many hours to arrive. Work may overlap because one part of a team moves ahead without realizing that the other part of the team has already completed some work. Time differences that cross the international dateline mean that some team members are enjoying their weekends when others are struggling through Fridays or Mondays, limiting the time for collaboration to four days rather than five.

As project manager, you need to work out a time-based system and get feedback about its success or failure. You will find it most important not to let problems fester until they become crises of confidence. If members of your team do not use instant messaging or email effectively, work in some quick training activities. Consider adding a chat area to the team website so that everyone can see all the discussions about a topic. Even consider adding to the communications subject-matter experts who are project stakeholders.

Time zones should not be an impediment to collaborative success but they will require the team to take them into account in managing communications.

Collaborating with team members working from home

Many project teams today include people who work all or part of the time at home. Telecommuting team members also present challenges for the management of a collaborative team. Most companies have effective policies that require telecommuters to maintain email or instant message contact, be available by phone, and be free of distractions during work hours. On the other hand, remember that telecommuters often agree to work at odd hours to accommodate time zone differences with other team members.

As project manager, your job is to ensure that telecommuting does not present a problem for other members of the team. Once again, be certain to listen for signs of tension. Telecommuters need to work hard to ensure they are up to date on project activities and be willing to attend team meetings as necessary to ensure that they understand how the project is developing. Build time into the schedule when the telecommuters are present at meetings rather than on the phone so that communication problems can be corrected.

Collaborating across cultures

Collaborating across diverse cultural environments is actually quite similar to collaborating across teams formed through mergers and acquisitions. It is certainly possible to keep the project work completely separate, assigning different projects to information developers in each geographic location. However, unless you have completely different products and no interest in creating an information brand for your organization, you will need to work toward a collaborative culture.

In our benchmark studies of information development in India and China, we have identified several best practices that help to ensure that projects are successful and that help you build a sustainable environment for global information development. To develop your project in collaboration with information developers in other cultural environments, consider the following:

- ✔ If at all possible, schedule a face-to-face meeting with the team at their location. Personal relationships are extremely important in establishing rapport, more so in high-touch cultures outside North America and Western Europe. If you don't meet personally with your team members, you will face communication problems throughout the project.

- ✔ If you expect the new team members to interact with your existing team, send a senior information developer or editor to train the new staff on authoring guidelines and style requirements and to help build personal relationships.

- ✔ Ensure that everyone has the resources they need to work productively. The team members may need training on your information-development tools and technology. You may have to insist that they have equipment that is compatible with your home team members and of equal quality. If everyone is using the same content management system, you need to ensure that backups or maintenance activities are not done in the middle of the night in your time zone when others need the system to be up and working.

✔ While everyone is learning the authoring guidelines, style guides, and new processes you use to conduct your project, assign mentors from your existing team to work with the new members. Mentoring will help to build confidence and trust.

✔ Be certain to involve everyone on your existing team in building relationships. In a collaborative environment, everyone must work together, not just the project manager. Involving all the staff in the collaborative effort helps to dispel the feeling among existing team members that they are threatened by the new team members. One organization has been particularly successful mixing the members of special project teams so that they include representatives from diverse locations.

✔ Ensure that the new team has a local manager who is an information developer and dedicated to the success of your projects. Assigning new information developers to be managed by someone who is not part of the information-development process invites conflicts of interest.

✔ If possible, avoid having one or two information developers working in isolation alongside product developers. There needs to be a critical mass of six to ten people to form a viable organization at the new site, even if only one or two people are working on your project.

✔ Just as you would for any team members who are not co-located, develop a project website, provide pictures of everyone, and invite all the team members to write brief introductions.

✔ Be certain that managers are aware of holidays in each location and take them into account when setting deadlines.

✔ Implement a practice to handle language difficulties. Allow more time for conference calls until people adjust to accents. Ask all team members to speak more slowly to ensure that they are understood. At regular intervals, ask all participants if they fully understand the points being made.

✔ Catalog the education, experience, and special skills of your new team members so that you know how best to involve them in your project. If you are part of the hiring team, be careful in your selection process. You may not be able to find people with the experience or education that you would expect in your home location. Work with local managers to develop a closer match.

Managing your cross-cultural project

In terms of the management of the project itself, you may have to plan your project in much more detail than you would for a team in your home location, especially if many of your new team members are newcomers to information development. Consider their needs as you develop the project plan and go over the details of the plan carefully with everyone. Ensure that your information architect or others responsible for planning the content do the same.

You will find that new team members may need far more detailed instructions than you are accustomed to provide. They often want to know exactly what is expected of them and understand all the interim and final deadlines. It would be best to have your entire process mapped out in a process guide so that the newcomers understand every project phase and all the activities that are required before the interim deliverables are acceptable. Explain the quality assurance activities of the team. Explain the nature of the collaborative environment you have created and the expectations you have for their participation.

Once your new team members gain experience and expertise, remember that they don't appreciate always being assigned the simplest, most boring work. They will expect to be treated as equal team members with their share of the interesting and challenging assignments. Recognize that some of your new team members may already have experience in the field and knowledge of the subject area. Reward their expertise by giving them the most challenging new assignments.

In the same way that you have already worked with team members in multiple locations, plan your communication carefully. Ensure that the new team members get lots of feedback from their mentors, the other team members, and you. If the new information developers are new to information development, they may need to have more feedback than you are accustomed to providing. No one wants to continue to create unacceptable work. Find ways to communicate about the quality and timeliness that you expect. Leverage the knowledge and experience of your senior information developers to work with new team members.

Consider how you are going to ensure adequate knowledge transfer between the product developers or other subject experts and your new team. If they are co-located with product developers, they may become primarily responsible for understanding the product and communicating with developers, some of whom may not be fluent in English. If they are not co-located with developers, plan how the communication of subject knowledge will occur. You may need to brief the product developers about the new processes required to help the team work effectively. They may be accustomed to providing information in person. In the new collaborative environment, most of their communications may have to take place electronically. You may need to monitor the communications until you're confident that a good exchange is occurring. In some cases, I have found product developers who become increasingly frustrated working with new team members who have little background in the subject area. A mentor arrangement will help, but you may have to provide more subject education to the new team members and intervene in their initial communications with the subject experts.

Recognize that the cultural expectations of your new team members may be different from your existing team members. In our benchmark studies, we learned that many cultures are concerned about losing "face," which reflects on the way they are viewed by their fellows and their peers in the larger organization.

You may find, for example, that your new team members are afraid to tell you that they have been assigned more work than they can accomplish. They may be likely to work extreme numbers of hours to complete everything asked of them and without informing you. Extra hours that continue for too long do nothing for the team's morale.

You may find that team members will not tell you about problems they may be experiencing on the project. That might include problems with tools, access to subject experts, difficulty in understanding the subject area, or lack of clarity about the processes. You will have to rely upon a local manager to help uncover problems or require more detailed and frequent feedback. You may have to inquire directly about areas in which problems are likely to occur. If you don't, you may discover at the deadline that work is not complete because of problems that could have been solved had you known about them sooner.

You may also experience differences of opinion about appropriate writing styles. If you are working in American English, you may find yourself or your editors in conflict with people trained in British English. At one company, the writers refused to correct errors marked by the editors because they considered the required changes to be incorrect. The editors had to communicate the expectations about working in American English and explain the use of the imperative voice in task instructions. The imperative is considered rude in some cultures.

You may also experience issues with culturally specific gender expectations. In some cultures, women never supervise men. If you are a female project manager, you may have male team members who will not cooperate with you. Cultural training may help correct the problem, or you may have to communicate through a local manager whose direction is accepted. You may also find that women on your team are likely to defer to the opinions of male product developers with whom they interact. The product developers or other on-site managers may ask them to perform secretarial or other tasks that are not part of their jobs. Again, you will need to work with the on-site managers to ensure that your team members work for you and follow your direction.

Despite all the differences I've mentioned, most of the cross-cultural projects we have tracked turn out to be successful after the initial problems are worked out. However, don't feel as if you have to impose your own culture on someone else. You're probably not out to change the world but rather to deliver your project on time and on budget with the quality that your customers deserve. Find ways to accommodate cultural differences. If your team has not had training in accommodating cultural differences, ask your manager that the training be arranged for all team members. The newcomers need to learn about your culture, and your existing team members need to learn about theirs.

Recognize that with all the planning, communicating, and problem solving you will have in accommodating new team members in a different culture, you may have to return to your initial project estimates and account for the extra management time, as well as the extra time it will take for new information developers to create usable content. Your hours per topic metrics may increase significantly, as will the percentage of time you will have to devote to project management tasks. If you thought that you could do some writing your-self on the project, think again. More likely, you will find yourself spending far more than 40 hours per week to ensure the success of your project. Track the time everyone spends carefully. Your senior management may have assumed that lower personnel costs would reduce project costs by the same percentage. Given the amount of start-up work you will encounter, the first several projects are likely to cost more than if you had done them with your original team.

Collaborating with both regular and temporary employees

Many project teams include both direct employees of the company and temporary employ-ees hired for particular projects or time periods. If your project team includes both, you need to consider the long-term consequences of project assignments. You should plan to involve temporary employees in the work of the team, including planning, design, development, and production. In fact, temporary employees may bring skills to the team that regular employees lack. If they are knowledgeable about information architecture, new tools and technologies, subject areas, usability, structured writing, minimalism, or other areas that you want to add to your project, their presence enhances the quality of the work.

By creating a collaborative team, you can better take advantage of the skills that temporary employees bring to educate your regular team members, especially in areas where training opportunities are limited. Consider setting up mentoring relationships that facilitate the transfer of knowledge from temporary to regular team members. The regular employees will need to learn enough to take responsibility for the innovations introduced. However, it may be easier for them to continue practices already established than to be innovative on their own.

Because using temporary employees is ubiquitous in information development, I don't generally find resentment between the regular and temporary team members. However, experienced temporary employees may bring with them work habits that don't support a collaborative environment that you want to establish. I find that interviews generally don't reveal problems with collaboration. Most people will tell you that they favor teamwork, but, in practice, you discover that they avoid it and tend to maintain their status as independent contributors. People who have been contract workers for many years have generally learned that they need to work independently to thrive in many corporate environments. Others genuinely enjoy the opportunity to work with a team and happily contribute their expertise.

If you sense that temporary employees are going to resist working closely with team members, replace them quickly. You may face enough challenges with recalcitrant regular team members than having to endure an uncooperative temporary person. During the interview process and at the first opportunity in planning meetings,

- ✔ explain the need for a collaborative environment and the benefits that will accrue to everyone on the team

- ✔ ask people to articulate their own understanding of how the team will function and how they believe they should work with others

- ✔ discuss ways to build communication and trust among all the team members and include the ideas in your own planning

The more you are able to make every team member responsible for the success of the collaboration, the easier your own task will be.

Best Practice–Recognizing that collaborations may fail

Successful collaboration is all about communication. When communication fails, collaboration fails as well. Building a collaborative team means setting up the means for communication and then making certain that it occurs successfully.

If your team members do not learn to trust one another and feel secure that everyone is working effectively and responsibly, they will revert to independent ways of working, especially if that seems to be the only way to meet deadlines.

Collaboration can fail if team members feel they owe their allegiance to other stakeholders. They may feel obligated to follow the dictates of technical experts or remain loyal to their previous colleagues, managers, and corporate cultures. They may believe their ways of working are superior to those that you are trying to sponsor.

You may find that some individuals are simply not suited temperamentally to working in teams. If there happens to be work that they can do on their own without affecting the success of the project, you may be able to keep them. In some cases, however, you'll find that they will be happier in another work environment.

Remember that you have already decided that a collaborative environment is necessary in your organizational structure and have made a business case to that effect. You are certain, I hope, that your project will be more successful if team members work together effectively. You believe you can deliver a better product to your information customers at lower costs than if you work in any other way. If collaboration does not take place, the project will suffer.

Use all the tactics that you can assemble, with a focus on communication, to create the working environment you have envisioned. If some members of your team seem dedicated to undermining your efforts, work with them individually or suggest that they find other opportunities.

Summary

Collaboration is essential to contemporary information-development projects. When your team is developing topics designed to be used in multiple deliverables, they must plan together, establishing and enforcing standards for information types, information content units, writing style, and even XML markup if appropriate. They must develop a common development process and communicate continuously to avoid duplication of effort and consistent branding of content.

As a project manager, you must

- ✔ make a business case to your team members, stakeholders, and senior managers about adopting a collaborative environment for your project

- ✔ communicate your vision for the project to everyone involved, ensuring that your team members understand that they will gain in creativity and quality what they might believe they are losing in independence

- ✔ establish a plan for creating a collaborative environment that fosters communication at all levels of your organization

✔ consider the challenges that you may face if your team members are geographically dispersed, come from different corporate cultures brought together by merger and acquisition, work in different time zones, work at home, are temporary employees, or are entirely new to information development because they come from low-cost emerging economies

If you have never worked in a collaborative environment yourself, you may want to review information available on collaboration or attend a workshop on the subject. You may also need to find others who have already worked collaboratively to serve as informal mentors for your new venture. Despite the learning curve you may experience, I believe that you will find the process worthwhile and rewarding. If you are moving your team to a topic-based, structured authoring environment in which you intend to increase productivity and quality by using content across deliverables, collaboration is essential.

"We all know that Quality is Free."

Chapter 21

Managing Quality Assurance

> Be a yardstick of quality. Some people aren't used to an environment where excellence is expected.
>
> —Steve Jobs
> US computer engineer & industrialist

Why shouldn't every team member be responsible for the quality of his or her own work during the information-development life cycle? Isn't it enough to make certain that people are following a sound process and are conscientious? Why does anyone have to manage quality assurance?

In fact, information developers should be responsible for the quality of their work. However, many people involved in the information-development life cycle have roles to play in assuring that the information conveyed to the customers provides value. Product developers, those involved in defining services or processes, information architects, usability professionals, editors, and project managers have responsibility for understanding the customers' information needs and providing that information to the best of their abilities. Without the involvement of the entire team, you cannot ensure that you are doing the best work.

Under pressure of cost reductions and reductions in force, many project managers and information-development organizations as a whole find themselves compromising on quality assurance. Many organizations no longer employ editors to review content for consistency and readability. Most organizations do not perform usability tests on information. Even basic copyediting seems to be turned over to spelling and grammar checking software. Yet, all of these processes help to ensure that information is complete and consistent with the organization's standards.

Although information reviews by subject-matter experts remains a typical part of the information-development process, many information developers complain about the inadequacy of this process for uncovering problems with technical accuracy, let alone problems with readability and general usability.

Information developers themselves are responsible for understanding the customer for their information and developing content that complies with the information architecture. In developing the content, they are responsible for ensuring that the information provided is complete enough to serve the customers' needs and is both accurate and usable.

In many organizations, some information developers seem to believe that they are not really responsible for the quality of the content. Rather they are responsible for meeting their deadlines no matter how inadequate the content is that they produce. Information developers sometimes seem happy to copy information from development documents, leave legacy content unexamined, or even write content that they do not understand. Then, they expect that subject-matter experts will find the time and attention to identify and correct all problems with the information. Subject-matter experts are often dismayed by the poor quality of the draft information they are given to review. Not only is the information often full of basic writing errors, but it often appears to reflect a lack of understanding on the writers' part.

Information developers who are content to copy content from specifications or marketing documents are likely to deliver content that is either irrelevant to the customers or out of date before it is published. In such cases, subject-matter experts may not react negatively because they have often written the content themselves. They might ask, however, what job the writers are being paid to do.

Developing high-quality content is not simple. It takes analysis and care. It requires that the information developer become knowledgeable about the customers and understand what they are trying to achieve in working with a product, service, or other information. Producing high-quality information requires competent writers who are intent upon understanding the content they are working with and transforming the source material they work with into something genuinely useful.

As a project manager, you need to ensure that everyone on your team is pursuing content that brings value to the customer. You may need to schedule time for team and editorial reviews of content. You may have to work hard to ensure that technical reviews are handled competently and carefully. You may need to organize your team so that peer reviews are an integral part of the process rather than an afterthought.

Finally, you may have to become involved in the review process yourself to ensure that the project is going in the right direction, especially if you are working with information developers who are unfamiliar to you.

Best Practices in Assuring Quality

As you proceed through the five phases of the information-development life cycle, you and your team members will find many opportunities for assuring the quality of the content you are developing. Review the six best practices for quality assurance:

✔ Assuring quality throughout the information-development life cycle

✔ Facilitating expert reviews

✔ Establishing developmental editing

✔ Validating content accuracy

✔ Obtaining customer feedback

✔ Scheduling copyediting

Best Practice—Assuring quality throughout the information-development life cycle

The project manager's responsibility is to ensure that activities that help to assure quality and value in the work are part of the information-development process and receive the attention they deserve. Without quality assurance in place, you can almost guarantee that the information delivered to the customer will have problems that could have been corrected.

Clearly, quality assurance begins with good project planning. Without adequate time set aside in the project plan for the quality assurance activities, they will either not take place at all or will be addressed inadequately. Information architecture itself must address issues of quality in determining at the beginning of the project what content is required to support tasks the user wants to perform. Without a sound architectural foundation for the content, achieving a good outline of tasks and supporting conceptual and reference information will be difficult. Information architecture that includes user and task analysis will ensure that the information developed is appropriate for the various audiences.

Beginning with project planning and information architecture, you can plan a number of review activities during the life cycle to assure quality, depending upon your resources and time and the importance you place on quality in your organization:

✔ content reviews by subject-matter experts and other stakeholders in your project

✔ structure reviews by the information architect, developmental editor, and other team members

✔ style reviews by an editor and other team members

✔ copyediting and final deliverables editing

Reviews like these are designed to identify errors, misunderstandings, and problems in complying with standards. They are typically performed by colleagues and stakeholders in the organization. They most often include individuals who are considered experts in the subject area of the larger project. For example, in an engineering organization, the development engineers and software developers ordinarily review content for technical accuracy. In a process-centered organization, specialists in a particular subject area review for accuracy and completeness. These experts are asked to identify errors or misunderstandings in the subject area of the content.

Although team members who are working on related content in a project may also provide valuable feedback on the accuracy and completeness of the content, team members also play an important role in assuring that the information complies with organizational standards. An information architect ensures that the content follows the structure standards laid out in the information architecture. A development editor ensures that the content follows standards for readability and consistency. Copyeditors ensure that the content follows language and style standards. Publication editors look for layout problems and adherence to format standards in the final deliverables.

In each case, the reviews may be done by peers among your team, by the project manager serving in more than one role, or by specialists who assist the team in quality assurance activities.

Other quality assurance activities may be handled by other external or internal information users:

✔ usability testing by actual product end users

✔ validation of the content by a testing group or by team members

You may decide to include customers or users of the information being developed in the quality assurance process. When information is being developed for internal use in an organization, end users may be asked to review the initial design of the information. During the Development Phase, end users are often asked to review the content and the final deliverables. They may look for problems in readability, accuracy, or usability and usefulness of the information. They may be asked to verify that the design and content of the information matches their needs and the environment in which the information will be used.

Both internal and external customers may also be asked to verify the usability of the content through informal or formal usability testing. During a usability test, the intended users of the information are asked to perform actual tasks using the information as a guide.

Customers, team members, subject-matter experts, and members of formal testing teams may be responsible for validating the content, ensuring that the tasks can be accomplished with a product or process by someone following the instructional steps. Although validation helps to identify errors in the content, it does not substitute for usability testing. An instruction may be completely accurate but nearly impossible for the intended customers to use.

Figure 21-1 illustrates how quality assurance activities are often associated with the phases of the information-development life cycle.

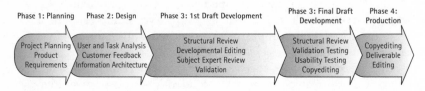

Figure 21-1: Quality assurance opportunities in the information-development life cycle

Depending upon the design of your project and the resources available to you, you should be able to schedule several quality assurance activities. Even if you can afford only to conduct stand-up reviews among your team members of one another's work, that activity will help to ensure that your team has a common understanding of the content as well as the structure and style standards that must be applied. It is not enough for content to be reviewed by subject-matter experts, who are often unaware of the customer requirements or the standards adopted by your organization or your project. Only through the involvement of peers or specialists such as information architects or development and copyeditors can you ensure that your team is working toward the same goals at the same level of quality.

Table 21-1 defines the quality assurance activities suggested in Figure 21-1.

Table 21-1: Quality Assurance Activities

Quality Assurance Activity	Description
Project planning	Project Plans are generally subject to review by various stakeholders, including product marketing, engineering, sales and marketing, and others. Project Plan contain the strategy for addressing user needs and the specification of the required information-development activities, schedules, roles, and deliverables. Project managers should consider presenting their plans directly to stakeholders rather than relying upon passive reviews of emailed content.
Product requirements	Product requirements should provide some detail with respect to user information requirements. These requirements will be translated into project plans. They should be reviewed by information-development management and project managers for completeness and relevance to information-development activities and schedule.
User and task analysis	User and task analysis provides the critical basis for determining the content required by the various user communities. A thorough user and task analysis for a project will have a direct effect on the selection of content, the level of detail of the content, and the usability of the information to promote successful implementation and use of a product or process.
Customer feedback	Customer feedback is a first and a last step in quality assurance. At the beginning of a project, customer feedback can guide change in information architecture and content development. At the end of a project, customer feedback can help to evaluate the success of the project in delivering customer quality. Customer feedback may come directly from surveys, print and online feedback forms, and customer interviews and site visits. It may come indirectly from customer support, sales, product managers, and training.
Information architecture	A sound information architecture, adequately informed by a user and task analysis and customer feedback, provides the basis for creating and delivering information that provides value to customers. Information architecture determines the design of deliverables, the design of topic-based content, and the relationship between topics of content.

Continued

Table 21-1: Quality Assurance Activities (Continued)

Quality Assurance Activity	*Description*
Structure review	A structure review of draft topics or other content ensures that the information developers are following the information architecture determined for the project. A structure review examines the sequence of content within topics and adherence to the topic architecture guidelines for the project's content. A structure review may also review the appropriate use of XML or other tag systems, including format tags in desktop publishing systems. A structure review ensures that the processing of final deliverables will proceed with few structure problems.
Developmental editing	Developmental editing reviews the draft topics to ensure that the content follows the guidelines established by information architecture and the style guidelines established for content readability and retrievability. Developmental editing is typically conducted by a senior staff member and focuses on new information developers, although it eventually includes all information developers to ensure consistency in approach to the content audiences.
Subject-matter expert review	Members of development and service teams provide expertise that ensures that information is accurate. They are responsible for avoiding errors in the content and providing insight into the subject matter that is required to implement and use products and procedures effectively.
Validation testing	Validation testing ensures that the content matches the user interface of a product or corresponds with the way a process or procedure is executed. Validation testing may be done by the information developer directly, by a product testing group, or by subject-matter experts. The information is used to exercise the product, process, or procedure and corrected for accuracy.
Usability testing	Usability testing ensures that the content allows the information user to perform tasks successfully, in addition to be able to locate the appropriate content in support of a task or to understand a concept. Usability testing must be conducted by actual product or process users and be observed by people trained in usability testing.
Copyediting	Copyediting ensures that the content adheres to a style guide and to domain-specific rules for spelling, punctuation, and grammar. Copyediting may be done by information developers themselves, through peer editing, or by a dedicated copyeditor. Copyediting ensures correctness of the content.
Production editing	Production editing ensures that the deliverables are produced correctly and follow the style and format guidelines for final production of content into print and electronic outputs. Production editing is done by information developers themselves or by dedicated production editors prior to the final release of the documents to the users.

Reviewing the project plans and design documents

I find that reviewing project plans and design documents is often a frustrating process. Project managers assume that all they need to do is send copies of the plans and designs to various stakeholders and wait for their comments to come back. In most cases, no comments are forthcoming, or the comments are stated in vague generalities such as "looks good to me." Only later, when the initial drafts are being reviewed, do you discover that you did not have the level of agreement for your planning documents that you had assumed.

In Chapter 16: Planning Your Information-Development Project, you learned how to put together your project plan, with an emphasis on the planning rather than the writing of the plan. Too often, project managers spend too much time making the plan an elegant, polished document. You will find your time better spent in communicating the plan to the appropriate stakeholders so that they understand the direction your project will take and provide feedback if they disagree. You want to know about any disagreements in the overall plan and design of the project architecture before you become embroiled in disputes about the draft content.

Consider planning how best to review the planning documents in the first two phases of your project. You may find it best to prepare slide presentations of the key points or mockups of the designs your information architect has proposed, rather than assuming that anyone outside of your team will actually read long design documents. Even reviewers who are supposed to sign the project plans and design documents often sign without really understanding what they are agreeing to.

If you have a slide presentation prepared, you can ask for time to present your plans to key stakeholders. The presentation might be in person or through an electronic conference. The opportunity to present directly is crucial to the success of your planning. In fact, your slide presentation may be more important to the project than the written plan. The written plan is for you to work out your ideas for the project. The presentation is to gain consensus.

Best Practice—Facilitating expert reviews

Reviews by individuals who are experts in the subject area of your project are perhaps the most common type of review to occur in information development. Project managers generally agree that engineers, software developers, and other experts in the subject area addressed by the content should have an opportunity to review drafts of the content before it is ready to be released. Experts are asked to ensure that the content matches their understanding of the product, service, or other subject matter.

The reviewers for the content being developed for your project should have been listed in your project plan. If you have divided review responsibilities among different subject areas, that information should also be in your project plan.

Depending upon the organization of your project, the expert reviews may be informal or formal or both. Informal reviews are usually handled by the individual information developers and occur throughout the development phase. Information developers ask subject-matter experts to read early drafts and provide immediate feedback. For example, an information developer may lack adequate information about a product function and ask the expert for clarification. The information developer may be unsure about the intent of a service, a policy, or an internal process and ask for additional resources to provide additional insight into the subject. Informal reviews and feedback often take place through email, phone conferences, or in person.

During the informal review process, the project manager needs to ensure that the subject-matter experts are responsive to the information developers' requests. It should be clear to the experts that they are simply trying to ensure that the early drafts of content are as complete and accurate as possible. Timely feedback, answering questions and commenting on early drafts, facilitates the development of sound content initially and means less time later in the process to correct errors and misunderstandings that could easily have been identified and corrected through early feedback. Ensuring timely and thorough feedback on small pieces of content means less time wasted going in the wrong direction with an increasing volume of information.

A formal review process is often a required part of an organization's product and information-development life cycle. During formal reviews, responsible subject-matter experts are asked to ensure that the content being developed is complete and is ready to proceed to the next milestone. Depending upon your organization's policies, formal reviews by subject-matter experts may take place at the following milestones:

- ✔ at the end of the Planning Phase with a review of the project plan
- ✔ at the end of the Design Phase with a review of the project architecture
- ✔ during the Development Phase when first drafts of content are complete
- ✔ during the Development Phase for second and subsequent drafts of content as necessary
- ✔ at the end of the Development Phase for final content review

Formal reviews may be conducted in several ways: access to draft documents by individual reviewers for comment; review meetings in which all reviewers meet to present, discuss, and resolve comments; electronically supported simultaneous reviews to which all reviewers contribute. Each of these methods has advantages and disadvantages to ensure the accuracy and quality of the content.

Conducting individual reviews

Individual reviews are the most common way of conducting expert reviews. Experts receive copies of draft documents and add comments manually or electronically. Individual experts work alone, usually unaware of the comments of other experts. The information developer takes responsibility for reviewing the expert comments, reconciling differences among experts, asking for clarification of comments, and making changes to the content.

The process of individual reviews may be facilitated by automated workflow systems that automatically send out drafts suitable for review when the information developer indicates they are ready. In some organizations, the drafts are posted to a common area in a repository or collaboration website where all reviewers can find them. Often the reviewers are notified of the readiness of the content for review and are provided with links to the content.

The advantage of the individual review process is that reviewers can work whenever they have time available and take the time they need to add comments. The disadvantage is that reviewers often neglect their review responsibilities, do not provide their feedback in a timely or complete manner, and are unaware of the comments and concerns of other reviewers. Because each reviewer comments on the same content, reviews end up with considerable duplication of effort.

In many cases, when the information developers receive review comments from individuals, they have to contact the reviewers for clarification or additional information or to reconcile disagreements among reviewers.

Conducting review meetings

Review meetings are designed to bring all the reviewers together at one time and place. In most cases, the reviewers receive preliminary copies to read and come prepared to the review meeting with comments. However, in practice, many reviewers come unprepared to review meetings, using the meeting itself as the time to read draft documents.

Reviewers often complain about the time required to attend review meetings. Review meetings for extensive content may take several days, much more time than the reviewer might devote to the review process working alone. However, you will find that review meetings, while time-consuming, are much more effective at uncovering and resolving problems with the information than are individual reviews.

At review meetings, reviewers get to hear what other reviewers have to say about the content. They often discover that one reviewer finds a problem with content that another reviewer had overlooked. Reviewers also learn that they often do not agree about what is correct in the content. You will discover at review meetings that reviewers have different conceptualizations of the purpose, audience, or requirements of the project itself.

During review meetings, you need to ensure that disagreements about the content are resolved either on the spot or soon afterward. Reviewers may need to take action items to resolve questions about the content which, as project manager, you need to track and close. You need to ensure that questions raised during the review meeting do not remain unanswered or incomplete.

Using electronic review systems

New electronic systems are in place to simulate the positive effects of review meetings without requiring everyone to be present in the same physical location. In some cases, common websites are set up that allow reviewers to enter comments that can be seen and comments on further by other review team members. Electronic review tools make the draft documents available so that each reviewer can enter individual comments at the same time seeing all the other comments being entered. The ability to see all the comments helps to reduce review time (you don't have to comment on a problem that someone else has already noted) and reconcile differences of opinion among reviewers (reviewers can add comments to other comments to clarify points of disagreement). Some systems allow reviewers to conduct electronic chat sessions to discuss a point or send email messages to others outside the review team for additional information.

At the end of the review process, the original information developer can import all comments into the source files to facilitate making changes. All the review comments are archived in the content repository. The information developer may even be able to note the resolution of each comment in the final draft of the information, facilitating traceability.

Overcoming problems with expert reviews

Despite all the new tools available to facilitate the information-review process, reviews by subject-matter experts remain problematic. Review problems probably cause project managers and information developers more grief and frustration than any other single activity during the information-development life cycle. Reviewers are notoriously inattentive to their review responsibilities, often ignoring them completely. Without the support of the senior managers to whom the reviewers report, you may have little recourse except your art of persuasion to get adequate feedback from reviewers.

In your project estimate, you may want to include estimates of the amount of time needed for thorough reviews so that time can be factored into the reviewers work schedule. In many cases, reviewers are expected to meet all of their own deadlines at the same time they have been asked to review information. If possible, provide a schedule of review times or review meetings early in the project planning so that everyone gets the dates onto the schedule.

One problem with the review process may be solvable as your project moves to a topic-rather than a book-based architecture. Small groups of topics can be sent for review by one or a small number of experts in the content, instead of asking everyone on the review team to review everything. Reviewers can be asked to provide feedback on individual topics or small topic sets on a focused part of the product and the content. Reviewers who receive a few topics to review may be more prompt and thorough than if they receive hundreds of pages to read in a few days, much of which has nothing to do with their areas of expertise.

If you are not already using electronic tools to support the review process, consider adding them. Although some reviewers may baulk at having to review on line rather than on paper, the time saved by avoiding duplication of effort and being able to see what others have said may overcome the initial hesitation.

If your project includes the development of topics that will be used in more than one deliverable, communicate to reviewers that after they review a topic once, they will not have to review it when it appears in multiple contexts. Topics that are judged complete and correct should no longer be part of the review process, again decreasing the total time needed for reviews.

Best Practice—Conducting structure reviews

During the Design Phase of your project, the information architect and other team members have developed a structure for the information to be created. With legacy content, of course, the existing structures may simply be maintained with the addition of new and changed content. However, if your project is new or undergoing a major redesign, you need to plan for a review of the implementation of the new design during the Development Phase of the project.

If you are using a book-based architecture, the initial structure will be developed as a set of tables of content for the various deliverables. If you have instituted content planning, you will have an annotated table of contents in addition to a statement of the audience and objectives for each deliverable. If you are using a topic-based architecture, you will have a comprehensive content plan for all the topics to be developed, in addition to an annotated topic list and content maps that show how the topics will be organized in the final deliverables.

As your team members work through the content, they are inevitably going to discover that the content plans need to be revised. Tables of content and annotated topic lists will have to be revised to reflect new content or content that turns out to require a somewhat different structure than originally imagined. I find that a topic area needs to be expanded from one topic to several or several topics need to be compressed into a single topic. Such changes to the high-level structure of the content need to be reviewed with the team and those primarily responsible for the architectural decisions.

Stand-up or table top structure reviews by team members, perhaps including some additional stakeholders, should be regular events during the Development Phase. Weekly discussions of new or changing content that affects the overall structure of final deliverables will help to keep the content plans and topic lists up to date. They also help to ensure that all team members know when a project is changing and help the project manager to recognize when schedules may be compromised and scope adjusted.

However, high-level structure is not the only structure that needs to be reviewed by the information architect and the team members. If your team is using a structured writing method, you need to schedule reviews in which all team members have an opportunity to review each other's work. You will find that team members easily drift away from the common structure that the architect put into place at the beginning of the project. Individuals make decisions about content that may reflect an incomplete understanding of the original structure. Developers in the subject area may require adjustments to the structure or new structural elements to be introduced.

For example, some teams that may not agree among themselves on the proper use of content tags. One information developer may label an item as a <uicontrol> while another labels the same item as <bold>. Review sessions that allow everyone to look at text markup for structured documents can quickly resolve misinterpretations so that everyone agrees upon a common structure for the same types of information. Even at the level of information types, you will find confusion and disagreement. One information developer may consider a list of definitions to be a concept while another decides a similar list is a reference information type. Early reviews of the internal structure among the team members will resolve differences and facilitate a common understanding early in the Development Phase of the project before the number of problems has multiplied beyond your capacity to fix them.

Resolving structure differences of opinion among team members will help to educate everyone about the common architecture and decrease the amount of time spent on editorial reviews, copyediting, and production edits.

Scheduling structure reviews

Structure reviews should be conducted early in the Development Phase to avoid significant restructuring late in the project, especially when a new information architecture is being introduced, when significant changes are made to the architecture, when new members join the team, or when individual team members have had problems with structure in the past.

Editors and information architects should be responsible for reviewing sample files of each team member before too much content has been written or revised. The reviewers must open the sample files to check the proper application of templates or markup using XML or other tag systems.

Editors and information architects should read sample files closely to ensure that content units are being handled according to authoring guidelines. Information developers need to know if they are properly implementing the content units in the information architecture. If someone begins by misunderstanding a content unit, that misunderstanding may require the restructuring of hundreds of topics. For example, if an information developer begins to include step-by-step instructions in the middle of concept topics, it's best to find the problem early and be certain that the idea of separating concepts from tasks is well understood.

Best Practice—Establishing developmental editing

Despite the fact that developmental editing has long been established as an integral part of information development and I advocated its use in 1994 in *Managing Your Documentation Projects*, it is still not practiced on many information-development projects. Organizations are much more likely to include copyediting in their information-development process than they are to include developmental editing. Yet, in terms of quality assurance, developmental editing is one of the best ways to ensure consistency among the work produced by multiple authors working on content that customers will use as a set.

If you are serious about ensuring that information coming from the same corporation or organization or department looks, reads, and is structured consistently, then you need developmental editing. If you have on your team, individuals who are more and less experienced in information development, in the subject area, in your house style, or in your information architecture, you need developmental editing. If you have information developers spread over the globe, including new team members in developing economies, you need developmental editing.

A developmental editor, typically an experienced team member with special skills in reviewing draft content and working with individual developers, is responsible for ensuring that a uniform level of quality is maintained across the information set. He or she is also responsible for ensuring that the standards established by the team for the structure of the content are adhered to by everyone equally.

A developmental editor often assumes the role of educating new, less experienced team members about the standards and the expectations for quality. A developmental editor may, for example, review the work of a new information developer to ensure that he or she understands how to implement the standards and does so effectively in the draft content. The editor may work with new team members on writing style or may review content to be certain that the writer has assigned the correct information type (concept, task, or reference).

First-rate developmental editors can review drafts and ask critical questions about the content, even if the editors are not themselves subject-matter experts. Skilled developmental editors find gaps and sequence problems in task instructions without ever having tried to perform the task. They suggest ways in which the task can be written with fewer, clearer steps, sorting out what content must be included and what content is nice, but not necessary. They read conceptual information and question whether it is necessary for the audience to know. They ask how the conceptual information is related to task performance. They look at tables of data and other reference information and find inconsistencies. They even test equations to find out if the results reported make sense.

Developmental editors can quickly catch small problems that could become big problems with the content if left unchecked. They work effectively to reduce the technical debt of a project, ensuring that the content developed for the release is the best it can be.

If you are managing a project with a global team, many of whom are new to your projects, organization, or even to information development itself, you should insist that you have a developmental editor on the team. The editor serves as the coach for the new team members, helping to educate them to the practices and processes of the project and guide them through the learning process.

You may find it practical to find one person who assumes the role of both information architect and developmental editor. Both roles are responsible for ensuring that the architecture is implemented in the individual topics. Whether one person can do both jobs depends in part on temperament. Although your architect may fully understand the structure for the project, he or she may not be a good coach or teacher. Your developmental editor often has the most important education role among your team members.

If you do not have someone in the role of developmental editor, you may have to assume that responsibility yourself. Although the project manager often focuses on the nuts and bolts of the project, on many teams, the project manager wears many hats, including information architect and developmental editor. The key is not to assume that everyone on the team is automatically fully engaged with the project vision and plan. Even experienced information developers may be moving to a topic-based architecture for the first time and need support. Years of experience are insufficient. People with many years of experience may have grown accustomed to taking technical source content and dumping it into their topics without examination or revision. Even people with many years' experience may find it difficult to say "no" to a senior product developer when he or she wants inappropriate information added to a topic. If the content developed by your team members is never reviewed by anyone, it is unlikely to be the best content that the team can produce.

In the absence of a developmental editor, you may have to assume the role yourself or use a peer review process, such as a stand-up or table-top review, that gives all the team members an opportunity to read and comment upon the work of the team as a whole.

Best Practice–Validating content accuracy

Reading through technical content to check for accuracy and completeness certainly works to discover problems with the information. However, read-throughs are no substitute for actually performing the task by following the step-by-step instructions. For this reason, you must schedule time for task validation in your project quality assurance process.

In many projects, information developers, who are quite diligent about trying to understand how tasks should be performed by customers, have little access to the products they are writing about during the early part of the Development Phase. The products may not be completed as specified. Early prototype designs are being tested and changed in response to customer and stakeholder feedback. Design documents are often incomplete or never updated as the designs for the product are modified. Hardware may not be available at all for testing until late in the Development Phase. Software may change daily or even more frequently, especially if the user interface is being designed by programmers who are not experienced interface developers.

In some organizations, a product testing group conducts a formal validation of task instructions. Typically, the testing group validates the content at the same time that they are validating the functionality of the product. If done well, this testing process can help ensure that the instructions match the product that is delivered to the customer. Unfortunately, the testing process does not always work as intended. In some cases, the product is changed after the testing, and the information is not revalidated. In other cases, the testing group never gets around to testing the instructions because they are too busy finding bugs in the product.

As a result, information project managers should consider moving the validation process into the project team. Individual information developers are often quite able to test their own instructions against the product, if and when the product becomes available. Product developers may test instructions with their own builds of the products. Even better, team members can test each other's instructions, meaning that no one has to test his or her own work. Some project managers hire interns or other temporary workers to help test the instructions, an especially useful process since the temporary staff bring a fresh perspective to the task and often find errors that are overlooked by information developers who have been working with the product and the information for months.

In organizations developing internal policies and procedures, a validation test is often performed by end users, supervisors, or even trainers. They review the procedures to ensure that the process can be done in the working environment. Unfortunately, sometimes this validation step includes only a read-through rather than a walk-through of the instructions. Although reading may uncover problems, until the instructions are used in a working environment, they may still have flaws.

During a validation test, no one, especially the information developer, should coach the testers. If the instructions are not self-sufficient for a tester, they will probably not be adequate for a typical user. If someone has to assist the person performing the validation, the test itself is flawed.

Validation testing can approach usability testing in the value it provides for quality assurance of instructional text. However, most validation testing is done to ensure accuracy and completeness, not to guarantee usability. Validation testing should be supplemented with usability testing, especially in situations where content is being newly architected or rearchitected to better meet the needs of the customers.

Scheduling validation

Validation testing can occur after first drafts are complete, after reviews by subject-matter experts. Testing content before reviews are completed seems pointless because problems that might have been easily identified and corrected interfere with the testing.

In the best case, the information developer is able to test the content against the product during the draft development, either for the first or second drafts. But, too often, no product is available for testing until much later in the information-development life cycle, requiring the information developers to guess how the product might actually work. One of the primary frustrations for information developers is the inability to validate content because products are unavailable in time or interfaces change after the information has gone into production.

Testing done by a formal testing group must often be scheduled to accompany product testing. The testers are supposed to use the content to test the product for defects. Unfortunately, the testing may be scheduled before the content is ready, or the testers may exercise the product without paying attention to the content. I know of many instances in which testers added comments about the information after they had completed their product testing. Under those circumstances, key problems are often forgotten and never fixed.

Information developers often choose to ignore validation testing because they claim they are too busy. It is up to the project manager to ensure that this quality assurance activity takes place in the schedule in sufficient time to correct the content. The project manager is also responsible for finding products to test, for pressuring development to "freeze" product interfaces in enough time to correct the information, and for ensuring that information developers have the assistance they need to run the products through the testing process. Information developers often argue that they cannot test their information because they do not know how to use the products.

Best Practice—Obtaining customer feedback

Feedback from information customers provides us with a measurement of quality that is unequalled by any of the internal quality assurance activities. After all, customers are the real judges of the usefulness and usability of the information you provide them. Without their feedback, you are always left hoping that their needs are met.

Information developers have historically used mostly passive instruments to obtain feedback from customers. When most information was delivered in print, we asked customers to fill out feedback forms put at the end of their manuals. Unfortunately, most organizations rarely received any of the forms, and when they did, it was almost impossible to decide if the information represented the viewpoint of more than a single customer.

Using electronic feedback systems

Now that much information is delivered electronically to customers, you can use direct feedback more effectively than you did when only paper feedback forms were available. A number of organizations include feedback mechanisms on their websites, often attached directly to a topic or web page of content. When customers respond, the location of the information they are viewing is automatically added to the response, freeing them from having to describe the content that has presented a problem. The information collected from online feedback forms is, however, valuable to information developers only when it informs decisions about quality improvements in the information design and delivery. If the information is not easily accessible or is difficult to interpret, it is often ignored. A well-designed customer feedback system, like the ones used at Microsoft and Apple, provide a constant flow of information about customers that is invaluable in guiding the work of the information developers.

Customer Feedback System at Microsoft

The Office Applications group at Microsoft has developed a customer feedback system that has been successful in helping information developers revise content to increase its value to the customer. Customers using the web-based help system are invited to rate the topic they are consulting and provide a comment about its usefulness or lack thereof. The comments are delivered daily to the information developers who have been assigned topics to monitor. Each topic is rated according to the value assigned by customers, and all the comments are attached.

The information developers monitor the topics that are receiving low ratings and decide how they might be modified or completely recast to improve their usefulness. As soon as a rewritten topic is reviewed and approved, it is posted to the help website. New and rewritten topics are added to the help system daily.

Not only are the information developers improving the value of the content of individual topics, they report that they feel much more in touch with customers' information needs. They find that the constant feedback has led them to rethink how help topics should be written, which has influenced the information developed for the next release of the products.

Conducting customer surveys

Some information-development organizations conduct customer surveys, most often electronically, although occasionally include telephone calls to follow up. More often, a general customer survey includes a single question about information quality. Questions typically ask customers to rate the quality of the information they receive and may ask them to rank the importance of information among other evaluations of product and service included in the survey.

Responses to general survey questions provide no information that can be used to improve quality. Rather, they provide a baseline for a periodic comparison. For most product-centered information, I usually find a direct correlation between the customers' assessment of information quality and their assessment of product quality. If customers find the product to be of low quality, they often evaluate the information in the same way. However, when a product is well received, information quality that is rated low may reflect quality problems that require attention.

Survey managed by the information-development organization itself is usually better designed to elicit information that you might use to improve quality. Such surveys might ask what parts of the information customers use most frequently and what parts they never use at all. You might be able to learn information that is considered essential or which information causes most problems. However, remember that survey feedback is only a starting point. Once you have identified a trend in customer responses to a survey, you may be able to plan a more in-depth analysis of quality problems.

Working with customer service

Customer service personnel are often asked to serve as a conduit for customer feedback. Customers calling into help lines, for example, might be asked if they were able to find information they needed in the documentation. The support person might conclude that the information the customer was looking for was, in fact, not in the documentation or, if it was there, was incorrect or insufficient to answer the question. Customer service is an excellent partner for information development in learning about customers, but the formal reports from many customer service organizations do not provide enough information to foster good decisions about changes to the documentation. Only by working closely with customer service can information developers learn about the responses of customers to the information and make sound decisions to improve the quality.

As a project manager, you need to establish a dialogue with the customer service organization. Review how data are collected about customer information and discuss how a better understanding of information problems might be achieved. Interview customer service personnel who are interested in improving information delivery. They can often tell you more about the problems that customers experience than is revealed by formal reports.

Adding usability assessments

Although usability testing appears to be used increasingly to validate the usability of software user interfaces, usability professionals do not often perform tests on customer information. Nonetheless, usability testing provides us with a way to measure the effectiveness of information delivery to our customers in a way that is nearly impossible in any other way.

Usability testing means direct observation of a customer using the information to perform actual tasks. The customer is usually observed by a usability professional or an information developer or both. During the observation, the customer is asked to comment on the text and the product or process and how useful the text is to performing the task. The observers may ask probing questions but are generally expected to refrain from interfering with the customer's progress.

As a result of such direct observation, information developers learn how the customer interacts with information, how effective the information is in helping the customer perform the tasks successfully, how much of the information seems unnecessary, how much is misinterpreted or even incomprehensible. The feedback is extensive and always leads to improvements in the information or even to a complete revamping of the information to improve its effectiveness.

Although usability testing of this sort has been and continues to be done in usability laboratories, more often today the testing is done at the customer's workplace. Usability professionals have learned that the customer's environment affects the usability of the information. The key to measuring customer quality is the direct observation of the customer at work. If you rely upon retrospective feedback to measure usability, you expect the customer to recall the problem and report upon it accurately long after the problem occurred. Most retrospective feedback is unreliable because customers rarely remember exactly what was wrong with the information.

Conducting user site visits

Any opportunity you and the team members have to observe customers using information in their environment is invaluable to assessing quality. In *User and Task Analysis for Interface Design* (Wiley, 1998), Ginny Redish and I outlined in detail how to plan and organize a customer site visit. At a typical site visit, you may not, of course, find customers using information by chance. You may have to ask them to find information related to the work they are performing and review it with you observing. Although their observations may not be as timely as they might have been without you observing, you can get great insights into the problems they experience as well as learning what they find useful.

Learning how customers are affected by the quality of the information you deliver is the best measurement of the information's value. Without direct information from customers, all your other quality assurance activities may still not ensure that customers are getting the information they want and need.

Best Practice—Scheduling copyediting

Ensuring that the information your team produces is accurate, complete, and usable provides you with the most important assessments of the quality of your work. Copyediting activities add an extra element of quality by ensuring that the information you are delivering to customers follows company standards and contains no embarrassing mistakes in language, spelling, grammar, or format.

I believe it is unfortunate that so often, copyediting is the only quality assurance activity that is pursued with rigor, except for reviews by subject-matter experts. Information developers should never be publishing content that is perfect from the perspective of copyediting but is full of technical errors or is unusable and unsuitable for the customers. That is not to say that copyediting is unimportant. Information that contains copy errors does not help customers to be confident in other aspects of its quality. A document replete with spelling or grammar errors reflects badly on the standards of an organization.

At the most basic level, you should expect all your team members to use the electronic copyediting tools available to them through their authoring software. Spelling and grammar checking software, although not foolproof, provides at least a basic level of correctness. Unfortunately, such systems cannot find words that are spelled correctly but are the wrong words in context, and the systems frequently miss many errors that could be easily found if someone other than the author read through the text. If you find that a team member is delivering draft content that has not been run through spelling and grammar checks, ask that he or she check the content and resubmit it. Make it clear that everyone is responsible for basic copy quality. If you believe that writers have not read over their own content and made basic corrections, establish a policy that writers must read their own work before submitting it for review by others.

Decide where in the information-development life cycle copyediting should occur, either by the original writer, by a peer, or by an editor:

- ✔ immediately after the information developers have completed a draft and before it is submitted to external reviewers

- ✔ before a draft is released to production or translation

- ✔ after the final production tasks have been completed and before the final deliverables are released

Copyediting before a draft is submitted to reviewers is critical because it reflects upon the quality standards of your team and your organization. Reviewers are often distracted by copy errors, spending their time fixing spelling and punctuation rather than reading the text for technical accuracy. At this early phase, you may ask the information developers to take responsibility for copyediting or use informal peer reviews if people have time. It is always easier for another person to catch errors than for the original writer.

Copyediting before the draft content is released to the Production Phase of the process is the most obvious place to check for final content quality. Many information-development organizations schedule copyediting during the Production Phase, if not immediately before, especially if content has changed extensively during the Development Phase. If content is moving to translation during the Development Phase or in the Production Phase, you should schedule a copyedit before a translator has to deal with copy errors. Misspellings will increase your translation costs because they will not match translation memory. Misspelling and grammar and punctuation errors can also lead to mistranslations.

During the Production Phase, you should schedule a production edit, which is a form of copyediting that also includes final checks of the format of the deliverables. Production editors look for bad page breaks, misplaced illustrations, badly reproduced illustrations, overprinting, and other typical errors that occur when final formats are rendered from draft content. If you are developing book-centered information, the information developers may be responsible for final production and require time to check final desktop-published content for errors. Output redirected to web pages or help systems also need to be checked for layout problems.

When copyediting takes place at the end of the information-development life cycle, you may find team members tempted to make corrections in the final deliverables. All corrections must be made in the source documents and the deliverables rerun. If you don't institute this policy, you will be left with a legacy of error-filled source material that will have to be checked again, or you may find that errors persist for years.

Controlled writing tools

Well-designed, controlled writing tools can help to assure basic copy quality in your deliverables and may even absorb some of the work of your editors. The best tools available are designed following an analysis of your style standards and terminology. The tools are customized to meet the needs of your information. They can also be customized to address problems with written English among team members who are not native speakers or writers of English.

Information developers run the controlled writing tool against their draft content. They then review the results and decide which errors need to be corrected. If the tools report many errors that are rejected by the information developers as incorrect, the tools may need further customization. However, controlled writing tools can become useful in maintaining a standard terminology in your content and maintaining your style standards. By using standard terminology, you not only decrease your translation costs, but you also maintain a level of consistency that your customers will benefit by.

Summary

Quality assurance in information development is everyone's job, from the project manager and information architect through the information developers, editors, and those who handle production and localization. At every step of the information-development life cycle, quality assurance activities need to be scheduled and carefully implemented.

Despite the shared responsibility for quality, many project managers discover that team members have difficulty reviewing and editing their own work. For that reason, project managers must find ways that reviews and editing can occur in a team, even if there is no designated editor:

- ✔ Plan for quality assurance activities through the information-development life cycle.

- ✔ Ensure that reviews by subject-matter experts take place in a timely and thorough manner.

- ✔ Either establish a team-based review process or select someone to serve as the team's developmental editor, or do both. Developmental editors work on ensuring that information standards and expectations are met from the beginning of the project.

- ✔ Ensure that content is validated against the product or process so that it is correct.

- ✔ Recognize that copyediting is not developmental editing. Copyediting should focus on compliance with the corporate style guide and the basics of correct writing, graphics, and layout.

- ✔ Find opportunities for customers to become involved with the quality assurance process through feedback, usability testing, and user site visits. Recognize that some of these activities can occur prior to or as part of the information-development life cycle.

- ✔ Add a production edit at the end of the process. See Chapter 23: Managing Production and Delivery, for more information on conducting thorough production edits.

Work closely with your team members to establish a series of best practices for quality assurance. Build quality into your process from the first Planning Phase, well before any content development begins.

Chapter 22

Managing Localization and Translation

> Think like a wise man but communicate in the language of the people.
>
> —William Butler Yeats

Information-development projects increasingly deliver content in multiple languages and content designed to serve the needs of customers in many different international environments. Both *localization* of content, which means preparing it for use in different national or cultural environments, and *translation,* which means preparing content in the local languages of the customers, play increasingly significant roles in information development. The term localization is often used to include translation, but it may be simpler for you to consider the two activities separately.

Localization often refers to the process of converting a software product, especially its user interface, into a version appropriate in another country. The process involves more than translation of the words. It requires keyboards, menu selections, and myriad other issues to be taken into account. In your information, localization may include adding different government regulatory text required by different countries. It may include changing measurements from the English standard used in the United States to the metric standard used in most of the rest of the world. Other changes may be required to ensure that your content is acceptable elsewhere. In one project, I had to change the names of the people in an example scenario because the names might offend people in a particular religious group.

If your organization is experienced with producing content for a global market and you already prepare content in multiple languages, you may have a series of best practices already in place to help you manage these activities. You may even have entire departments of specialists in localization and translation, fully prepared to work with outside service providers.

Note: Most organizations today do not have inhouse translators but either work with one or more service providers to coordinate the work of translators located in many countries or do this coordination themselves.

If you are new to localization and translation activities, as a project manager or because your organization has just begun the process, you may find yourself responsible for managing the entire effort. Even if you work with experienced localization and translation managers in your organization, you have an opportunity to make their work easier and more effective if you prepare your project team members properly.

Adding translation-friendly activities to your project, including writing for translation and performing pre-translation edits of the source content, can greatly reduce translation costs and the time required for translations to be completed. Writing for translation can also improve the quality of the translated content, as well as improving the quality of the original content in the source language. Translation-friendly writing represents, after all, effective and consistent information development. The better written the content is in the source language, the easier it will be to translate effectively and efficiently.

Best Practices in Localization and Translation

Managing localization and translation in the information-development life cycle requires the collaboration of the entire information-development team, an internal localization coordinator, a localization service provider, one or more translators, and you as the information-development project manager. Depending upon the roles and responsibilities in your organization, you may coordinate activities with a department or individual responsible for managing the localization process and interacting with the service provider or you may be responsible for coordinating the service providers yourself. These three best practices will help you succeed in preparing content and delivering it effectively to your global customer community:

- ✔ Preparing your content for localization and translation
- ✔ Including localization and translation requirements in the project plan
- ✔ Selecting and working with a localization vendor

Best Practice—Including localization and translation requirements in the project plan

The starting point for a successful global information-development project is during the Planning Phase of the life cycle. Without a sound Project Plan that accounts for localization and translation, accompanied by a sound information architecture and a team of information developers who create content with a global audience in mind, your efforts to support a global development process are likely to be too little and too late. Your project plan is your best starting point, in which you describe the localization and translation requirements for your project and schedule time for the activities in the plan.

Including localization in your project plan

The template for your project plan, outlined in Chapter 16: Planning Your Information-Development Project, provides information about localization and translation requirements.

As you prepare your deliverables list for the project plan, you should consult with your marketing organization or other's responsible for defining customer requirements. You want to understand the global reach of your project and how your colleagues are preparing for a global market.

Translation requirements include all the countries and languages that will be needed for the final deliverables. You want to include in your project plan's deliverables list all the languages into which each deliverable must be translated. At the same time, you want to be certain that the translation requirements have been decided correctly. For example, the European Union countries require that end-user documentation be translated into the appropriate local languages. If your organization expects to sell product in Estonia, you will need to translate end-user information into Estonian. You may not be required to translate administrative or highly technical instructions into the customer's language, but it is always a good idea to find out before making a decision to avoid translation.

In some countries, highly technical customers prefer to read the information in the original language, arguing that the source language is more accurate than the translation. In other countries, even highly technical customers may be more comfortable reading their own language. In either case, you or someone in your marketing organization should find out what is the best course of action.

If you are selling your product in South America and in Spain, you will have to develop two separate Spanish translations, because the Spanish spoken in South America is substantially different from the Spanish spoken in Spain.

If your product is going to mainland China and Taiwan, you may need to translate into what is called Simplified Chinese for the mainland and "traditional" Chinese for Taiwan, depending upon the analysis of customer requirements. Simplified Chinese, introduced by the government of the People's Republic of China to promote literacy, uses fewer characters than traditional Chinese.

I recommend that not only do you consult with colleagues in marketing and sales and other parts of your organization, but that you do your own research. Sometimes people make assumptions about translations that are simply incorrect and expensive to fix later. One organization, for example, hired a Spanish translator who worked in Latin American Spanish for a document that was being prepared for dissemination in Spain. The work had to be completely redone before it was accepted by the customer. In another case, an organization sent a Spanish documentation set to Brazil, not realizing that Brazilians speak a Brazilian version of Portuguese.

Translations into Multiple Dialects

One well-known fast food franchise organization translated instructions for preparing and serving food items into 29 different versions of Spanish for their global end users. They found that the local languages spoken by their employees were different from the standard for Latin American Spanish, especially regarding work in a kitchen and food items. They decided to translate the instructions into language that could be understood by staff with minimal education.

By asking questions, you may also find that others in our organization assume that localization and translation can be done cheaply and easily inside. For example, they assume that someone in the support organization in Europe will be able to translate all the documentation into French, German, Italian, and Spanish. Or, the product manager has a cousin from Poland who will be happy to translate your manual into Polish.

Both of these are extremely foolish and expensive decisions. You should know enough to make a strong case for professional translations. Non-professionals cannot produce responsible translations. If they have lived outside of a country for many years, they no longer know the common jargon. If they are not experts in your technology, they will not know the common terminology in the field. If they are support professionals, they may not be good writers, introducing grammar and other errors into the text. Professional translators prepare for a translation by researching the terminology to be used in a particular field and are skilled writers in their native language, as well as skilled interpreters of the source language. They know how to maintain consistency in the document and ensure that the correct words are used in the context.

Remember that you can always ask your support staff or other in-country experts in your organization to review the translation and to contribute to building a multi-language glossary for your technical field.

Scheduling time for localization and translation

If you have managed a traditional information-development project, you may expect to schedule time for localization and translation as part of Phase 4 of your project. At that phase, the content is fully developed, further changes to the content are less likely, and the production process has begun. You may even decide to complete the production of the information in English or another source language before sending anything to the localization service provider. As long as it doesn't matter when the translated information is available to the customer, scheduling localization after the source content is finished may be perfectly reasonable. Just recognize that the translated content may not be available for several months.

If you hope to have translated content available at the same time as your source-language content, you will have to schedule localization and translation activities much earlier in the life cycle. Simultaneous shipping of product and information in all languages is challenging but with careful planning, you may be able to achieve these results or come close to them.

To plan for simultaneous shipping or a close approximation, you need to schedule several events early in the information-development life cycle.

- ✔ Prepare a glossary of terms specific to your industry or service and have these terms researched in each target language

- ✔ Review the terminology with representatives from your organization in each of the target countries

- ✔ Enlist individuals who will review the translated content for accuracy once it is complete

- ✔ Develop a process that enables your information developers to produce content that supports localization

- ✔ Communicate with your localization service provider to develop a process for handing off files

- ✔ Meet with the service provider to develop a workable schedule for delivery of files, return of content for review, and production of final deliverables

It is important to start work early enough, especially on a glossary of terms in multiple languages and writing for translation, that you will be ready when your localization service provider needs to begin. Traditional schedules for first translations could be as long as six months to complete. With the use of topic-based authoring, development in XML, reduction of desktop publishing, and excellent preparation of the source content, these schedules can be reduced to a few weeks.

Budgeting for localization and translation

If you have never contracted for localization and translation before, you may be surprised by the cost. You should contact your service provider to obtain preliminary cost estimates for each language you need. You will find that the basic European languages (French, Italian, German, and Spanish, abbreviated FIGS) are much less expensive than Japanese, Chinese, or other languages for which fewer translators are available or with character sets or right-to-left orientation that require special processing.

The cost of localization and translation typically includes the actual translation time, project management, and production of final deliverables in the target languages. The cost may also include quality assurance reviews, reviews of the source language content for ease of translation, and the costs associated with localizing and translating the product itself.

You can expect to pay for translation by the number of words in the source language, typically between 20 cents and 50 cents per word, depending on the language. You will pay additional fees for desktop publishing tasks, from $15 to $20 or more per page. You will also find that there are fees for project management, quality assurance, and other activities that must be completed as part of a translation project, leading to costs that might be as high as $1 per word.

Best Practice—Supporting localization and translation with content management and workflow

Traditional localization and translation practices can become quite expensive when you pay by the word. Of course, reducing the number of words to be translated will help to reduce costs, a strategy discussed in the next best practice. However, developing a strategy to create and manage content for translation will help to reduce costs as well. Adding a content-management system that is translation-ready will provide you with additional cost- and time-reduction activities. Single sourcing, or reusing content in more than one deliverable, will further reduce the number of words to translate.

Developing a single-source strategy to reduce translation costs

A single-source strategy for the development of your information helps to decrease translation costs at the same time that you reduce information-development time. Modules of content that are reused in more than one deliverable need to be written once and translated once rather than multiple times. Using the same content multiple times also ensures that both the original text and all the translated versions will have identical content. Even miniscule changes to the original content, including changes to punctuation, can affect translation costs.

Creating a multiple language repository

Adding a multiple language repository or database to your single-source strategy will further reduce your costs. With a multiple-language repository, you can store all the content in every language. By maintaining links between the multiple-language modules, you can manage the cost of revisions. Only those modules that have changed from the original versions need to be retranslated, drastically reducing the number of words that need to be translated following the first version of the information.

Establishing a translation memory

The use of translation memory tools has become ubiquitous in translation management. On your first project, your localization service provider will build a database referred to as a translation memory for each target language. This database is then available for the next iteration of the translation and has the potential to greatly reduce subsequent costs. Translation memories, also known as translation databases, are collections of entries where a source text is associated with its corresponding translation in one or more target languages.

Translation memories are databases that relate the source content with the corresponding translated content in each of your target languages. The translation memory for a particular project becomes available to the translator, usually through a desktop system. When the translator begins the translation of a section of text, he or she first searches the database for equivalent source content.

If the source content and the "new" content is exactly the same, that translator accepts the existing translation for a 100% match. If the source content and the "new" content are similar but not exactly alike, the translator has a "fuzzy" match. The translator can then work with the fuzzy match to translate the text without having to invent the entire translation from scratch.

By using the translation memory, the translator can work more quickly and be more assured that the terminology and phrasing is correct, that is, if the original translation was correct. However, the translator has to be aware of the context in which a match occurs. The new context may be somewhat different from the original context, requiring a different translation. Even a single term may be translated differently, depending on the context. A discussion about a bridge over a river and a subsequent discussion about bridging a gap in knowledge are likely to require different translations.

As project manager, recognize that you will pay at a reduced rate per word for 100% and fuzzy matches. If the content produced by your team has changed drastically, you may find that the translation memory for your project has to be updated. If someone did not update the translation memory after the last translation or if the translator does not install the correct version of the translation memory, the number of 100% and fuzzy matches will also be reduced.

Just be certain that the correct translation memory is always applied. It is best to maintain the translation memory as part of your repository or file system. Always ensure that the service provider has returned to you the most up-to-date version of the memory. Also discuss the importance of beginning with the most up-to-date translation memory for every project. Occasionally a translator does not install the latest version of the translation memory before beginning the translation, thereby increasing your costs. You pay much less per word if the translation memory recognizes the text in your deliverables.

Using content management to support pre-translation

The combination of translation memory and content management can work to decrease both the costs and time of translation. If you have a multiple-language content-management system, you can avoid entirely sending out modules that have not changed since the previous translation. Because applying translation memory will cost you by the word, although less than new text, to translate, avoiding translation costs altogether is a distinct advantage.

Your content-management system should track the changes and alert you to modules that must be retranslated. However, if you maintain the translation memory as part of your content-management system, you may be able to apply it to modules with minor changes and control the number of new text that goes to the translator.

If you are using topic-based authoring and controlling the content units with XML markup inside the topics, you can further reduce costs by applying pre-translation techniques to content units that are smaller than entire topics and thus less likely to change.

Integrating automated workflows with your localization service provider

If your content-management system includes an automated workflow, you may be able to avoid the frustration of transmitting the wrong files to your localization service provider or forgetting to send the latest version of files that have been revised. A workflow system can be designed to automatically forward the appropriate files to the service provider as soon as they are ready to be translated, even before the entire project is complete. Many content-management systems provide a browser-based interface that can be directly accessed by the translators so that they can pick up and deliver files easily. Your service provider may have an internal workflow system that you can link to, seamlessly providing the correct information at the right time in the process. Modules of content can flow back and forth between the various process steps, including reviews by your in-country experts.

Integrating the workflows implies that you are maintaining the translating files in your repository as they move back through the workflow to you. It is essential for subsequent translation that you maintain the files in your content-management system. Note that when translations are handled by local sales and support groups at your company, you often have no electronic copies of the translated content to leverage for subsequent translations.

Automating publishing

Desktop publishing tools such as Word, FrameMaker, Quark, or InDesign require that the localization service provider disassemble the original documents for translation and reassemble them in each target language. Such repetitive desktop publishing becomes quite expensive. By using automated publishing systems, you can avoid desktop publishing costs entirely.

If you move to standard markup languages using XML or SGML in your source files, most localization service providers can automatically strip out the tags, send the content to their translators, and reassemble the translated content automatically into the original markup.

Given XML or SGML file, you will find tools available that make it possible for you to handle the final production process in multiple languages inhouse. If you are using such tools, you should ensure that all the translated content modules are returned to your content-management system or file system. Then, you can run the scripts needed to process the content and produce final deliverables.

Even if you bring the final publishing inhouse, you should arrange for a quality assurance review by your service provider. It is difficult to tell if the text is laid out correctly and if there are no missing sections, when you cannot read the target language. Especially for double-byte languages like Japanese and Chinese, meaning that the character set is large and complex, or bidirectional content like Arabic and Hebrew, you will need experts to review the final deliverables to be certain that all the processing has worked correctly.

Recognize that text expansion also requires that content be reviewed in final form. Most languages require more space than English, either horizontally or vertically or both. That means that tables and graphic labels may have to be rearranged to be readable. It also means that page breaks must be rechecked to avoid unreadable text.

In any event, work closely with your localization service provider to establish a process that works efficiently and effectively for your content. By keeping the costs of translation as low as possible, you have an opportunity to add additional languages to the mix for customers without breaking the bank and exceeding your budgets.

Best Practice–Preparing your content for localization and translation

Keeping translation costs under control and reducing time to market for multiple languages means more than streamlining the production processes. After you have implemented a single-source strategy, translation memory, and a content-management system, the next area to address to reduce costs and increase the quality of translation is the original source content. By writing for translation, you not only make translation easier and more accurate, you make the original text more consistent and readable as well. Writing for translation means following good writing practices with a few more ideas included that specifically affect translations.

If your content has been previously translated, also consider minimizing the number of cosmetic changes you make to the text. If your localization service provider has a translation memory database for the previous translation, even tiny changes may require a re-translation, increasing your costs. Refrain from "tweaking" the previously translated content, especially if the corrections do not add demonstrable value to the customers.

Developing global standards for your content

The style guide that you invoke to support your team's information-development work should reflect the requirements for global content. The place to begin is consistency of writing style. The style guide should embody the need for consistency in terminology, heading styles, and sentence structure.

As the first strategy, your style guide should include a glossary with terms and their official definitions. With your quality assurance process, you can support the use of the terms in the glossary, ensuring that they are used in every case and that the terms are not used in the text with different definitions. Try to keep newly invented terms out of the content, especially if ordinary, well-established terms already exist with the same meaning. Product developers often invent terms in the user interface that have no equivalent in other languages, even when perfectly acceptable words are available. Your information developers should be encouraged to track new terms and suggest variants that are already in your translation memory. The same process holds, of course, for the information developers themselves. The more consistent their use of terminology throughout the content, the simpler the translation work will be.

Styles used for topic titles, depending upon the information type, should be consistently applied. For example, task titles should begin with verbs, and concept and reference titles might be restricted to noun phrases.

Sentences should be reasonably simple with obvious relationships between subjects and verbs. Information developers should avoid long, complex sentences with many prepositional phrases. Strings of prepositional phrases are always difficult to read and even more difficult to translate accurately. Readers often have difficulty figuring out how the phrases modify one another. Consistency in word choice and sentence structure will help the English reader as well as the translator.

Expressions that are understandable to Americans may not be understandable by people in other countries or cultures. Avoid Americanisms in the text; remember that although English is spoken worldwide, American English is not. Even native English speakers in the United Kingdom, Australia, or South Africa may not understand American expressions. Even a football field is a different length outside the United States.

For more guidelines on writing for translation, see http://www-306.ibm.com/software/globalization/topics/writing/style.jsp developed by translation coordinators at IBM.

Localizing the content ahead of time

Much of the work of localizing content can be handled during the information development. Alternative measurements for English and metric can be included in the original information. If you don't want to display both measurements in the text, use a method for marking the conditions and processing only the one appropriate for a particular language and culture.

Research the regulatory content required in every country in which your information will be used and find the appropriate texts. As you construct maps of your content for your final deliverables, you can point to or insert the appropriate content for each language and country.

Explore ways of making localization simpler by avoiding icons, graphics, names, and other items that are understood only in your country and language. The American symbol for mail as a rural-delivery mailbox is unknown in other countries.

Recognize that most languages take up more space on the page or screen than English. Leave sufficient space around graphic labels or in tables to accommodate longer character strings.

Using controlled language tools

Controlled language tools have been available for a number of years and can be quite helpful in ensuring that information developers follow the style guide and write more consistently. Some of the tools use English or another source language version of your translation memory. If someone introduces a term that does not exist in the memory, it is flagged for revision or inclusion in the glossary.

Other controlled language tools build various checks into the word-processing system, essentially creating an electronic version of your style guide, fully incorporated into the information developers' working environment. The tools may, for example, flag sentence structures that violate the guidelines or are especially awkward or confusing for non-native readers of the language. The controlled language tools are especially useful when you have writers for whom English is not a native language. Rather than relying on expensive editing to correct basic problems with sentence structure, grammar, or spelling, you can have the controlled language system configured to address the most common problems.

Preparing graphics for translation

Text developed inside graphics tools can present special problems for translation. To get at the text, the localization service provider has to open the graphic with the tool originally used to create it, extract the text, send it to the translator, and place it back into the graphic correctly. Since most text is longer in languages other than English, placing the text back into the graphic can cause problems. The translated words may not fit in the same places as the English words.

One way to handle the problem is to place the text on a separate layer so that it is easily removed for translation, although the text size problem is still not solved. Another way organizations handle the problem of translating text in graphics is to remove the text completely, reference graphics with numbers or letters and providing a table that converts the reference items into names and explanations. Such indirect labeling is easier to translate but not easy to reference for readers.

Some organizations are using XML to facilitate the translation of text in graphics. The graphic is converted to a Scalable Vector Graphics (SVG) format. Because the entire image is presented in XML, the text can be extracted using the XLIFF standard for transmitting text for translation. After the text is translated, it can be returned to the graphic in the target languages. Problems may still exist with text expansion, but the need to work directly in the graphics development tool is eliminated, making the process faster and less expensive.

When you are sizing graphics for a print page, remember that paper sizes are not the same worldwide. A graphic that works well in the US's paper size of 8.5 inches x 11 inches, will not necessarily work in the ISO standard A4, which is 210mm x 297mm (longer and slightly narrower).

Localizing the product interface

In some organizations, localizing and translating the product interface is handled by another group than information development. If, however, you are responsible for working with your localization service provider to localize the interface, you have much to keep in mind:

- ✔ The better the product is designed to support localization and translation, the faster and less expensive the process will be. You may want to work with the service provider to design the product to be ready for localization.

- ✔ The localization process often requires that the product be taken apart so that all the translatable text can be extracted. Once the translations are done, the product will have to be reassembled. In addition, certain aspects of the product may have to be changed to accommodate using it in the target language.

- ✔ All the screens and other images in the user interface will have to be translated and otherwise restructured to accommodate the target languages. Screen shots in the documentation will have to be resaved in the target languages.

- ✔ Once a product is being localized, changes that affect localization become increasingly expensive the later they arrive. Small changes may lead to many changes throughout the interface. It's best to keep changes to a minimum after a product is being localized and the screens translated.

- ✔ Your glossary of terms must include all the terms in the interface, including menus, dialog boxes, and error messages. Too many stories occur of user interfaces in one language with error messages in another language because they were left out of the localization process.

Product localization is a process fraught with difficulties. If you have never managed this process before, and neither has anyone else in your organization, start early in discussions with your service provider. They will assist you in finding ways to make the process go more smoothly.

Best Practice—Selecting and working with a localization service provider

Some information-development organizations contract with individual translators for each language they need, managing the entire localization project inhouse. I even find an occasional information-development organization that employs its own translators inhouse. For the rest of the organizations, the best route is to contract with a localization service provider. A localization service provider will partner with you to provide a complete range of services, from recruiting and managing the in-country translators to handling all the project management and quality assurance activities of the localization project.

Selecting a professional localization service provider

One of the best ways to find localization service providers that can meet your needs is to use your network of other managers. When you ask colleagues for recommendations, however, be certain that you understand their localization needs. If your needs are considerably different, the recommended vendor may not be appropriate.

As you prepare to evaluate vendors, list your requirements:

- ✔ Languages you will need to have produced, especially double-byte and bidirectional languages

- ✔ Graphics and screen shots that will be included in your information

- ✔ The tools you use to produce information, especially if you are using XML-based authoring and graphics development

- ✔ Formats you will need for the final deliverables

- ✔ Systems you will use to transmit files, especially if you are developing individual topics

- ✔ Responsibility for final production of complete deliverables

- ✔ Responsibility for maintaining translation memories

- ✔ Quality assurance activities and responsibilities

- ✔ Schedule considerations

- ✔ Number and locations of in-country reviewer

With these requirements in hand, you are ready to begin discussions with three to five potential service providers who appear to have the capabilities you need. You may want to consider at least one large, international and at least one smaller, locally based provider so that you can evaluate a range of capabilities and costs.

Some questions you may want to address in your selection process include the following:

- ✔ Which languages do they specialize in, especially considering less well-known languages that you may need?

- ✔ Do they have experience with your technology?

✔ How do they recruit translators with the technical skills you need?

✔ Are their translators located in their native countries? Do they always use native speakers?

✔ How do they manage localization and translation projects? Do they assign you a project manager who handles all communications?

✔ How do they handle quality assurance activities and who is responsible for them?

✔ What tools and systems do they use for translation memory, content management, software localization, graphics translation, and so on?

✔ How do they manage XML files and individual topics? If you deliver partially translated topics, how will these be handled?

✔ What are the fees based upon? How are they calculated?

Many more questions are likely to arise as soon as you begin your discussions. Once you and the potential services providers have a mutual understanding of the requirements of your projects, you should ask for formal proposals and price quotations for your immediate project.

In your selection process, be certain to consider more than the lowest price. You will need to be comfortable working with the people in the company, especially the people who will be directly managing your project. You probably should ask for and consult with references who have projects similar to yours, asking about day-to-day relationships, project management, and quality assurance activities.

Establishing a working relationship early in the life cycle

As soon as you have selected a service provider, ask for a planning meeting. This meeting should be scheduled early in your information-development life cycle, hopefully in the Planning Phase. During the planning meeting, review the proposal to ensure that you have a clear mutual understanding of the project. Ask for a review of the project management and quality assurance process, and go over the service provider's requirements for the delivery and return of files. Convey as much information as you have about the schedule, especially your assessment of the project risks. If you are confident that the project will proceed on schedule, make that clear. If you are confident that the project will fall behind and that late changes are a regular part of the product-development process, explain what you think is likely to happen to the schedule.

As you discuss process, ask the service provider how they will report to you on the progress of the localization project. Make sure you receive regular reports that include budget information. You need to understand how you will be informed about budget increases and what choices you may have in controlling costs.

Developing a multiple-language glossary of terms

As part of your initial planning meeting or soon afterwards, establish a plan for developing the multiple-language glossary. If your service provider uses terminology software to manage the glossary in multiple languages, ask for a demonstration so that you understand how the software works. Also find out if they have software that allows them to "mine" your existing information for potential glossary terms. Software to create draft glossaries can save you considerable time coming up with the initial term list. It will also help find terms that you don't recognize as special because they are so familiar. Your glossary should include terms that your industry uses in special ways, even if they are otherwise ordinary parts of the language.

You may also have a glossary of terms and definitions, which will be a good starting point. However, if you already know that your legacy information does not use terminology consistently, make that information available to the service provider. They may have suggestions about establishing a synonym list or otherwise working with you so that the translation is consistent. They are likely to suggest that you develop a process to make the writing more consistent to keep translation costs under control.

You may be asked to clarify the meanings of words in your English glossary as the translators begin their research. They will research the terminology used in your field or with your technology in the target countries. If they encounter terms that don't seem to exist, they will ask for clarification. If you have access to up-to-date materials in your field that are already in the target languages, especially materials that are verified publications like scientific papers, provide them to the service providers as a terminology resource.

Once the service provider has drafted the glossary in the language you need, they should ask you to review the terminology with people in your organization who are native speakers of the target languages. Make sure you have subject-matter experts in-country who will contribute to the review process. They may find errors in the glossary because they are experts. Have them provide explanations about the word choice so that the translators better understand the technology.

Recruiting internal reviewers and establishing an internal review process

Not only will you need reviewers in-country when you are developing a multiple-language glossary of terms, you will need reviewers for the translated versions of the information. Look for people who are subject-matter experts in engineering locations in the target countries. If your information is designed for end users, work with in-country trainers or people on the support team who are more likely to know the language used by the end user rather than the technical customers. If you only have sales organization in-country, find people in the sales organization who are willing to review the translated information from the perspective of the customer and who know enough about the technical content to be competent reviewers.

Consider the need for legal review of the content in the target languages. One company discovered too late that the French version of a policy manual did not say the same thing as the English version. An employee who was fired for a policy infringement won a lawsuit against the company. He followed the policy in the French version of the manual, which had been mistranslated and was therefore different from the English version of the policy. Some organizations with sensitive information have the translated text retranslated back into the source language so that the two versions can be compared.

As you recruit in-country reviewers, provide them with information about how the process will work and how important it will be to meet schedules. If the reviewers are slow, they can hold up the release of the product in their market, which is usually a considerable disadvantage for them. Understanding the relationship between their review tasks and product release can help motivate them to work quickly.

Provide your reviewers with the original language copy of the information so that they are able to compare the translations with the source language.

Inform your service providers about the arrangements with in-country reviewers and establish a process for transmitting the completed information. If the service provider uses an automated workflow system, the review process and the reviewers should be included.

Providing product training for translators

Unless they have worked with you before, the translators cannot be expected to be experts in your particular product and information, although they may have related experience. Consider how you will help the translators come up to speed on the product. Do you have a basic e-learning course available? Can you set up a web presentation by your information developers or others in your organization? Are there trainers or scheduled training courses near the translators' locations? Is there literature available that explains the product in general terms?

Even if you are able to arrange for training in the product, you should also plan for time that your information developers will be available to answer questions from the translators. Technical terms, ambiguous and inconsistent writing, and information that is not exactly the same in multiple locations all present problems for translators. Establish a line of communication between the translators or the service provider's editors or project managers and your team members so that questions can be addressed quickly. The communication will help to increase the quality of the translations.

To assist the translation process, you may want your information developers to include notes to the translators in the text. They should indicate which words are not to be translated. If you are using XML authoring tools, you can use the XML translate attribute set to "no" to identify a word or phrase that must not be translated. However, use this attribute sparingly and carefully to avoid confusion, such as a problem when you have set a word or phrase to translate inside a passage that is set for no translation.

Managing translation memory

By contract, your organization should own the translation memory even if it is maintained by your service provider. If you are using a content-management system that allows you to run the translation memory against the files in the repository, you may ask that the revised memory be transmitted back to you. You can then make the memory available for the next translation round or to a different service provider. Work with your service provider to ensure that the correct version of the translation memory is always available to the individual translators and that the memory was properly updated at the end of the previous project.

Evaluating the localization project

After the project is complete, schedule a meeting with your localization service provider to review the project. Find ways to improve the timeliness and quality of the process and the final products. You may want to make changes to the information-development process in creating the source content so that it becomes easier to translate. You may want to consult with a translation expert to find ways to make your text more consistent and unambiguous. You may want to research the use of controlled language tools to enforce style standards throughout your team.

Provide feedback to the service provider. If you work together to solve communication problems now, the next round of localization will be smoother. Remember that communication problems are unlikely to be one-sided; both you and the vendor project managers should be able to find ways to improve.

Review the entire process step-by-step so that you uncover any misunderstandings that led to delays or mistakes. Consider ways to make the process more effective for both teams. If you are dissatisfied with the work of someone on the vendor team, let more senior managers know. They may be able to find an individual who can work with you more effectively.

Go over the entire budget for the project so that you understand where all the costs come from. Discuss ways that you can work together to reduce costs in the future. Some of that cost reduction may be up to your team through more consistent writing and the exact reuse of more content. Your process may benefit from a move to a content-management system that is compatible with the system used by your service provider.

If you intend to change to XML- and topic-based authoring, inform your service provider during your planning and implementation. The change will necessitate changes in the translation process as well. It is best to be prepared well in advance of the next project. If you find that your service provider is unable to work in the new environment you are creating, you may have to look elsewhere.

Summary

Localizing product and translating information can be expensive and time-consuming if you do not plan for it carefully and establish a process that includes localization- and translation-friendly actions through the information-development life cycle. If you are responsible for the localization and translation of your project information, remember the following recommendations as you put your project plan together:

- ✔ Know exactly what your project requirements are for localization and translation. List all the deliverables that must be translated with every target language.

- ✔ Establish translation activities as part of your project schedule, working closely with your translation coordinator or your localization service provider to determine the optimal schedule.

- ✔ Include writing for translation as part of your information design and development process. Create a style guide that promotes consistent, jargon-free writing.

- ✔ Plan graphics and text so that translation is as simple as possible.

- ✔ Use topic-based authoring to ensure that information that occurs in more than one context is exactly the same in each instance unless there is a customer-driven reason for a difference.

✔ Develop a glossary of terms in your source language and facilitate the translation of the glossary into the target languages.

✔ Establish a cadre of in-country reviewers who are knowledgeable about your technology and customers.

✔ Work closely with your localization service provider to develop a smooth process for transferring files and keeping lines of communication open.

✔ Standardize and automate as many of the tasks as possible to reduce your costs and decrease time to market.

✔ Solicit feedback about the process by conducting a review session with your service provider.

✔ Solicit feedback from in-country employees and customers about the quality of the translations and share the feedback with your service provider to improve the translation for the next release.

"Thank goodness we planned.
We actually made our deadline."

Chapter 23

Managing Production and Delivery

> Production is like planning an elegant dinner; timing is everything.
>
> —Diane Davis

The Production Phase of your project is your last opportunity to ensure quality. A successful Production Phase begins with your project plan, in which you determine what final deliverables are required and when they have to be completed so that they are available to customers in a timely manner.

When I wrote *Managing your Documentation Projects* in 1994, delivering printed manuals was the overwhelming standard. Help systems were still relatively new and the web was unknown to the general public. As a result, I suggested that project managers allot 20% of total project time to preparing final deliverables. That often resulted in calendar time of six to eight weeks, if not a full quarter (13 weeks) before completed information would be available to customers. The result was reduced quality to the customers; the information was often out of date before it was published.

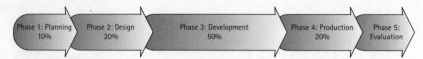

Figure 23-1: Information-development life cycle

Since then, information-development organizations have come to depend much more heavily on electronic delivery, primarily through help systems and websites, although handheld devices like cell phones and personal digital assistants are increasingly included in the deliverables planning. Even when you decide to print and ship paper to customers, the methods you now have to handle print content require far less time than they once did. Delivery of camera-ready copy to printers is handled electronically, with most of the final preparation completed before the files are sent (see Fig. 23-1).

As a result, the time needed for final production activities has been shortened. Information-development managers report Production Phases for their projects decreasing from 6 to 8 weeks to 2 to 3 weeks. Production tasks that used to take days, now take minutes to complete. And, you need fewer people to handle the production tasks. For these reasons, I estimate that the Production Phase for most projects that incorporate content-management systems or use automated methods to handle production tasks is 10% or less of the total project time. Figure 23-2 shows the new life cycle phase diagram with more time allocated for development as a result of decreased production time.

Figure 23-2: New Production Phase time allocation

In many cases, information developers handle more of the production process directly, sometimes with help from dedicated inhouse production specialists. Information developers using high-level desktop publishing tools may complete most of the final production reviews and prepare the final copy for delivery to printers or electronic systems. Organizations using advanced content-management systems, accompanied by XML and topic-based authoring systems like DITA, prepare scripts that assemble complex documents automatically from scores of independent modules of text and graphics. And, they do so in multiple languages and for multiple output formats.

Production specialists in many groups have merged with or become the tools specialists, because so much of final production is handled by tools. With single-sourced production, for example, you can easily produce multiple output media from the same source, including PDFs, HTML, various help systems, and other formats needed for an increasing number of special devices.

At the same time, the most critical quality activities during the Production Phase remain the same. As project manager, you still need to ensure that final approvals and signoffs occur, production edits are completed, deliverables are assembled in multiple languages with appropriate localizations, and final outputs are properly distributed. So, although you have many new and powerful production tools at your disposal, some of the primary tasks remain the same. Your role is to ensure that your team does not assume that the tools will take care of everything.

Best Practices for Managing Production

In managing the production process, you will once again find yourself coordinating with people inside and outside your organization. Not only will you work with your own team members to ensure that all files are ready for production, but you may also work with a production coordinator in your own or another department. You are also likely to work with others in the organization responsible for making content available to users by embedding help systems into the product or making information available through various websites. Finally, you may be responsible for selecting and coordinating with outside vendors. These five best practices lead you through the production management process:

- ✔ Planning for production and delivery
- ✔ Preprocessing final deliverables
- ✔ Performing production edits
- ✔ Handing off final deliverables
- ✔ Working with vendors

Each of these best practices will help you ensure that the final deliverables of your project are properly produced. You do not want the final result of a creative effort, done with dedication and hard work, to be done poorly. The quality of your deliverables reflects not only upon your team but also on the larger organization you work with.

Best Practice—Planning for production and delivery

Good production actually starts long before Phase 3: Development. If you have established a sound information architecture, templates that support the structure, and you have designed a careful production process, you do not have to squeeze production activities into the last few days of your project. By building production-supporting activities throughout the information-development life cycle, you have sufficient time to ensure that you have achieved the quality of work your customers deserve.

The life-cycle processes that support production during the project include the following:

✔ Instituting a strong, standardized information-architecture support by well-designed templates or Document Type Definitions (DTDs) in XML

✔ Using quality assurance activities to ensure that your team members adhere to the information architecture and the templates

✔ Establishing production reviews or checks of XML markup during the Development Phase rather than only in the Production Phase

✔ Ensuring that the graphics comply with standards. If you have professionals handling graphics, rather than writers, they are more likely to be ready for production.

✔ Moving legal reviews and approvals far enough upstream in the process that the do not produce bottlenecks at the end of the project

In your project plan, all the way back in Phase 1 of your project, you listed the deliverables you expected to produce for your project. Table 23-1 illustrates a typical deliverable plan. Note that you should have taken into account the delivery method for each output type. It is especially important that you know far in advance exactly what is expected by all the various parts of the delivery process.

Table 23-1: Deliverable Schedule Example

Deliverable Title	Part#	Delivery Method	Languages	%Change	Page/Topic Count
STEW Reference Information System V5.1 Administrator Guide	2559-001	PDF on CD; HTML on web	fr, gr	0	410
STEW Reference Information System V5.1 Release Notes	2559-009	PDF on CD	fr, gr	0	18
Reference Information System V5.1 User Guide	2559-007	PDF on CD; HTML on web	fr, gr	0	215
Customer Information Worksheet	2559-012	HTML on web	fr, gr	75	2
Document Manager Online Help	2559-017	.chm help files		20	35

In the deliverable table, provide all part numbers. They are often the primary reference for retrieving the correct titles from inventory. Specify the delivery method, including print, CD, web, and help systems. Specify the languages required, and note the final page count or topic count as appropriate. Other information, such as the area or region of the world associated with a deliverable, should be included, depending upon the needs of your organization.

Knowing what you're supposed to produce

You have noted the delivery format that you need to produce in your project plan. However, you may need to uncover more details about the formats. For example, if you are developing HTML pages for a website, discuss with those responsible for building the web pages how they want the files delivered. Find out if they need both HTML and PDF output for the web. Many information websites provide PDF copies that customers can use for printing, rather than printing multiple web pages.

In determining the full extent of the deliverables, you should identify all the organizations you need to coordinate with and the key individuals who can provide you with details about the delivery methods. If you are responsible for delivering content to training sites, find out what needs to be shipped in what form. Classroom instructors may need source material well in advance of scheduled training to use for their own learning.

Scheduling backward

It's a long-established practice for information project managers to schedule back from delivery dates. Delivery dates in your project plan may be specific calendar dates or relative dates, showing dependencies on other events in the larger project schedule. A relative delivery date might read:

> One day before Block Product Test Stability Phase entrance. File sent to print at least 2 weeks before CD gold to allow for printing and kitting with CDs.

It may be useful to include information about who determines when the relatively scheduled delivery date actually occurs.

Once you have determined the specific or relative delivery dates, you should begin to count backward. If you've handled production tasks before, that counting will be fairly straightforward, especially if you haven't changed the process. But if you're new to production estimating or you are working with new processes and tools, counting backward may be difficult and anxiety provoking. Consult with as many people internally as you can find who have worked with the production process before. If you are changing your processes and tools, the first time through will require extra time, perhaps even more than you might have originally estimated, to handle the inevitable problems that occur.

If you're finding it difficult to estimate production time, list all the possible activities you can think of, with help from your team members. Leave enough "wiggle room" to allow for solving problems and for last minute changes coming from the larger project team. Attach either specific or relative dates and indicate how many hours or days you think should be allotted to each task. Don't forget to take into account production activities for multiple languages, especially if you don't expect to receive translated content until late in the Production Phase.

Table 23-2 illustrates a typical production schedule.

Table 23-2: Development Schedule, Including Production Dates

XYZ STEW Agent Installation Instructions

Start date	Submit 1st review	Submit 2nd review	Submit final review	Begin final edits	To production	RTM*	To translation	From translation	To web admin	Launch date	To archive
07/15/2004	08/15/2004	08/20/2004	08/20/2004	08/25/2004	08/31/2004	08/31/2004			10/20/2004	10/22/2004	10/25/2004

STEW Director Release Notes

Start date	Submit 1st review	Submit 2nd review	Submit final review	Begin final edits	To production	RTM*	To translation	From translation	To web admin	Launch date	To archive
07/15/2004	08/17/2004	08/20/2004		08/25/2004	08/31/2004	08/31/2004			10/20/2004	10/22/2004	10/25/2004

*RTM = Release to Manufacture

Different types of deliverables may have to be delivered on different schedules. For example, you may need to deliver an online help system to the product-development team or to the configuration management coordinator long before you have to deliver content for the web or for print. You may be asked to deliver draft help files for testing before other content is in final stages of development.

As you plan the delivery schedule, focus on giving your information developers as much time as possible to complete the content and ensure that it is up to date. If they have to turn in content four weeks before the software code is frozen, you can guarantee that the software and the documentation will not match. By extending the time for information development and shortening production schedules, you improve the chances that the information you deliver will be accurate.

Defining roles and responsibilities

Once you have your deliverables determined and your schedule outlines, you need to assign responsibilities to various people on your team. You may want your information developers to do part or all of the production tasks themselves, especially if you do not have a dedicated production individual or team. If the information developers are expected to handle the production edits and final reviews of the output, you need to plan time for these activities in their schedules.

Consider who you want to make responsible for testing the production files, making certain that they are intact and not corrupted. Consider who will ensure that all the production outputs are transferred to the appropriate delivery channels. Think about who will coordinate the printing and work with the printing and fulfillment providers. Know what tasks you and your team members are responsible for completing and what tasks are the responsibilities of others in your organization. Table 23-3 illustrates possible roles and responsibilities during the Production Phase. Note that you may have one or more people in these roles and that some responsibilities will be handled by the same individuals.

Table 23-3: Production Phase Roles and Responsibilities

Role	*Responsibility*
Production specialist	Responsible for transforming content into delivery formats, such as HTML for web pages or various help systems. May also run or monitor scripts that massage content into different formats such as UNIX man pages or quick reference cards.
Production editor	Responsible for developing the production checklist and ensuring that all production edits are completed. Prepared to decide what should and should not be changed at the last moment.
Indexer	Responsible for either completing the indexes for all deliverables or reviewing the indexes for completeness and accuracy. May also prepare a list of common index terms across deliverables.
Vendor coordinator	Responsible for negotiating contracts and schedules with the outsource vendors, handing all required files and systems to the vendors, and ensuring that all deliverables meet requirements.
Production coordinator	Responsible for the management of all Production Phase activities after deliverables have been handed off from the information developers. May also coordinate all activities with the vendors.

If you are coordinating activities between your team of information developers and a production team, negotiate which group is responsible for final changes to the documents. Consider which types of change need to be reviewed by the original author and which can be handled by the production team. Develop a policy about making changes in the source files versus making changes in the production files. If you make changes in the production files, develop a procedure for ensuring that the changes are made to the source files before they are forgotten.

Scheduling indexing

In Table 23-3, I introduced the need for someone to be responsible for final indexes for your deliverables. In some organizations, the information developers enter index terms while they are writing. In others, indexing is done during the Production Phase by a specialist in indexing. In either case, you need to allocate time in your schedule for the indexes to be produced and edited. You will find that it is impossible to generate a good index on the first round. Even well-developed indexes will contain many inconsistencies: you will find both singular and plural forms of nouns, nouns and verbs used for the same item, spelling differences among words that should be the same, or problems with index levels that require rearrangements among primary, secondary, and tertiary index items. All of these problems and more appear in unedited indexes. If you do not have someone on your team who is knowledgeable about indexing, consider using an outside specialist or getting training in indexing techniques and best practices.

Too often, information developers give short shrift to indexing, even though customers tell us that the index is essential to the usability of a document.

If you are delivering electronic content without indexes, remember that keywords attached to web pages play the same role for the user as index terms. The same practice should be in place to ensure consistency in keywords that will facilitate search. You may want to develop a glossary of keywords, just as you would develop a list of master index terms. Then, you will need to leave time in the schedule for the keywords to be reviewed.

Including a legal review

If you have not already included a legal review during earlier review activities in the Development Phase, schedule a legal review as early in the Production Phase as possible. During the review, ask the legal representatives to check copyright pages and trademarks to ensure that the information conforms to requirements. Note that legal requirements change often. Guidelines that were correct for your last release may need to be updated.

You ought to have a policy established through your style guide for handling danger, warning, and caution notices in your documents. In some organizations, all warnings must be collected from the document and repeated in a special section either at the beginning or at the end of the document. Legal notices may have to be included, with text coming directly from your legal representatives or from government documents. Ensure that you have all the appropriate pieces that you will need for final production before the last moment. If necessary, ask your legal representatives to perform final reviews on the completed deliverables.

At this point in your schedule, delays in key reviews can have a deleterious effect on the final deliverables. Be certain that legal representatives and anyone else who has to approve final copy is alerted to the schedule in advance. Ask that the approvers appoint surrogates who can approve the documents in their absence. If you are aware of approvers who may want to make changes, try to schedule a time during the Development Phase for their input rather than wait until production is underway. If they won't cooperate until they can see final copy, try to schedule a specific time for you to meet with them to conduct their reviews rather than have the documents languish in their inboxes.

If at all possible, work with legal and management representatives to develop an acceptable workflow for approvals. If they are aware of the procedures and have agreed to them in advance, they may be more likely to comply.

Best Practice—Preprocessing final deliverables

If you wait until the last moment in your schedule to first produce the final deliverables for your project, you are inviting serious trouble. Inevitably, something will not work properly, especially if you have introduced new tools into the process. As you work backward from your final due dates, allow time to preprocess the draft deliverables before you conduct your production edit and in enough time to run the processes more than once before they produce the correct output. Be sure that your information developers understand that the documents do not need to be completed for preprocessing.

Developing an automated process

Many teams continue to rely on manual processing steps long after they have the opportunity to automate the processing. Review the process used by your production team, looking for steps that can be automated. If your process is taking too long, consider adding new tools to the process that allow you to write processing scripts that run automatically.

Automated processes can accommodate changes in output format, converting XML to PDF, HTML, and a variety of help systems. Automated processes can convert graphics to appropriate formats and sizes for online and print or PDF renderings. You can automatically import graphics, develop tables of contents, and run indexes.

The more you can automate, the easier it will be for your production team to prepare copy for review and signoff.

Running test content

If you are working with chapters of books, run each chapter individually and then assemble them and run them again. Include the front- and backmatter, especially the table of contents and the index. You will need all of the pieces in place for the production edit, but you may have to correct problems during processing before you have a version that can be edited.

If you are working with individual topic-based modules, run them against the final assembly maps to ensure that all of the pieces assemble correctly. You may find that errors in tagging will cause the production process to break down. You may need to reapply templates in desktop publishing systems or correct errors in the source files before the pieces come together correctly.

If you are using new tools or introducing a new architecture for your information, run test content on a small scale before trying to preprocess your deliverables. Work through test content on all the deliverables, including websites and online help systems.

As you work through test content or preprocessing of final deliverables, include editorial reviews to make sure everything is working as intended.

Ensuring that all files are available and correct

Ensuring that all the files are available and in the correct versions may not be as straightforward as you might think. Unless you have a content-management system with version control, you will find it remarkably easy to include the wrong versions of files in final deliverables.

If you are working with a number of remote information-development teams who are supplying files of final content, ask the remote team members to transfer preliminary files early even if they are not quite complete. Your production team can preprocess the files and detect problems early. Recognize that large file sizes may slow transfer or transfer may cause errors in the files. If you don't plan for testing of remote files, you are likely to find yourself with unusable content when your deliverables are due.

If you have early copies of files, you have something to work with even if problems occur with final deliverables.

If you are handling the final production of content in multiple languages, your test and preprocess steps are critical. Surprises of all sorts occur: fonts that will not process, missing text elements, problems processing double-byte languages like Chinese and Japanese, problems with text and graphics that must run right to left rather than left to right. Plan for quality checks by your localization service provider even if you handle production the multiple language assembly and production inhouse. If your production team cannot read the languages of the final deliverables, they will not detect problems.

Using content management to facilitate production

A content-management system can be a great advantage during the Production Phase of your project. The version control technology in most content-management systems ensures that you know exactly which versions of files are the most recent. Version control is especially important when you are working with scores of individual topics that will be assembled in a variety of maps that define individual deliverables. Every topic must be delivered in its most current version.

By creating references to standard product names and version numbers stored in your content-management system, you can more easily verify that content meets your delivery requirements. By creating references to separate files that contain standard, repeated content, you need check the accuracy of that content in one place rather than in hundreds.

Content-management systems can also facilitate the production of deliverables in multiple languages by storing all language versions of files (from book chapters to individual topics) in parallel (through links between the files). As a result, you will know that if a topic in the source language is changed late in the Development Phase, you must send the language files for retranslation to your service provider.

Many content-management systems also facilitate final production of multiple formats, automating the production of the source files into PDF, PostScript, HTML, help, and other formats. Once you have designed the production process into the content-management workflow, final deliverables can often be produced with the single push of a button (at least figuratively). Production cycles that took weeks to complete can be reduced to hours or minutes.

Reducing production time and costs is one of the benefits of a content-management system, providing a significant return on investment to your organization. It also helps ensure that your customers get content that is up to date and aligned with the product at release.

Best Practice—Performing production edits

Perhaps the one most critical quality assurance activity during the Production Phase is the production edit. The production edit is your final opportunity to make sure that the deliverables do not contain errors that will interfere with their usability or create a negative impression with the customer. During the production edit, you will look for errors in the deliverables that will affect their quality.

A typical production edit means that your production-edit team checks for problems that were not caught during the final copyedit and problems that occur or can be detected only after a deliverable has been assembled. Some of the checks you might perform include the following:

- ✔ Correct versions of all the components of the deliverables
- ✔ Correct assembly of the chapters into a final manual
- ✔ Correct assembly of topics into a final deliverable (PDF, HTML, help, etc.)
- ✔ Title and copyright pages with the correct information
- ✔ Page numbering, including areas with Roman numerals
- ✔ Headers and footers
- ✔ Bad page breaks
- ✔ Correct placement of graphics
- ✔ Incorrect tagging of elements
- ✔ Introduction of new tags (creative tagging by authors)

These are only a few of the possible items on your checklist; in fact, your list could probably go on for pages.

Establishing a production-edit team

Even if your information developers are responsible for all or some of your production activities, organize them into a production-edit team. Of course, the best qualification for a production editor is attention to detail. Information developers, editors, and even managers can contribute.

I recommend that you place one person in charge of the production edit. The "production editor in chief" is responsible for coordinating the activities, developing the editing checklist, and training everyone on the team in using the checklist properly.

The best choice for a chief production editor is someone who already knows the process, having done production edits successfully on previous projects. An experienced individual is able to make a judgment call about what is necessary to change and what can wait until the next release. At the eleventh hour of a project, some problems will have to wait until later to be corrected, particularly if they do not affect the accuracy of the information or create a negative impression on the customers.

The chief production editor should be someone who is aware of the problems that ordinarily occur with your information, especially problems that are caused by the tools you use to produce your final deliverables. If you are using a well-known desktop publishing tool that regularly introduces errors to numbered lists, the editor will know that numbered lists must be checked carefully. If you are using a topic-based architecture, your editor will be aware that topics may be incorrectly ordered in a final deliverable and include that problem on the checklist.

Watching for Problems

Diane Davis, who has managed technical publications departments and projects for many years, mentioned that she always reads the last line of each page and the first line of the following page. She wants to be certain that lines of text have not been dropped at page breaks. The habit of checking for missing lines of text grew out of experience with a tool that regularly caused that problem.

There are other, internal, problems to watch for as well. I recall a project in which one of the editors decided to measure the margins of all the pages because she was concerned that they might be slightly less than expected. After discovering that she had already spent 8 to 10 hours with a ruler and had found no margin errors, I suggested that the problem was not worth worrying about. You'll find that some individuals don't know when to let something go.

An experienced production editor may decide not to correct all the problems found. He or she might decide that a problem has to go back to the information developer to fix in the source files. In some cases, he or she might decide that a last minute glitch must be fixed in the output, with a note to go back and correct the source as soon as production is complete.

Your goal should be to place someone in charge of the production-edit team who knows what is really important, what you should take the time to correct, and what you can afford to postpone.

Information developers should not complete production edits on their own documents. Peer production edits are fine.

Preparing a production checklist

The first responsibility of the production editor is to prepare a production checklist. The checklist should be based on the experience of knowing what is likely to go wrong with a final deliverable. It should embody, to the extent needed at this point in the process, the requirements of the inhouse style guide. The checklist should include both text and graphics. You want to be certain that you have the correct version of the graphics, the correct release of the software, and that the screen captures and other illustrations are clear in the final renderings. A graphic that looks perfectly fine on screen may be unreadable in print.

Once the checklist is ready, the editor reviews the process with the whole team, not only the newcomers. Given the review and training in the process, everyone should be able to contribute to the quality of the production edit. Nevertheless, the chief production editor spot-checks the work of the team members, especially those who have not participated in a production edit before.

Figure 23-3 is an example of a production checklist prepared by the production editor in one organization. Note that checklists will be different depending upon the type of deliverable you are checking. The items you might check for a quick reference card are likely to be different from the ones you would check in a help system or website. If everyone on your team is familiar with the production checklist from the beginning of the project, many of the items on this list can be covered during the Development Phase as information developers are creating content. Also schedule some of the activities for the production editor to complete during Development. The production editor can check for correct title and copyright pages and be sure that the page numbering is set up correctly. Sample files can be checked to make sure that markup tags are applied correctly before correcting markup problems becomes onerous. You don't want to spend two to four weeks at the end of the project correcting problems that should have been caught much earlier.

Being Efficient

Amy Witherow, who established the production processes at Cadence Design Systems, tells us not to pay a production editor to clean up bad markup tags when the writers could have learned to use them correctly. She points out that the more production editing you include in the development cycle, the more time you can give the writers to work on content.

How to Do a Production Edit

Check the Documents In and Out

See the production editor to pick up a document (and turn it back in). She will give you a printed copy of the book (or take the edited book from you) and log the information into the production database. Please do not take books on your own.

Use the Production Database to Verify the Title and Document Order Number

This is the URL:
    http://www.????

Flag Only Critical Corrections

In a production edit, we focus on fixing major errors only.

* Mark and flag these corrections to be made in production:

* Incomplete or incorrect chapter introductions

    Chapter introductions should include a complete bulleted list of the level-1 heads in the chapter. If not, mark for fixing in production. (Tip: Compare the bulleted list with the level-1 heads for that chapter in the TOC.)

* Items in a table that have footnote symbols but no footnote (Ask the writer what needs to be done.)

* Typographical errors

* Discrepancies between text mentions and what appears in text

    Scan to ensure that items mentioned in text actually appear.

    Example: You might see a mention of "Figure 12-1," but you can't find this figure in the text.

    Example: The text cross-reference refers to "Table 3-1," but the table says "Table 3-2."

Do not flag these types of corrections to be fixed now, but do mark them (using a different ink color) for the writer to correct next time:

* Tables that continue to another page but do not include "Continued" on the second page

* Inconsistent hyphenation, word style, and so forth

* Figure and table mentions that include a page number when the figure or table is on the same page as the mention

    Example scenario: Figure 8-12 appears on page 8-20. The text mention for Figure 8-12 says, "The flow is shown in Figure 8-12 on page 8.20." In the future, the page number should be deleted.

* Missing figure and table mentions (All tables and figures should have text mentions.)

* Inconsistent table styles for similar content

    Example scenario: Sometimes tables containing options and descriptions have titles and sometimes they don't.

* Cross-references to sections that don't include page numbers
    * Terms that should (or should not) be wrapped as syntax
    * Any content changes, without checking with the writer first

Check With the Writer on Any Content Changes

Before making any content changes, check with the writer.

* Deleting even a single stand-alone sentence that might seem unnecessary or repetitive, for example, is a content change.
* Don't assume a strange-looking word is just a typo: In one production edit, the LEF construct VIARULE (correct as all caps and one word) was incorrectly changed to be two words, lowercase.

Copy the production editor on Corrections to Quick References

Most, if not all, quick references are automatically generated.

To ensure that the corrections we mark in the quick references are also eventually made in the source main pages, give copies of changed pages to the production editor. She will determine which writer is handling the collection of products the original page belongs to and pass the information to that writer.

Mark Page Breaks Extremely Sparingly

Awkward page breaks usually don't affect content; readers will get the information anyway. So at this late stage in the process, it's best to leave all but the page breaks that might look like mistakes. Also, when you mark a page break, all the pages that follow in that chapter are affected, possibly disrupting a writer's intended layout and introducing bad breaks later in the chapter.

Do mark this type of page break to be made in production:

* Sometimes a table with only one row of entries appears at the bottom of the page, with the table continuing to the next page.

Do not flag this type of page break, for example:

* Sometimes text introducing a list or code example is at the bottom of the page and the list or code example appears at the top of the next page. Leave the page break as is.

    Users:

    Any staff helping with the production process, which can include production staff, editors, writers, and managers.

    Purpose:

    A production specialist uses this checklist to verify the final appearance of the document. A production proof and check is a page turn, not a reading of each sentence.

Title page

Note: Though the title and copyright pages are now automatically generated, please continue to check them.
　　Check for the following:

* Correct title (Check the production database)
* Version number (**Example:** X-2005.09)
* Date (**Example:** September 2005)
* Appropriate TM or R symbol with product name (compare with current copyright page to see which symbol, if any, to use)

Copyright page

Note: Though the title and copyright pages are now automatically generated, please continue to check them.
　　Check for the following:

* Current copyright page (Example: 2005)
* Correct version of the copyright page
* Correct document order number at bottom, which should end with the appropriate two letters (example: ZA), and correct title (the same as on the title page) followed by a comma and, for example, version X-2005.05.

    Example:

    Document Order Number:　37656-000 ZA

    XYZ User Guide, version X-2005.09

Table of contents

* Make sure all headings in the table of contents have a minimum of three leader dots, and page numbers are aligned properly. Long headings in the table of contents should have the second line indented.
* Be sure TOC begins with the three 1-heads in the preface ("What's New in This Release," "About This <Manual, Guide, whatever>," and "Customer Support") and doesn't include "Preface."
* Check that "Index" is the last item in the contents if the book has an index.
* Check all chapters and appendixes listed in the TOC to be sure they are included, with the right number (or letter), in the book

Preface

* Check that current boilerplate has been used.

    Tip: In most books, the section "What's New in This Release" will contain only a reference to the release notes. (Note, however, that some books will still include detailed information--the choice is up to the writers and their teams.)
* Check that the "Conventions" table is on one page.

Text

* Run your eye over pages; don't read—about 30 seconds per page.
* Page breaks: Mark as few as possible (mark only the most serious breaks, such as a table that begins with only one line at the bottom of a page).
* Check footers:

> Even-numbered pages have Chapter X: Chapter Title above the page number.
>
> **Example:**
>
> Chapter 2: Running Automated Workflow
>
> 2-2
>
> Odd-numbered pages have only the number 1 section head above the page number.
>
> **Example:**
>
> Preparing Your Design
>
> 2-3

* Check that chapter intros include a complete bulleted list of the level-1 heads in the chapter. If not, mark for fixing in production. (Some books, such as the Library reference manuals, do not include these lists at this time. See the production editor if you have questions.)

  > ? **Tip:** Compare the bulleted list with the level-1 heads for that chapter in the TOC.

* Introductions to lists: Be sure lists deliver the promised content.
* Look for dropped (accidentally missing) graphics and text.

Index

* Spot-check four page numbers or so.
* Mark only very bad column breaks (for example, if a letter is at the bottom of a column and the first entry under that letter is at the top of the next column).

Figure 23-3: Production checklist

Another organization uses two checklists: one for production editing and the other for quality assurance. The production editing checklist includes checks for the following:

✔ Tags are consistent with the standard set.

✔ Content has been spell checked.

✔ Variables are correctly set up as required by the tools.

✔ Conditional text options are correctly set up as required by the tools.

✔ Version numbers are correct.

✔ Book files or DITA maps are correctly set up.

✔ Index items and keywords are properly marked.

The quality assurance checklist includes checks for different items:

✔ Titles and version numbers are correct.

✔ Tables of contents are properly linked to the text.

✔ Graphics are legible.

✔ Cross-reference links work.

✔ Search works.

✔ Fonts are legible and appropriate for the output format.

✔ Software code has appropriate line endings and indents for readability.

The checklists can be handled in one of two ways. Some editors use the checklist for a single pass over each page of a deliverable, with special checks for unique pages, such as the front- and backmatter. Other editors conduct a series of passes, selecting five to ten items from the checklist for each pass. When the production editors have finished checking an item on the checklist throughout the entire deliverable, they mark the checklist item as complete. During an edit pass, editors often find a new problem. The best course of action is to add the new item to the checklist, rather than going back to the beginning of the deliverable. Otherwise, you'll never finish.

Production editors should spend no more than 30 to 45 seconds on a page. It's important not to begin reading the text, duplicating the copyedit. It's also important to spot-check the table of contents and the index. It's amazing how often you will find errors in these. In one example, the writer had attached the wrong index to a document. Look for bad page breaks, but consider making changes only if they are really bad and the change does not require the entire text to be repaginated.

Many production editors work from print copies of the deliverable if possible, although some are comfortable working electronically. Many editors will tell you that they still find more errors in print than on the screen. You might try both methods, evaluating their comparative effectiveness.

Best Practice—Handing off final deliverables

Because you are likely to collaborate in managing final deliverables with other parts of your organization, you need to be prepared to hand off the deliverables to these groups at an appropriate time and in an appropriate manner. You do not want your handoff to cause problems downstream in the production process.

Planning for handoffs

From the earliest phases of your project, you should be planning for the final phases. You find it essential to communicate among all the parties associated with the project, including those responsible for final production. Hold production planning meetings between the information developers and the people receiving the content to ensure that everything will go smoothly.

Anticipating potential problems

Every production process will encounter problems. Nothing as complex as getting an entire release synchronized properly works without some flaws in the process or unforeseen errors. The more different problems you or your team members have encountered in the past, the more you will be aware of the areas in which things are likely to go wrong.

In one organization, the web designers had changed the interface for the technical manuals without informing the publications organization. As a result, all the frontmatter for various documents had to be updated. It is important to establish and keep open the lines of communication with other organizations. It may end up being your responsibility to coordinate with groups that fail to keep you informed. Call other groups or their managers to find out what may be changing.

If your information must be attached to a software or hardware system in the form of embedded help, you will need to coordinate with the group responsible for the final configuration of information and product. Be certain that the correct versions of the information files have been attached in the final product deliverables. Use your content-management system to ensure that version control is in place and that the people responsible for embedding files know where to find the correct versions. If necessary, include these handoffs in an automated workflow process to avoid errors.

If the operations or configuration groups or those responsible for building the information websites do not have access to your file or content-management system, work with them to know how they want you to deliver your files. Establish the details of the schedules for delivery so that the correct information files are in place when they are needed.

Handling final production and delivery inhouse

Your organization may find that it is cost-effective and efficient to handle all the final production of printed material inhouse. If you are considering this move, calculate the cost of setting up a print production facility with staff and equipment and compare that cost with proposals from outside vendors. In some cases, your internal costs may be less.

You may also find it feasible, especially if you have a small number of well-defined customers, to handle order-taking and shipping in your own group. Once again, compare your costs with proposals from vendors to ensure that you are taking the most cost-effective direction.

If your print production is especially complex, you may find it simpler to handle it rather than outsourcing it to a supplier.

Verifying the deliverables

Your production team may be responsible for print or PDF content, but you may need to rely on another organization to test HTML output. If you work with a group responsible for producing the web content, plan in advance how the new information will be organized on the website. Become involved in testing the navigation schemes so that you know how customers are supposed to find your content. You will often find that web developers have made assumptions about the nature of your content that are incorrect, leading to usability problems for the customers.

In one case, the information-development manager discovered that all but the latest versions of documentation have disappeared from the support website. Another manager discovered that the release notes were listed together on one web page, rather than associated with the products they discussed.

With websites, you may work with a separate production organization that has its own process to follow. Understand their requirements in advance. Schedule a meeting with the production manager to understand their process and deadlines.

You may be responsible for coordinating with outside vendors or distribution channels. Once again, meet well in advance of the production schedule so that you understand thoroughly what is expected of your team.

Best Practice—Working with vendors

In many organizations, your team will be responsible for coordinating with outside vendors during the production process. In other organizations, a separate production group may handle this responsibility. If you are responsible for working with vendors, select them carefully and establish a communication plan that ensures quality.

Selecting outside vendors

Selecting appropriate outside vendors is a detailed process beyond the scope of this discussion. The choices will depend upon the reach of your delivery of information to customers, especially for print and CD reproduction and shipping. If your customers are global, for example, you may need vendors operating in many international locations. You may want one international vendor to handle all global production and delivery, or you may want to work with local vendors in a number of different areas to respond best to local requirements.

In any case, if you are starting with outside vendors, use the recommendations of colleagues in other companies in your area or those with similar customer needs. Work through your personal networks to develop a short list of appropriate vendors.

Be aware that not all vendors are equal. I recall receiving a very high quote for a full-color print project from a vendor with whom I had worked with for many years. When I inquired about the quote, I learned that they were planning to subcontract the color printing to a provider who specialized in color reproduction, adding their own fees to the outsourcing. My vendor was a black-and-white specialist and did not have the equipment to handle the new project. I learned that I could get a better price by going directly to a printer who could handle my color-printing needs.

Begin by developing your requirements carefully, consulting with your sales and marketing colleagues and your support organization so that everyone's needs are included. Once you have reviewed the requirements with all the appropriate stakeholders, develop a Request for Information. In such a request, you have an opportunity to learn how vendors will handle your account and work with you to develop a creative solution before you ask them for cost quotations. Requests for Information are less formal than Requests for Proposal, allowing you to evaluate responses and find vendors who seem most compatible with your needs. You may even discover that vendors will suggest solutions that you never knew existed.

After you have narrowed the selection, you can then issue a more formal Request for Proposal so that you can associate costs with services. Be careful not to consider only the low bid. Generally, you get what you pay for. A low bid may also mean low-quality customer service.

Many publication organizations select vendors offshore because costs may be substantially lower. If you are considering an offshore vendor, discuss thoroughly their requirements for handing off files. Learn how they ensure the quality of the deliverables, asking especially about your role in ensuring that all the deliverables are accurately assembled before they are shipped to customers.

Maintaining good working relationships

As in every relationship, you will find it to your advantage to maintain open lines of communication with your vendors. Consider them to be partners rather than adversaries. Learn how best to communicate with them, particularly the names and contact numbers of the people who are responsible for various aspects of your project.

Ask how best to deliver the correct files to them and when they will be needed. Learn how much flexibility is in their schedules. Sometimes you can delay a handoff by a few days without affecting the final due dates if you alert the vendor in advance that a problem might be brewing.

If your vendor is handling distribution, work through the process of order taking thoroughly. Conduct spot checks from time to time to ensure that the correct versions of documents are being shipped, especially if they are shipped with products. Vendors ordinarily work with part numbers. As long as your part numbers for documents are correct, the fulfillment activities should also be correct. However, be careful about the shipment of documents toward the end of a product release. If a customer happens to place an order between releases, they may receive an outdated set of documents.

One organization put a notice out on their public website alerting customers that the documents were going to be updated on a particulate date.

Be certain that internal stakeholders also know about your vendors and how the process works. Customer support, in particular, will need to know how documents are ordered and what versions are the most up to date for a particular product release.

Finally, assume responsibility for ensuring that your vendors are paid in a timely manner. Project managers discover too late that a trusted vendor will no longer work with them because the larger department has failed to pay its bills. You must follow through by signing-off on invoices quickly and reviewing problems with the invoices as soon as you discover them. Follow through with your accounts payable department to make sure they have received all the documentation they need for payment to be scheduled. Let your vendors know that they can contact you if they have a payment problem, and then take responsibility. If your company has a reputation for late payment, discuss this with the vendors so that they are warned. But, also work with accounts payable to escalate your trusted vendors to the top of the list. I know of at least one publications manager who lost all of her trusted vendors because they were paid six months to a year late.

Evaluating the delivery process

After your customers have received their new documents, CDs, help systems, or website deliverables, consider surveying them to ensure that your production and delivery processes are working effectively. Several days after a customer has ordered and presumably received their information products, have an email sent that asks about the success of the ordering and delivery process:

✔ Was the ordering experience quick and easy?

✔ Did they receive the information products in a timely manner?

✔ Did they receive what they ordered?

✔ Did they discover that something they needed was not included in the ordered set?

Review the survey results with your vendors at the end of each release. Chapter 24: Evaluating the Project, presents an evaluation process called "lessons learned." Use this same process to discuss improvements with your vendor. Include all the stakeholders who are involved in the delivery process in the lessons learned.

Adding to your project management file

At the end of the Production Phase, you have considerable information to maintain in your project management file. Since the next project is likely to pick up where your project has ended, you want to have all materials in the best state possible. If you have a content-management system, you will probably archive the final versions of the source files and the deliverables so that you can recreate the deliverables at any time. The final versions of the source files, including the last minute fixes you made to the deliverables during production, should be in perfect condition for the next project iteration.

With a content-management system, you can archive the release version of files and deliverables at the same time that you move the last version into the proper folders for the new project. The last versions of the files become the first versions of the new project. Remember that the graphics files are just as important to be archived and brought forward as the text files.

Check your project management files. Make sure you have all copies of project documents in the proper folders:

✔ Planning documents and their updates

✔ Project spreadsheets with estimates and actuals

✔ Copies of proposals, contracts, and invoices for outsourced work

✔ All project correspondence

✔ All project meeting minutes and phone call logs

✔ Physical artifacts such as print, CD-ROM, or DVD masters and any materials from outside your organization, such as photographs and drawings that you received in hardcopy

✔ Any paper review copies dated and organized by deliverable

Make a list of everything you have archived or placed in the project management files for the project and include the list as a table of contents of sorts for the records. Remember that someone else may need information about your project when you are no longer available.

Summary

The successful completion of the Production Phase of your project is essential to the successful completion of your entire project. If production fails to deliver quality information to customers, all the rest of your work has been to no avail. You do not want a failed production process to be all that remains of a million dollar project.

✔ Plan your production process at the beginning of your project, including lists of final deliverables and production schedules (to the extent you know them) in your original project plan.

✔ Update your production plan as necessary throughout your project.

✔ Be certain to test your production process long before the final deliverables have to be run, especially if you are changing the process or introducing new tools and technology.

✔ Don't assume that something new will work at the last minute. Ensure that you have a process in place to assemble the correct parts within the needed schedule.

✔ Don't assume that a new production process that you've tested on a small project will necessarily run smoothly on a large project. Allow time for things to go wrong and be fixed.

✔ As soon as you have assembled content, introduce a thorough production edit, following the rules and practices established in a production checklist. Train and involve enough team members in the production edit to ensure that you meet your deadlines and produce quality output.

✔ Know to whom you need to hand off your final content assemblies and coordinate with them to understand their requirements.

✔ Select well-qualified outside vendors with whom to work and establish communication lines with them to ensure quality in the final deliverables.

✔ Evaluate the final result of your production process by asking customers if they view the process as successful.

Following these best practices will help to ensure the success of your project.

"Good Job!"

Chapter 24

Evaluating the Project

> The intelligent man is one who has successfully fulfilled many accomplishments, and is yet willing to learn more.
>
> —Ed Parker

The project is finished, and the final deliverables have arrived in the hands and on the computers of the customers. You're ready for a month in Tahiti. But wait, you're not quite through. As project manager, you need to tie up loose ends, some of which you did at the end of the Production Phase. Now you're at Phase 5: Evaluation (see Figure 24-1). You still have things to do that are as important to the success of future projects as all your previous activities were essential to the smooth running of your completed project.

Although the first four phases account for 100% of the total project time, Phase 5 remains, although I have not formally allocated time to this phase. If you have kept good records and issued progress reports throughout the project, your final reporting about the project should take minimal time. However, you may want to allocate five percent of total project time to the evaluation activities if you can formally include them in the project budget.

The major activity of the Evaluation Phase is a meeting with your team and other stakeholders to discuss lessons learned from the project and planning improvements in the process and the deliverables. In addition, you may want to write personal evaluations for your team members and begin the process of collecting feedback from customers that will provide a basis upon which to plan the next project.

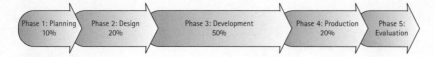

Figure 24-1: Phase 5 is the culmination of your project

Best Practices in Project Evaluation

Four best practices are associated with Phase 5: Evaluation. You need to gather all the data you have been keeping throughout the information-development project and summarize it in a final report. Then, you need to schedule a project review meeting with your team and other relevant stakeholders in the project. You will write reviews for your team members before you forget their contributions to the project. Finally, you need to begin the process of obtaining customer feedback.

✔ Reviewing project data

✔ Conducting a "lessons learned" review with team members and stakeholders

✔ Evaluating the team

✔ Collecting customer feedback

By devoting even a small amount of time to the final Evaluation Phase of the project, you will achieve closure and collect ideas for improving your next projects.

Best Practice—Reviewing project data

Throughout the information-development life cycle, you have been collecting data about the project. You began with estimates and added actuals, and you produced periodic progress reports, especially for long projects. Collect this data so that you have a complete record of your projects as you begin your review. During your review, consider the following questions:

✔ How well were you able to accomplish the goals laid out in your project plan? Were you able to provide all the information required by customers? Did you complete all the activities in your schedule, especially your quality assurance activities? Were there any areas where you ended up cutting corners?

✔ How much did the scope of the project change from your initial estimate? Did you add deliverables or increase the topic list?

✔ Did the project dependencies at the end of the project correspond to your assessment of dependencies at the beginning of the project? What changed and why?

✔ Did you keep to the original schedule? Was the final schedule shorter or longer than expected? How did schedule changes affect your milestone schedule?

✔ How well were you able to estimate hours per topics or hours per page?

✔ Did your project require more or less time for project management or quality assurance than you have originally estimated?

✔ Were you able to estimate the number of graphics accurately?

✔ Were you able to estimate time for production and localization and translation accurately?

✔ Did your team remain constant during the project? Did you lose members from or add members to the team?

As you think through these and other issues, document the changes from your original plan and explain why they occurred in your draft final report. You don't need pages of explanation, but having notes about the changes in the project will help you evaluate your estimating metrics more effectively. It will be important to think about what happened on the project that caused your original estimates to change. The better you understand the variations, the more likely you are to take them into account in future projects.

Unforeseen Issues Can Affect the Schedule

Most projects differ from the original estimates because the product scope or design changes. More work must be done than originally estimated or work that was completed has to be done again. But sometimes the time and cost of the project increase because of stakeholder interference. In one project, the engineer insisted that we include hundreds of pages on the theory of light to an installation and configuration manual. We had to edit, manage, and publish material that doubled the size of the manual. Not only was the information useless to the customer, it was technically wrong.

Comparing estimates to actuals

As you compare your original estimates to your actual project data, build a table to illustrate the results. Table 24-1 shows typical changes to a project.

Table 24-1: Project Statistics Compared

Project Activity	Estimated	Actual	Percent Change
Total hours	2,000	2,456	+23%
Project management	200	232	+16%
Writing	1,500	1,720	+15%
Editing	180	240	+33%

Continued

Table 24-1: Project Statistics Compared (Continued)

Project Activity	Estimated	Actual	Percent Change
Graphics	69	199	+188%
Production	51	65	+27%
Topic count	308	352	+14%
Hours per topic	6.5	7.0	+8%

Note that everything in this project has increased, including the number of topics produced. However, the hours to produce the topics have increased significantly more than the increase in topics might have justified. Obviously, something else has affected the efficiency of this project than a straightforward change of scope. You'll need to include information about the changes and why they occurred to explain why your original estimate was 23% lower than the actual project numbers show.

The most interesting change in this project is the graphics time. That might indicate a change from a design that was primarily text to one that was primarily graphics. A change in design of this sort could account for most of the increase in hours per topic. The editing time was a third higher than estimated. That might indicate that the information developers were not as experienced as you had originally thought or that the design style was changing sufficiently to require a new kind of writing.

You may also want to look at the changes that have occurred in the project life-cycle phases. If you are using standard percentages to estimate your schedule, examining the actual percentages may give you insight into how projects are likely to proceed in your organization. Your real phase data will be more useful for scheduling future projects than industry averages.

Table 24-2 shows the phase estimates for the same project described in Table 24-1.

Table 24-2: Project Phase Estimates and Actuals

Project Milestone	Estimate	Percent of Total	Actual	Percent of Total
Total hours	2,000		2,456	
Planning	200	10	200	8
Design	400	20	491	20
First draft	600	30	842	34
Second draft	300	25	547	22
Production	300	15	376	15

Clearly, the primary change in this project occurred during the Design and Development Phases, especially the completion of the first draft of the topics. It looks like there were design changes that increased the amount of time required to produce the first draft of the topics. The percentage of the total for the second draft is actually lower than originally estimated.

These results may indicate that the project went through a design change. Such a change might not be typical of future projects, making it difficult for you to assume that these percentages will hold for new projects without the same design changes. You'll have to look at more projects to decide if your schedule estimating requires some modification.

Evaluating project dependencies

Although scope and design changes will account for many of the cost overruns, changes in the way the project proceeds are equally important to review. In your original project plan, you evaluated 10 project dependencies and used the result of your analysis to influence your basic estimating metric, hours per topic or hours per page. During your analysis of the actual project, reexamine the dependencies to discover if changes in the nature of the project had an effect.

Table 24-3 shows the results of your analysis.

Table 24-3: Reevaluation of Project Dependencies

Dependency	initial	final
Product stability	4	5
Information availability	3	3
Prototype availability	2	2
Review process	4	2
SME availability	3	2
Technical experience	2	3
Writing experience	2	2
Audience experience	3	3
Team experience	2	2
Tools experience	3	4

Several interesting observations can be made about this project analysis. Five dependencies did not change from your initial evaluation (information and prototype availability and the writing, audience, and team experience of your team). Three dependencies increased (product stability and the technical and tools experience of your team), meaning that these factors were more problematic than you had originally imagined. The product was less stable, and your team members were not as experienced with the product technology or the new tools as you had hoped. Fortunately, these increases were countered by the decrease of two dependencies (review and subject-matter expert availability). Apparently, the developers or other experts your team was working with were much more available and cooperative than you had expected.

If you began with an average of 5 hours per topic, your original dependencies calculation would have given you 4.7 hours per topic for this project. With the new dependencies in place, the project becomes slightly more difficult at 4.87 hours per topic.

The last question to ask yourself about your dependencies calculation is whether or not the original 5 hours per topic is a good starting point. Many project managers begin with this industry average, only to discover that their typical projects are much shorter or longer than they estimated. The solution is to recalculate the average hours per topic.

The best way to discover your average hours per unit metric for projects in your organization is to accumulate the results of many projects and calculate the actual average. If you have only one project or a few projects, you may have a difficult time calculating an average, especially if these projects are untypical. Under these circumstances, you can use the dependencies calculator backward to arrive at a possible new average hours per topic.

What if you had estimated your project to be 5.2 hours per topic? At the end of the project, you find that the actual hours per topic was 7.5. Was this project that much more difficult than expected, or did you begin with an average that was too low?

Go back to the dependencies calculator and set all the rankings to 3. That's the neutral, average position. Valued at one in the calculations, this ranking does not change the starting hours per topic. If you begin with 5 hours per topic, you'll end up with 5 hours per topic if all your dependencies are ranked at 3.

Next, put in the final dependencies for the project that was 7.5 hours per topic. Then, adjust the original hours per topic at the top right of the calculator until the calculated hours per topic at the bottom right equals 7.5. If I use the values in my previous example (5, 3, 2, 2, 2, 3, 2, 3, 2, 4), the average hours per topic for a typical project in your organization becomes 7.3. That means that for your next project, you should begin with 7.3 and have an estimate that is much more accurate.

Remember that if you have identified factors that contribute to the ease or difficulty of projects in your organization, you can easily add them to the dependencies calculator.

Preparing a project final report

To prepare your project final report, use the results of your data collection and analysis to draw conclusions about the successes and challenges of the project. Information you might include in the final report are included in the following template:

Final Report Template

[Enclosed in brackets you will find instructions for completing the plan. Replace or delete all of the instructions and add your own content to the template.]

[Project Name] Final Report

Project Manager:
Start and end dates:
Team members:

Project Summary

[Summarize what the project covered and what deliverables were produced. If the actuals are significantly different from the estimates (5% or more), use this summary statement to explain why the project changed from your original expectations. Refer to the following tables or others you devise to illustrate where and why the changes occurred.]

Estimates and Actuals Table

Project Activity	Estimated	Actual	Percent Change
Total hours	2,000	2,456	+23%
Project management	200	232	+16%
Writing	1,500	1,720	+15%
Editing	180	240	+33%
Graphics	69	199	+188%
Production	51	65	+27%
Topic count	308	352	+14%
Hours per topic	6.5	7.0	+8%

[Note: You may want to include in a note to this table the reasons for the changes from the estimates to the actuals in each area of the project.]

Estimates and Actuals by Milestone

Project Milestone	Estimate	Percent of Total	Actual	Percent of Total
Total hours	2,000		2,456	
Planning	200	10	200	8
Design	400	20	491	20
First draft	600	30	842	34
Second draft	300	25	547	22
Production	300	15	376	15

[Note: You may want to include in a note to this table the reasons for the changes to the schedule and milestone calculations for the project.]

Project Dependencies Comparison

Dependency	Initial	Final
Product stability	4	5
Information availability	3	3
Prototype availability	2	2
Review process	4	2
SME availability	3	2
Technical experience	2	3
Writing experience	2	2
Audience experience	3	3
Team experience	2	2
Tools experience	3	4

[Note: You may want to include in a note to this table the reasons for the changes to the initial dependencies.]

What we learned from this project

[Summarize the lessons learned from this project. You may want to wait to complete this section until after you have conducted a "lessons learned" session with your team members and stakeholders.]

Add to your Project Management folder

Be certain that you include this final report in your Project Management folder. If your organization hopes to compare projects and learn how to improve project management and performance as a whole, you may want to include all the final reports and relevant spreadsheets for each project in a central project management folder.

Best Practice—Conducting a "lessons learned" review

As soon as you have completed your final report, schedule one or more review meetings. The first meeting should be with your own team members and any stakeholders from outside the team that you would be comfortable including when your team discusses the challenges of the project. If you have stakeholders that you would not be comfortable including, schedule a second meeting with them or meet with them individually. In either case, summarize the results at the end of your final report.

Setting the agenda

Set an agenda for the review meetings, especially those involving outside stakeholders. You want to be sure that everyone attending knows the purpose of the meeting and knows what it will not include. Try to avoid having some participants use the meeting to blame others about the project. The goal is to find ways to improve, not to assign blame.

In planning your meeting, you may find these logistics steps helpful. This way of handling the meeting was suggested by Diane Davis.

Logistics for your review meeting

1 Reserve a large conference room.

2 Request that several (at least two) flip charts are in the room.

3 Order food.

4 Take markers, masking tape, and a laptop computer to the conference room.

5 Request administrative assistance to type responses into a computer file.

6 Provide opening guidance:

   a The purpose is to improve the quality of documentation for customers by improving Technical Publications processes and efficiency.

   b There are two overriding issues: education and process. Process, not people, is the problem.

   c Make statements that refer to a job description, not to a specific person.

   ```
   For example, the technical publications manager did not understand the
   core team concept to help make positive changes.
   ```

7 Explain the questions that the meeting will address:

   What did we do right?

What could we do better?

Where did the process not work?

How can we build on the actions that led to success?

How can we change the actions that caused problems or failures?

What would we do if we had no barriers?

8 Explain the brainstorming method:

Everyone participates

No criticism

No comments

No discussion

Ok to clarify

Members can pass at any time.

Each person speaks in turn.

Each person should use paper and pen to record ideas in between turns.

No one censors or interrupts anyone else.

All ideas are acceptable.

Ideas can be developed by hitchhiking off of someone else's ideas.

9 Ask for any questions.

10 Remind everyone that this is not a venting session.

11 After brainstorming, allow participants to select the issues that are most important to them by assigning lucky bucks to specific topics.

12 Provide each person with the results.

Solving Problems with Lucky Bucks

Diane Davis uses lucky bucks to help her team identify the problems they want to fix most urgently. After the brainstorming, each person assigns lucky bucks to the problems that they are most interested in fixing. Each person has 100 lucky bucks to distribute among the problem topics in increments of 5. One person can assign all 100 bucks to one problem or spread the bucks out among no more than five topics.

After everyone assigned the lucky bucks, they total them. The five top problems become the objectives for improving the process for the next project. They track the solutions and rate the team on how well they have met their goals.

The lucky bucks process helps ensure that problems are solved with everyone assuming responsibility for the process improvements.

Inviting stakeholders to a review

If your team wants to focus on internal improvements, you may not want to invite outside stakeholders to the same review meeting. However, you do need to learn from the stakeholders, including representatives from the development team, product marketing, training, support, and others that have contributed to the success of the project.

You may find it most helpful to meet with these stakeholders in a separate meeting, although you may want to include your team members. You may want to meet with some stakeholders independently so that you can learn about their concerns without creating an uncomfortable atmosphere. However, I find that meetings with stakeholders and team members can be extremely helpful, although sometimes sensitive. Set the ground rules for behavior, asking people to avoid personal attacks. Once again, focus on how to improve for the future rather than rehashing the problems of the past.

Taking action

The review meetings should not turn into gripe sessions or personal attacks on individuals. Unless you can find ways to improve your processes as a result of the review, you should not hold the meeting at all. After you and your team members and stakeholders have identified challenges and discussed possible solutions, ask people to take responsibility for action items.

Here is an example of a feedback memo as the result of a "lessons learned" meeting.

Sample Feedback Memo

Dear Participant:

Thank you for participating in the Lessons Learned meeting. Your input is valuable and useful. Now, to make your input effective, we must act on it.

The first five pages that follow contain your input grouped in the categories we used at the meeting. Where appropriate, I combined similar input into one objective. I then listed each objective in descending order of the number of "lucky bucks" you assigned them. The last page contains the seven objectives that received the most lucky bucks.

I will own the bulk of the second objective (resource issues), the first part of the fifth objective (including a freeze date on the program manager's release schedule), and the seventh objective (planning better).

However, we need volunteers to own the other objectives. For example, for the first objective, I think we need a group of three to define the writers' roles and responsibilities, another group of three to define the editors' roles and responsibilities, and a third group of three to define production's roles and responsibilities. The other objectives also need volunteers.

If you are interested in working on one of the objectives to improve our process for the next release, please contact me or your manager.

Note that the project manager takes responsibility for some of the changes and asks team members or stakeholders to take responsibility for others.

Once you have completed the "lessons learned" meetings and assigned action items to implement the recommendations, include the conclusions, recommendations and action items in your final project report. Send the report to the appropriate stakeholders. If possible, schedule individual meetings with certain stakeholders to discuss the report if you need their assistance or support to implement any of the recommendations.

The final project report and the "lessons learned" meetings are designed to evaluate the processes used to conduct the project so that they can be improved for future projects. Better estimating, better scheduling, better analyses of project dependencies all help on the road to a more mature information-development process.

Best Practice—Evaluating the team

I recommend that the project manager provide brief personal reviews for the team members as soon as the project is complete and add these to each person's personnel file for reference when yearly performance appraisals occur. Exemplary work on projects or project-related performance problems that need to be addressed can be forgotten by the time yearly performance appraisals come around.

The formal project review activities mentioned earlier in this discussion are designed to avoid pointing to challenges experienced by particular individuals. They also do not encourage special praise for individuals who contributed more than their share to the project's success. In most organizations, such contributions are relegated to yearly personnel reviews and may be easily forgotten if they're not addressed immediately following the project.

You might also want to consider conducting a 360° review for the project, which means that team members review each other's contributions as well as the contributions of the project manager.

Evaluating team effectiveness and efficiency

Evaluating the team's work together is different from evaluating individual performance but just as valuable to future project success. You may want the team to evaluate its own performance in addition or in place of your review. Consider the following questions:

- ✔ Did the team work together effectively as a whole? If disagreements occurred, were they settled quickly and amicably?

- ✔ Did team members respect each other's contributions? Did people treat each other professionally?

- ✔ How well did the team members work with other teams, especially the product developers or other subject-matter experts? How could the relationships be improved?

- ✔ Did each team member contribute to the project's success? If some members decided to act independently, were they brought back into the team?

✔ Did team members support each other's work? As deadlines approached, did everyone contribute by putting in extra time, or were some people willing to leave the extra work to others?

✔ Did the team work together effectively through all phases of the project?

These or other questions will help you and the team members identify areas in which collaboration can be improved. Collaboration is increasingly important in maintaining the efficiency of projects and the quality of deliverables.

If you or others believe the team was dysfunctional, you may have to challenge the team during your review meeting to find ways to work together more effectively. People accustomed to working independently may find the transition to teamwork difficult. You may simply need to give them more time to adjust.

Writing project performance reports

By completing your evaluation of each team member, you ensure that their contributions to the project are not lost in the annual performance appraisals. First, be certain to recognize outstanding contributions to the project's success. You may have team members who added innovative design ideas, worked hard to understand customer needs, or were especially adept at handling difficult subject-matter experts. Record these achievements so they get the credit they deserve.

If you are concerned about an individual team member's performance, share that information as well in the report. However, you must review the reports with them and give them the opportunity to disagree or provide explanations. If you think a problem is serious enough, discuss it with the department head before writing the review. You may need help in working out a solution to a performance problem.

 ## Best Practice—Collecting customer feedback

Improving process and the performance of a project team is vital to the internal success of projects. You want a project that proceeds smoothly, keeping on schedule and budget, and delivering effective information projects. A good process should help to ensure that you are effective as well as efficient. But, until you actually measure the success of the project in delivering value to customers, you may be spending a lot of time and effort doing the wrong things.

Customer feedback allows you to judge whether your process is getting the results you want. If your information is not accessible and usable for your customers, all your attention to the details of a project is for naught.

Surveying customers

Customer surveys are a quick but not always satisfying method for garnering feedback about information products. Most customer surveys address many issues, with information products represented by one or two questions. Unless you conduct specific surveys of customers about information, you are unlikely to get much information.

The single question on a general customer satisfaction survey can, at most, provide a useful baseline. If you are improving the quality of the information delivered to customers, you hope to see the positive responses to this question increase. However, a single question on a survey is inevitably biased by responses to other questions. In general, I find that when customers dislike a product experience they also dislike the information provided. It seems that you cannot create good information for a poor product.

In many cases, you may have difficulty finding out about the individuals who responded to the general survey unless the responses are monitored by whatever organization is responsible for the survey. Even if the respondents are known, you may have difficulty knowing the context in which they responded to the question about information. I have found that people respond positively or negatively based on no experience whatsoever with the information. They are simply answering questions. All you may obtain is a general impression about their general impression.

Specific surveys focused on information products can yield better results, especially if you can monitor who responds to them. Depending upon how well the survey is designed, you may get some useful information, although, once again, it is likely to represent a baseline for future comparison rather than a treasure trove of information.

With survey data as a starting point, however, you can follow up with the respondents to get more complete information. Surveys should ask respondents if they are willing to be emailed or called for additional information. If they are willing to help, you can gather more than a baseline impression and begin to identify areas that need improvement.

Be careful, as with any feedback, with the effect of self-selection. Those customers who respond to surveys and agree to follow-up may be different in some unknown way than customers who never respond.

Tracking support calls

Working with your customer support organization to gather information about customer problems with information products is an invaluable form of feedback. Many telephone support organizations keep logs of calls and can be asked to check off a box that associates a call with an information problem or shortcoming. Unfortunately, the check-off method, while useful, has shortcomings. Information problems that you might notice may never be checked off because the caller or the support person doesn't recognize it as a problem that could be solved through better information.

You may find it more valuable to establish a good working relationship with the call center personnel. Talking to them about typical customer questions can reveal areas in which information might be strengthened. Reviewing call logs as a whole can point to patterns of calls that indicate information problems or areas that require additional explanation or clarification.

Paying Attention to Customer Calls Can Save Money

On one project, our discussion with an interested support engineer revealed a pattern of calls associated with the installation of the operating system. We worked together to analyze the call data and design a new way of writing the installation instructions. As a result, the number of calls regarding installation decreased by more than 50%, an enormous cost savings for the company.

Receiving customer feedback reports

Other forms of feedback from customers will help you assess the success of your information-development project. Telephone surveys of people responding to surveys or calling customer support can lead to an in-depth understanding of information shortcomings. Feedback from websites has become increasingly valuable to many organizations, especially if the feedback is focused on particular topics of information rather than on information in general.

The Faster the Feedback Gets Implemented, the Better

Microsoft's Office Applications group encourages customers to comment on individual help topics. The comments and the accompanying ratings of the topics are delivered daily to the information developer responsible for each topic. The information developer monitors the responses, focusing on topics that seem to be creating the most problems for customers. When it is clear how to revise or completely rewrite a troublesome topic, the information developer creates a new topic. As soon as the topic is approved, it is immediately updated on the website. As soon as the customer asks for help again on that topic, he or she will find the new content.

Some of the most significant feedback comes from direct contact with customers in their native environments. Site visits, discussed in the section on usability testing, are an excellent way of getting feedback on innovations in information design and delivery.

Consulting other stakeholders

Many people in your organization are likely to refer to the information products in their own work or are in direct contact with customers. Sales, field support, training, and others may get comments from customers about your deliverables. Asking these stakeholders about the reception of your information adds to your assessment.

In many cases, stakeholders will not volunteer information themselves. I learned of a trainer who spent hours having workshop participants make corrections in their copies of the technical manuals without ever informing the publications' organization about the problems.

To obtain the feedback you need, you have to ask for it directly. You may have to interview individual stakeholders to get more than cursory comments about information quality. You also have to work hard to ensure that the comments you receive are judicious. Too often, stakeholders comment about information that was in place years before rather than what has been offered to customers most recently.

Summary

Each of the evaluation methods presented here provides you with ways to improve your process for conducting information-development projects and improve the value of the information you deliver to customers. None of the methods alone is sufficient, however. You need a combination of techniques to identify challenges and find ways to overcome them.

- ✔ Begin by reviewing all of your project data and analyzing the relationship between your initial estimate and the actual results.

- ✔ Create your final project report following the basic template and including your data, your analysis of the data, and your overall conclusions about the project.

- ✔ Schedule review "lessons learned" meetings with your team members and with other stakeholders to identify opportunities for improvement in processes.

- ✔ Based on the review meetings, create a set of action items to improve processes for the next project.

- ✔ Add the recommendations and actions to your final project report.

- ✔ Identify opportunities to learn if your information products have been successful with the customers. Include surveys, telephone interviews, site visits, and meetings with people in your organization who are close to the customer.

Once you have completed your final report, include it with the other materials in your Project Management folder.

Bibliography

Ament, Kurt. *Single Sourcing: Building Modular Documentation.* New York: William Andrew Publishing, 2003.

Andriole, Stephen J. *Rapid Application Prototyping: The Storyboarding Approach to User Requirements Analysis,* 2nd edition. Wellesley, MA: QED Technical Publishing Group, 1992.

Applehans, Wayne, Alden Globe, and Greg Laugero. *Managing Knowledge: A Practical Web-Based Approach.* Boston, MA: Addison Wesley Longman, Inc., 1999.

Baker, Mark. "What Makes an Authoring System Tip?" *Best Practices* of The Center for Information-Development Management, February 2004.

Barney, Matt, and Tom McCarty. *The New Six Sigma: A Leader's Guide to Achieving Rapid Business Improvement and Sustainable Result.* Upper Saddle River, NJ: Prentice Hall, 2003.

Bethke, F.J., W.M. Dean, P.H. Kaiser, E. Ort, and F. H. Pessin. "Improving the usability of programming publication," *IBM Systems Journal,* Vol 20 No 3, 1981.

Bonura, Larry S. *The Art of Indexing.* New York: John Wiley & Sons, 1994.

Brooks, Frederick P. *The Mythical Man-Month: Essays on Software Engineering, 20th Anniversary Edition.* Reading, MA: Addison-Wesley, 1995.

Buckingham, Marcus, and Donald O. Clifton. *Now, Discover Your Strengths.* New York: The Free Press, 2001.

Carroll, John M. *Minimalism Beyond the Nurnberg Funnel.* Massachusetts: The MIT Press, 1998.

Carroll, John M. *Scenario-Based Design: Envisioning Work and Technology in Systems Development.* Hoboken, NJ: John Wiley & Sons, 1995.

Carroll, John M. *The Nuremberg Funnel: Designing Minimalist Instruction for Practical Computer Skill.* Cambridge, MA: The MIT Press, 1990.

Christensen, Clayton M. *The Innovator's Dilemma: When New Technologies Cause Great Firms to Fail.* Cambridge, MA: Harvard Business School Press, 1997.

Christensen, Clayton M., and Michael E. Raynor. *The Innovator's Solution: Creating and Sustaining Successful Growth*. Cambridge, MA: Harvard Business School Press, 2003.

Christensen, Clayton M., Michael E. Raynor, and Scott D. Anthony. "Six Keys to Creating New-Growth Businesses." *Harvard Management Update*, January 2003.

Clark, Ruth Colvin. *Developing Technical Training*. Boston: Addison-Wesley Publishing, 1989.

Cooper, Alan. *About Face: The Essentials of User Interface Design*. New York: John Wiley & Sons, 1995.

Cooper, Alan. *The Inmates Are Running the Asylum: Why High Tech Products Drive Us Crazy and How to Restore the Sanity*. Indianapolis, IN: Sams Publishing, 1999.

Crosby, Philip B. *Quality Is Free*. New York: McGraw-Hill Book Company, 1979.

Datz, Todd. "Portfolio Management: How to do it right." *CIO Magazine*, May 1, 2003.

Davis, Diane. "Writing Performance Appraisals." *Best Practices* of The Center for Information-Development Management, June 2001.

DeLuca, Joel R. *Political Savvy: Systematic Approaches to Leadership Behind-the-Scenes*. Berwyn, PA: EBG Publications, 1999.

DeMarco, Tom. *Controlling Software Projects: Management, Measurement, & Estimation*. Englewood Cliffs, NJ: Yourdon Press, 1982.

Dicks, R. Stanley. *Management Principles and Practices for Technical Communicators*. New York.: Pearson, Longman, 2003.

Dillon, Andrew. *Designing Usable Electronic Text: Ergonomic Aspects of Human Information Usage*. London: Taylor & Francis, 1994.

Dreyfus, Hubert L. and Stuart E. Dreyfus. *Mind over Machine*. New York: The Free Press, 1986.

Drucker, Peter F. *People and Performance*. Burlington, MA: Butterworth-Heinemann, New Edition, 1995.

Drucker, Peter F. *The Practice of Management*, reissued edition. New York: HarperCollins, 1993.

Dumas, Joseph S. and Janice C. Redish. *A Practical Guide to Usability Testing*. Norwood, NJ: Ablex Publishing Corporation, 1993.

Eleder, Michael. "Making Project Management a Valiant Voyage." *Best Practices* of The Center for Information-Development Management, December 2005.

Ensign, Chet. *SGML: The Billion Dollar Secret*. Upper Saddle River, NJ: Prentice-Hall, 1997.

Fisher, Kimball, and Maureen Duncan Fisher. *The Distance Manager*. New York: McGraw-Hill, 2001.

Frame, J. Davidson. *The New Project Management: Tools for an Age of Rapid Change, Corporate Reengineering, and Other Business Realities*. San Francisco: Jossey-Bass Publishers, 1994.

Gau, JoCarol. "Putting Kotter's Ideas to the Test: Leading Change Through an Offshoring Effort." *Best Practices* of The Center for Information-Development Management, August 2005.

Gladwell, Malcolm. *The Tipping Point: How Little Things Can Make a Big Difference*. Little, Brown and Company, 2000.

Goleman, Daniel. "Leadership That Gets Results." *Harvard Business Review*, March–April 2000.

Guglielmetti, Krista. "Technical Communications and Customer Support: Partnering to Publish What Customers Want to Know." *Best Practices* of The Center for Information-Development Management, August 2000 and the *47th Annual Conference Proceedings* of The Society for Technical Communication, 2000.

Hackos, Bill. "The Information Developer's Dilemma." *Best Practices* of The Center for Information-Development Management, October 2004.

Hackos, Bill. "The Shackleton Way: Leadership Under Stress." *Best Practices* of The Center for Information-Development Management, October 2001.

Hackos, JoAnn T. *Content Management for Dynamic Web Delivery*. Hoboken, NJ: John Wiley & Sons, 2002.

Hackos, JoAnn T. *Managing Your Documentation Projects*. Hoboken, NJ: John Wiley & Sons, 1994.

Hackos, JoAnn T. and Janice C. Redish. *User and Task Analysis for Interface Design*. Hoboken, NJ: John Wiley & Sons, 1998.

Hammer, Michael. "Deep Change: How Operational Innovation Can Transform Your Company." *Harvard Business Review*, April 2004.

Hargis, Gretchen, Michelle Carey, Ann Kilty Hernandez, Polly Hughes, Deirdre Longo, Shannon Rouiller, and Elizabeth Wilde. *Developing Quality Technical Information: A Handbook for Writers and Editors*. Upper Saddle River, NJ: IBM Press, 2004.

Harkus, Susan. "Ignore Context and Risk Your Project." *Information Management*, the e-newsletter of The Center for Information-Development Management, April 2004.

Hartman, Peter J. *Starting a Documentation Group: A Hands-On Guide*. Peabody, MA: Clear Point Consultants Press, 1999.

Herzberg, Frederick. "One More Time: How Do You Motivate Employees?" *Harvard Business Review*, January–February 1968.

Highsmith, Jim. *Agile Project Management: Creating Innovative Products*. Boston: Addison-Wesley Publishing Company, 2004.

Holtzblatt, Karen and Hugh Beyer. *Contextual Design: A Customer-Centered Approach to Systems Designs*. San Francisco, CA: Morgan Kaufmann Publishers, 1997.

Hope, Jeremy and Robin Fraser. "Who Needs Budgets?" *Harvard Business Review*, February 2003.

Horton, William K. *Designing Web-Based Training*. Hoboken, NJ: John Wiley & Sons, 2000.

Humphrey, Watts S. *Managing the Software Process*. Boston, MA: Addison-Wesley Publishing Company, 1989.

Kaplan, Robert S., and David P. Norton. *The Balanced Scorecard: Translating Strategy into Action*. Cambridge, MA: Harvard Business School Press, 1996.

Kaplan, Robert S. and David P. Norton. *The Strategy-Focused Organization*. Cambridge, MA: Harvard Business School Press, 2001.

Kim, W. Chan and Renee A. Mauborgne. "Tipping Point Leadership." *Harvard Business Review*, April 2003.

Kotter, John P. *Leading Change*. Cambridge, MA: Harvard Business School Press, 1996.

Kotter, John P. and Dan S. Cohen. *The Heart of Change: Real-Life Stories of How People Change Their Organizations*. Cambridge, MA: Harvard Business School Press, 2002.

Levy, Paul F. "The Nut Island Effect: When Good Teams Go Wrong." *Harvard Business Review*, March 2001.

Maslow, Abraham H. *Maslow on Management*, revised edition. Hoboken, NJ: John Wiley & Sons, 1998.

McLuhan, Marshall and Lewis H. Lapham. *Understanding Media: The Extensions of Man*. Cambridge, MA: The MIT Press, 1964.

Moore, Geoffrey A. *Crossing the Chasm: Marketing and Selling Technology Products to Mainstream Customers*. New York: HarperCollins, 1991.

Moore, Geoffrey A. *Inside the Tornado: Strategies for Developing, Leveraging, and Surviving Hypergrowth Markets*. New York: HarperCollins, 2004.

Morrell, Margot and Stephanie Capparell. *Shackleton's Way: Leadership Lessons from the Great Antarctic Explorer*. New York: Viking Press, 2001.

Norman, Donald A. *The Design of Everyday Things.* New York: Doubleday Publishing Group, Inc., 2002.

Patterson, Kerry, Joseph Grenny, Ron McMillan, and Al Switzler. *Crucial Confrontations*. New York: McGraw-Hill, 2005.

Pek, Elizabeth and Susan Harkus. "Building Success on the User Agenda." *Best Practices* of The Center for Information-Development Management, April 2002.

Pellak, Mary T. "Sustaining Motivation and Productivity During Significant Organization Change." *Performance Improvement*, November–December 2001.

Perkins, Dennis N. T. *Leading at the Edge: Leadership Lessons from the Extraordinary Saga of Shackleton's Antarctic Expedition*. New York: American Management Association, 2000.

Price, Jonathan and Henry Korman. *How to Communicate Technical Information*. San Francisco, CA: The Benjamin/Cummings Publishing Company, Inc, 1993.

Price, Lisa and Jonathan Price. *Hot Text: Web Writing that Works*. Indianapolis, IN: New Riders, 2002.

Redish, Janice (Ginny). "The Innovator's Dilemma: When New Technologies Cause Great Firms to Fail." *Best Practices* of The Center for Information-Development Management, June 2001.

Rogers, Everett M. *Diffusion of Innovations*. New York: The Free Press, 1962.

Rosenfeld, Louis and Peter Morville. *Information Architecture for the World Wide Web,* 2nd edition. Sebastopol, CA: O'Reilly & Associates, Inc., 2002.

Rubin, Jeffrey. *Handbook of Usability Testing: How to Plan, Design, and Conduct Effective Tests*. New York: John Wiley & Sons, 1994.

Schriver, Karen. *The Dynamics of Document Design*. New York: John Wiley & Sons, 1996.

Shackleton, Ernest. *South: A Memoir of the Endurance Voyage*. New York: Carroll & Graf, 1998.

Weinberg, Gerald M. *Quality Software Management: Systems Thinking*. New York: Dorset House Publishing, 1991.

Index

A

acquisitions and mergers, 53, 240
agile project environment, 342–344, 481
annotated task list, 402–404
annotated topic list, 409–411, 434
assertiveness issues for team members, 255–256
assessments
    benchmarking opportunities, 190
    budget characteristic, 107–108
    conducted by CIDM, 34
audience. *See* customers
automated deliverables process, 561
automated publishing for localization, 542
automated workflow systems, 458, 541–542

B

Balanced Scorecard
    adding value to customer deliverables, 100
    best practice, 95–104
    CIDM example, 103
    connecting layers of goals and tracking results, 93
    customer satisfaction, 12
    described, 96
    effective operations, 12
    efficient and knowledgeable employees, 13
    Employee Growth and Development as foundation, 27
    example for CIDM, 102–104
    financial perspective, 96, 97, 98–100
    financial success and profitability, 12
    if-then sequence for information development, 95
    illustrated, 97
    information-development activities related to
        company success, 95
    "intangible and intellectual assets", 95
    internal business process perspective, 96, 97, 101
    layers of goals, 93
    learning and growth perspective, 96, 97, 101–102
    outcome measures, 102

performance measures, 97
performance measures for meeting customer needs,
    100–101
portfolio management, 95–104
reasons for using, 97–98
relating work done to financial goals, 108
representative example, 93–94
Strategy-Focused Organizations', 85
benchmark studies, 308–310
    budget practices, 111
    increasing efficiency and effectiveness, 189–191
    project costs, 112
    quantifiable data, 191
    understanding competitors, 76
best practices
    Balanced Scorecard, 95–104
    budget management, 112–119
    change management, analyzing ongoing project risk,
        483–488
    change management, communicating, 488–490
    change management, initiating change, 481–483
    change management, of the team, 466–470
    change management, responding to change,
        472–481
    change management, tracking, 470–472
    collaboration, 78–79
    collaboration management, business case, 496–498
    collaboration management, questions to ask,
        494–496
    collaboration management, recognizing possible
        failure, 508–509
    collaborative environment, creating, 498–508
    communication plan, developing, 348
    cost control, 73–74
    customer feedback collection, 589–591
    customer information requirements analysis, 140
    customer partnerships, establishing, 140–149
    defining new roles and responsibilities, 229–232

continued

597

C

continued

continued

continued

continued

continued

continued

continued

continued